Periodic Table of the Elements

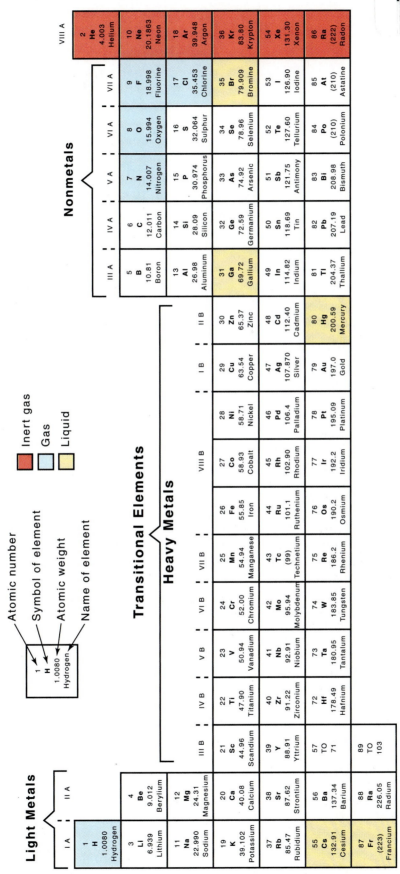

Light Metals

Transitional Elements

Heavy Metals

Nonmetals

Legend:
- Atomic number
- Symbol of element
- Atomic weight
- Name of element

1
H
1.0080
Hydrogen

- Inert gas (red)
- Gas (blue)
- Liquid (yellow)

IA	IIA	IIIB	IVB	VB	VIB	VIIB	VIIIB			IB	IIB	IIIA	IVA	VA	VIA	VIIA	VIIIA
1 **H** 1.0080 Hydrogen																	2 **He** 4.003 Helium
3 **Li** 6.939 Lithium	4 **Be** 9.012 Beryllium											5 **B** 10.81 Boron	6 **C** 12.011 Carbon	7 **N** 14.007 Nitrogen	8 **O** 15.994 Oxygen	9 **F** 18.998 Fluorine	10 **Ne** 20.1863 Neon
11 **Na** 22.990 Sodium	12 **Mg** 24.31 Magnesium											13 **Al** 26.98 Aluminum	14 **Si** 28.09 Silicon	15 **P** 30.974 Phosphorus	16 **S** 32.064 Sulphur	17 **Cl** 35.453 Chlorine	18 **Ar** 39.948 Argon
19 **K** 39.102 Potassium	20 **Ca** 40.08 Calcium	21 **Sc** 44.96 Scandium	22 **Ti** 47.90 Titanium	23 **V** 50.94 Vanadium	24 **Cr** 52.00 Chromium	25 **Mn** 54.94 Manganese	26 **Fe** 55.85 Iron	27 **Co** 58.93 Cobalt	28 **Ni** 58.71 Nickel	29 **Cu** 63.54 Copper	30 **Zn** 65.37 Zinc	31 **Ga** 69.72 Gallium	32 **Ge** 72.59 Germanium	33 **As** 74.92 Arsenic	34 **Se** 78.96 Selenium	35 **Br** 79.909 Bromine	36 **Kr** 83.80 Krypton
37 **Rb** 85.47 Rubidium	38 **Sr** 87.62 Strontium	39 **Y** 88.91 Yttrium	40 **Zr** 91.22 Zirconium	41 **Nb** 92.91 Niobium	42 **Mo** 95.94 Molybdenum	43 **Tc** (99) Technetium	44 **Ru** 101.1 Ruthenium	45 **Rh** 102.90 Rhodium	46 **Pd** 106.4 Palladium	47 **Ag** 107.870 Silver	48 **Cd** 112.40 Cadmium	49 **In** 114.82 Indium	50 **Sn** 118.69 Tin	51 **Sb** 121.75 Antimony	52 **Te** 127.60 Tellurium	53 **I** 126.90 Iodine	54 **Xe** 131.30 Xenon
55 **Cs** 132.91 Cesium	56 **Ba** 137.34 Barium	57 TO 71	72 **Hf** 178.49 Hafnium	73 **Ta** 180.95 Tantalum	74 **W** 183.85 Tungsten	75 **Re** 186.2 Rhenium	76 **Os** 190.2 Osmium	77 **Ir** 192.2 Iridium	78 **Pt** 195.09 Platinum	79 **Au** 197.0 Gold	80 **Hg** 200.59 Mercury	81 **Tl** 204.37 Thallium	82 **Pb** 207.19 Lead	83 **Bi** 208.98 Bismuth	84 **Po** (210) Polonium	85 **At** (210) Astatine	86 **Ra** (222) Radon
87 **Fr** (223) Francium	88 **Ra** 226.05 Radium	89 TO 103															

Lanthanide series

| 57 **LA** 138.91 Lanthanum | 58 **Ce** 140.12 Cerium | 59 **Pr** 140.91 Praseodymium | 60 **Nd** 144.24 Neodymium | 61 **Pm** (147) Promethium | 62 **Sm** 150.35 Samarium | 63 **Eu** 157.25 Europium | 64 **Gd** 158.92 Gadolinium | 65 **Tb** 158.92 Terbium | 66 **Dy** 162.50 Dysprosium | 67 **Ho** 164.93 Holmium | 68 **Er** 167.26 Erbium | 69 **Tm** 168.93 Thulium | 70 **Yb** 173.04 Ytterbium | 71 **L** 174 Lute |

Actinide series

| 89 **Ac** (227) Actinium | 90 **Th** 232.04 Thorium | 91 **Pa** (231) Protactinium | 92 **U** 238.03 Uranium | 93 **Np** (237) Neptunium | 94 **Pu** (242) Plutonium | 95 **Am** (243) Americium | 96 **Cm** (247) Curium | 97 **Bk** (249) Berkelium | 98 **Cf** (251) Californium | 99 **Es** (254) Einsteinium | 100 **Fm** (253) Fermium | 101 **Md** (256) Mendelevium | 102 **No** (256) Nobelium | 103 **L** (25) Lawrei |

Essentials
of Oceanography

Essentials
of Oceanography

FIFTH EDITION

Harold V. Thurman

Mt. San Antonio College

PRENTICE HALL
Upper Saddle River, New Jersey 07458

Library of Congress Cataloguing-in-Publication Data

Thurman, Harold V.
    Essentials of oceanography / Harold V. Thurman. — 5th ed.
      p.   cm.
    Includes index.
    ISBN 0-13-360231-1
    1. Oceanography.   I. Title.
   GC11.2.T49   1996
   551.46—dc20   95-17384
                                     CIP

Acquisitions Editor: *Robert A. McConnin*
Editor in Chief: *Paul F. Corey*
Editorial Director: *Tim Bozik*
Assistant Vice President of Production and Manufacturing: *David W. Riccardi*
Executive Managing Editor: *Kathleen Schiaparelli*
Assistant Managing Editor: *Margaret Antonini*
Marketing Manager: *Leslie Cavaliere*
Manufacturing Buyer: *Trudy Pisciotti*
Creative Director: *Paula Maylahn*
Art Director: *Heather Scott*
Cover Designer: *Amy Rosen*
Cover Photograph: *Warren Bolster/Tony Stone Images*
Photo Editor: *Lorinda Morris-Nantz*
Photo Researchers: *Chris Migdol, Diane Kraut*

©1996 by Prentice-Hall, Inc.
Simon & Schuster/A Viacom Company
Upper Saddle River, New Jersey 07458

Previous editions copyright ©1993, 1990, 1987, 1983 by Macmillan Publishing Company, a division of Macmillan, Inc.

Photo credits: Chapter 1 opener (H. Thurman)/Chapter 2 opener (©1980 by Marie Tharp; Reproduced by permission of Marie Tharp)/Chapter 3 opener (H. Thurman)/Chapter 4 opener (H. Thurman)/Chapter 5 opener (H. Thurman)/Chapter 6 opener (Tony Stone Worldwide)/Chapter 7 opener (H. Thurman)/Chapter 8 opener (Tony Stone Worldwide)/Chapter 9 opener A & B (©Clyde H. Smith/Peter Arnold, Inc.)/Chapter 10 opener (Tony Stone Images)/Chapter 11 opener (NOAA)/Figure 11–4 (The Stock Market)/Chapter 12 opener (H. Thurman)/Chapter 13 opener (North Atlantic Productivity (flyer)—February and May)/Chapter 14 (Tony Stone Images)/Chapter 15 opener (H. Thurman).

Printed in the United States of America

10 9 8 7 6 5 4 3 2 1

ISBN 0-13-360231-1

Prentice-Hall International (UK) Limited, *London*
Prentice-Hall of Australia Pty. Limited, *Sydney*
Prentice-Hall Canada Inc., *Toronto*
Prentice-Hall Hispanoamericana, S.A., *Mexico*
Prentice-Hall of India Private Limited, *New Delhi*
Prentice-Hall of Japan, Inc., *Tokyo*
Simon & Schuster Asia Pte. Ltd., *Singapore*
Editora Prentice-Hall do Brasil, Ltda., *Rio de Janeiro*

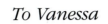

To Vanessa

Contents

CHAPTER 8

Waves *181*

CHAPTER 9

Tides *201*

CHAPTER 10

Coastal Geology *220*

CHAPTER 11

The Coastal Ocean *243*

CHAPTER 12

The Marine Habitat 265

CHAPTER 13

Biological Productivity and Energy Transfer 286

CHAPTER 14

Animals of the Pelagic Environment 311

CHAPTER 15

Animals of the Benthic Environment *336*

Appendixes *366*

Glossary *371*

Index *388*

Preface

The fifth edition of *Essentials of Oceanography* has been revised with the same basic goals as previous editions. It is designed for a course in oceanography taught to students with no formal background in mathematics or science. An effort has been made to present the relationship of scientific principles to ocean phenomena in a way that can be clearly understood.

Changes in this edition are designed to increase the readability and appeal of the book to students. The fifteen-chapter format is designed for easy coverage of the material in a fifteen- or sixteen-week semester. For courses taught on a ten-week quarter system, the teacher may need to select those chapters that cover the topic and concepts of primary relevance to the course.

The full-color format with nineteen improved as well as seven new line drawings and nineteen new photographs should continue to increase the appeal of the book. The clarity of the writing continues to be improved; material thought to contain too much detail by reviewers has been removed in an effort to further "essentialize" the content.

The order of presentation of material has not been altered. A short Introduction is designed to motivate students and help them appreciate the importance of the course to their future life. All of the chapters are introduced with a new motivational paragraph or two.

Chapter 1, "History of Oceanography," has had sections added on the peopling of the Pacific Ocean islands, remotely operated vehicles, and autonomous underwater vehicles. Chapter 2, "The Ocean: Its Origin and Provinces," has had the discussion of the origin of life updated and simplified, as well as the coverage of excess volatiles and continental margin depositional processes. Chapter 3, "Global Plate Tectonics," and Chapter 4, "Marine Sediments," received a minor amount of updating and simplification.

The physical oceanography section begins with Chapter 5, "Properties of Water." The coverage of the behavior of light and sound in water has been moved to Chapter 5 from Chapter 6, "Air-Sea Interaction." The greenhouse effect and distribution of solar radiation sections have been simplified in Chapter 6. Additional aspects of physical oceanography are covered in Chapters 7, 8, and 9, "Ocean Circulation," "Waves," and "Tides." All three chapters have had excess detail removed, and a section on rogue waves has been added to Chapter 8. The concepts presented to this point are incorporated into the discussion of their role in developing the nature of features discussed in Chapters 10 and 11, "Coastal Geology" and "The Coastal Ocean." The discussion of the Chedabucto Bay oil spill has been removed from Chapter 11 and replaced by coverage of the Boston Harbor sewage treatment project. The discussion of ocean pollution is concentrated in these chapters, which focus on those parts of the marine environment most affected by this problem.

Chapter 12, "The Marine Habitat," introduces marine biology with a discussion of the changing physical conditions throughout the marine environment and some of the general adaptations required for living in the oceans. Chapter 13, "Biological Productivity–Energy Transfer," has had the discussion of the microbial loop added and completes the introduction to marine ecology. Chapters 14 and 15 focus on the ecology of "Animals of the Pelagic Environment" and "Animals of the Benthic Environment," respectively.

As we crowd more and more of us onto the planet, there is increasing evidence that human actions are having a significant negative impact on essentially all components of Earth's ecology. It thus seems appropriate to broaden the scope of such survey courses as this to include a greater ecological focus. This edition continues to discuss the role of the oceans in the possible increased greenhouse effect of the atmosphere (Chapter 6) and the possible relationship between the El Niño-Southern Oscillation and changes in climate throughout the world (Chapter 7). Also considered are the problems of coastal development (Chapter 10) and exploiting marine resources such as hydrocarbons and other aspects of coastal pollution (Chapter 11).

Every effort is made to include only factual information and point out controversial aspects of any discussion of environmental considerations. However, the overall theme is that we must take much more care in activities that can modify the environment and keep that modification to a minimum.

Key terms are noted with bold print and defined when they are introduced. A summary reviews the major concepts discussed in each chapter, and the key terms are listed at the end of each chapter. The "Questions and Exercises" section provides the student with an opportunity to focus on the major concepts. "Suggested Reading" at the end of each chapter provides a guide to articles on relevant topics that have popular presentation in *Sea Frontiers* or a somewhat more demanding presentation in *Scientific American*.

To assist the student further, there are a number of appendixes and a glossary that includes the definition of all of the key terms (except names of individuals) as well as additional terms that students may wish to look up.

The Instructor's Manual provides not only answers to end-of-chapter questions and a selection of possible exam questions, but also special feature articles, covering a wide range of interesting topics, that may be duplicated and distributed to students. These features could be used simply to add breadth and interest to the coverage of the material in a chapter or serve as a basis for student reports. Other teaching aids available are 50 color transparencies for use with an overhead projector and a set of 100 slides for selected illustrations from the textbook.

Acknowledgments

Many individuals have provided advice and assistance for this work, including:

Richard D. Little, Greenfield Community College; Edward Ponto, Onodaga Community College; Wallace W. Drexler, Shippensburg University; Timothy C. Horner, California State University, Sacramento; Dr. M. John Kocurko, Midwestern State University; Stephen A. Macko, University of Virginia, Charlottesville; Charles H. V. Ebert, SUNY Buffalo; Dr. Arthur Wedweiser, Edinboro University of PA; Cathryn L. Rhodes, University of California, Davis; William W. Orr, University of Oregon; Walter C. Dudley, University of Hawaii; Jackie L. Watkins, Midwestern State University; Kenneth L. Finger, Irvine Valley College; Donald L. Reed, San Jose State University; James M. McWhorter, Miami-Dade Community College; B. L. Oostdam, Millersville University; Dr. William Balsam, University of Texas at Arlington; Hans G. Dam, University of Connecticut; Johnnie N. Moore, University of Montana; Curt Peterson, Portland State University; Richard A. Laws, University of North Carolina; Jill K. Singer, SUNY College, Buffalo; Laurie Brown, University of Massachusetts; and Matthew McMackin, San Jose University.

I N T R O D U C T I O N

We hope the world will be a better place for our having been here. As good world citizens, we have that responsibility. If you approach the study of the oceans with such a goal in mind, I'm sure the oceans of the world will benefit from your presence.

When I first began writing texts in the early 1970s, I wanted to help students develop an appreciation for the oceans by learning about oceanic processes, both physical and biological—that is, develop an *ecological awareness*. Some of you may even find a lifelong interest in the ocean and continue to study it formally or informally.

This book can help you grasp the basic principles underlying marine phenomena. Such an understanding is essential to comprehending the interrelationships between the oceans and the life forms they support. The population density in coastal areas is increasing, but the threat to coastal ecological systems can be greatly reduced if significant numbers of this population understand the need to protect the coastal environment. With the continuing increase in technological capability, our power to modify the marine environment will continue to expand. Only an informed populace can help decide whether this increased capability is used to increase the damage to our oceans or is used to greatly reduce the ecological threat.

Interrelationships of the Ocean

Since prehistoric times, people have used the ocean as a means of transportation and as a source of food. However, the importance of ocean processes has been studied technically only since 1930. The impetus for this study began in a search for petroleum, continued with the emphasis on ocean warfare during World War II, and more recently has been expressed in the concern for ocean ecology. Of course, fishermen have always known to go where the physical processes of the ocean offer good fishing. But how life interrelates with ocean chemistry, geology, and physics was more or less a mystery until scientists in these disciplines began to investigate the ocean with high technology.

For example, in the late 1970s geologists were investigating global plate tectonics. The process of plate tectonics involves the creation of new ocean floor as lava erupts along underwater mountain ranges. When they searched the vents that release hot water along a range in the deep eastern Pacific Ocean, they discovered many large tube worms, clams, mussels, and crabs. Normally the deep-ocean floor supports relatively small forms of life because there isn't much food. But the vents provided energy for bacteria to support abundant life. This is a striking example of how the physical nature of the ocean determines the distribution of life within it (see Figure I–1).

As you will discover in this book, the atmosphere and the ocean powerfully affect each other. Scientists have long been interested in this interaction. Our present technology uses satellites to view atmospheric circulation, ocean circulation, and life distribution in the sea. We are gathering data from every ocean every day. Such simultaneous data are showing how the water reacts to wind speed and air temperature, how the air responds to oceanic changes, and how the distribution of life in the sea relates to both.

Population Impacts on Coastal Ecology and the Open Ocean

Population studies now show that three-fourths of all Americans live within 80 kilometers (50 miles) of the ocean or the Great Lakes. This migration to the coasts will further mar the delicate natural balance between the ocean and the shore. Specifically, the migration is resulting in more harbor and channel dredging, industrial waste, dredge spoils, the use of ocean water for cooling of power plants and industries, and the filling in of marshes that are vital to the cleansing of runoff waters and to the maintenance of coastal fisheries (Figure I–2). In the open ocean, deep-ocean mining and nuclear waste disposal are planned. How do we deal with these increased demands on the environment? How do we regulate the ocean's use?

Further, in 1980 the National Academy of Sciences predicted a gradual rise in sea level worldwide. As carbon dioxide and other gases that increase the atmosphere's ability to hold heat reach higher concentrations,

1

Figure I–1

Deep-Sea Hydrothermal Vents. One of the first photos ever taken of deep-sea hydrothermal vents discovered in 1977 on the Galápagos Rift. It shows crabs and 25-centimeter (10-inch) mussels at a 2500-meter (8200-feet) deep site near the equator off the coast of Ecuador. (Photo courtesy of Scripps Institution of Oceanography, University of California, San Diego.)

worldwide temperature increases may occur. This increased "greenhouse effect" could hasten sea-level rise and damage to coastal areas. Even considering present erosion and deposition along the coasts, geologists would like to have us move back from the shore as much as possible and leave coastline management to nature. Engineers, on the other hand, say any coastal region can be stabilized for the lifetime of a development, in which case much public money will be spent stabilizing the coastal property of a few people (Figure I–3). If sea level continues to rise, perhaps engineers will be so busy protecting present development that they will abandon plans to develop new areas!

Figure I–2

Coastal Population. Heavily developed bay barrier running from Sea Bright to Monmouth Beach, New Jersey. (Photo © Breck P. Kent.)

Rational Use of Technology

Looking at these issues, we see cause for concern, but certainly not despair. The world ocean is a vast resource that hasn't yet been lethally damaged. We have been able to inflict only minor damage here and there along its margin. But as our technology makes us more powerful, the threat of irreversible harm becomes greater—or less, depending on how we use our tools.

The most negative impact humans have had on ocean resources surely must be the damage inflicted by the world fishing fleet on fish resources. From Canada to New Zealand, fishermen have lost their livelihood because of the poor management of fisheries worldwide. Whereas the fishing harvest quadrupled from 1950 to 1990, the fishing fleet increased at double that rate, or eight times. Despite the great value of the fisheries, government subsidies were required to support the greater cost of maintaining the fishing fleet of each country. Great damage now has been done. Economics will always drive the exploiters away when the greed that motivates their activities can no longer be gratified by the resource they misuse; this is the ocean's ultimate defense against human exploitation. Must it always resort to this defense?

Figure I–3

Coastal Erosion. Storm surge washes over breakwater during heavy storm. (Photo © Mark Sherman/Bruce Coleman, Inc.)

Which path will we take? All of you will need to evaluate carefully those you elect to public office, and some of you will have direct responsibility for making the decisions that affect our environment. It is my hope that you, as a new student of the marine environment, will gain enough knowledge from your teacher and this course to help our nation, as well as other nations, make rational use of the oceans in your lifetime.

C H A P T E R   1
History of Oceanography

With each passing year, human interaction with the ocean increases. The fishing effort is stepped up, and the search for petroleum and minerals intensifies. With our increasing use of marine resources, we are obliged to increase our knowledge of the impact on the marine environment.

This chapter summarizes events that have brought us to our present state of knowledge and discusses how the pressing need for greater knowledge of the oceans may be met.

We speculate about early human interaction with the oceans and describe historical milestones in our knowledge of Earth and sea. Then we present the marvelous technological capabilities for studying ocean depth and breadth that have been developed over the past half-century.

However, even with this new technology, we have much to learn: If we consider this page to represent the entire area of the ocean floor, people have *directly observed* only an area that is smaller than the period at the end of this sentence. Clearly, challenges remain in our quest to understand the workings of the oceans and the basins they occupy.

The Concept of Geography Develops

Early History

Humankind probably first viewed the ocean as a source of food. At some later stage in the development of civilization, vessels were built to move upon the ocean's surface, thereby making it a wide avenue over which distant societies could begin to interact.

European Navigators. The first westerners known to have developed the art of navigation were the **Phoenicians,** who lived at the eastern end of the Mediterranean Sea, in the present area of Syria, Lebanon, and Israel. As early as 2000 B.C. they were investigating the Mediterranean Sea, the Red Sea, and the Indian Ocean. The first recorded circumnavigation of Africa, in 590 B.C., was made by the Phoenicians, who had also sailed as far north as the British Isles.

The Greek view of the world in 450 B.C. is seen in the map made by the "father of history," **Herodotus** (Figure 1–1). It shows the Mediterranean Sea surrounded by three continents, Europe, Asia, and Libya, and bordered by three major seas. The northern and northeastern margins of the continents of Europe and Asia are indicated as unknown, but the Greeks believed that the oceans were a margin of water that surrounded all three continents.

The Greek astronomer-geographer **Pytheas** sailed northward to Iceland in 325 B.C. He worked out a simple method for determining latitude (one's position north-south) that involved measuring the angle between an observer's line of sight to the North Star and line of sight to the northern horizon (see appendix II). Mariners could determine the latitude of any point on the surface of Earth using the method introduced by Pytheas, but it was impossible for them to determine longitude accurately. Using astronomical measurements, he also proposed that the tides were a product of lunar influence.

Eratosthenes (276–192 B.C.), a Greek librarian in the Egyptian city of Alexandria, was the first known person to determine Earth's circumference. He calculated that its circumference through the North and South Poles was 40,000 kilometers (24,840 miles), which compares very well with the 40,032 kilometers (24,875 miles) determined by the more precise methods in use today.

In about A.D. 150, **Ptolemy** produced a map of the world that represented Roman knowledge at that time. He introduced vertical lines of longitude and horizontal lines of latitude. Ptolemy's map indicated, as did the earlier Greek maps, the continents of Europe, Asia, and Africa. It showed the Indian Ocean to be surrounded by a partly unknown landmass. Unlike Herodotus, Ptolemy considered the major oceans to be seas similar to the Mediterranean, having boundaries defined by unknown landmasses.

Pacific Navigators. It is not known who first developed navigation in the east, but it may have been the ancestors of Pacific Islanders. The peopling of the

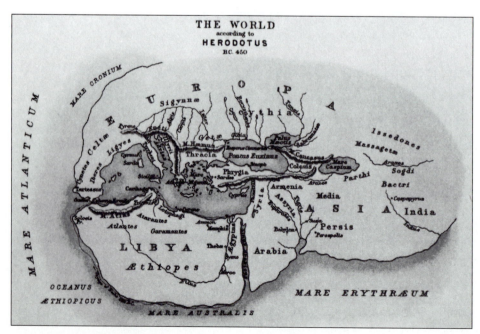

Figure 1–1

The World of Herodotus. The world according to Herodotus, 450 B.C. (From *Challenger* report, Great Britain, 1895.)

Pacific Islands is mysterious and may never be completely understood. Because there is no evidence that people directly evolved on these islands, their presence obviously required travel over hundreds or even thousands of miles of ocean from the continents, probably in double canoes, outrigger canoes, or balsa rafts. Only 0.7 percent of the Pacific Ocean is composed of islands, so probably only a fortunate few of the voyagers made landfall. Other than the Hawaiian Islands, all of the large islands in the Pacific are toward the western end. East of these large islands, in southeast Asia, the Pacific Islands have been divided into three groupings, as is shown in Figure 1–2: Micronesia ("small islands"), Melanesia ("black islands"), and Polynesia ("many islands").

Although no written record exists of Pacific human history prior to the arrival of Europeans in the 16th century, the movement of Asian peoples into Micronesia and Melanesia is easy to understand. It is the triangular Polynesia that must have presented the greatest challenge to ocean voyagers. This is especially true for Hawaii, which is 3000 kilometers (1860 miles) from the nearest inhabited islands, the Marquesas Islands. Easter Island, at the southeastern corner of the Polynesian triangle, is 1600 kilometers (1000 miles) from Pitcairn Island.

Archaeological evidence suggests that humans from New Guinea may have occupied New Ireland as early as 4000 or 5000 B.C. However, there is little evidence of human travel farther into the Pacific Ocean prior to 1500 B.C. By then, pottery makers called the "Lapita people" had traveled on to Fiji, Tonga, and Samoa. Over the next thousand years or so, the Polynesian culture developed in these islands, and Polynesian voyaging may have begun about 500 B.C.

The first archaeological evidence from the eastern Pacific indicates human occupation of the Marquesas Islands by A.D. 300. Although the Maori of New Zealand, the Hawaiians, and Easter Islanders of the 16th century were Polynesian, there is no clear evidence to explain *how* these peoples arrived at their present homes.

An adventuring biologist-turned-anthropologist, **Thor Heyerdahl,** believes that voyagers from South America may have reached islands of the South Pacific prior to the coming of the Polynesians. He demonstrated the possibility of such voyages with his 1947 voyage of the *Kon Tiki,* a balsa raft designed like those known to have been used by South American navigators at the time of European discovery. Heyerdahl sailed the raft 11,300 kilometers (7000 miles) in the South Equatorial Current to the Tuamotu Islands. He undertook this voyage to prove that early South Americans could have traveled to the Polynesian Islands more easily than canoe travelers from the west. Heyerdahl's views are not yet accepted by many anthropologists, but the final story of the peopling of the Pacific Islands is yet to be written.

The Middle Ages

After the fall of the Roman Empire in the 5th century A.D., the Mediterranean area was dominated by Arab influence; the writings of the Greeks and Romans passed into the hands of the Arabs, to be forgotten by the Christians in Europe. Subsequently, the Western concept of geography degenerated considerably; one notion envisioned the world as a disc with Jerusalem at the center.

The Arabs, meanwhile, were trading extensively with East Africa, Southeast Asia, and India and had learned the secret of the monsoon winds. They took advantage of these seasonal winds to make their trade voyages easier. During the summer, when the monsoon winds blew from the southwest, ships laden with goods would leave the Arabian ports and sail eastward across the Indian Ocean. The return voyage would be timed to take advantage of the reverse northeasterly trade winds that occur during winter, making the transit relatively simple for their sailing vessels.

In Europe, the nautical inactivity of the southern Europeans was offset by the vigorous exploration of the

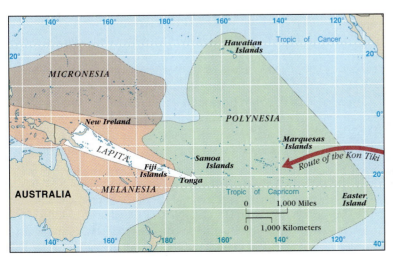

Figure 1–2

The Peopling of the Pacific Islands. The islands of the Pacific Ocean have been divided into Micronesia (brown), Melanesia (red), and Polynesia (green). The "Lapita people" present in New Ireland 5000–4000 B.C. can be traced to Fiji, Tonga, and Samoa by 1500 B.C. Polynesian culture may have been developed here, and by 500 B.C. voyages into the rest of Polynesia may have begun. The path of the *Kon Tiki* is also shown.

Vikings of Scandinavia. Late in the 9th century, aided by a period of climatic warming, the Vikings colonized Iceland. In 981 Erik the Red sailed westward from Greenland and discovered Baffin Island. In 995 Erik's son, **Leif Eriksson,** discovered what was then called Vinland (now Newfoundland) and spent the winter in that portion of North America (Figure 1–3).

Age of Discovery

During the 30-year period from 1492 to 1522, known as the **Age of Discovery,** the Western world came to a full realization of the vastness of Earth's water-covered surface. The continents of North and South America were discovered. The globe was circumnavigated for the first time. And it was learned that human populations existed elsewhere in the world. Human cultures were found throughout the newly discovered continents and islands, although they were vastly different from the cultures familiar to the voyagers.

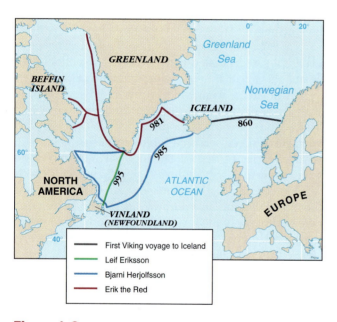

Figure 1–3

The Vikings Reach the New World. The expansion of Viking influence spread from Europe and probably reached Iceland by 860. Erik the Red was banished from Iceland and sailed to Greenland in 981. He returned to Iceland in 984 and led the first wave of Viking colonists to Greenland. Bjarni Herjolfsson sailed from Iceland to join the colonists but sailed too far south and is thought to be the first Viking to have viewed what is now called Newfoundland. Bjarni did not land but continued trying to find his way to Greenland. He eventually did land at the settlement Herjolfsness, named for his father, who waited for him there. In 995 Leif Eriksson, son of Erik the Red, bought Bjarni's ship and set out for the land that Bjarni had seen to the southwest. Leif set up a camp at Tickle Cove in Trinity Bay and named the land Vinland after the grapes that were found there.

The Age of Discovery did not just happen. Precipitating these voyages was the capture of Constantinople, the capital of eastern Christendom, in 1453 by Sultan Mohammed II. Constantinople is at the eastern end of the Mediterranean, providing access to the east—India, Asia, and the East Indies (modern Indonesia). This event isolated the Mediterranean port cities from the riches of the East and caused the Western world to search for a new trading route to this area.

Prince **Henry the Navigator** of Portugal had established a marine observatory to improve Portuguese sailing skills, but its attempts to reestablish trade by ocean routes around Africa met with failure for years. This need to travel around the tip of Africa was a great obstacle to establishing an alternative trade route. Cape Agulhas (the southern tip of Africa) was finally rounded by **Bartholomeu Diaz** in 1486. He was followed in 1498 by **Vasco da Gama,** who continued his trip around the tip of Africa to India.

The idea for a voyage such as the one **Christopher Columbus** undertook was initiated by an astronomer in Florence, Italy, named Toscanelli. He wrote to the king of Portugal, suggesting that a course be charted to the west in an attempt to reach the East Indies by crossing the Atlantic Ocean. Columbus later contacted Toscanelli and was given a copy of the information indicating that India could be reached at a certain distance. (Today, we know that this distance would have carried him just west of the continent of North America.)

After his well-known difficulties in initiating the voyage, Columbus set sail with 88 men and three ships on August 3, 1492, from the Canary Islands west of Morocco. During the morning of October 12, 1492, the first land was sighted, which is generally believed to have been Watling Island in the Bahama Islands southeast of Florida. Based on the inaccurate information he had been given, Columbus was convinced that he had arrived in the East Indies. Consequently, he called the inhabitants "Indians." Upon his return to Spain and the announcement of his discovery, additional voyages were planned, and the Spanish and Portuguese explored the coasts of North and South America.

The Atlantic Ocean became familiar to the European explorers, but they did not see the Pacific Ocean until 1513, when Vasco Núñez de Balboa attempted a land crossing of the Isthmus of Panama and sighted the Pacific Ocean from atop a mountain.

The culmination of this period of discovery was the circumnavigation of the globe initiated by **Ferdinand Magellan** (Figure 1–4). In September of 1519 Magellan left Sanlucar de Barrameda, Spain, and traveled through a passage to the Pacific at 52°S latitude, now named the Strait of Magellan. After discovering the Philippines on March 15, 1521, Magellan was killed in a fight with the inhabitants of these islands. Juan Sebastian del Caño completed the circumnavigation by taking

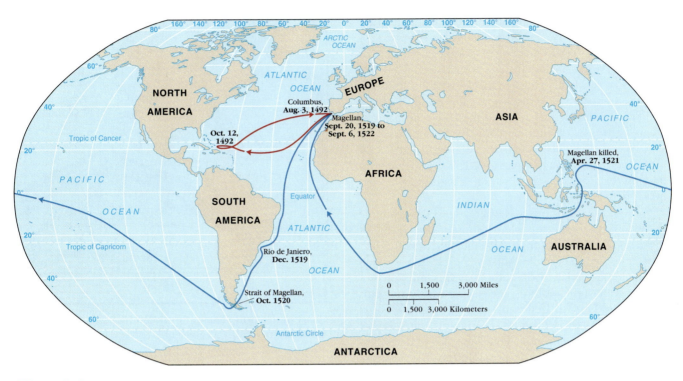

Figure 1–4

Voyages of Columbus and Magellan. Searching for a western route to the East Indies, Magellan culminated the Age of Discovery with a voyage (1519–1522) that ended for him with his death in the Philippines. Juan Sebastian del Caño returned to Spain in the *Victoria,* one of five ships that began the voyage, to complete the first circumnavigation of Earth. (Base map courtesy of the National Ocean Survey.)

one of the ships, *Victoria,* across the Indian Ocean and back to Spain in 1522. On this trip Magellan had attempted to measure the depth of the Pacific Ocean with a weighted line, but he was unable to reach bottom.

After these voyages, the Spanish initiated many others to take the gold that had been found in the possession of the Aztec and Inca cultures in Mexico and South America. While the Spanish concentrated on plundering the Aztecs and Incas, the English and Dutch plundered the Spanish. The political dominance of Spain came to an end with the defeat of the Spanish Armada by the British in 1588. With this squelching of an attempted invasion of the British Isles, the English became the dominant maritime power, and they remained so until early in the 20th century.

The Search to Increase Scientific Knowledge of the Oceans

Captain James Cook

The next major focus of interest in the oceans was more scientific, as the English determined that increasing their knowledge of the oceans would help them maintain their maritime superiority. The more successful

early voyages that sought to learn about the physical nature of the oceans were conducted by the English navigator **James Cook** (1728–1779), son of a farm laborer (Figure 1–5).

Figure 1–5

Captain James Cook, Royal Navy. (Courtesy of U.S. Navy.)

From 1768 until his death in 1779, Captain Cook undertook three voyages of discovery. He searched for the continent Terra Australis ("Southern Land," or Antarctica) and concluded that, if it existed, it lay beneath or beyond the extensive ice fields of the southern oceans. Captain Cook discovered the South Georgia and South Sandwich islands and the Hawaiian Islands, where he was killed after searching for a northwest passage from the Pacific Ocean to the Atlantic Ocean. Cook also developed a diet that prevented his crew from contracting the disease scurvy, which is caused by a vitamin C deficiency.

Cook's expeditions added greatly to the scientific knowledge of the oceans. He determined the outline of the world's largest ocean, the Pacific, and was the first person known to cross the Antarctic Circle in his search for Antarctica. Cook also led the way in sampling subsurface water temperatures, measuring winds and currents, sounding (to determine sea floor depth), and collecting data on coral reefs. By proving the value of **John Harrison's** chronometer as a means of determining longitude, Cook made possible the first accurate maps of Earth's surface (see appendix VII).

An American contemporary of Captain Cook was an early contributor to greater understanding of the ocean's surface currents. **Benjamin Franklin,** deputy postmaster general for the American colonies, determined that it took mail ships coming from Europe two weeks longer to reach the colonies by a northerly route than it did ships that came by a more southerly route. To find out why this occurred, he asked captains who came into port for information concerning the movement of surface waters in the Atlantic. Franklin inferred that there was a significant current moving northward along the eastern coast of the United States. The current then moved out in a more easterly path across the North Atlantic. He concluded that this east-flowing current, which the ships traveling a northerly route from Europe had to combat, was responsible for increasing the time of their voyage. He subsequently published a map of the Gulf Stream based on these observations (Figure 1–6).

Matthew Fontaine Maury

An even greater contribution was made by **Matthew Fontaine Maury** (1806–1873). This career officer in the U.S. Navy was placed in charge of the Depot of Naval Charts and Instruments after suffering an injury early in his career. In the depot were log books containing a large amount of information about currents and weather conditions in various parts of the oceans. Maury's systematic analysis of these logs produced a compilation of wind and current patterns that proved very useful to the navigators of the 19th century.

Maury helped organize the first International Meteorological Conference in Brussels in 1853 for the pur-

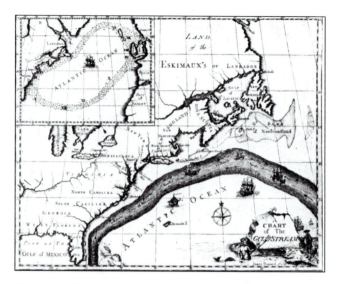

Figure 1–6

Chart of the Gulf Stream Compiled by Benjamin Franklin in 1777. (Courtesy of U.S. Navy.)

pose of establishing uniform methods of making nautical and meteorological observations at sea. This standardization greatly increased the dependability of such data, which Maury summarized in *The Physical Geography of the Sea* (1855). Maury is often referred to as the "father of oceanography" (Figure 1–7).

Charles Darwin

In the early 19th century, an English naturalist named **Charles Darwin** (1809–1882) entered the scientific scene. Darwin's interest, which was investigating the whole of nature, led him to make one of the most outstanding contributions to the field of biology: the theory of evolution by natural selection. Much of the background on which he based his conclusions was gained from observations made during his voyage aboard HMS *Beagle*.

The *Beagle* sailed from Devensport on December 27, 1831, under the command of Captain Robert Fitzroy with the major objective of completing a survey of the coast of Patagonia (Argentina) and Tierra del Fuego.

Darwin spent the next five years aboard the *Beagle,* which continued to sail around the world for the purpose of carrying on chronometric measurements (Figure 1–8). The voyage allowed the young naturalist the opportunity to study the plants and animals throughout the world. He studied closely the changes that occurred in the animal populations living in different environments and concluded that all animals change slowly over the immense period of time represented by the geologic past.

Darwin's observations led him to conclude that birds and mammals must have evolved from the reptiles

Figure 1–7

Lieutenant Matthew F. Maury. An engraving by Lemuel S. Punderson, modeled upon a daguerreotype (early photograph) autographed and inscribed by Lt. Maury. (Courtesy of Naval Historical Center.)

and that the similar skeletal framework of the human, the bat, the horse, the giraffe, the elephant, the porpoise, and other vertebrates required that they be grouped together. Darwin felt that the superficial differences that could be observed between populations were the result of adaptation to different environments and modes of existence.

On his return to England, Darwin published his controversial book, *The Origin of Species,* which dealt not so much with the origin of life but with the evolution of living things into their many forms. Although Darwin's ideas were highly controversial, many now can be indisputably proven by an impressive array of scientific facts.

The Rosses—Sounders of the Deep

Two of the earliest successful sounders of the deep oceans were Englishmen **Sir John Ross** and his nephew **Sir James Clark Ross.** Baffin Bay in Canada was explored by Sir John Ross in 1817 and 1818, and he was able to measure depths. Ross also collected samples of bottom-dwelling organisms and sediments with a device of his own design called a *deep-sea clamm.* He recovered starfish and worms in mud samples from a depth of 1.8 kilometers (1.1 miles).

Sir James Clark Ross extended the soundings to greater depths on voyages to the Antarctic during 1839–1843 (Figure 1–9). A 7-kilometer (4⅓-mile) sound-

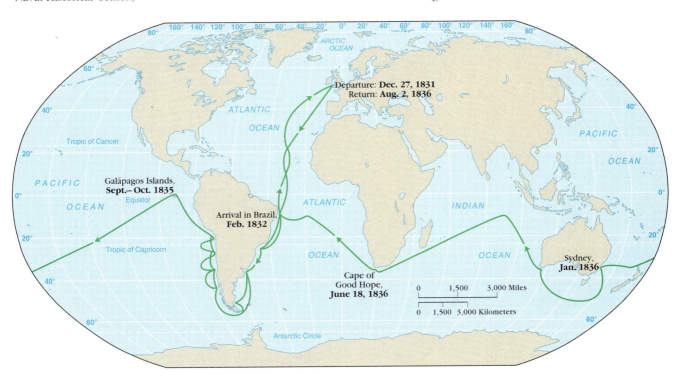

Figure 1–8

Voyage of the *Beagle.* Sailing as a naturalist aboard HMS *Beagle,* Charles Darwin gathered the evidence that enabled him to develop his theory of biological evolution through natural selection. (Base map courtesy of the National Ocean Survey.)

Figure 1–9

Ross Expedition. Members of the Sir James Clark Ross expedition to Antarctica after landing on Possession Island. The penguins seem to be at least as curious about the visitors as the members of the expedition are about them. (Reprinted by permission of Bettman Archive.)

ing line was used on these voyages, and even this great length was at times insufficient to reach the bottom of the ocean. Sir James observed that the animals he recovered from the cold waters of the Antarctic were the same species his uncle had recovered from the Arctic area, and they were found to be very sensitive to temperature increase. This discovery led him to the conclusion that the waters making up the deep ocean must be of uniformly low temperature.

Edward Forbes

Also making an important contribution to the study of marine biology was the British naturalist **Edward Forbes** (1815–1854). Forbes was interested in determining the vertical distribution of life in the ocean, and after repeated observations he divided the sea into specific life-depth zones. He came to the conclusion that plant life is limited to the zone near the surface. Forbes also found that the animal concentration was greater near the surface and decreased with increasing depth until only a small trace of life, if any, remained in the deepest waters.

Forbes's followers either misinterpreted his statement of the condition of life's distribution in the ocean or decided to apply their own logic to it, and they concluded that no life existed in the deep ocean, as life would be impossible under the conditions of high pressure and absence of light and air. Ignoring the findings of the Rosses in the high southern and northern latitudes, they persisted in their belief that no life existed in the deep ocean.

The *Challenger* Expedition

The controversy over the distribution of life in the oceans helped stimulate interested in the first large-scale voyage with the express purpose of studying the subject.

In 1871 the Royal Society recommended that funds be raised from the British government for an expedition to investigate the following subjects:

1. Physical conditions of the deep sea in the great ocean basins;
2. The chemical composition of seawater at all depths in the ocean;
3. The physical and chemical characteristics of the deposits of the sea floor and the nature of their origin;
4. The distribution of organic life at all depths in the sea, as well as on the sea floor.

A staff of six scientists under the direction of C. Wyville Thompson left England in December 1872 aboard **HMS *Challenger.*** A 2306-tonne corvette that had been refitted to conduct scientific investigation, the *Challenger* returned in May 1876 after having traversed large portions of the Atlantic and Pacific oceans (Figure 1–10). The achievements of the *Challenger* expedition, which covered 127,500 kilometers (79,223 miles), included 492 deep-sea soundings, 133 bottom dredges, 151 open-water trawls, and 263 serial water temperature observations.

The more outstanding aspects of the voyage were the netting and classification of 4717 new species of marine life and the measurement of a water depth of 8185 meters (26,850 feet) from the Mariana Trench in the western North Pacific. Seventy-seven samples of ocean water collected during the *Challenger* expedition were analyzed in 1884 by C. R. Dittmar. This first refined analysis contributed greatly to our understanding of ocean salinity, which will be discussed in Chapter 5.

Fridtjof Nansen

One of the most unusual of the voyages was initiated by a Norwegian, **Fridtjof Nansen** (1861–1930). Dr. Nansen developed a great interest in exploring the North At-

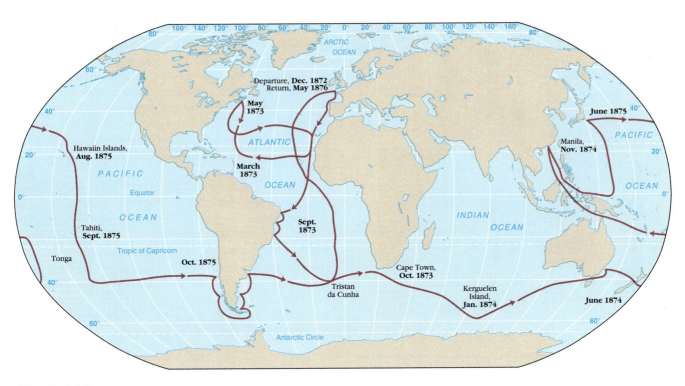

Figure 1–10

Voyage of HMS *Challenger*. The route traveled by HMS *Challenger* during the first major oceanographic voyage (December 1872 to May 1876). (Base map courtesy of The National Ocean Survey. Right, HMS *Challenger* from the *Challenger* Report, Great Britain, 1895.)

lantic and the Arctic area. To aid him in his planning, the results of earlier attempts to explore the Arctic during the latter part of the 19th century were compiled. Of particular interest to Nansen was the ill-fated American voyage of George Washington DeLong and his crew on the *Jeanette* in 1879. This expedition had attempted to sail through the Bering Strait between North America (Alaska) and Asia (Russia) to Wrangell Island, from which it planned to go overland to the North Pole. Scientists believed that Wrangell Island was possibly the southern tip of a peninsula extending southward from the great, but yet undiscovered, Arctic continent.

The *Jeanette* became stuck in the ice on September 6, 1879, and drifted north of Wrangell Island, proving that the island was not a peninsula of the Arctic continent. After two years of drifting, the *Jeanette* sank off the New Siberian Islands. DeLong and many of his crew perished in the Lena Delta region of eastern Siberia.

In 1884, five years later, a number of articles that could be traced to the *Jeanette* were found frozen in the pack ice off the eastern coast of Greenland. On the basis of this evidence, Nansen considered that the articles must have drifted from the wreck of the *Jeanette* to the observed location by following a path that would have taken them over or near the North Pole. He saw no reason why this ice flow that transported these articles

could not be used to transport an expedition across the North Pole as well.

Eventually, Nansen was able to raise funds for building the ***Fram,*** a ship designed so that the expanding ice would not crush it but rather would force it up to the surface, free of the grip of the growing ice sheets.

Provisions for 13 men for five years were stored on the small ship, and the crew set sail from Oslo, Norway, on June 24, 1893. On September 21, after fighting their way along the northern coast of Siberia, they had not yet sighted the New Siberian Islands when the ice captured them at a latitude of 78°30'N. They were 1100 kilometers (683 miles) from the North Pole. They had now begun what would be a long and lonely endeavor (Figure 1–11).

On November 15, 1895, the *Fram* reached its northernmost point, 394 kilometers (244 miles) from the pole. On August 13, 1886, the little *Fram* broke ice and was again unbound in the open sea. She had drifted a total of 1658 kilometers (1028 miles) during her three-year entrapment.

The drift of the *Fram* proved that no continent existed by the Arctic Sea. It also showed that the ice that covered the polar area throughout the year was not of glacial origin but was a freely moving pack ice accumulation that had been formed directly upon the ocean

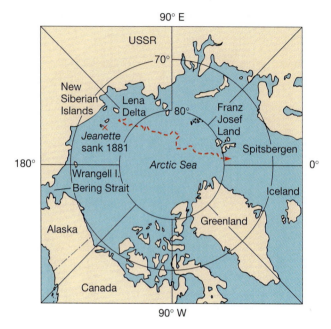

Figure 1–11

Voyage of the *Fram*. *A:* The course followed by the *Fram* after becoming frozen in ice near the New Siberian Islands, September 21, 1893. On August 13, 1896, the little ship was released by the ice off Spitsbergen. *B:* Fridtjof Nansen, Norwegian scientist and statesman, is using a sextant to determine his ship's position. He was a pioneer in the development of physical oceanography.

surface. During the voyage, the crew had found that the depth of this ocean exceeded 3000 meters (9840 feet). They also discovered a rather surprising body of relatively warm water with temperatures as high as 1.5°C (35°F) between the depths of 150 and 900 meters (about 500–3000 feet). Nansen correctly described this water as being a mass of Atlantic Ocean water that had sunk below the less saline Arctic water.

Nansen's realization of the need for more accurate measurements of water salinity and temperature led him to develop the *Nansen bottle,* a device once widely used for sampling the ocean depths. His observations of the direction of ice drift relative to the wind direction helped **V. Walfrid Ekman,** a Scandinavian physicist, to develop the mathematical explanation of this phenomenon. Nansen, who was awarded the Nobel Peace Prize in 1922, Ekman, and other Scandinavian scientists led the development of *physical oceanography* in the early 20th century.

20th-Century Oceanography

Voyage of the *Meteor*

Oceanography entered a new era with the voyage of the *Meteor* during 1925–1927. This German expedition

was the first to use an *echo sounder,* making it possible to obtain a continuous depth recording as a vessel proceeded along it course. This was a great advancement over reliance on scattered soundings and interpolation of the water depth between these points of known depths. The *Meteor* gathered data day and night for 25 months and concentrated on increasing the knowledge of the South Atlantic Ocean. It was this expedition that first revealed the true ruggedness of the ocean floor (Figure 1–12).

Many contributions from a number of nations have been made to the understanding of the oceans subsequent to the voyage of the *Meteor*. We must emphasize that the knowledge we now possess concerning the world's oceans is still far from complete. Recall the analogy made in the introduction: If this page represents the entire ocean floor, the period that ends this sentence represents how much we have observed directly. The effort to increase our knowledge of the oceans will continue for many years.

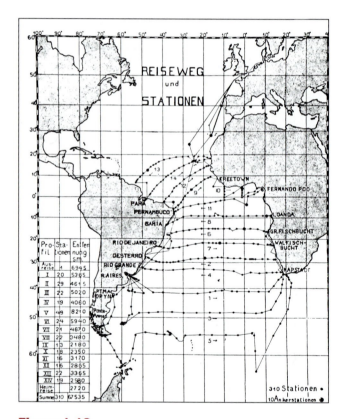

Figure 1–12

Voyage of the *Meteor*. This 1925 cruise crisscrossed the South Atlantic, where marine data were sparse, to enable a comprehensive study of the water masses and circulation of the Atlantic Ocean. (From Wüst, G., 1935. The stratosphere of the Atlantic Ocean. Scientific results of the German American Expedition of the Research vessel *Meteor* 1925–27, Vol. VI, sec. 1. W.J. Emery, trans. and ed. Amerind Pub. Co., New Delhi, 1978.)

Oceanography in the United States

The first large-scale contribution to oceanography by a U.S. citizen was the effort of Lieutenant Matthew Maury of the U.S. Navy, which culminated in his book, *The Physical Geography of the Sea*. However, it was not until later in the 19th century that the United States became active in organizing voyages to increase our knowledge of the oceans. This phase actually began in 1877 with the voyages of the *Blake* under the direction of **Alexander Agassiz.** This young scientist, the son of the Swiss naturalist Louis Agassiz, contributed greatly to the development of oceanographic research in the United States. Following his voyages, many more were initiated with the primary objective of increasing the scientific knowledge of the world's oceans. The following paragraphs discuss briefly some projects undertaken by major U.S. institutions concerned with the investigation of the world's oceans.

Government Research Agencies. The U.S. Navy Oceanographic Office exists to improve the combat readiness of our naval fleet. It conducts basic research toward this objective. However, its research results, as well as those of the Office of Naval Research and the U.S. Coast Guard, are of basic value to all oceanographers.

In 1807 the Congress created the U.S. Coast and Geodetic Survey after President Thomas Jefferson requested a survey of the coast of the United States. This organization, now called the National Ocean Survey, conducts bathymetric (ocean depth) surveys in the waters adjacent to the United States and its territories. Although the primary work of this agency is underwater mapping, its vessels are outfitted to conduct a wider variety of oceanographic research, gathering a broad range of oceanographic data.

The Commerce Department's National Oceanic and Atmospheric Administration (NOAA) works to ensure wise use of ocean resources through its National Ocean Survey, National Marine Fisheries Service, and Office of Sea Grant.

Marine Laboratories. Not directly under the control of the federal government, but very important in the achievement of research of national interest, are three marine laboratories that have pioneered much of the marine research in the United States. The oldest of these institutions is **Scripps Institution of Oceanography** of the University of California at San Diego. Originated at La Jolla, California, in 1903, Scripps investigates a wide spectrum of marine problems.

Woods Hole Oceanographic Institution originated at Woods Hole, Massachusetts, in 1930 as a private nonprofit organization devoted to the scientific study of the world ocean. Their ship *Atlantis II* was one of the first modern oceanographic research vessels.

The **Lamont-Doherty Earth Observatory** of Columbia University was founded at Torrey Cliffs Palisades, New York, in 1949 under the direction of **Maurice Ewing.** He was one of the great pioneers of geophysical methods for the investigation of the geology of the ocean bottom (Figure 1–13).

These institutions have led the way in 20th-century oceanographic exploration, including development of new devices for the study of the ocean depths:

- The **Floating Instrument Platform** (FLIP), shown in Figure 1–14, was designed for making acoustical and other measurements in the open sea that require a stable platform on the surface, despite waves, currents, and storms.
- Camera systems and instrument capsules have been developed for free-fall emplacement on the ocean floor. They can be left to operate for days or months. Automatic timers or acoustical signals release these devices to float to the surface for recovery.

Cooperative Studies. The need for cooperation was recognized in the **Deep Sea Drilling Project** led by the Scripps Institution of Oceanography with the

Figure 1–13

Maurice Ewing. Dr. Maurice Ewing (1906–1974) pioneered seismic techniques for studying sediments on the ocean floor. He was director of the Lamont-Doherty Earth Observatory at Columbia University from 1949 to 1972. (Photo courtesy of Columbia University.)

Figure 1–14

Floating Instrument Platform (FLIP). FLIP is the 108-meter (356-feet) Floating Instrument Platform, designed and developed by the Marine Physical Laboratory at the Scripps Institution of Oceanography. In the vertical position (*right*), it gives scientists an extremely stable platform from which to carry out underwater acoustic and other types of oceanographic research. FLIP has no motive power and must be towed to a research site in the horizontal position (*left*). Once on station, the ballast tanks are flooded and the vessel flips vertical, leaving 17 meters (56 feet) of the platform above water. When the work is completed, water is forced from the tanks and FLIP resumes its horizontal position for transport. (Courtesy of Scripps Institution of Oceanography, University of California, San Diego.)

cooperation of a number of educational institutions that have developed oceanographic programs. In 1963 the National Science Foundation announced a national program for drilling cores in ocean sediment to obtain samples for analysis.

The Scripps Institution of Oceanography united with three other major oceanographic institutions—the Rosenstiel School of Atmospheric and Oceanic Studies at the University of Miami, Florida; the Lamont-Doherty Earth Observatory of Columbia University; and the Woods Hole Oceanographic Institution in Massachusetts—to form the Joint Oceanographic Institutions for Deep Earth Sampling (JOIDES). They were later joined by oceanography departments of the University of Washington, Texas A&M, the University of Hawaii, Oregon State University, and the University of Rhode Island.

To carry on this program, a ship capable of drilling the ocean bottom while floating 6000 meters (3.7 miles) above the ocean surface had to be designed. The **Glomar Challenger** was constructed and launched in 1968, commencing the first leg of the Deep Sea Drilling Project. The *Glomar Challenger* is 122 meters (400 feet) long and 20 meters (66 feet) in beam (widest point), and its displacement (the amount of water the ship displaces) is 10,500 tonne loaded. The vessel is a self-sustained unit capable of remaining at sea for 90 days (Figure 1–15).

Initially, the oceanographic research program was financed by the U.S. government, but in 1975 it became international. Financial and scientific support was received from West Germany, France, Japan, the United Kingdom, and the Soviet Union. In 1983 the Deep Sea Drilling Project became the **Ocean Drilling Program** (ODP) with the broader objective of drilling the thick sediment layers near the continental margins (the area between the shore and the deep ocean), supervised by Texas A&M University.

Accompanying this change was the decommissioning (retirement) of *Glomar Challenger* after 15 highly successful years. The ship had made possible a major advancement in the field of Earth science: confirmation of sea-floor spreading, a process in which new ocean floor is created from lava along submarine mountain ranges. Following in the wake of the *Glomar Challenger* as the ODP drill ship is the much larger **JOIDES Resolution,** which conducted its first scientific cruise in 1985.

Since 1972 the Geochemical Ocean Sections (GEOSECS) program has been sampling and analyzing water from all oceans to study ocean circulation patterns, mixing processes, and biogeochemical cycling. The GEOSECS program is discussed further in Chapter 5.

Diving and Submersibles. Putting humans into the marine environment to observe it directly has always been a dangerous business. Our bodies are

Figure 1–15

Old and New Drilling Ships. *Above:* The DSDP's *Glomar Challenger.* D/V (drilling vessel) *Glomar Challenger* could produce high horsepower (8800 continuous or 10,000 intermittent) for propulsion and for operating drilling equipment. To hover over the drill site, the ship used dynamic positioning to move in any direction. The achievements of this vessel helped revolutionize geologic science. (Photo courtesy of Victor S. Soleto, Deep Sea Drilling Project.) *Right:* The *JOIDES Resolution,* drilling ship of the Ocean Drilling Program. This modern vessel replaced the *Glomar Challenger* in the important work of sampling the world's sea floors. (Photo courtesy of the Ocean Drilling Program.)

adapted to living in a different fluid—air—at a pressure of 1 kilogram per square centimeter (15 pounds per square inch). The effects of greatly increased pressure underwater limit the depth and duration of dives. The record ocean dive is 457 meters (1500 feet), but the researchers in this field believe that humans will eventually work for extended periods at depths below 600 meters (1970 feet). Table 1–1 provides the dates of some significant events in the history of human progress toward deep-sea descent.

Submersibles now take humans deep into the ocean. A submersible is simply a chamber in which the pressure remains the same as at the ocean surface. Some are suspended on cables; others are unattached and motorized for exploring the sea floor. Submersibles have been successful as a means of directly observing

the deep-sea environment. In 1934 zoologist William Beebe descended to a depth of 923 meters (3027 feet) off Bermuda in a tethered *bathysphere* to observe deep-sea life. After World War II the Swiss scientist Auguste Piccard designed his *bathyscaphe* (deep boat), *Trieste,* which was bought by the U.S. Navy. This vessel, which operates much like a balloon with a gondola, submerged to the record depth of 10,915 meters (35,801 feet) in the deepest ocean trench, the Mariana Trench off Guam, in 1960 (Figure 1–16).

Two untethered vessels are now widely used in deep-sea research. ***Alvin*** can descend to a depth of 4000 meters (13,120 feet), and ***Sea Cliff II*** can descend to 6000 meters (20,000 feet). The deepest-diving manned submersible is *Shinkai 6500.* This Japanese vessel can dive to 6500 meters (21,320 feet) (Figure 1–16).

Table 1–1
Some Important Achievements in the History of Human Descent into the Ocean

Date	Event
360 B.C.	Aristotle records the use of air trapped in kettles lowered into the sea by Greek divers in his *Problematum.*
330 B.C.	Alexander the Great descends in a glass berry-shaped bell during the siege of Tyre.
1620	Dutch inventor Cornelius van Drebbel tests first submarine in the Thames River. Drebbel descended to a depth of 5 m (16.5 ft) with King James I of England on board.
1690	Sir Edmond Halley descends in a lead-weighted diving bell. Air was replenished from barrels lowered into the Thames River.
1715	Englishman John Lethbridge tests a leather-covered barrel of air with viewing ports and waterproof arm-holes to a depth of 18 m (60 ft). This was the first armored diving suit.
1776	Colonist David Bushnell's *Turtle* attempts the first military submarine attack on the British ship *H.M.S. Eagle.* the attempt failed, but the British moved their fleet to less accessible waters.
1800	Robert Fulton builds the submarine *Nautilus,* the first of several to bear this name. It was powered by a hand-driven screw propeller.
1837	Augustus Siebe invents a prototype of the modern helmeted diving suit.
1913	A German company, Neufeldt and Kuhnke, manufactures an armored diving suit with articulating arms and legs.
1930	William Beebe and Otis Barton descend in a bathysphere to a depth of 923 m (3027 ft) off Bermuda.
1943	Jacques-Yves Cousteau and Émile Gagnan invent the fully automatic, compressed air aqualung.
1960	Jacques Piccard and Donald Walsh descend to a depth of 10,915 m (35,801 ft) in the bathyscaphe *Trieste.*
1962	Hannes Keller and Peter Small make an open-sea dive from a diving bell to a depth of 304 m (1000 ft) using a special gas mixture. Small dies during decompression.
1962	Albert Falco and Claude Wesly live for one week in Cousteau's Continental Shelf Station off Marseilles, France.
1964	*Sealab I*—U.S. Navy team lead by George F. Bond stays 11 days at 59 m (193 ft) off Bermuda.
1966	Submersibles *Alvin, Aluminaut,* and *Cubmarine,* with the aid of the unmanned remote-controlled vehicle *CURV,* locate and recover an atomic bomb.

Unmanned Remote-Controlled Vehicles. The **Seabeam/*Argo-Jason*** system combines many technologies to "put eyes on the sea floor." As Figure 1–17 shows, the system includes a research vessel, underwater vehicles, and a satellite uplink to control centers. Seabeam is a complex sonar system on the research vessel that maps the sea floor. *Argo* is towed by the research vessel; it contains an array of sonar and television systems and is a hangar for *Jason.* (*Argo* was being towed above bottom by the R/V *Knorr* in 1985 when its television cameras discovered the remains of the RMS *Titanic*—Figure 1–18). *Jason,* a self-propelled vehicle, can be tethered to *Argo,* a manned submersible, or a surface ship. With its stereo color television and two manipulator arms, *Jason* can be lowered to the ocean floor to investigate areas of interest.

Future plans call for using both manned submersibles and **remotely operated vehicles (ROVs)** such as *Jason* (Figure 1–19). Because they are unmanned, ROVs are cheaper to operate and can stay beneath the surface for months if necessary. Because of the need for oxygen, manned submersibles can stay submerged for only 12 hours, and it may take 8 of those hours to descend and ascend.

The Japanese are presently testing an ROV named *Kaiko* that will be the deepest-diving submersible, capable of investigating even the deepest trenches. In fact, it was in the Mariana Trench at a depth of 10,911 meters (35,788 feet) during sea trials in May 1994 that *Kaiko's* video signal failed. Researchers hope the problem can be fixed quickly so they can begin investigating deep-sea biology, sedimentation, and fault activity.

Another class of vehicle being developed for underwater surveys is the **autonomous underwater vehicle (AUV).** The *ABE* (Autonomous Benthic [sea floor] Explorer) is being developed at Woods Hole Oceanographic Institution. Another AUV that has undergone sea tests (Figure 1–19) beneath the ice floes of Antarctica is the *Odyssey.* It is being developed by the Massachusetts Institution of Technology Sea Grant Underwater Vehicles Laboratory. Although this technology is still developing, many researchers have high hopes for AUVs as the submarine information-gathering system of the near future.

Sea Satellites. Remote sensing from satellites is becoming an important tool in ocean studies. Although they orbit hundreds or thousands of kilometers above

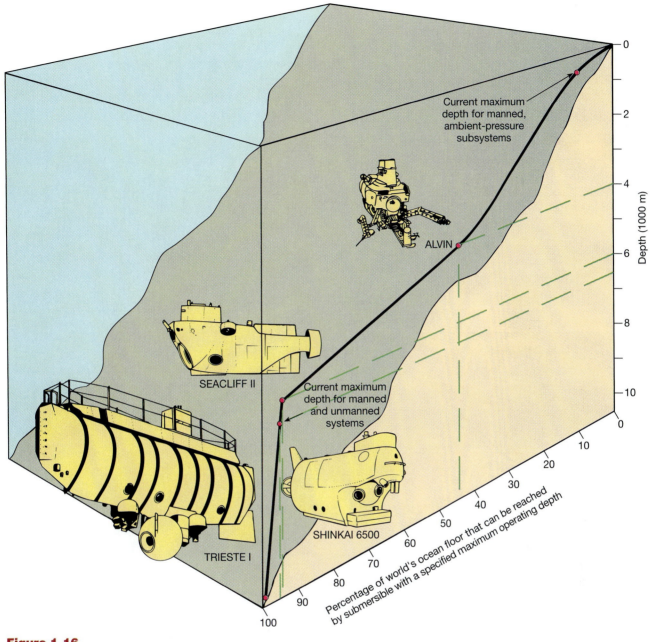

Figure 1–16

Manned Submersibles and the Ocean Depths. This figure shows the percentage of ocean floor that is reachable by four submersibles of importance in the history of ocean exploration. *Alvin,* which is the first submersible to be used extensively in marine research, can explore about 44 percent of the ocean floor with its maximum operating depth of 4000 meters (13,120 feet). *Sea Cliff II* has an operations limit of 6000 meters (20,000 feet). The *Shinkai 6500* is the world's deepest-diving manned research submersible. It is operated by the Japanese government and can explore 97 percent of the ocean floor. The 1960 dive of *Trieste* to a depth of 10,915 meters (35,801 feet) shows that it has the capability of exploring any part of the deep ocean. However, given its immobility, it is not practical for use in deep-sea exploration.

the ocean, instruments aboard satellites can measure the temperature, ice cover, color, and topography of the ocean surface, and even give us much information about the topography of the ocean floor. Figure 1–20 shows *Seasat A,* the first truly oceanographic satellite, launched in 1978. Unfortunately, it failed after only three months, but it still provided enough data to prove the great benefit of remote satellite sensing of the ocean's surface.

Many ingenious uses of today's technology in the study of the oceans are discussed in later chapters.

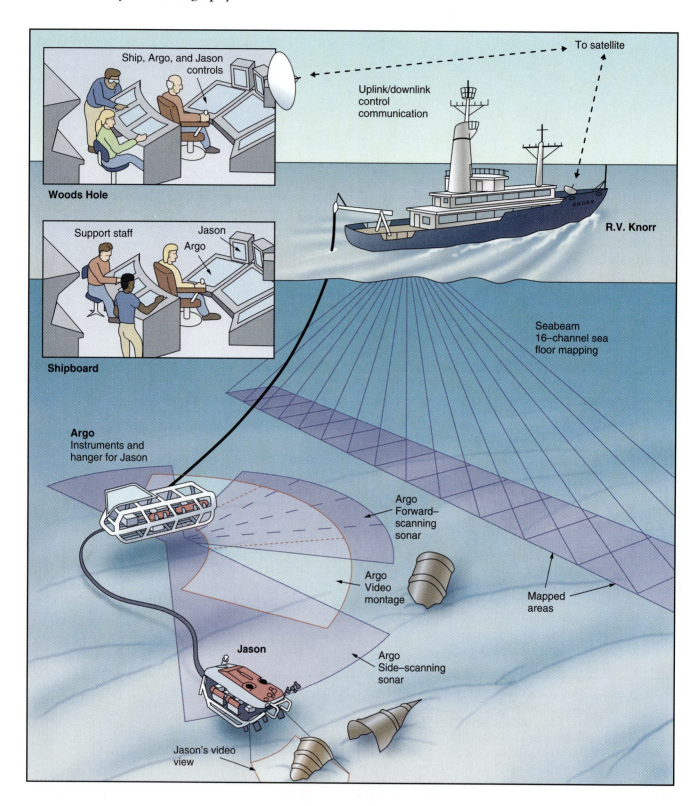

Figure 1-17

Seabeam/Argo-Jason. Seabeam, with its computerized system, makes very accurate sonar maps that can be converted to three-dimensional models of the ocean floor. When the *Argo-Jason* system reaches the ocean floor that has been modeled, the shipboard operator enjoys a view of the ocean floor equivalent to being inside the *Jason*. Both *Argo* and *Jason* use television eyes to guide the shipboard operator across the ocean floor. *Argo* sends out sonar signals to help determine its position on the three-dimensional model created by Seabeam.

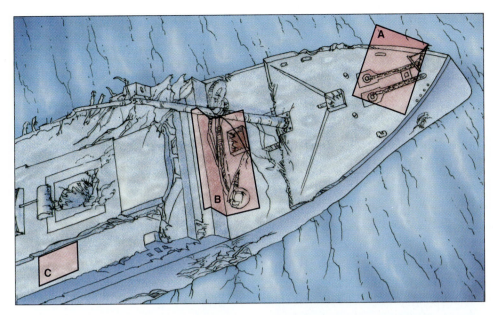

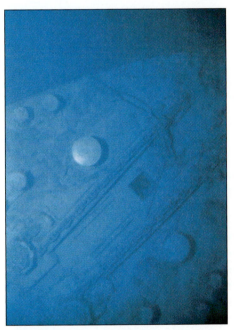

A.

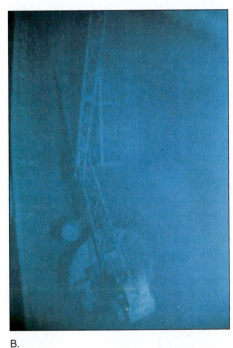

B.

Figure 1–18

RMS *Titanic* Is Photographed.
RMS *Titanic,* unseen by human eyes since it sank in 1912, 600 kilometers (373 miles) south of Newfoundland, is at last visible in television and photographic images. The ship was discovered during sea trials of the new deep-sea investigative instrument *Argo* on September 1, 1985. Photos *A* and *B* were taken by a towed photographic unit, ANGUS (Acoustically Navigated Geological Underwater Survey System), that takes high-resolution still photographs. (Note that the locations of photos are shown on the ship diagram.) *A:* Anchor chains, capstans, and windlasses. *B:* Loading cranes on the bow of the *Titanic.* Photos *C* and *D* were taken the next year when *Alvin* returned for the sea trials of *Jason. C:* Electric winch on boat deck photographed by prototype of *Jason* called *Jason Jr. D: Jason Jr.* photographed from *Alvin. Jason Jr.'s* "garage" on *Alvin* is in the foreground. (Courtesy of Woods Hole Oceanographic Institution.)

C.

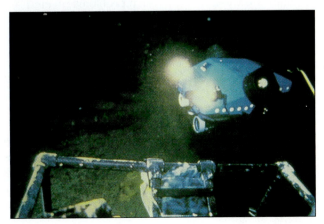

D.

A.

B.

Figure 1–19

Jason and *ABE.* *A: Jason* is the remotely operated vehicle (ROV) used in conjunction with the manned submersible, *Alvin.* It is operated on a tether and can also be operated from a surface ship. *B: ABE* (Autonomous Benthic Explorer) represents a new class of underwater vehicles that are untethered. They can remain beneath the surface for months, gathering data continuously. (Courtesy of Woods Hole Oceanographic Institute.)

These include the use of sound waves to image horizontal slices of ocean temperature by use of **ocean acoustical tomography.** This technique is similar in design to the use of X rays to image slices through the human body with computerized axial tomography, or CAT, scans.

Law of the Sea

Who owns the ocean? Who owns the sea floor? If people want to drill for oil 14 kilometers off the Saudi Arabian coast, who must they deal with? Extensive exploitation of the ocean floor for minerals and petroleum is imminent. Much of this activity will occur beyond the jurisdiction of the country that has the nearest coastline. Furthermore, problems of overfishing and pollution are worsening. Are such problems covered by long-established laws? Yes and no.

In 1609 Hugo Grotius, a Dutch jurist and statesman whose writings were important in the formulation of international law, established in *Mare liberum* the doctrine that the oceans are free to all nations. But controversy continued over whether nations could control *portions* of the oceans, such as the ocean adjacent to the coastlines.

Cornelius van Bynkershonk solved this problem in *De dominio maris,* published in 1702. It provided for national domain over the sea out to the distance that could be protected by cannons from the shore, an area called the **territorial sea.** How far was that? The first

official width of this territorial sea was probably established in 1672 when the British determined that cannon range encompassed a distance out to one league (three nautical miles) from shore. Thus, every country with a coastline had a *three-mile territorial limit.*

Because of the rapidly developing technology for drilling beneath the ocean, the first **United Nations Conference on the Law of the Sea,** held in Geneva, Switzerland, in 1958, produced a treaty relating to the *continental shelf,* the sea floor from the coastline out to where slope increases markedly. The treaty stated that prospecting and mining of minerals on the shelf are under the control of the country that owns the nearest land. Unfortunately, the seaward limit of the continental shelf varies widely and was not well-defined in the treaty.

The second United Nations Conference on the Law of the Sea was held in Geneva in 1960 and achieved little. A third conference held meetings during 1973–1982.

The Third Law of the Sea Conference came to a disappointing conclusion when a large majority voted in favor of a new Law of the Sea treaty. The vote was 130 to 4, with 17 abstentions. Most developing nations that could benefit significantly from the treaty voted for its adoption. The opposition included the United States, Turkey, Israel, and Venezuela, led by the United States. These countries support private companies that plan seabed mining operations and believe that certain provisions of the treaty will make it unprofitable. Among the abstainers were the Soviet Union, Great Britain, Belgium, the Netherlands, Italy, and West Germany, all of whom are interested in seabed mining. The treaty was

Figure 1–20

Seasat A. This satellite, which pioneered the remote sensing of the oceans, was launched July 7, 1978. It shorted out on October 10, 1978, but provided sufficient data to prove the value of satellite remote sensing to the study of the oceans. The ghostly image below *Seasat* is the HMS *Challenger,* which pioneered the study of the oceans from the ocean surface in 1872. (Courtesy of Jet Propulsion Laboratory, NASA.)

signed by the sixtieth nation and became international law in 1994. Recent modifications of the mining provisions have made it more likely that the treaty will be ratified by countries with interest in deep-sea mining.

The primary features of the treaty are as follows:

1. *Coastal nations jurisdiction.* The treaty established a uniform 12-mile territorial sea and a 200-mile **exclusive economic zone** (EEZ). Each coastal nation has jurisdiction over mineral resources, fishing, and pollution within its EEZ. If the continental shelf (defined geologically) extends beyond the 200-mile EEZ, the EEZ is extended up to 350 miles from shore.

2. *Ship passage.* The right of free passage on the high seas is preserved. It is also provided within territorial seas and through straits used for international navigation.

3. *Deep-ocean mineral resources.* Private exploitation of seabed resources may proceed under the regulation of the International Seabed Authority (ISA), within which a mining company will be chartered by the United Nations, called Enterprise. This provision, which caused industrialized nations to oppose ratification, required mining entities to fund mining at two sites—their own and one operated by Enterprise—before modifications in favor of free market principles.

4. *Arbitration of disputes.* A United Nations Law of the Sea tribunal will perform this function.

SUMMARY

In the Western world, the Phoenicians were the first great navigators. In the Pacific, people who populated the Pacific Islands played that role. Later the Greeks, Romans, and Arabs made significant contributions, but they lay dormant during the Middle Ages for over 1000 years. The Age of Discovery renewed the Western world's interest in exploring the unknown. It began with the voyage of Columbus in 1492 and ended in 1522 with the first circumnavigation of Earth by a voyage initiated by Ferdinand Magellan.

Captain James Cook, who sailed in the late 18th century, was one of the first navigators to sail the oceans with the primary purpose of learning their natural history. He also made the first accurate maps using the chronometer developed by John Harrison. Matthew Fontaine Maury was the first American to make a large contribution to ocean study with publication of *The Physical Geography of the Sea* in 1855. By the mid-19th century, the English biologists Charles Darwin, Sir John Ross, Sir James Ross, and Edwin Forbes had increased our knowledge of life in the oceans.

The voyage of HMS *Challenger,* from December 1872 to May 1876, was the first major full-scale oceanographic voyage. The significance of this exploration can be appreciated by contemplating the fact that 4717 new species of marine life were identified during the voyage. An unusual voyage aboard the *Fram,* which spent three years entrapped in Arctic ice, was initiated by the Norwegian explorer Fridtjof Nansen in 1893. It was with this voyage that the first oceanographic knowledge of the Arctic Ocean was gained. For 25 months during 1925–1927, the German ship *Meteor* crisscrossed the South Atlantic Ocean, making the first large-scale continuous depth record with an echo sounder. This voyage first revealed the true ruggedness of the ocean floor.

Recent oceanographic investigations have been numerous. In the United States, the early ones were conducted by government agencies and three pioneering oceanographic institutions: the Scripps Institution of Oceanography, the Woods Hole Oceanographic Institution, and the Lamont-Doherty Earth Observatory. They have since been joined by many educational institutions. Major projects such as the Deep Sea Drilling Project, Ocean Drilling Program, and Geochemical Ocean Sections have resulted from this widespread cooperation among institutions and nations.

The use of manned submersibles such as *Alvin* and *Sea Cliff II* has aided new discoveries in the deep ocean. Remote unmanned systems such as the Seabeam/*Argo-Jason* and ABE will be able to replace more expensive submersibles for much deep-ocean research. Many new techniques have been developed for studying the oceans; ocean acoustical tomography and remote sensing from satellites will become increasingly important.

Because of imminent ocean overfishing, pollution, and exploitation, the United Nations has conducted conferences on the Law of the Sea to develop international agreements on the resources of the oceans. These efforts were certainly set back, and may have failed, in April 1982, when developed and developing nations could not agree. Recent modifications in the treaty have significantly reduced the legal barriers to seabed resource exploitation.

KEY TERMS

Agassiz, Alexander (p. 15)
Age of Discovery (p. 8)
Alvin (p. 17)
Autonomous underwater vehicle (AUV) (p. 18)
Challenger, HMS (p. 12)
Columbus, Christopher (p. 8)
Cook, James (p. 9)
Da Gama, Vasco (p. 8)
Darwin, Charles (p. 10)
Deep Sea Drilling Project (p. 15)
Diaz, Bartholomeu (p. 8)
Ekman, V. Walfrid (p. 14)
Eratosthenes (p. 6)
Eriksson, Leif (p. 8)
Ewing, Maurice (p. 15)
Exclusive economic zone (p. 23)
Floating Instrument Platform

(FLIP) (p. 15)
Forbes, Edward (p. 12)
Fram (p. 13)
Franklin, Benjamin (p. 10)
Glomar Challenger (p. 16)
Harrison, John (p. 10)
Henry the Navigator, Prince (p. 8)
Herodotus (p. 6)
Heyerdahl, Thor (p. 7)
JOIDES Resolution (p. 16)
Lamont-Doherty Earth Observatory (p. 15)
Law of the Sea (p. 22)
Magellan, Ferdinand (p. 8)
Maury, Matthew Fontaine (p. 10)
Meteor (p. 14)
Nansen, Fridtjof (p. 12)
Ocean acoustical tomography (p. 22)

Ocean Drilling Program (p. 16)
Phoenicians (p. 6)
Ptolemy (p. 6)
Pytheas (p. 6)
Remotely operated vehicle (ROV) (p. 18)
Ross, James Clark (p. 11)
Ross, John (p. 11)
Scripps Institution of Oceanography (p. 15)
Seabeam/*Argo-Jason* (p. 18)
Sea Cliff II (p. 17)
Territorial sea (p. 22)
Vikings (p. 8)
Woods Hole Oceanographic Institution (p. 15)

QUESTIONS AND EXERCISES

1. Construct a time line showing the major events of human history that have resulted in a greater understanding of our planet in general and the oceans in particular.
2. List the Mediterranean cultures that appear on your time line in the probable order of their dominance in marine navigation.
3. Using a diagram, describe the method used by Pytheas to determine latitude in the Northern Hemisphere.
4. How did the Greek concept of geography provided by Herodotus differ from that of the Romans as given by Ptolemy?
5. While the Arabs dominated the Mediterranean region during the Middle Ages, what were the most significant ocean-related events taking place in northern Europe?
6. List some of the major achievements of Captain James Cook.
7. What was Matthew Fontaine Maury's major contribution to an increased knowledge of the oceans?
8. Define in your own words the process of evolution by natural selection.
9. What controversy developed over the observations of the Rosses and Edward Forbes?
10. List some major achievements of the voyage of HMS *Challenger*.
11. What new knowledge of the Arctic polar region did the voyage of the *Fram* provide?
12. The voyage of which ship first revealed the true relief of the ocean floor?
13. List the names and locations of three pioneering oceanographic research institutes.
14. Name the major ocean research programs designed to drill into the ocean floor and study ocean circulation.
15. Describe the functions of the Floating Instrument Platform (FLIP), Seabeam/*Argo-Jason,* and ABE.
16. Discuss possible reasons why less-developed nations believe that the open ocean is the common heritage of all, whereas the more developed nations believe that the open ocean's resources belong to those who recover them.

REFERENCES

Allmendinger, E. 1982. Submersibles: Past—present—future. *Oceanus* 25:1, 18–35.

Bailey, H. S., Jr. 1953. The voyage of the *Challenger. Scientific American* 188:5, 88–94.

Borgese, E. M., and Ginsberg, N. 1993. *Ocean yearbook 10.* Chicago: The University of Chicago Press.

Brewer, P., ed. 1983. *Oceanography: The present and future.* New York: Springer-Verlag.

Duxbury, A. C. 1971. *The earth and its oceans.* Reading, Mass.: Addison-Wesley.

Heyerdahl, T. 1979. *Early man and the ocean.* New York: Doubleday.

Knauss, J. A. 1974. Marine science and the 1974 Law of the Sea Conference: Science faces a difficult future in changing Law of the Sea. *Science* 184:1335–1341.

Mowat, F. 1965. *Westviking.* Boston: Atlantic-Little, Brown.

Ryan, P. R. 1986. The *Titanic* revisited. *Oceanus* 29:3, 2–17.

———. 1985. The *Titanic:* Lost and found, 1985. *Oceanus* 28:4, 1–112.

Sears, M., and Merriman, D., eds. 1980. *Oceanography: The past.* New York: Springer-Verlag.

SUGGESTED READING

Sea Frontiers

Baker, S. G. 1981. The continent that wasn't there. 27:2, 108–114. A history of the search for Ptolemy's *Terra Australis Incognita,* the unfound southern continent.

Charlier, R. H., and Charlier, P.A. 1970. Matthew Fontaine Maury, Cyrus Field, and the physical geography of the sea. 16:5, 272–281. A biography of Maury with emphasis on his accomplishments as superintendent of the Department of Charts and Instruments. His role in laying the first transatlantic cable and a discussion of his most famous work, *The Physical Geography of the Sea,* are included.

Engle, M. 1986. Oceanography's new eye in the sky. 32:1, 37–43. Photographs taken by Paul Scully-Power, a U.S. Navy oceanographer, from the space shuttle add to our knowledge of ocean currents and waves.

Loeffelbein, B. 1983. Law of the Sea treaty: What does it mean to nonsigners (like the United States)? 29:6, 358–366. An overview of the Law of the Sea Convention and a discussion of the implications of the treaty not being signed by all nations.

Maranto, G. 1991. Way above sea level. 37:4, 16–23. A discussion of Navstar Global Positioning System satellites and their potential for helping us map changes in sea level.

McClintock, J. 1987. Remote sensing: Adding to our knowledge of oceans—and Earth. 33:2, 105–113. The role of remote sensing, primarily in aiding us in understanding weather and climate, is discussed.

Rice, A. L. 1972. HMS *Challenger:* Midwife to oceanography. 18:5, 291–305. An interesting account of the achievements recorded during the first major oceanographic expedition.

Schuessler, R. 1984. Ferdinand Magellan: The greatest voyager of them all. 30:5, 299–307. A brief history of the voyage initiated by Magellan to circumnavigate the globe.

Van Dover, C. 1987. *Argo* Rise: Outline of an oceanographic expedition. 33:3, 186–194. A description of the first scientific use of the *Argo* system to survey hydrothermal vent activity on the East Pacific Rise off the coast of Mexico.

———. 1988. Dive 2000. 34:6, 326–333. Events related to the 2000th dive of DSRV *Alvin* on the East Pacific Rise are recounted.

Scientific American

Bailey, H. S. 1953. The voyage of the *Challenger*. 188:5, 88–94. A summary of the accomplishments of the English oceanographic expedition of 1872.

Borgese, E. M. 1983. The law of the sea. 248:3, 42–49. One hundred nineteen nations sign a convention aimed at establishing a new international scheme of regulating the oceans.

Herbert, S. 1986. Darwin as a geologist. 254:5, 116–123. Before publication of *Origin of species,* Charles Darwin considered himself to be primarily a geologist. This article outlines his contributions in this field with special attention to his theory of coral reef formation.

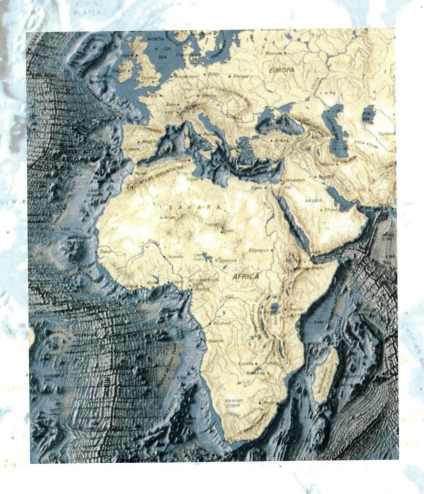

CHAPTER 2
The Ocean: Its Origin and Provinces

Learning about the world's oceans can be a wonderful and rewarding experience. The wonder is heightened when you come to understand the place of the ocean in the cosmos. We begin our discussion with aspects of the universe that we can readily observe and describe.

Then we will discuss some theories of the origin of the universe, our planet, and the sea.

We conclude the chapter with a discussion of the distribution of Earth's present ocean basins and an overview of oceanic provinces.

The Universe We See

As astronomers looked into the sky surrounding the planet Earth, they gradually developed an understanding of Earth's position in what is now called the solar system (Figure 2–1). The solar system is composed of the sun and its nine planets, which occupy a disc-shaped portion of the universe about 13 billion kilometers (8 billion miles) in diameter. Although this is an enormous structure to us, we find that it represents a minute portion of the total universe. Our sun is one of perhaps 100 billion billion stars that constitute the known universe. In terms of distances with which we are familiar, the solar system is incomprehensibly isolated. The distance from the sun to the nearest star is 40.7 thousand billion kilometers (25 thousand billion miles).

While the planets revolve around the sun, the entire system is revolving at about 280 kilometers/second (174 miles/second) around the center of the Milky Way, the **galaxy** or group of stars to which our solar system belongs. Estimates of the dimensions of the Milky Way indicate that it is a disc-shaped accumulation of stars about 100,000 light-years in diameter and 10,000 to 15,000 light-years thick near the center, thinning gradually toward the edges.

A **light-year** is a unit of measure used by astronomers to describe distances within the universe. It is equal to the distance traveled by light, during one Earth year, at the speed light travels in a vacuum—300,000 kilometers/second (186,000 miles/second). This works out to almost 10 trillion kilometers (6.2 trillion miles). Within the Milky Way are some 100 billion stars, tens of millions of which very probably possess families of planets, millions of which may be inhabited by intelligent creatures, according to the estimates of some astronomers.

All the stars you see with your unaided eye belong to the Milky Way (Figure 2–2). If you have exceptional vision and live in the Northern Hemisphere, you may see a hazy patch of light in the constellation (star group) named Andromeda. This is the faint light from some 1.5 million light-years away. It appears that galaxies are exceedingly numerous, and within a distance of a billion light-years from our galaxy, at least 100 million galaxies have been observed by telescope.

By observing light energy that radiates from these distant galaxies, astronomers have been able to determine that most are moving away from us. The velocity with which the galaxies recede from our solar system is directly proportional to their distance from it. Velocities of nearly 240,000 kilometers/second (149,000 miles/second) have been calculated for the most distant galaxies as they recede from the Milky Way. This is in excess of three-fourths of the speed of light!

You may be left with the impression that all the galaxies are being repelled by our solar system, but a more reasonable view is that galaxies are moving away from one another as would fragments from an explosion. The universe appears to be rapidly expanding, its component galaxies moving ever farther apart. If all these galaxies are moving away from one another in a manner similar to that of fragments created by explosion, we might ask if they all belonged originally to one large mass. We do not know. But if that was the case, the time required for them to have reached their present distribution may have been about 15 billion years.

Origin of Earth

This book is about the oceans, but to understand where the oceans came from, you first must read the story of how Earth, and then its atmosphere, formed. Then the story of the oceans will make sense.

As insignificant as our planet is in the overall composition of the cosmos, it is fascinating to learn what we can of its origin. No complete understanding of Earth's origin has been developed, but we can outline a probable sequence. As galactic matter revolved around its center of rotation, the stars began to coalesce as individual glowing masses due to the concentration of particles under the force of gravity. These concentrations may have occurred in local eddies, like little whirlpools in a stream, and over a vast period of time took their present form. Many of the resulting stars collected assemblages of matter that revolved around them, which we call planets.

Such a general sequence may have occurred in the formation of our sun. In its early stages, its volume may have at least equaled the diameter of our solar system today. As this large accumulation of gas and dust began to contract under the force of gravity, a small percentage of it was left behind in smaller eddies with features similar to the large eddies in which the stars developed.

As this material was left behind, it flattened itself into a disc that became increasingly dense. The stage was set for the formation of the suite of planets that revolve around the sun. Because of the increase in density of this disc, it became gravitationally unstable and broke into separate small clouds. These were protoplanets, which later consolidated into the present planets.

Protoearth was a huge mass, perhaps 1000 times greater in diameter than Earth today and 500 times more massive. The heavier constituents of the planet Earth, as with all the other protoplanets, were migrating toward the center to form the heavy core surrounded by lighter materials. The satellites we observe around the planets, such as our moon, began to form in a similar fashion.

A.

Figure 2–1

The Solar System. *A:* The orbits of the planets of our solar system are drawn to scale. *B:* Relative sizes of the sun and the planets are shown accurately, although distance is not. (Reprinted by permission from Tarbuck, E. J., and Lutgens, F. K., *Earth science*, 5th ed. (Columbus: Merrill, 1988), figs. 19.1 and 19.2.)

B.

During this early formation of the protoplanets and their satellites, the sun condensed into such a concentrated mass that forces within its interior began releasing energy through a process known as "hydrogen burning." This is the first step in a nuclear reaction called the *fusion reaction*. In the fusion reaction, when temperatures reach tens of millions of degrees, hydrogen atoms are converted to helium atoms, and large amounts of energy are released. In addition to light energy, the sun also intensely emits ionized (electrically charged) particles. In the early stages of our solar system, this emission of ionized particles served to blow

Figure 2–2

The Milky Way Galaxy. The sun and its planets are located about two-thirds the distance from the center of the Milky Way to the edge. The Milky Way completes a rotation every 200 million years. To maintain its position in the Milky Way, our solar system travels around the center of the galaxy at a velocity of about 280 kilometers/second (174 miles/second). (Reprinted by permission from Tarbuck, E. J., and Lutgens, F. K., *Earth science*, 5th ed. (Columbus: Merrill, 1988), fig. 20.17.)

away the nebular gas that remained from the formation of the planets and their satellites.

Meanwhile, the cold protoplanets nearer the sun were being warmed by the solar radiation, and their atmospheres began to boil away. The combination of ionized solar particles and the internal warming of the planets drastically shrank the planets closer to the sun. Because the gaseous envelope that surrounded the planets was composed mostly of hydrogen and helium (we may consider this to be Earth's initial atmosphere), the heating easily energized atoms of this small mass to escape the gravitational attraction of their planets. As the protoplanets continued to contract, the heat that was produced deep within their cores from the spontaneous disintegration of atoms (radioactivity) became more intense.

Earth ultimately became molten, allowing the heavier components, primarily iron and nickel, to migrate to and concentrate in the **core.** The lighter materials segregated according to their densities in concentric spheres around this core. Thus, surrounding the dense core is a zone composed of heavy iron and magnesium silicates—the **mantle.** The thin **crust** overlying the mantle varies in thickness. Oceanic crust composed of the black volcanic rock basalt is 4 to 10 kilometers (2.5 to 6.2 miles) thick. The continental crust of lighter granite is 35 to 60 kilometers (22 to 37 miles) thick.

The crust is separated from the mantle by a zone of distinct density change called the **Mohorovicic discontinuity,** or the **Moho** for short (see Figure 2–3). It is named after the Yugoslavian seismologist who discovered it. (Please note that the internal structure of Earth is much more complex than this simplified discussion has indicated.)

Origin of the Atmosphere and Oceans

The solidification of Earth's crust marks the beginning of geological history. The dating of rocks from Earth and the moon indicates that this event occurred about 4.6 billion years ago (Figure 2–4). At this time Earth had lost all but a small fraction of its original gas envelope.

Where did the material to make up the present atmosphere and oceans come from? Although Earth's surface was relatively cool by this stage in its development, there was still much volcanic activity, and we know from craters on the moon that intense meteor bombardment occurred until 3.9 billion years ago. As this bombardment continued, volcanic gases surged from the upper mantle up through the crust to form Earth's new atmosphere. This phenomenon is appropriately called *outgassing,* and it continues in volcanoes today.

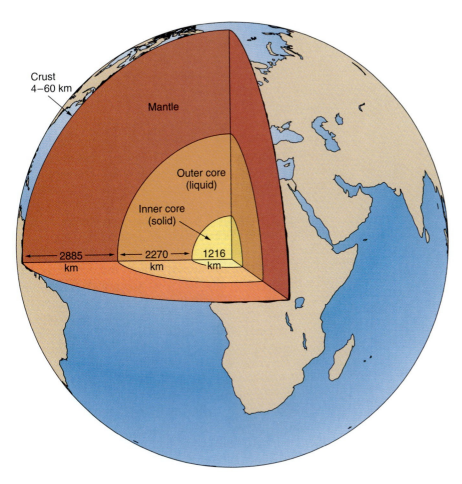

Crust
4–60 km

Mantle

Outer core
(liquid)

Inner core
(solid)

2885
km

2270
km

1216
km

Figure 2–3

The Earth's Structure. The cross section shows the three major subdivisions of Earth: core, mantle, and crust. The inner core, outer core, and mantle are drawn to scale, but the thickness of the crust is exaggerated about five times. The boundary between the crust and the mantle is the Mohorovicic discontinuity (Moho). (Reprinted by permission from Tarbuck, E. J., and Lutgens, F. K., *The earth: An introduction to physical geology,* 3d ed. (Columbus: Merrill, 1990), fig. 1.11.)

Although there is some dispute about the nature of this atmosphere, it was different from the initial one that Earth had lost (hydrogen-helium) and different from the one it has at present (nitrogen-oxygen). This formative atmosphere probably contained little free oxygen and had large percentages of carbon dioxide and water vapor. By at least 4 billion years ago, most of the water vapor had condensed to form the first permanent accumulations of liquid water on Earth's surface, and this was the beginning of the oceans (Figure 2–5).

The relentless rainfall on Earth's rocky surface weathered it, eroding particles and dissolving elements and compounds, carrying them into the newly forming oceans. This has given our oceans their present chemical composition.

How Components of Eroded Surface Rocks Were Distributed

Water entering the ocean from land dropped its load of particles on the sea floor as sediment. The dissolved elements and compounds generally stayed in solution. The components of surface rocks that were freed by chemical weathering today are found (1) dissolved in the ocean,

(2) as components of Earth's atmosphere, or (3) chemically bound into the sediments on the ocean bottom.

Estimates that have been made for these various amounts cannot be exact. The estimate for the ocean is probably one of the most accurate, because the ocean volume is well known and its composition is comparatively uniform because it is continually mixed. (Earth's oceans freely interconnect around the continents. This is why Magellan was able to sail around the world.) The estimates are less accurate for rock and the sediments, which are quite heterogenous. Nevertheless, geochemists have drawn up balance sheets for the various elements, and those calculated by different individuals working on the problem agree fairly well. For those elements that are most common in the surface rocks, the geochemical equations balance well (Figure 2–6).

There are some components that do not balance in any of the attempts that have been made. These elements and compounds are too abundant in the atmosphere, the oceans, and the sediments to be accounted for by the volume of surface rocks that have undergone chemical weathering. All these substances are volatile (gaseous) and are therefore called *excess volatiles*. Most abundant are water vapor and carbon dioxide. The answer is

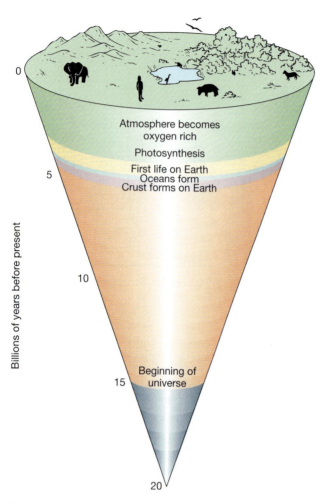

Billions of years before present

0

Atmosphere becomes
oxygen rich

Photosynthesis

First life on Earth
Oceans form
Crust forms on Earth

5

10

Beginning of
universe

15

20

Figure 2–4

Major Events in the Evolution of Earth. Based on best estimates from present evidence, the universe began to form about 15 billion years ago. About 4.5 billion years ago, Earth's solid crust solidified and the oceans are believed to have existed by 4 billion years ago. The earliest fossil evidence of life—bacteria-like cells—dates back 3.8 billion years. Although fossils that resemble present-day photosynthetic bacteria exist in rocks 3.5 billion years old, an oxygen-rich atmosphere is not indicated until 2 billion years ago. Photosynthesis was certainly well established by then.

Earth's continued outgassing. Volcanic activity is producing these gases and venting them into the atmosphere.

Development of the Oceans, Their Basins, and Their Saltiness

Due to gravity, the oceans occupy lower areas on Earth. But what formed these lower areas, and why are they lower?

When Earth began to form its solid crust about 4.6 billion years ago, the mass of the oceans was zero. At present, ocean mass is 14 trillion trillion grams. We do not know what the rate of volcanic release of gases from the mantle was throughout the period of 4.6 billion years. However, if we assume it to have been equal to the present rate, we can visualize the formation of Earth's oceans as a gradual process.

Does Earth's continuing volcanism mean that the oceans are gradually covering more and more of the surface? Not necessarily, because the *area* that the oceans cover is directly determined by the *volume that the basin in which they form is able to accept*. During the initial solidification of Earth's crust, there may not have been distinct continents and basins. These may have formed gradually as the oceans themselves were formed.

If we assume that the continents formed gradually, we can see that the capacity of the ocean's basins could have gradually increased along with the volume of water that was being produced during the same period. Thus, it is quite possible that the ocean's surface area has been similar for a long period of geological time, and that the only major change in the ocean's character has been an increase in depth. This process will be discussed further in Chapter 3.

Did the oceans start out salty? They certainly did, for many of the compounds eroded from surface rocks contained the elements that make up salt—chlorine, sodium, magnesium, potassium, and so forth. Considering the history of the oceans you might ask whether they have possessed the same salinity throughout their history, or whether they are growing more or less saline. By far the most important component of salinity is the chloride ion, Cl^-, which is produced by the same process that produces the water vapor forming the oceans: outgassing. Thus it boils down to this: Has the proportion of water vapor to chloride ion outgassed remained constant throughout geological time or has it varied?

We have no indication of any fluctuation in this ratio throughout geologic time. Therefore, we must consider, on the basis of the present evidence, that the oceans' salinity has been relatively constant, despite the increase in volume throughout the 4.6 billion years since Earth's crust began to form. In Chapter 5, we will further explore the ocean's salinity.

Life Begins in the Oceans

Recent research has offered several new ideas on the origin of life on Earth: The organic building blocks of life may have arrived embedded in meteors, comets, and cosmic dust. Life may have originated in association with hydrothermal vents deep in the ocean. A primitive form of photosynthesis (by which plants manufacture food) may have occurred in conjunction with the iron-sulfur mineral called pyrite. The ability to reproduce may be associated with the structure of clay minerals.

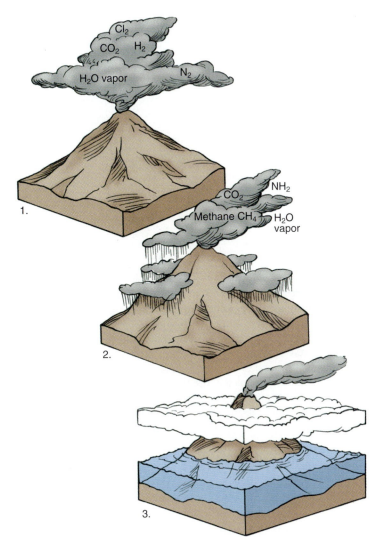

Figure 2–5

Formation of Earth's Oceans. Widespread volcanic activity released water vapor and smaller quantities of carbon dioxide (CO_2), chlorine gas (Cl_2), nitrogen (N_2), and hydrogen (H_2). This produced a cloudy water vapor atmosphere that also contained carbon dioxide, methane, and ammonia. As Earth cooled, the water vapor (1) condensed into thick clouds and (2) fell to Earth's surface. There it (3) accumulated to form the oceans.

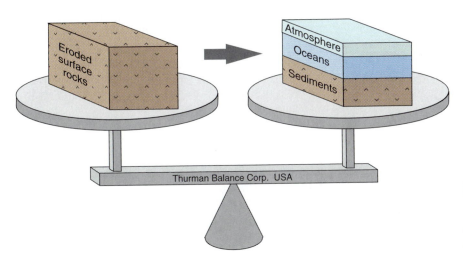

Figure 2–6

All the material in eroded surface rocks is accounted for by material in the atmosphere, oceans, and sediments.

However, we will emphasize events as they are most strongly supported by the fossil record. This record indicates that the basic building blocks for the origin of life were provided by materials already present on Earth, and that the most likely place in which they could interact and produce life was in the oceans.

Oxygen—Essential to Modern Life

Oxygen makes up almost 21 percent of our present atmosphere. It is essential to modern life for two reasons. First, oxygen is what we must inhale so that our bodies can "burn" (oxidize) food, releasing energy to our cells. Second, we depend on oxygen for protection from the ultraviolet radiation from the sun that constantly bombards our planet. Much of the ultraviolet radiation is absorbed in our upper atmosphere by the ozone layer (ozone molecules consist of three oxygen atoms). This allows very little of the ultraviolet energy to reach Earth's surface.

However, we believe that oxygen was essentially absent from Earth's early atmosphere. Why? Oxygen may well have been outgassed, but oxygen and iron have a strong affinity for each other (consider how common rust—iron and oxygen—is). Iron in Earth's early crust would have bound up outgassed oxygen immediately.

Of course, if oxygen was missing from the early atmosphere, the ultraviolet radiation would have readily penetrated the atmosphere and reached the surface of the young oceans. These oceans and the atmosphere contained the gases methane (CH_4), ammonia (NH_3), nitrogen (N_2), and carbon dioxide (CO_2).

Laboratory experiments have shown that exposing a mixture of hydrogen, carbon dioxide, methane, ammonia, and water to ultraviolet light, plus an electrical spark readily available in the form of lightning, will produce a large assortment of organic molecules (Figure 2–7).

The exposure of the mixture to ultraviolet radiation causes photosynthesis. **Photosynthesis** is a chemical reaction in which energy from the sun becomes stored in organic molecules (basically sugars). The production of organic substances in the shallow oceans that developed early in Earth's history must have resulted in a vast amount of "organic" material, as do the laboratory experiments, but this material did not yet represent life on planet Earth.

Carbon—The Organizer of Life

Let us now consider the element most critical to the buildup of the organic compounds we have just mentioned: carbon. At one time, the chemistry of carbon compounds was synonymous with organic chemistry.

Chemical study of organisms is called "organic chemistry," and because the compounds in living things contain carbon, we are constantly concerned with very

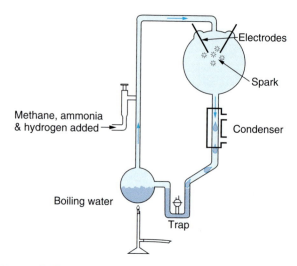

Figure 2–7

Synthesis of Organic Molecules. The apparatus used by Stanley L. Miller in the 1952 experiment that resulted in the synthesis of the basic components of life, amino acids. A mixture of water vapor, methane, ammonia, and hydrogen was subjected to an electrical spark that provided the energy for synthesis. This mixture is thought to resemble the composition of the atmosphere and oceans that existed when "organic" molecules first formed on Earth.

large molecules. The greater percentage of inorganic (noncarbon) molecules contain fewer than 10 atoms, but in organic chemistry, molecules that contain 100 atoms are common.

This fact arises out of the "combining power" of the carbon atom. The combining power of an atom is its ability to bond with other atoms. Chemists specify this ability for every kind of atom by stating how many hydrogen atoms it can bond with. This property is called *valence*; carbon has a valence of four, which means it can combine with four atoms of hydrogen.

To demonstrate how this ability gives carbon an unusually high combining power that makes large molecules, let us compare the carbon atom with other common atoms that have valences less than four. If we consider chlorine (a gas with a valence of one), you can see that a single atom of chlorine combines with a single atom of hydrogen to produce a molecule of hydrogen chloride, HCl:

$$Cl—H$$

The gas oxygen has a valence of two. Thus an atom of oxygen combines with two molecules of hydrogen to form water, H_2O:

Another gas common in our atmosphere, nitrogen, has a valence of three. One atom of nitrogen will combine with three atoms of hydrogen to produce the gas ammonia, NH_3:

$$
\begin{array}{c}
\text{H} \\
| \\
\text{N} \\
\diagup \quad \diagdown \\
\text{H} \qquad \text{H}
\end{array}
$$

We return to the carbon atom, with a valence of four, and show how one carbon atom will combine with four atoms of hydrogen to produce the gas methane, CH_4:

$$
\begin{array}{c}
\text{H} \\
| \\
\text{H--C--H} \\
| \\
\text{H}
\end{array}
$$

You can readily see that the higher the valence of an atom, the greater the complexity of molecules it can form. Carbon has a valence of four and chlorine has a valence of one, so carbon must have a much greater combining power than chlorine or any other element with a valence of one. Consider chlorine as it combines with the metal sodium (Na), which also has a valence of one.

NaCl

You can see that these two elements combine to form only one compound, sodium chloride (common table salt). But when the valences increase, atoms can combine in multiple ways to form different compounds.

For example, methane, an important constituent of Earth's early atmosphere, is only the simplest of the group of carbon compounds called *hydrocarbons*—those containing multiple combinations of carbon and hydrogen atoms. Somewhat more complex are the *carbohydrates,* which are composed of carbon and water molecules. Belonging to this group are the sugars and starches. Even more complex carbon compounds are known as *lipids* (*fats*) and *proteins*. Aiding carbon further in forming large molecules is the ability of carbon atoms to attach to one another and produce long chains and rings.

We do not know precisely how the organic material took on the characteristics that allowed it to become living substance. Through some process, however, the organic material became chemically self-reproductive, developed the ability to actively seek light, and began to make food and grow toward a characteristic size and shape dictated by internal molecular codes. Thus, we call carbon the organizer of life.

Plants and Animals Evolve

The very earliest forms of life must have been *heterotrophs*. These organisms depend on an external food supply. That food supply was certainly abundant in the form of molecules of the nonliving organic material from which the simplest organisms originated.

The *autotrophs,* organisms that do not depend on an external food supply but manufacture their own, eventually evolved. The first autotrophs may have been similar to our present-day bacteria of the *anaerobic* type, which live without atmospheric oxygen. They may have been able to use inorganic compounds to release energy for producing their own food internally. This process is called **chemosynthesis.**

At some later date, the more complex single-celled autotrophs probably evolved. They developed a green pigment called *chlorophyll,* which captures the sun's energy. Through photosynthesis, these organisms produced their own food from the carbon dioxide and water that surrounded them.

Fossils that may be the remains of photosynthetic bacteria have been recovered from former sea floor rocks dating back almost 3.5 billion years. However, an oxygen-rich atmosphere is not indicated until about 2 billion years ago, following the appearance of complex cells with nuclei that first appear about 2.1 billion years ago. By 1.8 billion years ago, the oxygen content of the atmosphere had reached one-tenth of the present concentration. This addition of free oxygen to the atmosphere surely caused the extinction of many anaerobic single-celled species that had evolved in an oxygen-free environment. The survivors of this environmental crisis are the progenitors of Earth's present dwellers.

Photosynthesis and Respiration. Photosynthetic cells had developed security in the form of their built-in mechanism to manufacture a food supply. Heterotrophs, not so endowed, were at a considerable disadvantage because they had to search for food. To meet this need, animals eventually evolved, with their inherent mobility and an awareness or a consciousness that gave them a much more active role in the biological community than that of the plants.

Photosynthetic organisms and animals developed in a beautifully balanced environment where the waste products of one filled the vital needs of the other. We can see this expressed chemically by looking at the complementary process of *photosynthesis* and *respiration.* In the photosynthetic process, energy is captured and stored in the form of organic compounds (food sugars). This is an *endothermic* chemical reaction, one that *stores* energy. **Respiration** is the reaction by which energy stored in the foods produced by photosynthesis is extracted by the plants and animals to carry on their life processes. It is the opposite of photosynthesis; it is an *exothermic* reaction, one that *releases* energy. The process is shown in Figure 2–8.

Every living organism that inhabits Earth today is the result of an evolutionary process of *natural selec-*

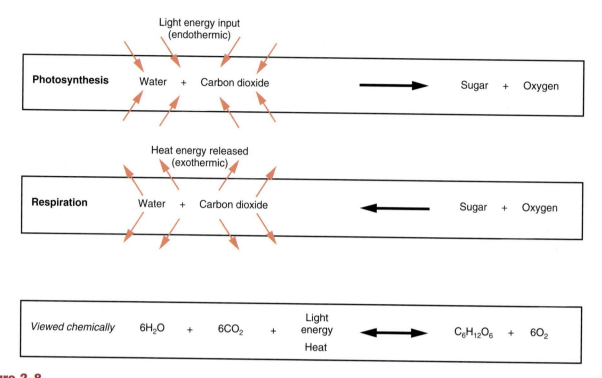

Figure 2–8

Photosynthesis and Respiration.

tion that has been going on since these early times and by which various life forms (species) have been able to inhabit increasingly numerous niches within the environment. As these diverse life forms adapted to various environments, they also modified the environments in which they lived.

As plants emerged from the oceans to inhabit the terrestrial environment, they changed the landscape from a harsh, bleak panorama (which we may envision as being typical of the lunar surface) to the soft green hues that cover much of Earth's land surface today. Other changes were manifested in the ocean itself as vast quantities of the hard parts of dead organisms accumulated on the bottom. Some of these accumulations we now see exposed on the continents, sometimes high in mountains. We take this as evidence that these rocks formed from sea floor sediments that were elevated to their present position by great forces within Earth. Because over half of the rocks exposed at the surface of continents originally formed on the ocean floor, we can learn much about the oceans of the past by studying these rocks.

Probably the most important environmental modification of the environment by its living inhabitants involved the atmosphere. This change, the release of large quantities of oxygen, made it possible for most present-day animals to develop. A byproduct of photo-

synthesis in the early oceans was the free oxygen that makes up 21 percent of our present atmosphere. Carbon dioxide, which at one time must have made up a large portion of our atmosphere, was removed by photosynthesis to produce a tiny (but important) concentration of 0.035 percent in Earth's present atmosphere. These changes in the atmosphere must have had a great effect on climates and organisms (Figure 2–9).

In petroleum and coal deposits, we see the remains of plant and animal life that became buried in an oxygen-free environment. This allowed their energy to remain in storage for millions of years. These deposits, which we call fossil fuels (coal, oil, and natural gas), provide us with over 90 percent of the energy we consume today. Not only do we, as animals, depend upon the present productivity of plants to supply the energy required by our life processes, but we also depend very heavily on the energy stored by plants during the geological past.

Because of increased burning of fossil fuels for home heating, industry, power generation, and transportation over the past century, the atmospheric concentration of CO_2 and other gases that help warm the atmosphere has increased. Many people are concerned that Earth is warming to a degree that could cause serious problems for your generation and those that follow. This phenomenon, referred to as the increased *greenhouse effect,* is discussed in Chapter 6.

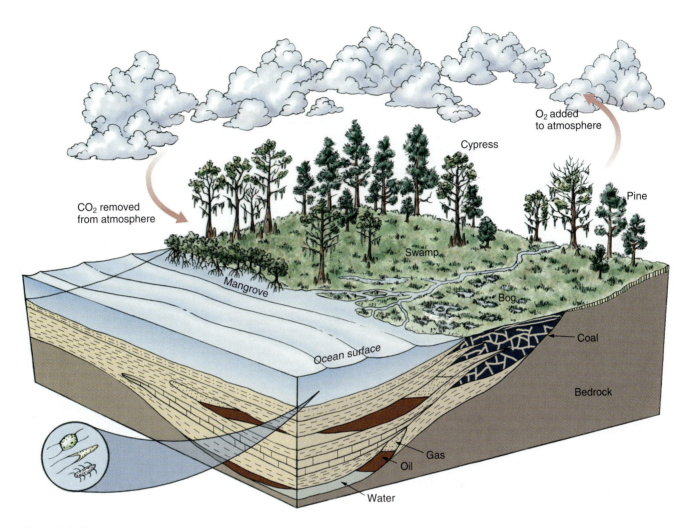

Figure 2-9

The Effect of the Evolution of Plants on Earth's Environment. As photosynthetic cells became established in the ocean, Earth's atmosphere began to enrich in oxygen and to decrease in carbon dioxide as a result of photosynthesis. As these cells and the microscopic animals that ate them died and sank to the ocean floor, their remains were incorporated into the marine sediments. Some of the remains were buried before they could be completely decomposed and were eventually converted to petroleum (oil and gas). When plants migrated onto the continents, they evolved into the various forms that support the animals that subsequently evolved there. In the low-lying, highly productive wetlands, plant remains accumulated in the sediment to be converted to peat and eventually to coal.

Bathymetry

Throughout the passage of time, the shape of the ocean basins has changed as continents have ponderously migrated across Earth's surface in response to forces in Earth's interior—a process that will be discussed in Chapter 3. The ocean basins as they presently exist reflect the long process of Earth history we have been discussing (Figure 2–10).

Bathymetry is the study of ocean depths. Because bathymetry examines the vertical distance between the essentially "flat" ocean surface and the

mountains, valleys, and plains of the sea floor, bathymetry is analogous to topography of the dry land surface. Instead of measuring height above sea level, as on land, *bathymetry measures depth below sea level.* An understanding of the general bathymetric features of the ocean is important because these features are related ultimately to the origin of the ocean basins, as well as to physical and biological phenomena to be discussed in following chapters.

As investigations into the depth of the ocean have proceeded, it has become apparent that a broad shelflike feature has generally developed around the

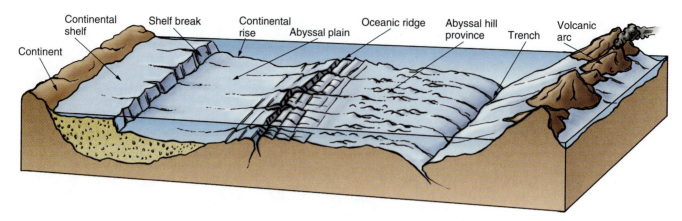

Figure 2–10

Marine Physiographic Provinces. A generalized longitudinal profile showing submarine physiographic provinces. To the left of the oceanic ridge is a continental margin typical of the Atlantic Ocean, with a well-developed continental shelf, slope, and rise at the surface of a thick sedimentary deposit. The features to the right of the ridge are more typical of the Pacific Ocean, where trenches and associated volcanic arcs are common.

continents. At varying distance from shore, it steepens and slopes off into deep basins. Quite commonly, linear mountain ranges run through the basins. In all oceans, but especially around the margin of the Pacific Ocean, linear trenches up to 11 kilometers (6.8 miles) in depth separate the continental slopes from the deep-ocean basin.

Figure 2–11 is a *hypsographic curve* (from the Greek *hypsos,* meaning height). It shows the distribution of Earth's solid surface both above and below sea level. The curve indicates that 29 percent of Earth's surface is above the oceans (land), and 71 percent is be-

neath the oceans. The mean elevation of the continents above the sea (840 meters, or 2755 feet) and mean depth of the oceans (3800 meters, or 12,460 feet) result from the different densities of continental crust and oceanic crust. The ocean crust is denser.

Isostasy

The rock called *basalt* is dark and dense, and it originated as molten magma beneath Earth's crust. The magma is extruded through openings on the sea floor

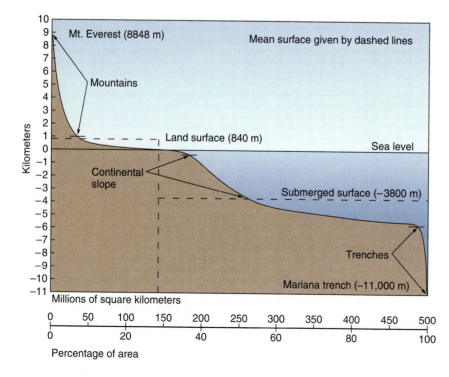

Figure 2–11

Elevations of Earth's Crust. The horizontal dashed lines indicate that the average height of the continents is 840 meters (2755 feet) above sea level and that the average depth of the oceans is 3800 meters (12,460 feet) below sea level. The vertical dashed line marks the division between land and sea. About 24 percent, or 150 million square kilometers, of Earth's surface is land above sea level. It is clear that the area of Earth below sea level is many times greater than the area of the continents above sea level. (Reprinted with permission from Tarbuck, E. J., and Lutgens, F. K., *The earth: An introduction to physical geology,* 3d ed. (New York: Macmillan, 1990), fig. 1.14.)

(essentially underwater volcanoes) and hardens to form *oceanic crust*. It averages about 8 kilometers (5 miles) in thickness. The basaltic oceanic crust has an average density of about 3.0 grams per cubic centimeter (three times the density of water). The continents are composed of a lighter-weight and lighter-colored crustal material called *granite*. It ranges in density from 2.67 grams per cubic centimeter near the surface to 2.8 grams per cubic centimeter deep beneath the continental mountain ranges. The continental crust averages about 35 kilometers (22 miles) in thickness but may reach a maximum thickness of 60 kilometers (37 miles) beneath the highest mountain ranges.

The key fact here is that continental crust is less dense than oceanic crust. Both "float" on the denser mantle beneath, but the less dense continental crust floats higher. This concept of crustal flotation, comparable to the flotation of ice on water, is called **isostasy** (Figure 2–12).

If we take a block of ice with a normal ice density of 0.91 and float it on water with a density of 1.0, the ice will sink into the water only until it displaces a volume of water equal to the ice's mass. When this displacement, called **buoyancy,** occurs, 91 percent of the ice mass will be submerged. Similarly, about 55 percent of the mass of the continents is submerged in the mantle. Actually, we would expect about 91 percent of the continents to be submerged because they are, on the average, about 91 percent of the density of the upper mantle. However, where high heat flow in the mantle heats and reduces the density of the mantle beneath the continents, this low-density mantle provides isostatic compensation. In such situations as exist beneath the southwestern United States and eastern Africa, the low-density mantle material provides the buoyancy that would normally be provided by the deep root of continental crust.

Techniques of Bathymetry

As described in Chapter 1, the first systematic bathymetry of the oceans was made in 1872, when the HMS *Challenger* undertook its historic three-and-one-half-year voyage. Every few thousand kilometers, *Challenger's* crew measured the depth with a sounding line (a weight on the end of a line marked off in fathoms; a fathom is 6 feet). These measurements indicated that the deep ocean floor was not flat but had significant topographic relief, as dry land does.

It wasn't until 1925 that the *Meteor,* using an **echosounder,** identified a mountain range running through the center of the South Atlantic Ocean. In fact, until recently, most of our knowledge of ocean bathymetry has been provided by the echosounder (Figure 2–13).

Echosounding lacks detail because, from a ship 4000 meters (13,100 feet) above the ocean floor, the sound beam emitted from the ship widens to a diameter of about 4600 meters (15,000 feet) at the bottom. Because the earliest sound to return from the bottom will likely be from the nearest (highest) peak within this broad area, one may obtain an inaccurate view of the topography of the deep sea floor.

Oceanographers who wish to know more than just the depth must use strong low-frequency sounds produced by air guns that will penetrate beneath the ocean floor and reflect off the contact zones between different layers of sediment. This provides *seismic reflection profiles.*

The *precision depth recorder (PDR)* was developed in the 1950s. With a more focused high-frequency sound beam, it could provide depths to a resolution of about 1 meter (3.3 feet). Throughout the 1960s, PDRs provided a reasonably good view of the ocean floor of the North Pacific and North Atlantic oceans.

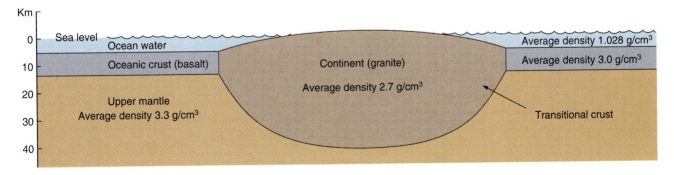

Figure 2–12

Isostasy. Isostasy is a state of equilibrium, or balance, reached by different components of Earth's crust and the mantle. The lithosphere (the outer layer of Earth's structure) contains the continental and oceanic crustal units as well as the upper mantle. For the lithosphere to maintain a uniform average density, continental crustal units that extend up to 14 kilometers (8.7 miles) above the floor of the ocean basins must extend deep roots into the mantle.

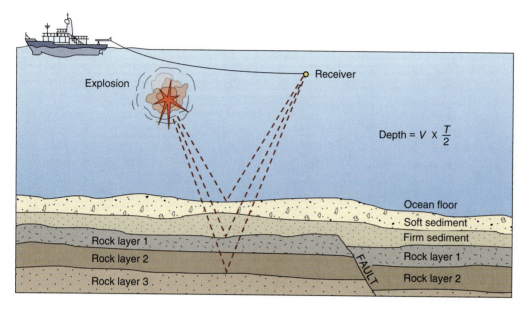

Depth = $V \times \dfrac{T}{2}$

Ocean floor
Soft sediment
Firm sediment
Rock layer 1
Rock layer 2

Figure 2–13

Seismic Profiling. Low-frequency sound that can penetrate bottom sediments is emitted by an explosion. It reflects off the boundaries between rock layers and returns to the receiver. The depth of each reflecting layer is equal to the velocity of sound travel (V) times one-half the time (T) required for the sound to travel from the source of the reflecting layer and back to the receiver.

A great leap forward occurred in the 1970s when the multibeam sonar and side-scan sonar began to give us a more precise picture of the ocean floor. The first multibeam echosounder, *Seabeam,* made it possible for a survey ship to map the topography of the bottom along a strip of ocean floor up to 60 kilometers (37 miles) wide (Chapter 1). The system uses sound emitters directed away from both sides of the ship. The map is developed with the aid of computers. A more recent version, called *GLORIA,* is illustrated in Figure 2–14.

Deeply towed side-scan systems such as *Deep-Tow,* operated by the Scripps Institution of Oceanography, also became available in the 1970s. Towed about 100 meters above the ocean floor, it provides high-resolution images of the seafloor up to 1 kilometer (0.6 miles) wide. It can image features less than 10 meters (30 feet) in size. A recently developed deep-tow side-scan system called *TOBI* (Towed Ocean-Bottom Instrument) can resolve features as small as 2 meters (6.5 feet) across-track and 12 meters (40 feet) along-track. It is towed

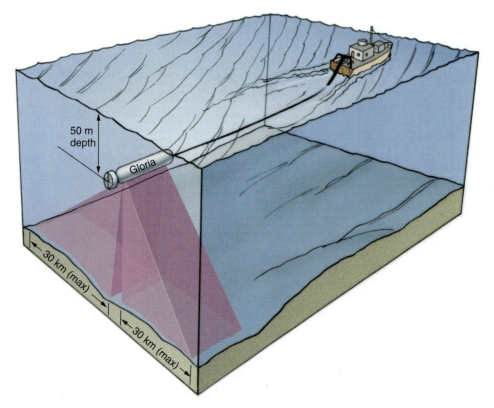

Figure 2–14

GLORIA System for Side-scanning Sonar Surveys. GLORIA is an acronym for Geological Long-Range Inclined ASDIC. ASDIC is an acronym for Acoustical Side Direction. The neutrally buoyant GLORIA is towed 300 meters (1000 feet) behind a research vessel. At water depths of 5000 meters (16,400 feet), it can map a strip of ocean floor 60 kilometers (37 miles) wide. At shallower depths, the width of the strip is reduced.

about 500 meters (1600 feet) above the seafloor and provides a swath image 6 kilometers (10 miles) wide.

Due to the gravitational effects on the ocean surface, sea level rises or falls 4 meters (13 feet) with each decrease or increase in ocean depth of 1000 meters (3300 feet). Altimeters aboard the short-lived *Seasat A* remotely sensed these differences and provided data to produce a map of the ocean surface. The maps in Figure 2–15 show how these changes in ocean surface elevation reflect major bathymetric features of the ocean floor.

The Shape of the Seafloor

Figure 2–10 showed the general features of the sea floor. We will now examine these as two broad groups, the *continental margins,* which form the aprons around all continents, and the *deep-ocean basins.*

Continental Margin

The continental margins comprise the continental shelf and shelf break, the continental slope with its submarine canyons and turbidity currents and the continental rise that extends toward the deep-ocean basins.

Continental Shelf. Extending from the shoreline is a shelflike feature called the continental shelf, which is geologically part of the continent (it is continental crust, covered with marine sediment). During the geologic past, much of it was exposed above the shoreline when colder climates prevailed during the ice ages, freezing more of Earth's water, which lowered sea level. Over time, as sea level has fluctuated, the shoreline has migrated back and forth across the shelf. Thus, its general bathymetry can usually be predicted by looking closely at the topography of the adjacent coastal region. With few exceptions, this coastal topography can be expected to extend beyond the shore and onto the continental shelf.

We define the **continental shelf** as a shelflike zone extending from the shore beneath the ocean surface to a point at which a marked increase in slope occurs. This latter point is referred to as the *shelf break,* and the steeper slope beyond the break is known as the **continental slope** (shown in Figure 2–10).

The width of continental shelves varies from a few tens of meters to about 1300 kilometers (800 miles). You will find it helpful to look at Figure 2–22, a remarkable painting of the seafloor, as you read this description. The broadest shelf developments occur off the northern coasts of Siberia and North America in the Arctic Ocean and in the North and West Pacific from Alaska to Australia. Note that, very generally, the eastern coasts of continents have broader shelves than western coasts (for example, see North and South America).

The average width of the continental shelf is about 70 kilometers (43 miles), and the average depth at which the shelf break occurs is about 135 meters (443 feet). Around the continent of Antarctica, however, the shelf break occurs much deeper, at 350 meters (2200 feet). The mean slope of the continental shelf is quite gentle, only about a tenth of a degree (0°7' or 1.9 meters/kilometer or 10 feet/mile). Sediment is transported across the continental shelf by current flow and submarine landslides, which are often generated by earthquakes.

Continental Slope. The continental slope beyond the shelf break has features similar to mountain ranges on the continents. The break at the top of the slope may be from 1 to 5 kilometers (0.6 to 3 miles) above the deep-ocean basin at its base. In areas where the slope descends into submarine trenches, even greater vertical relief is measured. Off the west coast of South America, the total relief from the top of the Andes Mountains to the bottom of the Peru-Chile Trench is about 15 kilometers (9.3 miles).

Continental slopes vary in steepness from 1° to 25° and average about 4°. Around the margin of the Pacific Ocean, the continental slopes are steeper than in the Atlantic and Indian oceans, because the slope is associated with the processes that form coastal mountain ranges and submarine trenches. Slopes in the Pacific Ocean average more than 5°, whereas those in the Atlantic and Indian oceans are about 3°. The differences in slope steepness are related to tectonic processes, which are discussed in Chapter 3.

Submarine Canyons and Turbidity Currents. The continental *slope* is cut by large **submarine canyons** that resemble the largest of canyons cut on land by rivers. To a lesser extent, the continental *shelf* exhibits these canyons as well (Figure 2–16). Like the canyons cut by rivers on the continents, submarine canyons have tributaries and steep V-shaped walls. Exposed in the walls of the canyons are rocks of widely varied age and type.

Submarine canyons are in some cases related to *land* river systems, because the canyons lead right into the mouths of the rivers. The majority of submarine canyons, however, do not tie so nicely with land drainage systems. Many are confined exclusively to the continental slope. The most obvious objection to explaining them as "drowned river valleys" is that they continue to the base of the continental slope, which on average is some 3500 meters (11,500 feet) below sea level. Because rivers lose their ability to erode shortly after reaching the ocean, it seems impossible that rivers could have cut canyons this far below sea level.

Side-scan sonar surveys indicate that the continental slope is dominated by submarine canyon topography along the Atlantic coast from Hudson Canyon near New York City to Baltimore Canyon. Canyons confined

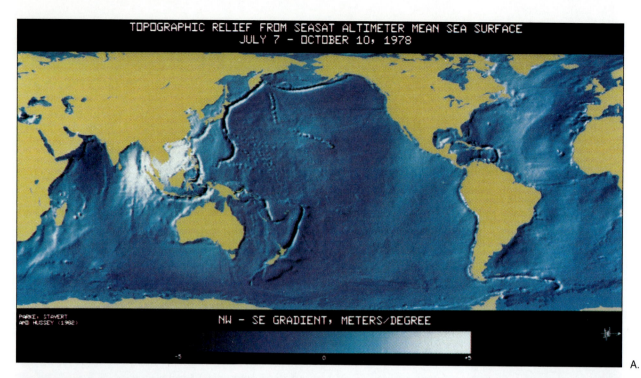

TOPOGRAPHIC RELIEF FROM SEASAT ALTIMETER MEAN SEA SURFACE
JULY 7 - OCTOBER 10, 1978

PARKE, STAVERT
AND HUSSEY (1982)

NW - SE GRADIENT, METERS/DEGREE

A.

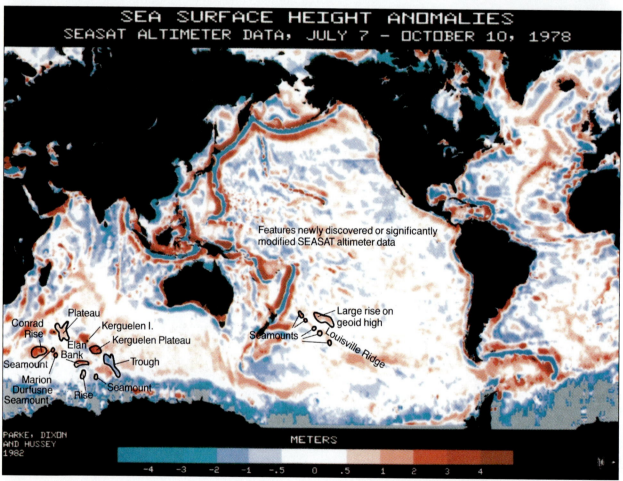

SEA SURFACE HEIGHT ANOMALIES
SEASAT ALTIMETER DATA, JULY 7 - OCTOBER 10, 1978

Features newly discovered or significantly
modified SEASAT altimeter data

Large rise on
geoid high

Seamounts

Louisville Ridge

Plateau
Kerguelen I.
Conrad
Rise
Kerguelen Plateau
Elan
Bank
Seamount
Trough
Marion
Durfusne
Seamount
Rise
Seamount

PARKE, DIXON
AND HUSSEY
1982

METERS

-4   -3   -2   -1 -.5   0   .5   1   2   3   4

B.

Figure 2–15

Ocean Surface Mapped by SEASAT. *A:* Due to gravitational effects, the ocean surface rises over high volcanic peaks and drops over depressions such as trenches on the ocean floor. An altimeter on the *SEASAT* satellite that operated in 1978 was able to measure these elevation in the ocean surface with sufficient precision to map major bathymetric features on the ocean floor. *B:* In the southern Indian Ocean and Pacific Ocean, this remote sensing of the ocean surface was able to identify some new bottom features and provide information that significantly altered the shape or location of previously identified features (circled).

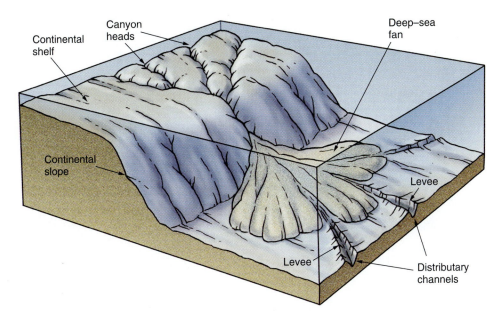

Figure 2–16

Submarine Canyons. Typical submarine canyons along the Atlantic coast, showing erosional and depositional features.

to the continental slope are straighter and have steeper canyon floor gradients than those that extend into the continental shelf. Their characteristics suggest that the canyons are initiated on the continental slope by some marine process and cut their way into the continental shelf as they age.

Probably the most widely supported hypothesis to explain the origin of submarine canyons is that which explains the erosion by **turbidity currents.** These currents are dense flows of sediment-laden water. Turbidity currents in many canyons confined to the continental slope apparently are initiated after sediment moves across the continental shelf into the head of the canyon and accumulates there. Initiation of the turbidity current may result from an earth tremor or other disturbance that makes the mass move down the slope under the force of gravity. The mass is deposited at the base of the continental slope in a fan-shaped structure (Figure 2–16).

Interesting evidence that turbidity currents move across the ocean bottom comes from trans-Atlantic cable breaks that occurred on the continental shelf and slope south of Newfoundland after an earthquake in 1929. The location is shown by a hexagon in Figure 2–17, left of center. These cables broke in a pattern: The cables closest to the earthquake broke first and those that crossed the shelf at greater depths and distances from the earthquake were broken later. This phenomenon could be explained by a dense flowing mass, moving down the slope and snapping the cables as it passed. The velocity would have been approximately 27 kilometers/hour (16.8 miles/hour). The possibility that such a mass may move with this velocity makes it easier to understand how the submarine canyons may have been formed on the continental slope and continental shelf.

Continental Rise. Deep-sea fans accumulate as deposits at the mouths of submarine canyons (Figure 2–16). The merging of these deep-sea fans along the base of the continental slope has formed the **continental rise.**

A major factor in shaping the continental rise is the strong **western boundary undercurrent (WBUC),** or slope current. It flows toward the equator at the base of the continental slope along the western boundaries of most ocean basins (Figure 2–17). This deep boundary current and the continental rise over which it flows have been extensively studied along the east coast of North America using submersibles, surface ships, and side-scan sonar.

The WBUC, which originates in the Norwegian Sea and flows into the North Atlantic through deep troughs between Greenland and Scotland, veers to the right and flows snugly against the base of the continental slope at all times. (The reason for this is the Coriolis effect, discussed in Chapter 6.) Picking up volcanic debris from Iceland and sediment from turbidity flows, and scouring sediment from the continental slope and rise, the WBUC flows at velocities that reach 40 centimeters/second (1 mile/hour). As the current negotiates the bends in the continental slope, its velocity decreases and sediment drops out, producing the deposits called *drifts* or *ridges* shown in Figure 2–17.

Summary. The continental margin and its components (shelf, slope, and rise) are of great interest to oceanographers. Increased knowledge of continental-margin features and the processes that create them are of importance in selecting ocean sites for disposal of toxic waste and for effective exploitation of oil, gas, and mineral resources. Now we will turn our attention further seaward, toward the deep-ocean basin.

Figure 2–17

Floor of the North Atlantic Ocean. Arrows indicate the path of North Atlantic Deep Water or western boundary undercurrent. Hexagon indicates cable breaks associated with the 1929 Grand Banks earthquake. (North Atlantic Panorama excerpted from *The world ocean floor* by Bruce C. Heezen and Marie Tharp. Copyright © 1977 by Marie Tharp. Reproduced by permission of Marie Tharp, 1 Washington Ave., South Nyack, NY 10960.)

Deep-Ocean Basin

The deep ocean is generally called the *abyss,* from a Greek word meaning bottomless. It is not literally bottomless, of course, but it seemed so to people thousands of years ago.

The ocean floor consists of oceanic crust covered by thinner sediment than occurs on the continental margin. Many volcanic peaks extend to various elevations above the ocean floor. Those that extend above sea level are, of course, islands. Those that are below sea level but rise more than 1 kilometer (0.6 miles) above the deep-ocean floor are called **seamounts.** If they have flattened tops, they are called tablemounts, or **guyots,** after the Swiss scientist A. H. Guyot. (Seamounts and guyots are illustrated in Figure 2–18.)

Guyots are less common than seamounts and are found in the Pacific Ocean. The tops of the guyots are between 1800 and 3000 meters (5900 and 9850 feet) below the present ocean surface. There is evidence that the flattened surface of the guyots was produced by wave action at the ocean surface, so significant subsidence of the ocean basin in these regions must have occurred since the guyots were at the surface. (An explanation of this subsidence is presented in Chapter 3.) Figure 2–18 shows many seamounts and guyots, especially in the Pacific.

Volcanic features on the ocean bottom that are lower than seamounts are called **abyssal hills** (Figure 2–18). They cover a large percentage of the entire ocean basin floor and have an average relief of about 200 meters (650 feet). Many such features are found buried beneath the sediments of abyssal plains of the Atlantic and Indian oceans. In the Pacific, the lower rate of sediment deposition has left extensive regions dominated by abyssal hills. These areas are called **abyssal hill provinces.** The evidence of volcanic activity on the bottom of the Pacific Ocean is particularly widespread—more than 20,000 volcanic peaks exist there.

Abyssal Plains.   Extending from the base of the continental rise into the deep-ocean basins are flat depositional surfaces with slopes of less than 1:1000 that cover extensive portions of the basins (Figure 2–18). These **abyssal plains** are particularly extensive and flat in the Atlantic Ocean, although they are occasionally interrupted by volcanic peaks. Most abyssal plains lie at depths between 4500 meters (15,000 feet) and 6000 meters (20,000 feet).

In the deep ocean, bottom current speeds average at least 8 centimeters/second (approximately 0.15 miles/hour). Such currents may play an important role in distributing fine sediment of the deep abyssal plains.

Trenches.   The continental rise commonly occurs at the base of the continental slope, where it meets the abyssal plain. However, sometimes the slope descends into long, narrow, steep-sided **trenches.** A dramatic example in Figure 2–18 is the Aleutian Trench off the coast of Alaska.

The deepest portions of the world's oceans are found in these trenches. Such features are characteristic of the margins of the Pacific Ocean along the coast of South America, Central America, the Aleutian Islands, Japan, and Australia (Figure 2–18). Figure 2–19 shows the great relief that occurs along South America's western coast, from the Andes Mountains to the Peru-Chile Trench.

The lowest point known on Earth's surface is the Challenger Deep of the Mariana Trench, with a depth of 11,022 meters (36,150 feet). Table 2–1 presents the dimensions of several trenches. The landward side of the trench rises as a volcanic arc that may produce islands (the origin of Japan) or a volcanic mountain range at the margin of a continent (the origin of the Andes Mountains).

Oceanic Ridges and Rises.   In Figure 2–18, observe the nearly continuous, fractured-looking ridges that extend through the middle of the ocean basins. These ridges form Earth's longest mountain chain, extending across some 65,000 kilometers (40,400 miles) of the deep-ocean basin. The features are called **oceanic ridges** where they are mountainous with steep and irregular slopes and **oceanic rises** where the slopes are more gentle. The best explored of these features are the Mid-Atlantic Ridge and the East Pacific Rise (Figure 2–22).

This system of mountains is entirely volcanic and is composed of lavas with a basaltic composition characteristic of the oceanic crust. The Mid-Atlantic Ridge was a focal point during the development of the theory of global plate tectonics because of its position, spearing the Atlantic Ocean into equal halves. Recently, the East Pacific Rise has been studied because of the hydrothermal vents and associated biological communities located in its central rift valley.

Table 2–1
Dimensions of Trenches

Trench	Depth (m)	Mean Width (km)	Length (km)
Peru-Chile	8,100	100	5,900
Aleutian	7,700	50	3,700
Middle America	6,700	40	2,800
Mariana	11,022	70	2,550
Kurile	10,500	120	2,200
Kermadec	10,000	40	1,500
Tonga	10,000	55	1,400
Philippine	10,500	60	1,400
Japan	8,400	100	800

Figure 2–18

The World Ocean Floor. (By Bruce C. Heezen and Marie Tharp, 1977. Copyright © 1977 Marie Tharp. Reproduced by permission of Marie Tharp.)

120° 140° 160° 180° 160° 140° 120° 100° 80° 60°

70°

60°

40°

20°

0°

20°

40°

60°

70°

CANADA
ABYSSAL
PLAIN

BAFFIN
BAY

WRANGEL
ISLAND

CHUKCHI SEA

NEW SIBERIAN
ISLANDS

BERING SEA

ALASKA

NORTH AMERICA

HUDSON
BAY

Aleutian Tr.

Emperor Seamounts

Kurile Tr.

Mendocino F. Z.

Japan Tr.

Hawaiian Chain

Clarion F. Z.

Mid. America Tr.

East

Clipperton F. Z.

Galápagos
Rift

Philippine Tr.

Mariana Tr.

Line Is.

Pacific

SOUTH AMERICA

Marquesas Is.

Peru-Chile Tr.

Tuamotu Is.

Society Is.

Rise

Cook Is.

Kermadec - Tonga Tr.

AUSTRALIA

Eltanin F. Z.

ANTARCTICA

Editions Pierre Charron, 51 rue Pierre-Charron. 75008 Paris., Draeger, Imp.

47

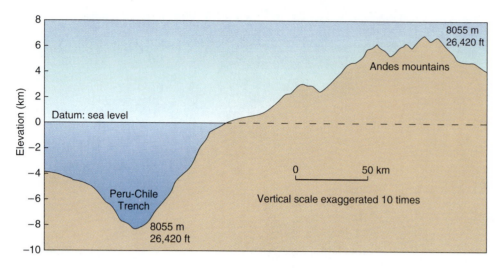

Figure 2–19

Profile across Peru-Chile Trench and Andes Mountains. This profile of the coast of Chile is typical of the deep trenches common around the margin of the Pacific Ocean. Note that over a distance of 200 kilometers (125 miles) a change in elevation occurs of more than 14,900 meters (48,870 feet).

Oceanic ridges and rises extend an average of 2.5 kilometers (1.5 miles) above the abyssal plains or abyssal hill provinces on either side. Oceanic ridges practically surround Antarctica, as careful study of Figure 2–18 reveals.

Fracture Zones. In Figure 2–18, note that the oceanic ridges and rises look as though they have been sliced like a meatloaf into hundreds of sections. The slices intersect all oceanic ridges at nearly right angles, forming linear bands of mountains and troughs. In the Pacific Ocean, where the scars are less rapidly covered by sediment than in other ocean basins, fracture zones extend visibly for thousands of kilometers and have widths of up to 200 kilometers (120 miles). The origin of these fractures is quite interesting and will be explained in Chapter 3.

S U M M A R Y

Our solar system, consisting of the sun and nine planets, belongs to the galaxy of stars we call the Milky Way. All the stars we see with the unaided eye belong to this galaxy, yet some 100 million galaxies are thought to exist in the universe. Most seem to be moving away from us as though all were set in motion by a single large explosion.

Stars like our sun coalesced into masses large enough to become fusion reactors and to give off energy at high rates. The planets were smaller, so they did not reach this state.

The massive protoearth was composed mostly of hydrogen and helium, but as it condensed and heated up, these elements were driven off into space. Earth became molten and developed a layered structure of core, mantle, and crust. The crust is of two types: a thin basaltic unit, which is denser oceanic crust, and the thicker, less dense granite crust of the continents. During this period, Earth also developed an atmosphere rich in water vapor and carbon dioxide produced by volcanic activity. As Earth's surface cooled sufficiently, the water vapor condensed and accumulated in depressions on the surface to give Earth its first oceans.

It is in these oceans that life is thought to have begun. As ultraviolet radiation fell on the oceans with their dissolved carbon dioxide, methane, and ammonia, inorganic molecules may have combined to produce carbon-containing molecules that are now formed naturally on Earth only by organisms. Chance combinations of these molecules eventually produced heterotrophic organisms (which can't make their own food) that were probably similar to present-day anaerobic bacteria. Eventually, autotrophs evolved that made their own food (chemosynthetic). Later some cells included chlorophyll in their makeup. Plants were the result.

The photosynthesis of plants extracted carbon dioxide from the atmosphere and released free oxygen at Earth's surface. This produced an oxygen-rich atmosphere in which animals as we know them could survive. Eventually, both plants and animals evolved into forms that could survive in the stark environment of the continents, producing the lush continental environments we enjoy today.

Seventy-one percent of Earth is covered by oceans. The continents stand high and bound the oceans because they are composed of granite, which is less dense than the basaltic crust that underlies the oceans. Both continental and oceanic "float" on the underlying mantle (isostasy).

Much of our knowledge of the ocean floor has been obtained using seismic surveys (topography of the ocean floor) and side-scan sonar (used to make strip maps of ocean floor topography).

The seafloor is divided into the continental margin and deep-ocean basin. Extending from the shoreline of continents are gently sloping continental shelves. Sediment is transported across the continental shelves by currents and slides. When the shelves reach an average depth of 135 meters they steepen into the continental slope. Cutting deep into the slopes are submarine canyons caused in part by turbidity currents. Turbidity currents deposit their sediment load at the base of the continental slope to produce deep-sea fans that merge to produce a gently sloping continental rise.

The continental rises gradually become flat, extensive, deep-ocean abyssal plains, often penetrated by volcanic peaks. In the Pacific Ocean where sedimentation rates are low, abyssal plains are not extensively developed, and abyssal hill provinces cover broad expanses of ocean floor. In some places, particularly the Pacific Ocean margin, the continental slope does not merge into a continental rise but continues down into deep linear trenches.

Oceanic ridges and rises are volcanic mountain ranges running through the deep-ocean basins. Volcanic peaks protruding from the abyssal plains are called seamounts and guyots. Rugged fault scars called fracture zones cut across vast distances of ocean floor and offset the axes of oceanic ridges and rises.

KEY TERMS

Abyssal hill (p. 45)
Abyssal hill province (p. 45)
Abyssal plain (p. 45)
Bathymetry (p. 37)
Buoyancy (p. 39)
Chemosynthesis (p. 35)
Continental rise (p. 43)
Continental shelf (p. 41)
Continental slope (p. 41)
Core (p. 30)

Crust (p. 30)
Echosounder (p. 39)
Galaxy (p. 28)
Guyot (p. 45)
Isostasy (p. 39)
Light-year (p. 28)
Mantle (p. 30)
Mohorovicic discontinuity (Moho) (p. 30)
Oceanic ridge (p. 45)

Oceanic rise (p. 45)
Photosynthesis (p. 34)
Respiration (p. 35)
Seamount (p. 45)
Solar system (p. 28)
Submarine canyon (p. 41)
Trench (p. 45)
Turbidity current (p. 43)
Western boundary undercurrent (WBUC) (p. 43)

QUESTIONS AND EXERCISES

1. Why is it thought that the observed motions of galaxies were initiated by an explosion?
2. What is the origin of Earth's atmosphere?
3. What is the origin of Earth's oceans?
4. New water is continually being released to the atmosphere by volcanic activity. Why does this not necessarily mean that the oceans will progressively cover an increasing percentage of Earth's surface?
5. Why is it believed that free oxygen was not released into the atmosphere by volcanic activity?
6. How does the presence of oxygen in our atmosphere help reduce the amount of ultraviolet radiation that reaches Earth's surface?
7. How did the photosynthesis that produced the first organic molecules in the early oceans differ from plant photosynthesis? How does chemosynthesis differ from photosynthesis?

8. When we say carbon has a greater combining power than oxygen, we are referring to their chemical valences. How is valence related to combining power?
9. Describe some basic characteristics of living things.
10. Discuss photosynthesis and respiration and explain their relationship.
11. As plants evolved on Earth, great changes in Earth's environment were produced. Describe some of these major changes caused by plants.
12. Compare the concept of isostasy with that of buoyancy, which explains the flotation of ice in water.
13. Describe the major features of the continental margin: continental shelf, continental slope, continental rise, submarine canyon, and deep-sea fans.
14. In which ocean basin are most ocean trenches found?
15. Describe the following: seamount, abyssal hill, guyot, oceanic ridge, fracture zone.

REFERENCES

Anderson, R. N. 1986. *Marine-geology: A planet earth perspective.* New York: John Wiley.
Damuth, J. E.; Kolla, V.; Flood, R. D.; Kowsmann, R. O.; Monteiro, M. C.; Gorini, M. A.; Palma, J. J.; and Belderson, R. H. 1983. Distributary channel meandering and bifurcation patterns on the Amazon deep-sea fan as revealed by long-range side-scan sonar (GLORIA). *Geology* 11:2, 94–98.

Field, M. E.; Gardner, J. V.; Jennings, A. E.; and Edwards, D. E. 1982. Earthquake-induced sediment failures on a 0.25° slope. Klamath River delta, California. *Geology* 10:10, 542–545.

Glaessner, M. F. 1984. *The dawn of animal life: A biohistorical study.* Cambridge: Cambridge University Press.

Gregor, B. C.; Garrels, R. M.; Mackenzie, F. T.; and Maynard, J. B., eds. 1988. *Chemical cycles in the evolution of the earth.* New York: John Wiley.

Hay, A. E.; Burling, R. W.; and Murray, J. W. 1982. Remote acoustic detection of a turbidity current surge. *Science* 217:4562, 833–835.

Holland, J. D. 1984. *The chemical evolution of the atmosphere and oceans.* Princeton, N.J.: Princeton University Press.

Rubey, W. W. 1951. Geologic history of seawater: An attempt to state the problem. *Geological Society of America Bulletin* 62:1110–1119.

Schopf, J. W. 1993. Microfossils of the early archean Apex Chart: New evidence of the antiquity of life. *Science* 260:5108, 640–646.

Strom, K. M., and Strom, S. E. 1982. Galactic evolution: A survey of recent progress. *Science* 216:4546, 571–580.

Sverdrup, H. U.; Johnson, M. W.; and Fleming, R. H. 1942. Reprinted 1970. *The oceans: Their physics, chemistry, and general biology.* Englewood Cliffs, N.J.: Prentice Hall.

Twichell, D. C., and Roberts, D. G. 1982. Morphology, distribution and development of submarine canyons on the United States Atlantic continental slope between Hudson and Baltimore canyons. *Geology* 10:8, 408–412.

Weirich, F. H. 1984. Turbidity currents: Monitoring their occurrence and movement with a three-dimensional sensor network. *Science* 224:4647, 384–387.

SUGGESTED READING

Sea Frontiers

Feazel, C. T. 1986. Asteroid impacts, sea-floor sediments, and extinction of the dinosaurs. 32:3, 169–178. A discussion of the evidence in marine sediments that may support the theory that the demise of the dinosaurs was caused by the impact of a large meteor.

Mark, K. 1976. Coral reefs, seamounts, and guyots. 22:3, 143–149. A discussion of the role of global plate tectonics in explaining the distribution of seamounts, guyots, and the evolution of coral reefs.

Rice, A. L. 1991. Finding bottom. 37:2, 28–33. Depth-sounding devices developed by ingenious early navigators are described.

Schafer, C., and Carter, L. 1986. Ocean-bottom mapping in the 1980s. 32:2, 122–130. The use of SeaMARK (Seafloor Mapping and Remote Characterization), a side-scan sonar device, in mapping the continental margin off the coast of Labrador is described.

Scientific American

Badash, L. 1989. The age-of-the-Earth debate. 261:2, 90–97. A history of the development of knowledge concerning Earth's age.

Barrow, J. D., and Silk, J. 1980. The structure of the early universe. 242:4, 118–128. Although the universe is inhomogeneous on the small scale of the solar system or a galaxy, it is very homogeneous on the scale of the universe as a whole.

Emery, K. O. 1969. The continental shelves. 221:3, 106–125. The nature of the continental shelves and the effect of the advance and retreat of the shoreline across them as a result of glaciation are discussed.

Frieden, E. 1972. Chemical elements of life. 227:1, 52–64. The roles of the 24 elements known to be essential to life are discussed, including some background on how they may have been selected from the physical environment in which life evolved.

Gott, J. R., and Gunn, J. E. 1976. Will the universe expand forever? 234:3, 62–79. Based on data regarding the recession of galaxies, average density of matter, and chemical elements, the authors suggest expansion will be reversed.

Heezen, B. C. 1956. The origin of submarine canyons. 195:2, 36–41. Theories explaining the origin of submarine canyons are presented along with data on the location and nature of such features.

Herbert, S. 1986. Darwin as a geologist. 254:5, 116–123. Before publication of *The Origin of Species,* Charles Darwin considered himself to be primarily a geologist. This article outlines his contributions in this field, with special attention given to his theory of coral reef formation.

Horgan, J. 1991. In the beginning. 264:2, 116–125. An overview of data relating to the validity of the big bang theory of the origin of the universe.

Kasting, J. F.; Toon, O. B.; and Pollack, J. B. 1988. How climate evolved on the terrestrial planets. 258:2, 90–97. A discussion of the possible sequence of events that culminated with the atmospheres that now exist on Mercury, Venus, Earth, and Mars.

McMenamin, M. A. S. 1987. The emergence of animals. 256:4, 94–103. A discussion on how the explosive diversification of animal forms 570 million years ago may have been related to the breakup of a single large continental landmass.

Menard, H. W. 1969. The deep-ocean floor. 221:3, 126–145. A summary of the dynamic effects of sea-floor spreading is presented with a description of related sea-floor features.

Stebbins, G. L., and Ayala, F. J. 1985. The evolution of Darwinism. 253:1, 72–85. New advances in molecular biology and new interpretations of the fossil record add to the knowledge of evolution.

Vidal, G. 1984. The oldest eukaryotic cells. 250:2, 48–57. Cells with a nucleus appear to have evolved as marine plankters 1.4 billion years ago.

Wilson, A. C. 1985. The molecular basis of evolution. 253:4, 164–175. Mutations within the genes of organisms play an important role in evolution at the organismal level.

CHAPTER 3
Tectonics and the Ocean

The fact that we live on a dynamic planet, on which movement is the rule rather than the exception, has long been understood. The massive amounts of energy released by volcanic eruptions and earthquakes indicate that processes are at work that constantly change the face of our planet. However, geologists and oceanographers—and virtually everyone else—clung to the precept that the continents themselves were static, remaining stationary in their position.

Many Earth scientists were not prepared until the 1960s to accept a dynamic concept of Earth that was of much broader scope, including the movement of continents over Earth's face and the changing shapes of ocean basins. In this chapter we briefly discuss the contributions of those who brought into general acceptance the plate tectonics concept of Earth's dynamics and describe the basic principles of this phenomenon, which is still incompletely understood.

Geologists sometimes refer to the movement of continental masses across Earth's surface as **continental drift.** Marine geologists who have studied the ocean floor and developed theories concerning why the continents are moving relative to one another may use the term **seafloor spreading.** This name derives from evidence indicating that new oceanic crust and rigid upper-mantle material are being formed along the oceanic ridges and rises. This newly formed material then moves at right angles away from the axes of the ridges and rises. The continents float on this denser material and are slowly carried along by the moving layer of denser rock.

The current term, which encompasses the totality of this process, is **global plate tectonics.** Earth's outer layer is divided into a dozen large slabs, or plates, like pieces of eggshell on an egg. The interaction of these plates as they move builds the structural features of Earth's crust that geologists observe. Thus, *tectonics* refers to the building of Earth's crustal structural and is derived from the Greek *tektonikos,* which means "to construct."

The concept of migrating continents is not new. Such a possibility was suggested early in the 19th century, and the first theory attempting to explain the movement was presented in 1912. With such a long history of awareness of this possible movement, why has its acceptance been so long in coming?

Plate Tectonics Theory

Alfred Wegener (Figure 3–1) is considered by most scientists to be the pioneer of the modern continental drift theory. This German scientist originally was drawn to the concept to explain the ancient climates that were evidenced in the rocks deposited in ocean basins and on landmasses. Wegener's theory was published in 1912 and encountered a great deal of resistance from the scientific community.

Wegener postulated that about 200 million years ago all the continental mass of Earth was one large continent, which he called **Pangaea** ("all Earth"). About 180 million years ago, the continent began to break up, and the various continental masses we know today started to drift toward their present positions. Since it was impossible at that time to cite a plausible mechanism for such movement, the theory did not receive wide acceptance.

Although many geologists in the Southern Hemisphere had accepted continental drift as a geologic reality, it was not until the 1950s that geologists of the Northern Hemisphere began to give it serious attention. The impetus for the renewed attention arose from the study of Earth's ancient magnetism. The British geophysicist S. Keith Runcorn used continental movement to explain his observations of the magnetic properties of the rocks of Europe and North America. (The details of this will be discussed later in this chapter.)

On the basis of this study of Earth's ancient magnetic fields, convincing arguments could be made that the continents had drifted relative to one another. As study continued, additional data suggested the mechanism by which the movement might have taken place. In the next section, we will look at the early observations that led to the initial interest in the theory of continental drift, and at more recent findings that led to the theory of global plate tectonics. We will first examine the evidence on the continents, and then on the seafloor.

Figure 3–1

Alfred Wegener. Shown waiting out the 1912–1913 Arctic winter during an expedition to Greenland, where he made a 1200-kilometer (745-mile) traverse across the widest part of the island's ice cap. (Courtesy of Bildarchiv Preussischer Kulturbesitz, West Berlin.)

Continental Jigsaw Puzzle

Before we look at the scientific evidence that supports plate tectonics, we should consider how well the continents fit together. Early investigators tried to arrange the continents to achieve a reasonable fit and support their data. Most of these attempts used the existing *shoreline* as the margin of the continent. However, sea level changes over time, so a shoreline fit would be good only if the shoreline at the time the continents separated could be simulated. We consider the continental shelf and continental slope to be part of the continental mass, based on magnetic intensity studies.

Sir Edward Bullard, an English geophysicist, constructed a computer fit of all the continents in 1965. The best fit of the continents was at the 2000-mile (6560-feet) depth contour. This contour is approximately halfway down the continental slope (Figure 3–2).

Evidence from the Continents

To test the fit of the continents, we may compare the rocks along their margins to identify those of the same type and age. Identification is not easy in some areas. During the millions of years since continental separation, younger rocks may have been deposited, covering those that might hold the key to the past history of the continents. However, there are many areas where such rocks are available for observation, and in these areas we can compare the ages by the use of (1) **fossils,** the remains of ancient organisms preserved in rocks, and (2) **radioactive dating** of the rocks.

The Fossil Record

The use of fossils became an important dating device in the early 19th century when it was realized that a particular layer of sedimentary rock would contain the remains of organisms that, as a group, were unique in time. Rocks laid down earlier and those laid down later would have a distinctly different assemblage of plant and animal fossils. Once these assemblages are recognized, a geologist can tell whether rocks exposed in one area are younger or older than in another (Figure 3–3A). The unique character of these assemblages is the result of evolution throughout geologic time. Appendix III shows the general pattern of this evolution and the geologic time scale.

Dating by such a method is referred to as **relative dating.** One can tell only whether or not an assemblage is relatively younger or older than another, but not its *actual,* or *absolute,* age in years. Such assemblages are abundant in sedimentary rocks of the last 600 million years. Note that this is but a little over one-eighth of the time represented by Earth's existence,

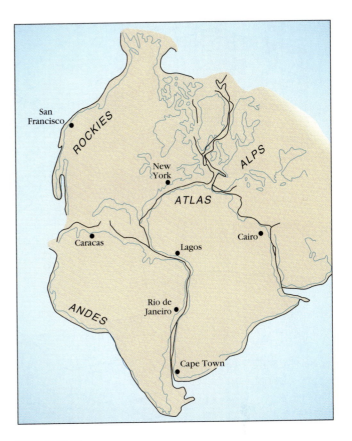

Figure 3–2

Computer Fit of Continents by Sir Edward Bullard. Sir Edward Bullard's fit of the continents was attempted in 1965 after a convincing fit pattern had been achieved in 1958 by the Australian geologist S. Warren Carey without the aid of computers. Bullard's fit was in complete agreement with that of Carey. (From *Continental Drift* by Don and Maureen Tarling, © 1971 by G. Bell & Sons, Ltd. Reprinted by permission of Doubleday & Co., Inc.)

which is thought to exceed 4.6 billion years. Therefore, this method cannot be used to compare the ages of the continental margins composed of rocks older than 600 million years.

Radioactive Dating

Most of the rocks we find on the continents contain small amounts of radioactive elements such as uranium, thorium, and potassium. Radioactive elements break down into atoms of other elements. Each radioactive **isotope** is an atom of an element that has an atomic weight different from that of other atoms of the element. Each isotope has a specific **half-life,** the time it takes for one-half of the atoms in a sample to decay to atoms of some other element.

By comparing the quantities of the radioactive isotope with the quantities of their decay products in rocks (Figure 3–3B), the age of the rocks may be determined

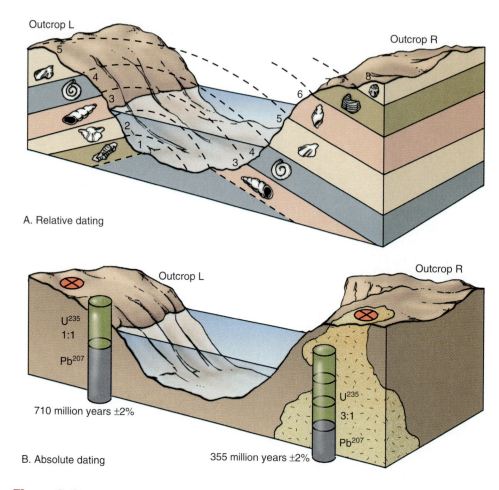

A. Relative dating

B. Absolute dating

U^{235}
1:1

Pb^{207}

710 million years ±2%

U^{235}
3:1

Pb^{207}

355 million years ±2%

Outcrop L

Outcrop R

Figure 3–3

Relative and Absolute Dating. *A:* Fossil assemblages used in relative dating occur in out-crops L and R. Assemblage 3 matches, telling us that the rocks in this segment of the out-crop formed at the same time. A similar correlation can be done for assemblages 4 and 5. We could not, however, tell how many years ago this formation occurred when fossil evidence was the only dating tool we had. The concept of superposition states only that younger sediments are laid down on older deposits. *B:* Outcrops L and R contain the ra-dioisotope uranium 235 (U^{235}). It has a half–life of 710 million years and decays to lead 207 (Pb^{207}). Dating with radioisotopes is possible for many types of rocks but is more broadly possible with igneous rocks like the basalt and granite that make up most of the oceanic and continental components, respectively, of Earth's crust. At outcrop L the ratio of U^{235} to Pb^{207} of 1:1 means that half of the U^{235} atoms have decayed to Pb^{207}, so the rocks are one half-life of U^{235} old, or 710 million years. In outcrop R, the ratio of U^{235} to Pb^{207} is 3:1. This means that one-fourth of the original U^{235} atoms have decayed to Pb^{207}, so the rock is one-half of a half-life old, or 355 million years.

within 2 to 3 percent. In very old rocks, this seemingly small error can be quite large—for example, 2 percent of 2 billion years is 40 million years. However, this method of age determination is a powerful tool and is the first possibility we have had of determining the *actual* age of rocks in years. We refer to such dating as **absolute dating.** As compared with the relative dating possible with the use of fossils, absolute dating tells us not only which rocks are younger or older but how much younger or older.

Ancient Life and Climates

The fossil record of plants and animals contained in sedimentary rocks can tell us much about the environments of the past. For instance, we can determine whether an organism lived in the ocean or on land by looking at its body structure and adaptations. Where it lived also can be told by the characteristics of the sedimentary rock itself.

Some fossil assemblages and rocks could not have formed under the climatic conditions where they exist today. These anomalous assemblages and rocks may be explained in two ways: (1) The climate changed or (2) the rocks moved.

The dominant factor controlling climatic distribution is latitudinal position on the rotating Earth. Assuming that Earth's axis of rotation has not changed significantly throughout its history, we may conclude that given latitudinal belts have possessed climatic characteristics that have not changed greatly during the evolution of life on Earth. This leaves us with the conclusion that these assemblages and rock types must have moved to their present position by migration of the continental masses.

Laurasia and Gondwanaland

Modern corals are animals whose accumulated skeletons build reefs where the water is clear and shallow and its temperature does not fall below 18°C (64.4°F). Although we cannot be sure that such conditions have been required throughout the long history of coral evolution, we may assume that the conditions required were similar.

Exactly the same species of coral are found in rocks 350 million years old in western Europe and eastern North America, as well as throughout the Alps and Himalayas. Other fossil evidence indicates that, throughout much of the 600-million-year fossil record, a major ocean separated two large continents, one composed of North America, Europe, and Asia to the north and the other composed of South America, Africa, India, Australia, and Antarctica to the south. This ocean has been given the name **Tethys Sea,** and the supercontinents were **Laurasia** to the north and **Gondwanaland** to the south. (Figure 3–4C). The record also indicates that the two supercontinents periodically came into direct or close contact across the Tethys Sea. It appears that both continents merged to form one great landmass, Pangaea, about 200 million years ago. Pangaea was surrounded by the ocean **Panthalassa** (one sea) (Figure 3–4B).

Fossils also support the existence of the Tethys Sea. From 350 to 285 million years ago there existed two distinct plant assemblages on the two supercontinents. The Laurasian assemblage included many tropical plants that were incorporated into the sediment to form the extensive coal beds mined throughout the eastern United States and Europe.

To the south, throughout Gondwanaland, existed an assemblage with a few species of plants thought to have grown in a cold climate. Supporting this belief are indications of glaciation in South America, Africa, India, and Australia, which at that time must have been very near the southern polar region (Figures 3–4A and 3–4C).

Continental Magnetism

We cannot reconstruct the relative position of continents prior to 200 million years ago with great accuracy. However, there is enough evidence to give us some idea of the paths followed by the continents after the breakup of Pangaea. The first clues to such movements came from the study of the magnetism of continental rocks. Recall that Earth is a large magnet (Figure 3–5A).

Igneous rocks are rocks that solidify from molten magma, either underground or after discharge from a volcano. All igneous rocks contain some particles of *magnetite,* a naturally magnetic iron mineral. The particles align themselves with Earth's magnetic field at the time of the rocks' formation. (They can do so because the magma is fluid, allowing the magnetite particles to rotate.) Volcanic lavas such as basalt are high in magnetite content and solidify from molten material in excess of 1000°C (1800°F). As they cool below 600°C (1100°F), the magnetite particles become oriented in the direction of Earth's magnetic field, permanently recording the angle of that field relative to the rock location.

Magnetite is also deposited in sediments. While the deposit is in the form of a sediment surrounded by water, magnetite particles have an opportunity to align themselves with Earth's magnetic field. This alignment is preserved when the sediment is buried and becomes solidified into sedimentary rock.

Although a number of rock types may be used to study Earth's **paleomagnetism** (ancient magnetism), the basaltic lavas and other igneous rocks high in magnetite content are best. The magnetite particles act as small compass needles, as shown in Figure 3–5B. They not only point in a north-south direction but also point into Earth at an angle relative to Earth's surface called the **magnetic dip,** or inclination, which is related to latitude. At the equator, the "needle" will not dip at all, but will lie horizontally. It will point straight into Earth at the magnetic north pole and straight out of Earth at the magnetic south pole. At points between the equator and the pole, the angle of dip increases with increasing latitude. It is this dip that is retained in magnetically polarized rocks. By measuring the dip angle, we can estimate the latitude at which the rock formed.

Apparent Polar Wandering. Earth's *geographic* North and South poles are determinal by its rotation. If Earth's axis of rotation were a rod, the poles are where the rod would poke through Earth. Its *magnetic* poles are different, determined by the distribution of Earth's internal minerals.

Because the present magnetic poles do not coincide with geographical poles, you might expect that determining latitude by magnetic dip would give incorrect results. However, for the last few thousand years, the *average* positions of the magnetic poles have nearly co-

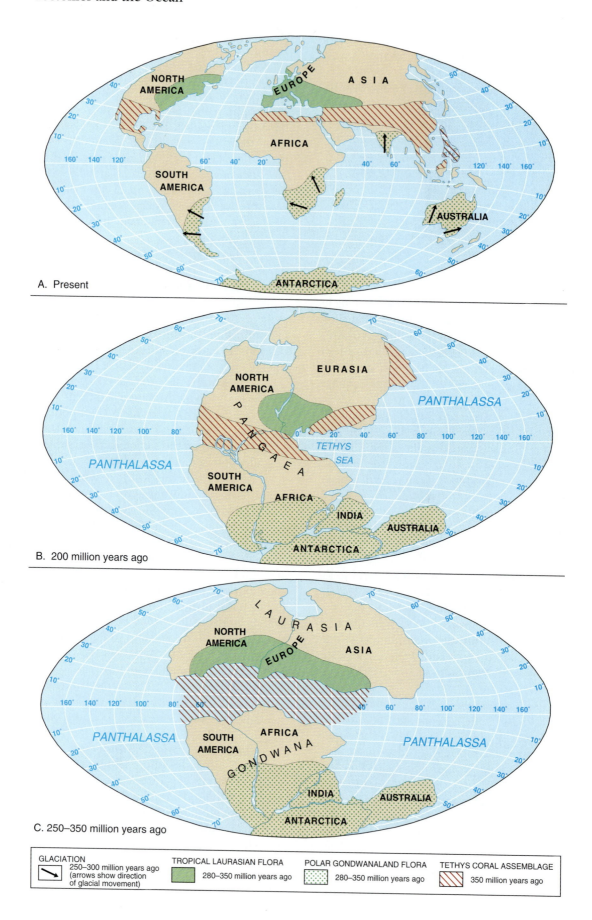

A. Present

B. 200 million years ago

C. 250–350 million years ago

GLACIATION	TROPICAL LAURASIAN FLORA	POLAR GONDWANALAND FLORA	TETHYS CORAL ASSEMBLAGE
250–300 million years ago (arrows show direction of glacial movement)	280–350 million years ago	280–350 million years ago	350 million years ago

Figure 3–4 (opposite)

Fossil and Glacial Evidence Supporting Continental Drift and the Existence of Pangaea.
A: The present distribution of continents contains three fossil assemblages that help in the reconstruction of past continental distributions. Assemblage A is a tropical assemblage found on Laurasia between 280 and 350 million years ago (see *C*). This assemblage helps support the idea that North America and Europe were combined at that time to form Laurasia. Assemblage B is of a cold-climate flora found in association with evidence of glaciation. This assemblage indicates that a large Southern Hemisphere continent, Gondwanaland, which contained present-day South America, Antarctica, Africa, India, and Australia, was in existence between 280 and 350 million years ago. Assemblage C is of fossil corals that are believed to have grown in the shallow Tethys Sea that separated the two large continents. *B:* By 200 million years ago Laurasia and Gondwanaland combined to produce the single large continent Pangaea. The ocean that covered the rest of Earth, Panthalassa, may be considered the ancestral Pacific Ocean, which has been decreasing in size since the breakup of Pangaea about 180 million years ago. *C:* Laurasia and Gondwanaland may have been in the relative positions shown here during the interval from 250 to 350 million years ago.

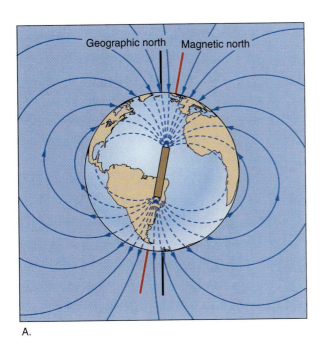

A.

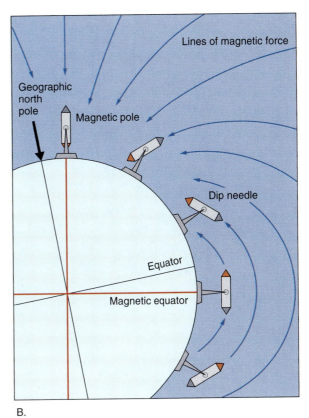

B.

Figure 3–5

Nature of Earth's Magnetic Field. *A:* Earth's magnetic field consists of lines of force like those a giant bar magnet would produce if placed inside Earth. Earth's magnetic poles seldom coincide with its geographic poles, but over long periods of time the average position of the magnetic poles approximately represents the position of the geographic poles. *B:* Magnetite particles in newly formed rocks are aligned with the lines of force of Earth's magnetic field. This causes them to dip into Earth. The angle of dip increases uniformly from 0° at the magnetic equator to 90° at the magnetic poles. When the rocks solidify, the magnetite particles are frozen into position and become fossil compass needles that tell today's investigators about the strength and alignment of Earth's magnetic field when the rocks of which they are a part formed. The angle at which they dip into Earth also tells at what latitude they formed. (Reprinted by permission from Tarbuck, E. J., and Lutgens, F. K., *The Earth: An introduction to physical geology*, 3d ed., 1990, figs. 18.8 and 18.9.)

incided with those of the geographic poles. If we can assume that this has been true in the past, we can determine average positions of the magnetic poles during a specific time and consider them to represent the geographic poles.

As these average positions for the magnetic poles were determined for rocks on the continents, it was found that their positions changed with time. It appears the magnetic poles were wandering. The wandering curve of the pole for North America shows an interesting relationship with that determined for Europe. Both curves have a similar shape, but for all rocks older than 70 million years the pole determined from North American rocks lies to the west of that determined by the study of European rocks. This difference implies that North America and Europe have changed position relative to the pole and relative to each other (Figure 3–6).

If there can be only one north magnetic pole at any given time and its position must be at or near the north geographic pole, the wandering curves can be resolved only by moving the continents. The two wandering curves can be made to coincide by moving the two continents together as we go back in time. This gives a single wandering curve, which shows the magnetic north pole to be much too far south during the interval of time from 200 million to 300 million years ago. But rotating the merged continents brings the pole into the proper position. The fact that moving the continents was the only solution to this problem was strong evidence in support of the movement of continents throughout geologic time.

Estimating the latitude of formation for rocks from many areas throughout the continents on the basis of paleoclimatic and paleomagnetic evidence produces similar results. The most logical explanation for the changes that have occurred throughout the past in both climate and magnetic dip of the rocks is that the continents have moved.

Magnetic Polarity Reversals. Not only have the magnetic poles seemed to wander throughout geologic time, but the polarity, or the direction of the mag-

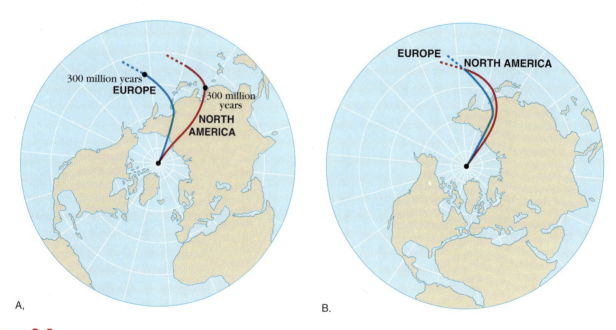

A,

B.

Figure 3–6

Apparent Wandering Curves of North Magnetic Pole Determined from Rocks of North America and Europe. *A:* Diverging paths of the apparent wandering curves of the north magnetic pole, as determined from North America and Europe, going back in time from the present to 300 million years ago. *B:* If one moves the apparent wandering curve of the pole as determined from North America to a position that makes it coincide with the curve as determined from Europe, the direction and distance equal that required to close the Atlantic Ocean between the two continents. This still leaves the apparent north magnetic pole out in the North Pacific Ocean about 40° south of the geographic north pole. Again, movement of the continent is required to solve this problem. Rotation of Laurasia, which is formed by closing the Atlantic Ocean and joining Europe and North America, can bring the apparently wayward north magnetic pole to a position coincident with the north geographic pole. (Reprinted by permission from Tarbuck, E. J., and Lutgens, F. K., *The Earth: An introduction to physical geology,* 3d ed., 1990, fig. 18.10.)

netic field, seems to have *reversed* itself periodically. These dramatic reversals in polarity can be described in the following way: A compass needle that would point to magnetic north today would, during a period of reversal, point south. It is not known why these reversals occur, but for the last 76 million years they have occurred once or twice each million years.

The period of time during which a change in polarity occurs lasts for a few thousand years. It is identified in magnetic properties of rocks by a gradual decrease in the intensity of the magnetic field of one polarity until it disappears, followed by the gradual increase in the intensity of the magnetic field with the opposite polarity. The time during which a particular paleomagnetic condition existed ("normal" or reversed) can be determined by the radiometric dating of the igneous rock from which the paleomagnetic measurements were taken.

Earth's present magnetic field has been weakening for the last 150 years, and some investigators believe that the present polarity will disappear in another 2000 years.

Evidence from the Seafloor

We have so far considered only data obtained from the *continents* in considering the theory of moving continents. This continental evidence did not convince many geologists that the continents had moved. For many years no other data were available because extensive sampling of the deep-ocean bottom did not become technologically feasible until the 1960s.

The Deep Sea Drilling Project, which began in 1968, and the Ocean Drilling Program, which followed it in 1983, have added greatly to our knowledge of the ocean floor. Extensive geophysical studies had given planners of the Deep Sea Drilling Project clues to where the drilling should be concentrated to gain the greatest new knowledge.

Paleomagnetism

A detailed magnetism study of the Pacific Ocean floor by the Scripps Institution of Oceanography identified narrow strips where the magnetic properties of oceanic crust differ from the properties of crust currently forming. These differences are called **magnetic anomalies.** The magnetic anomalies run parallel to the **Juan de Fuca Ridge** off the Washington-Oregon coast. Each anomaly represents a period when the polarization of Earth's magnetic field was *reversed* compared with today's. The anomalies are separated by bands of oceanic crust that display present-day polarity.

Researchers observed that the sequence of polarity changes on one side of an oceanic ridge is identical to the sequence of reversals on the opposite side of the ridge—one is a mirror image of the other. During dating of the reversal points on both flanks of the ridge, it became apparent that the rocks became older with increasing distance from the ridge axis. This evidence indicated that new oceanic crust is being formed at the oceanic ridges and is moving away from them simultaneously on opposite sides of the ridges. Having determined from continental studies the dates at which many of the more recent reversals occurred, it was possible to determine the age of the ocean floor at each strip boundary. Dividing the width of a strip by the number of years that polarity lasted produced the *rate* at which the ocean floor appeared to be moving away from the ridge (Figure 3–7).

Confirmation of the spreading of oceanic crust away from the oceanic ridges required actual samples of the crust and sediment at various locations. This was necessary to confirm three hypotheses:

- Radioactive dating would reveal that age increased away from the ridge.
- Because sediment could not fall upon crustal material until it formed at the oceanic ridges, fossil assemblages observed in sediments immediately overlying the oceanic crust should contain organisms that existed at the time of the crustal formation.
- Sediment thickness would be greater on older seafloor than on younger seafloor.

A significant task of the Deep Sea Drilling Project was to check the age of the ocean bottom by drilling through sedimentary sections into the oceanic crust. Although some attempts failed, enough data were obtained to confirm that the ocean floor is moving in the manner proposed. Oceanic crust was found at the axes of the oceanic ridges, with the age of the crust increasing at greater distances from the ridge crests.

Seafloor Spreading

The studies indicated that new ocean floor is forming at the oceanic ridges and rises. It then is carried away from the axes of such features, and volcanic processes continue to fill the void with new strips of ocean floor. Thus, the axes of oceanic ridges and rises are referred to as **spreading centers.** The term seafloor spreading is applied to this theory.

However, more than the seafloor is moving. The seafloor is just the surface of the outermost 700 kilometers (435 miles) that is moving, called the lithosphere (Figure 3–8). The **lithosphere** ("rock sphere") is a relatively cool, rigid shell that includes the crust and upper mantle (see the detail part of the figure). The lithosphere is broken into about a dozen major slabs, or **lithospheric plates,** that we observe moving across Earth's surface (Figure 3–9). Underlying the lithosphere

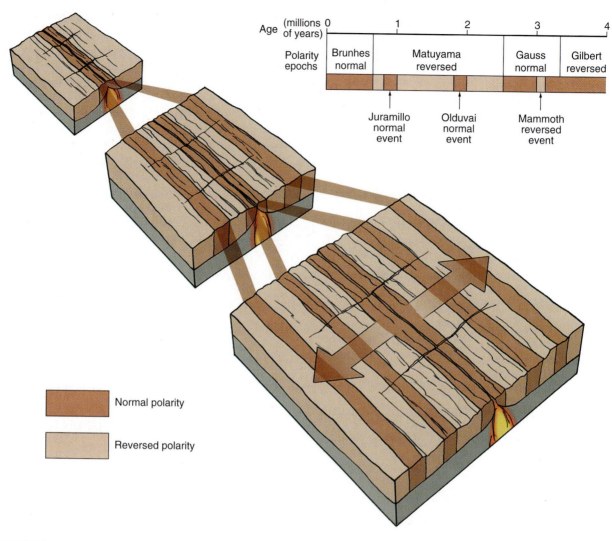

Figure 3–7

Magnetic Evidence in Support of the Seafloor Spreading Hypothesis. *A:* Time scale of Earth's magnetic field in the recent past. This time scale was developed by establishing the magnetic polarity for volcanic lavas of known ages. *B:* New seafloor records the polarity of the magnetic field at the time it formed. Hence it behaves much like a tape recorder, as it records each reversal of Earth's magnetic field. (Data from Allan Cox and G. B. Dalrymple. Reprinted by permission from Tarbuck, E. J., and Lutgens, F. K., *The Earth: An introduction to physical geology,* 3d ed., 1990, fig. 18.12.)

is a high-temperature layer within the mantle. This layer is the **asthenosphere** ("weak sphere") (Figure 3–8, detail). It can flow slowly—*plastically*—allowing the rigid lithospheric plates resting at its upper surface to move.

The ultimate fate of a lithospheric plate is **subduction**—a process by which it descends beneath another plate and is ultimately remelted into the mantle.

What Causes Spreading? We do not know for certain, but spreading seems to be caused by **convection cells** in the asthenosphere. The same connection

cells that stir Earth's atmosphere vertically, and which stir heated water vertically, also operate very, very slowly in heated plastic material (Figure 3–10).

Earth's interior is hot. It is believed that this heat moves to the surface through convection cells that carry the heat to the regions of the oceanic spreading centers. This further implies that cooler portions of the mantle descend somewhere to complete each convection cell.

Heat-flow measurements taken throughout Earth's crust show that the quantity of heat flowing to the surface along the oceanic ridges can be up to eight times

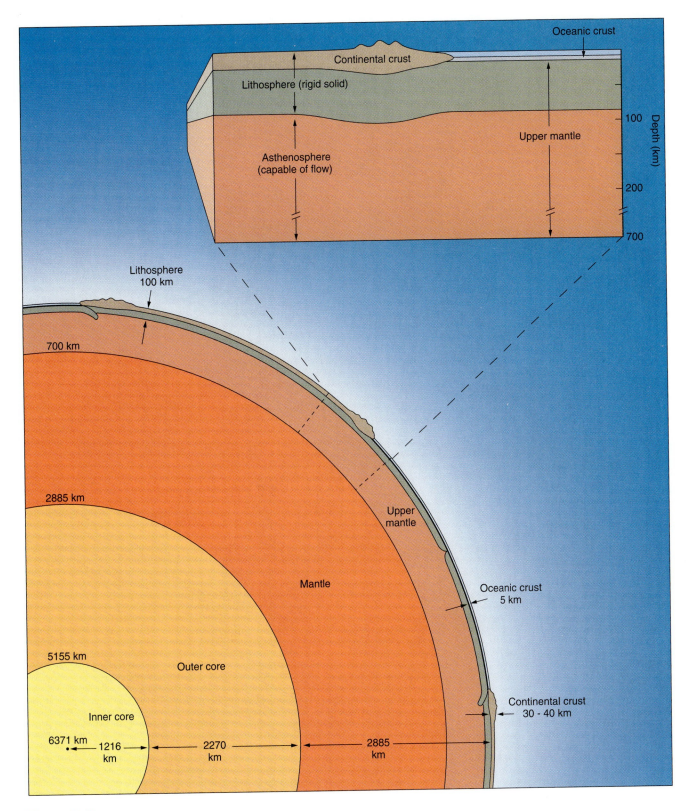

Figure 3–8

Lithosphere and Asthenosphere. The lithosphere is the rigid outer shell of Earth's structure that includes the crust and cooler upper mantle. It is supported and allowed to move across Earth's surface by flow within the warmer, plastic underlying portion of the mantle called the asthenosphere. This weaker, possibly partially molten layer is thought to extend to a depth of approximately 700 kilometers (435 miles). The three major subdivisions of Earth—core, mantle, and crust—are shown to clarify the position of the lithosphere and asthenosphere. (Reprinted by permission from Tarbuck, E. J., and Lutgens, F. K., *The Earth: An introduction to physical geology,* 3d ed., 1991, fig. 5.19.)

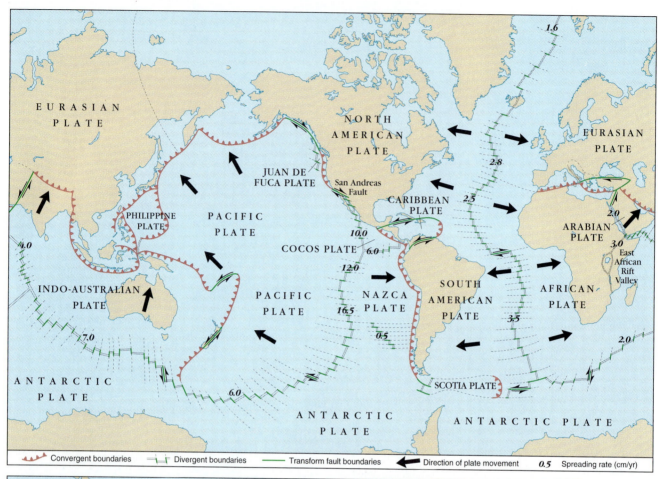

EURASIAN PLATE

EURASIAN PLATE

NORTH AMERICAN PLATE

JUAN DE FUCA PLATE

San Andreas Fault

CARIBBEAN PLATE

ARABIAN PLATE

PHILIPPINE PLATE

PACIFIC PLATE

COCOS PLATE

East African Rift Valley

1.6

2.8

2.5

2.0

3.0

4.0

INDO-AUSTRALIAN PLATE

10.0

6.0

12.0

SOUTH AMERICAN PLATE

AFRICAN PLATE

PACIFIC PLATE

NAZCA PLATE

16.5

0.5

3.5

7.0

2.0

ANTARCTIC PLATE

6.0

SCOTIA PLATE

ANTARCTIC PLATE

ANTARCTIC PLATE

ANTARCTIC PLATE

| ⌒⌒ Convergent boundaries | ⊢⊢ Divergent boundaries | — Transform fault boundaries | ⬅ Direction of plate movement | *0.5* Spreading rate (cm/yr) |

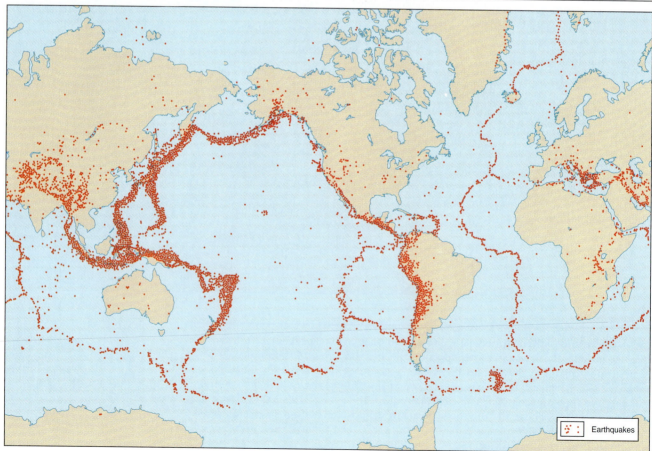

Earthquakes

Figure 3–9 (opposite)

Lithospheric Plates and Earthquakes. An early clue to the boundary locations of lithospheric plates (top) was the pattern of earthquakes (bottom). The plates form at the axes of submarine mountain ranges (divergent boundaries) and slowly plunge back into Earth's interior beneath deep–ocean trenches (convergent boundaries). The divergent plate boundaries are offset by transform faults, along which the plates grind past one another in opposite directions. The plates move from 1.5 centimeters (0.8 inches) per year (American plates) to more than 10 centimeters (5.4 inches) per year (Pacific Plate). (The bottom figure shows data from NOAA for a nine-year period; reprinted by permission from Tarbuck, E. J., and Lutgens, F. K., *The Earth: An introduction to physical geology,* 3d ed., 1991, fig. 5.11.)

greater than the average for Earth's crust. In addition, areas where ocean lithosphere subducts at trenches have heat flow as little as one-tenth the average. This pattern is shown in Figure 3–11.

It was once believed that most of the heat driving global plate tectonics was generated primarily by radioactivity in the upper mantle. However, now there is evidence that a significant amount of heat is "original" heat residue from Earth's formation, coming from the core. We do know that **plumes** arise from the base of the mantle, and their surface expressions are called **hot spots.** The high rates of volcanic activity associated with hot spots have produced such features as the Hawaiian Islands and Iceland.

Seismic data indicate that magma chambers providing the lavas for the spreading centers extend no deeper than 350 kilometers (220 miles) into the mantle. They may be passive features that result from the lithosphere's being pulled apart by the subducting slabs at the edges of these sheets (Figure 3–12).

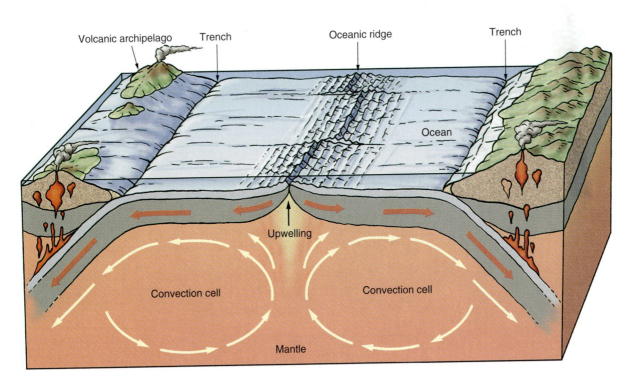

Figure 3–10

Movement in the Asthenosphere. Lateral flow within the asthenosphere is a possible mechanism for the lateral movement of the lithosphere, including the crust. The asthenosphere is a low-strength plastic zone within the mantle where pressure is low enough and temperature sufficiently near the melting point of the material to allow it to flow slowly. Some form of thermal convection within the asthenosphere appears to create new lithosphere at the oceanic ridges and may carry old lithosphere back into the mantle to be subducted (left and right). This subduction creates oceanic trenches. (Reprinted by permission from Tarbuck, E. J., and Lutgens, F. K., *The Earth: An introduction to physical geology,* 3d ed., 1990, fig. 18.11.)

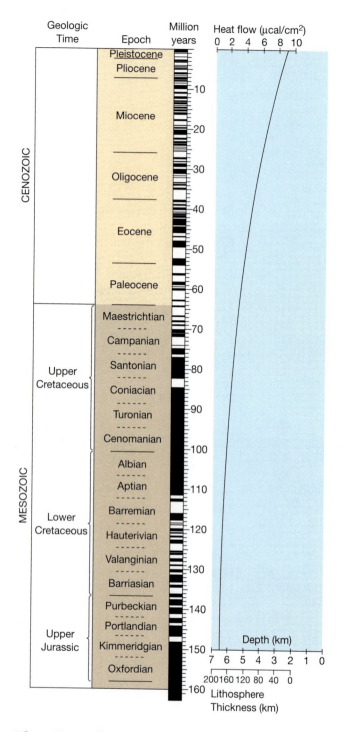

Figure 3–11

Geologic Time Scale, Magnetic Time Scale, Heat Flow, Ocean Depth, and Lithosphere Thickness. This diagram relates several important factors. The black and white bands (A) show changes in Earth's magnetic polarity during the past 160 million years. Black bands are periods when polarity *was the same as at present.* White bands show when it was reversed. The curve relates (B) heat flow, (C) ocean depth, and (D) lithosphere thickness to (A) increasing distance and age from the spreading centers. The ocean gets deeper, the lithosphere grows thicker, and heat flow diminishes with increasing age/distance from the spreading center. (This curve is very general and cannot be applied accurately to any specific area of the ocean; it is intended to show the *trend.*) The geologic time scale shown at left (E) is presented in greater detail in appendix III.

ridges are called **divergent boundaries.** Boundaries where plates are moving together and one subducts beneath the other are called **convergent boundaries.** Lithospheric plates also simply grind past one another along **transform boundaries** (Figure 3–13).

Divergent Boundaries

At spreading centers, deepening of the ocean floor correlates directly with how long the lithosphere has cooled. Thus, the rate of seafloor spreading significantly affects the steepness of the flanks of the seafloor mountain ranges associated with the spreading.

The term **spreading rate** applies to the total widening rate of an ocean basin resulting from the motion of both plates away from a spreading center. The faster the spreading rate, the broader the mountain range associated with the spreading. These gently sloping features are the oceanic rises; the best example is the **East Pacific Rise** between the Pacific Plate and the Nazca Plate, where spreading rates are as high as 16.5 centimeters (6.5 inches) per year. The **Mid-Atlantic Ridge,** with spreading rates in the range of 2 to 3 centimeters (0.8 to 1.2 inches) per year, displays the steeper slopes characteristic of slow spreading.

Features are significantly different at the axes of fast- and slow-spreading divergent plate boundaries. These differences probably are due to a much greater magma supply being available at fast-spreading oceanic rises than at slow-spreading ones. Figure 3–14 compares the sizes of axial rift valleys, which are much larger on the slow-spreading ridges. Project FAMOUS (French-American Mid-Ocean Undersea Study) in 1974 observed the rift valley along the axis of the Mid-Atlantic Ridge. These observations provide a clue to the spreading process (Figure 3–15).

Fissures parallel to the ridge axis range from hairline width near the center of the rift valley to more than

Plate Boundaries

Relentless seafloor spreading sets up stresses within the lithosphere. In fact, our first clue to the locations of plate boundaries was that dramatic tectonic events such as earthquakes and volcanic activity occur in linear belts (Figure 3–9, bottom).

Closer examination of these active regions identified three types of boundaries. The plate boundaries where new lithosphere is being added along oceanic

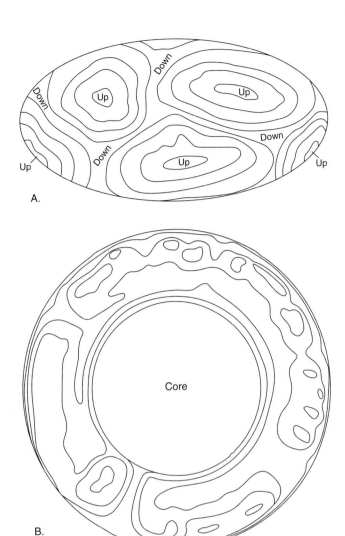

A.

B.

Figure 3–12

Three-Dimensional Spherical Model of Convection in Earth's Mantle. With half the heat coming from within the mantle and half from the core/mantle boundary, the circulation in the mantle according to some models appears as shown. *A:* At a depth midway between the surface and the core, mantle circulation shows plume-like upwelling surrounded by rings of downwelling. Upwelling is shown as high (red) or moderate (orange). Green shows little vertical motion. Blue and violet show increasing rates of downwelling. *B:* This longitudinal cross section of the mantle shows relative temperature. Blue represents the coldest near-surface mantle material. Green, yellow, and orange indicate increasing temperature, with red representing the hottest material associated with the upwelling plume and the core/mantle boundary. (After Bercovici, D., et al., "Three-Dimensional Spherical Models of Convection in the Earth's Mantle," *Science* 244, pp. 950–55, 26 May, 1989. Used with permission of the author and publisher. Copyright 1989 A.A.A.S.)

10 meters (33 feet) near both margins of the valley. This supports the theory that the plates are being *continuously pulled apart* rather than being pushed apart by upwelling of material beneath the ridges. Possibly, the upwelling of magma beneath the oceanic ridges is simply filling in the void left by the separating plates of lithosphere. Whatever the mechanism, a large mass of magma clearly must exist near the surface beneath the ridges. This process produces 20 cubic kilometers (4.8 cubic miles) of new ocean crust per year.

The magnitude of energy released by earthquakes along the divergent plate boundaries is closely related to spreading rate. Earthquake intensity is now measured on a scale called *seismic moment-magnitude* (which is replacing the well-known Richter scale for earthquakes). Earthquakes in the rift valley of the slow-spreading Mid-Atlantic Ridge reach a maximum magnitude of about 6, whereas those occurring along the axis of the fast-spreading East Pacific Rise seldom exceed 4.5. (Each unit increase of magnitude represents an increase of energy release of about 30 times.)

Hydrothermal Vents Recycle the Ocean. Near the axes of oceanic ridges and rises, where magma chambers may be less than 1 kilometer (0.6 miles) beneath the ocean floor, seawater seeps down fractures in the ocean crust and is heated by the hot underlying magma. It then rises back toward the surface to produce **hydrothermal vents** (Figure 3–16A). This rising water may exit into the ocean as *warm-water vents* (10° to 20°C or 50° to 68°F), as *white smokers* (30° to 330°C or 86° to 626°F, containing white particles of barium sulfate), or as *black smokers* (around 350°C or 662°F, containing black metal sulfides). A black smoker is shown in Figure 3–16B.

These hydrothermal vents are not just curiosities. The *entire volume of ocean water* is recycled through this hydrothermal circulation system in about 3 million years. Therefore, the chemical exchange between ocean water and the basaltic crust has a major influence on the nature of ocean water. And this process has major economic implications. The deposition of metallic sulfides around these vents is likely a major source of *continental* sulfide ore deposits because, over millions of years, they may be transferred to the continents during subduction of the oceanic plate.

The oceanic hydrothermal vents support unusual biological communities, first discovered in 1977 in the **Galápagos Rift** between South America and the Galápagos Islands (Figure 3–16C). Large clams, mussels, tube worms, and other animals abound around the vents. The reason is that the vents discharge hydrogen sulfide gas. Bacteria that oxidize the hydrogen sulfide gas provide energy to produce food for the community through chemosynthesis.

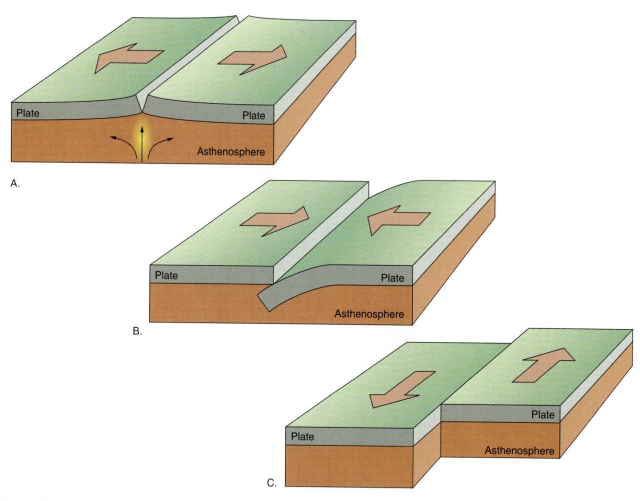

A.

B.

C.

Figure 3–13

Lithospheric Plate Boundaries. *A:* Divergent plate boundary. New material is added to plates at the spreading centers. The plates then diverge from one another as they move away from the spreading center. *B:* Convergent plate boundary. Where lithospheric plates converge, a trench may form as one plate subducts into the mantle and is remelted. *C:* Transform plate boundary. Spreading centers are offset by transform faults along which plates scrape past one another. No new lithosphere is formed, nor is any subducted. (Reprinted by permission from Tarbuck, E. J., and Lutgens, F. K., *The Earth: An introduction to physical geology,* 3d ed., 1991, fig. 6.7.)

Transform Boundaries

Beneath every ocean ridge or rise lies a series of separate magma chambers, not just a single continuous chamber (Figure 3–17). Each chamber creates its own *segment* of spreading seafloor, which may range from 10 to 80 kilometers (6 to 50 miles) in length and is centered over its magma chamber. These chambers and segments are not evenly aligned, and are offset from one another. The offset, called a **transform fault,** may be very minor, or it may be a major offset of hundreds of kilometers. You can see hundreds of them dissecting the ocean ridges and rises in Figure 2–22. One is shown in detail in Figure 3–18.

Along such faults, one plate slides past another, producing shallow earthquakes in the lithosphere. Magnitudes of 7 have been recorded along oceanic transform faults. An example of such a fault that has come ashore is the **San Andreas Fault,** which runs across California from the head of the Gulf of California to the San Francisco area. The San Andreas Fault runs through continental crust, so the lithosphere is much thicker than at oceanic sites. This can make earthquakes larger, and tremors along this fault have had magnitudes up to 8.5.

Fracture Zones. Continuing along the line of the transform fault away from the ridge or rise are *fracture zones.* The transform faults are *active* displace-

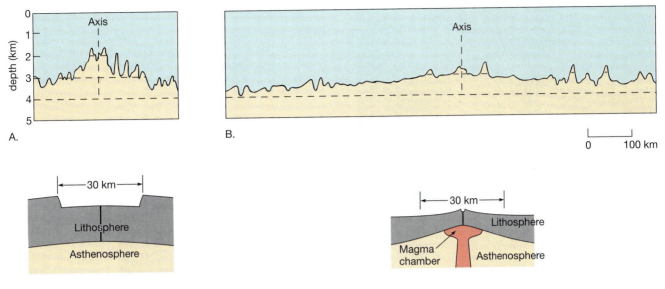

Figure 3–14

Oceanic Ridges and Rises. *A:* The Mid-Atlantic Ridge has a steeper and more irregular topography as a result of a low spreading rate. The axial valley of a slow-spreading ridge is typically 30 kilometers wide and may be from 100 to 3000 meters deep. The lithosphere may be from 3 to 10 kilometers thick at the plate boundary near the center of the axial valley. *B:* The high spreading rate of the East Pacific Rise has produced a more gently sloping and smoother bottom topography. (Vertical exaggeration 50X). At fast-spreading rises an axial ridge a few hundred meters high and from 2 to 10 kilometers in width typically overlies the plate boundary above a magma chamber that may be about 4 kilometers wide and 1 kilometer thick. An axial valley less than 1 kilometer wide and up to 100 meters deep may run along the axis of the axial ridge. The thickness of the lithosphere at the rise axis may be less than 2 kilometers. (*The ocean basins: Their structure and evolution,* The Open University and Pergamon Press.)

ments of the axes; the fracture zones are evidence of *past* transform-fault activity. On opposite sides of a transform fault, the lithospheric plates are moving in opposite directions, whereas no such motion is occurring along a fracture zone. In other words, the transform faults are actual plate boundaries, whereas the fracture zones are ancient fault scars embedded in a plate (Figure 3–18). Earthquakes shallower than 10 kilometers (6 miles) are common along the transform faults, whereas the fracture zones are earthquake free.

Convergent Boundaries

Island Arc Trench System. Figure 3–19A illustrates island arc trench system type of converging boundary. Both spreading centers and trench systems are characterized by earthquakes and volcanic activity, but in different ways. Spreading centers have shallow quakes, usually less than 10 kilometers (6 miles) in depth. But movements that cause earthquakes in the trench regions vary from shallow down to 670 kilometers (415 miles).

The earthquake *focus,* the actual point at which the movement that causes the quake occurs, is usually

shallow around a trench. Earthquake activity in the subduction zone deepens, down to 670 kilometers. The *Wadati-Benioff seismic zone* is a band 20 kilometers (12.5 miles) thick that dips from the trench region under the overriding plate, within which all of the earthquake foci are located.

What proof do we have that the lithosphere generated at the oceanic rises and ridges is subducted at the trenches? Measurements reveal that no oceanic crustal material more than 175 million years old has been recovered from the ocean bottom. This is only 4 percent of Earth's 4.6-billion-year history. Continental crust, which is very unlikely to be subducted because of its low density, remains buoyantly at Earth's surface, and continental rocks as old as 3.8 billion years have been found.

As the lithosphere is subducted into the asthenosphere, it becomes heated. Water and other volatiles are freed, producing a low-density mixture that very slowly rises to the surface through the overriding plate. This produces basaltic volcanoes that may become island arcs, as shown in Figure 3–19A. Such arcs usually develop about 100 kilometers (60 miles) above subducting

A.

Figure 3–15

Rift Valley Fissures. *A:* A fissure in the rift valley of the Mid-Atlantic Ridge photographed from the submersible *Alvin* during Project FAMOUS, 1974. (Courtesy of Woods Hole Oceanographic Institution.) *B:* A more readily observable fissure above sea level in the rift valley of Iceland. Here the Mid-Atlantic Ridge rises above the ocean surface as an island because of the high rate of volcanic activity in the Iceland hot spot. (Courtesy of Br. Robert McDermott, S. J.)

B.

lithosphere. Well-known examples are the Leeward and Windward Islands in the Caribbean and the Aleutian Islands in Alaska.

Continental Arc Trench System. Figure 3–19B shows the continental arc trench system type of converging boundary. Should an oceanic plate subduct beneath a plate with a continent at its leading edge, the melting of the subducting oceanic plate occurs beneath the continent. Consequently, the rising basalt melt passes through and mixes with the granite of the continental crust. This results not in an *island* arc but in a *continental* arc of volcanoes along the edge of the continent (Figure 3–19B). The volcanoes are composed of a rock called andesite, which is of a composition intermediate between those of basalt and granite and may form by the mixing of the two basic crustal rock types.

If the spreading center producing the subducting plate is far enough from the subduction zone, an oceanic trench becomes well developed along the margin of the continent. The Peru-Chile Trench is an example. It is associated with the Andes Mountains continental arc of volcanoes, from which the rock *andesite* takes its name.

No trench is visible where the Juan de Fuca Plate subducts beneath the North American Plate off the coasts of Washington and Oregon to produce the Cascade Mountains continental arc. Here the Juan de Fuca Ridge is so near the North American Plate that the subducting lithosphere is less than 10 million years old and hasn't cooled enough to produce deep-ocean depths. Also, the large amount of sediment carried to the ocean by the Columbia River has filled what trench structure may have developed. Many of the Cascade volcanoes of this continental arc have been active within the last 100

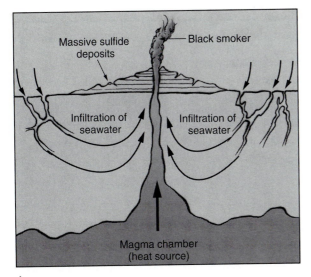

A.

C.

B.

Figure 3–16

Hydrothermal Vents along Spreading Center Axes. *A:* Cross section across the axis of a spreading center to illustrate the hydrothermal circulation in oceanic crust. Seawater infiltrates along a broad zone of fractured crustal rock, is heated as it approaches the underlying magma chamber, and rises in a narrow zone near the axis of the spreading center. The resulting vents usually lie within 200 meters (650 feet) of the ridge or rise axis. (Reprinted with permission from Tarbuck, E. J., and Lutgens, F. K., *The Earth: An introduction to physical geology,* 3d ed., 1988, fig. 10.17.) *B:* Black smoker. During a 1979 expedition to 21°N on the East Pacific Rise, this black smoker was observed spewing out 315°C (600°F) water. In the foreground is the submersible *Alvin's* sample basket. *C:* Large tube worms observed at East Pacific Rise warm-water vent. The tubes are about 1 meter (3.3 feet) long and the vestimentiferan worms that inhabit them have no mouth or digestive tract. In a symbiotic (mutually helpful) relationship between the tube worms and some chemosynthetic bacteria, the bacteria oxidize the hydrogen sulfide gas dissolved in the vent water and use the energy released to produce food for themselves and the worms. (Photos courtesy of Scripps Institution of Oceanography, University of California, San Diego; Photo *B* by Dr. Fred N. Spiess.)

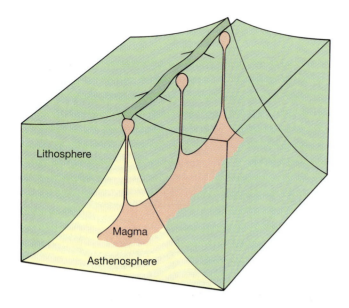

Figure 3–17

Segmentation of Oceanic Ridges and Rises. Beneath the ridge axis, the heat-bearing asthenosphere (yellow) rises between the separating lithospheric plates (green). A zone of partially molten asthenosphere (red) is less dense than the cold lithosphere above, so magma rises from it to form magma chambers at intervals along the ridge axis. The ridge axis rises over the magna chambers.

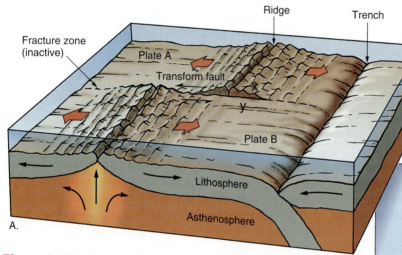

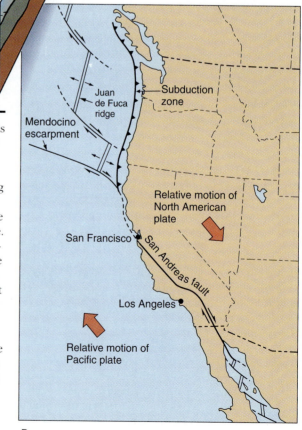

Figure 3–18

Transform Faults—Fracture Zones. *A:* The axis of an oceanic ridge is offset by a transform fault due to the stresses from the motion of rigid plates. They are called transform faults because they cut across the larger physiographic features of the ocean floor, oceanic ridges and rises. Earthquakes are common and of relatively great magnitude along transform faults because of their active movement. Extending beyond the offset ridges are fracture zones, where earthquakes are rare because there is no relative movement on the opposite sides of the fracture zone. Fracture zones are scars of old transform fault activity. They remain significant topographic features because the age of the ocean floor on one side of the fracture zone is older than the ocean floor immediately opposite it. For example, point X is younger than point Y. Therefore point Y has subsided more because of cooling and thermal contraction than point X. Thus there is an escarpment (cliff) along the fracture zone in Plate B that faces the older ocean floor on the side where point Y is located. There is also an escarpment on Plate A, but it faces the opposite direction. *B:* The San Andreas Fault is a classic example of a transform fault that offsets an oceanic ridge or rise and forms a boundary between plates that are sliding past each other. Along this contact, the Pacific Plate is moving north relative to the North American Plate. (Reprinted by permission from Tarbuck, E. J., and Lutgens, F. K., *The Earth: An introduction to physical geology,* 3d ed., 1990, figs. 18.22 and 18.23.)

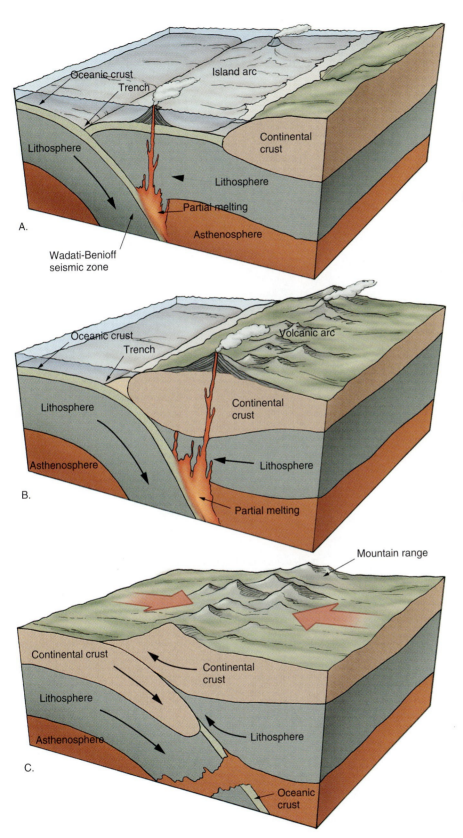

A.

B.

C.

Figure 3–19

Effects of Plate Collisions. *A:* Classic oceanic trench systems develop where oceanic lithosphere subducts beneath a plate with oceanic crust at its leading edge and a continent trailing some distance behind. A basaltic island arc-trench system develops. The sloping belt in which earthquake foci are located is the Wadati-Benioff seismic zone (an example is the Aleutian Islands). *B:* When an oceanic lithospheric plate subducts beneath a plate with continental crust at its leading edge, an andesitic continental arc develops. The volcanic rock andesite is named for the Andes Mountains, where a continental arc exists. Andesite has a composition between that of granite and basalt. It is thought to form as melted basalt from the descending oceanic plate rises through the continental granite to produce the volcanic continental arc (an example is the Cascade Mountains). *C:* As two plates carrying continents converge, trenches do not develop because sediments that accumulated in the sea between the continents are too light to be subducted. The shortening of Earth's crust in such regions is accommodated by the folding of the sediments into mountain ranges such as the Appalachians, Alps, Himalayas, and Urals. The lower lithosphere and asthenosphere may be involved in low rates of subduction beneath the mountains (an example is the Himalayas). (Reprinted by permission from Tarbuck, E. J., and Lutgens, F. K., *Earth Science,* 5th ed. (New York: Macmillan), 1988, fig. 6.11.)

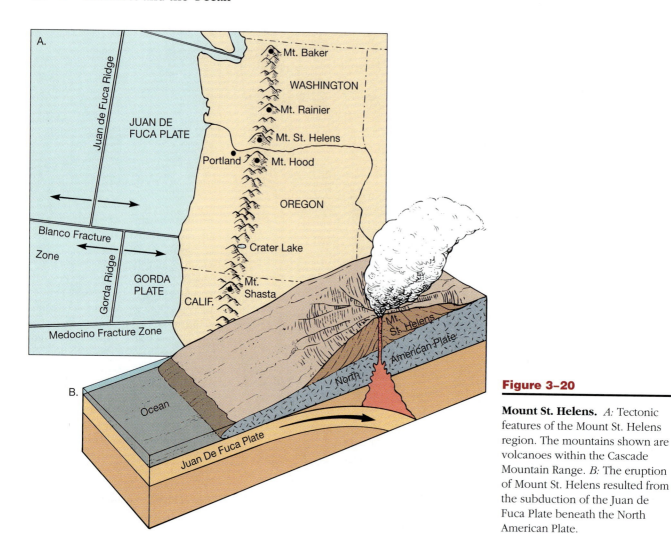

Figure 3–20

Mount St. Helens. *A:* Tectonic features of the Mount St. Helens region. The mountains shown are volcanoes within the Cascade Mountain Range. *B:* The eruption of Mount St. Helens resulted from the subduction of the Juan de Fuca Plate beneath the North American Plate.

years. Most recently, Mount St. Helens erupted in May of 1980, killing 62 people (Figure 3–20).

Continental Mountain System. Figure 3–19C shows the continental mountain system style of convergent boundary. If two lithospheric plates collide and both contain continental crust near their leading edges, the surface expression will be a mountain range. The mountains will be composed of folded sedimentary rocks derived from sediments on the seafloor that previously separated the continental blocks. For the most part, these sedimentary rocks do not subside in a zone of subduction because of their relatively low density, having been derived from the continents. The oceanic crust itself may, however, subside beneath such mountains. The classic example of this continental-continental crust merger is the Himalayan Mountains.

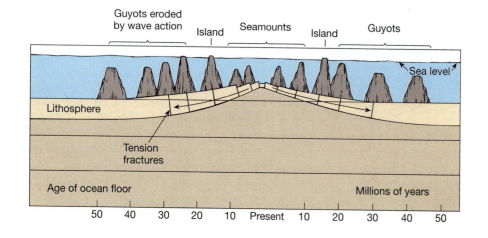

Figure 3–21

Formation of Seamounts and Guyots. Volcanic activity develops near the crest of the ridge because of tension fractures that develop parallel to the ridge as the plates are pulled apart. These fractures serve as conduits for lava but usually seal within 30 million years. Individual peaks may form instead of a continuous ridge because of the spacing of fractures or simply because the volcanism is periodic.

Intraplate Features

Features that exist in the *interior* of plates also provide important evidence of plate tectonic movements. These features include seamounts, coral reefs, and hot spots.

Seamounts and Guyots

Highly characteristic of the floor of the Pacific Plate are volcanoes called *seamounts* and *guyots*. Many appear to be related to oceanic ridges (Figure 3–21). Considering the oceanic ridge system worldwide, active volcanoes are characteristic of the axes of the ridges. Moving away from the ridge axis, the volcanic activity gradually decreases. This happens because the long cracks in Earth's crust that serve as conduits for rising lava near the axis of these ridges become sealed off as they move away from the axis and the active source of magma. Active volcanoes over 30 million years of age are rare.

Let us consider the life history of a theoretical volcano that begins life on an oceanic ridge and is carried away from the ridge by the lithospheric plate. The volcano may gradually increase in size as it moves downs-lope (Figure 3–21). By the time it is 10 million years old, if it has been very active, it may build enough height to become an island, and may exist for another 10 to 15 million years if the volcano remains active. The volcano will probably become dormant within 30 million years of its origin because it migrates away from its magma source. As it continues to move down the flank of the oceanic ridge, its top will become flattened by wave action and ultimately will be submerged. The flat-topped structure will sink deeper and deeper beneath the ocean surface as it descends the slope of the ridge. It then becomes a guyot. Most guyots are at least 30 million years old.

Coral Reef Development

Although a comprehensive theory of global plate tectonics was not to be achieved for another century, Charles Darwin hypothesized an origin of coral reefs that depended on the *subsidence of volcanic islands*. Darwin published his concept in *The Structure and Distribution of Coral Reefs* in 1842 (Figure 3–22). Drilling into reefs a half-century ago provided evidence to prove Darwin's hypothesis.

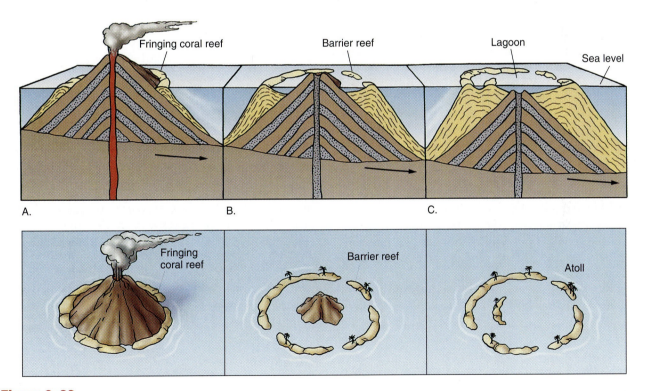

Figure 3–22

Darwin's Theory of Development of Reef Types. A volcanic island forms in warm-water latitudes. *A:* A *fringing coral reef* forms in the shallow, sunlit waters adjacent to the island shore. *B:* As the dormant volcano gradually subsides due to thermal contraction of the lithosphere, the reef maintains its elevation near the surface because new corals build on the skeletons of their ancestors. This produces a *barrier reef* where the actively growing reef is separated from the volcanic island by a well-developed lagoon. *C:* Continued subsidence submerges the volcano, but the reef may continue to grow upward to produce an *atoll,* with active reef growth confined mostly to the outer edge. Deposition of reef debris may produce low-lying islands inside the protective ring of coral reef. Within the atoll ring lies a broad, shallow lagoon. (Reprinted by permission from Tarbuck, E. J., and Lutgens, F. K., *Earth Science,* 5th ed (New York: Macmillan), 1988, fig. 10.19.)

Reef-building corals are colonial animals. They secrete a limestone framework in the sunlit waters on the flanks of volcanic islands and continents in warm, tropical waters. Once the corals have colonized all of the sunlit waters available to them, their primary direction of growth is upward. Each new generation is attached to the skeletons of its predecessors.

Three basic types of reef exist—fringing, barrier, and atoll (Figure 3–22). **Fringing reefs** develop along the margin of any landmass where the temperature, salinity, and turbidity (cloudiness) of the water are suitable for reef-building corals. Due to the close association between the landmass and the reef, at times runoff from the landmass will carry too much sediment for the coral to survive. The amount of living coral in a fringing reef at any given time is relatively small, with the greatest concentration on the side that is protected from sediment and salinity changes. If sea level does not change during the existence of the fringing reef, the process stops here, and the other two types of reefs, the barrier reef and the atoll, will not form.

Barrier reefs are linear or circular developments separated by a lagoon from the continent or island on whose margin they are growing. As the island or margin of the continent subsides, the fringe reef maintains its position at the optimum water depth by growing upward, producing the barrier reef (Figure 3–22B). Studies of reef growth rates indicate most have grown 3 to 5 meters (10 to 16 feet) per 1000 years during the recent geologic past. There is some evidence, however, that reefs in the Caribbean have grown at rates of over 10 meters (33 feet) per 1000 years.

The most outstanding example of a barrier reef is the Great Barrier Reef, which is 150 kilometers (90 miles) wide and extends for more than 2000 kilometers (1200 miles) along the northeastern coast of Australia (Figure 3–23). The effects of global plate tectonics are clearly visible in the structure of the Great Barrier Reef. It is oldest (around 25 million years) and thickest at its northern end. The reef thins to the south, and the base of the reef becomes younger. This is clearly the result of the Indo-Australian Plate moving toward the equator from colder Antarctic waters. Smaller barrier reefs are found around the island of Tahiti and other islands in the western Atlantic and Pacific oceans.

Atolls may form either on a subsiding continental shelf or around the margin of a sinking volcanic island in the open ocean. Most atolls are of the oceanic variety, and the greatest number occur in the equatorial Pacific. Atolls are generally circular and surround a lagoon that is usually 30 to 50 meters (100 to 165 feet) deep at the deepest part. The fringing reef generally has many channels that allow circulation between the lagoon and the open ocean. The reef flat is commonly broader on the windward side of the atoll, and channels are more

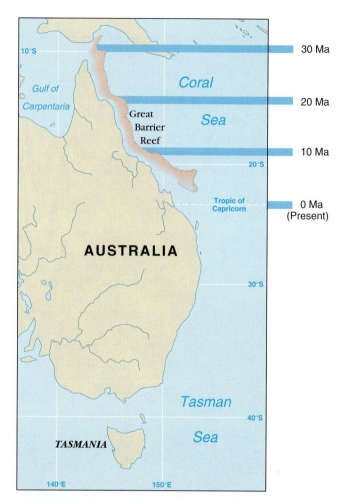

Figure 3–23

Australia's Great Barrier Reef and Global Plate Tectonics.
The Great Barrier Reef extends for more than 2000 kilometers (1200 miles) along the northeastern coast of Australia between latitudes 9° and 24°S. The reef first began to develop about 25 million years ago (Ma) when northern Australia was transported tectonically into tropical waters from colder Antarctic waters to the south. The southern end of the Great Barrier Reef has moved into tropical waters only recently.

numerous on the leeward side (Figure 3–24). Islands large enough to inhabit may develop.

Hot Spots

Throughout the Pacific Plate are many island chains that trend northwestward-southeastward. The most intensely studied of these is the chain that includes the Hawaiian Islands and the Emperor Seamounts. Refer for a moment to Figure 2–18 and locate this chain (labeled on the right-hand page, left of center). What caused this odd, boomerang-shaped string of islands and seamounts?

To understand, study the pattern in Figure 3–25. There are no active volcanoes in the entire chain north-

Figure 3–24

Atolls. View from space of a group of atolls in the Pacific Ocean. (Photo courtesy of NASA.)

west of Hawaii. All have long since gone dormant. To-day, the only active volcano is Kilauea, which is less than 1 million years old. A second volcano, Loihi, exists underwater southeast of Hawaii. Significantly, the age of islands and seamounts *increases* steadily northwest-ward from Hawaii. You can see this pattern in the inset map of the Hawaiian Islands in Figure 3–25.

Looking northwestward, the seamounts increase in age to a maximum of 70 million years.

Evidently, the Pacific Plate has moved steadily northwestward over a stationary hot spot in the mantle, as shown in the figure. This has generated a succession of volcanic islands and seamounts. At one point the plate shifted direction, which accounts for the bend in the island chain.

The Growth of Ocean Basins

So far we have discussed specific features of tectonic plates that tell us about seafloor creation at spreading centers, plate movement, and subduction of plates back into the mantle. In this section, we concentrate on the broader changes that have occurred since the breakup of Pangaea 180 million years ago. The study of histori-cal changes of continental shapes and positions is **pale-ogeography.** The term **paleoceanography** applies to changes in the physical and biological character of the oceans brought about by paleogeographical changes.

Figure 3–26A shows the changing positions of the continents during the last 150 million years for the North-ern Hemisphere. Figure 3–26B shows the same thing for the Southern Hemisphere. Observing the changes in the Northern Hemisphere, you can see that the biggest change is the separation of North America and South America from Europe and Africa to produce the Atlantic Ocean. During this process, Europe, Africa, and North America also moved a bit northward.

Historically, spreading rates are greater at lower latitudes. This shows up in the pattern of formation of the North Atlantic Ocean. Europe and North America separated at a much slower rate than Africa and North America, which began separating 180 Ma (million years ago).

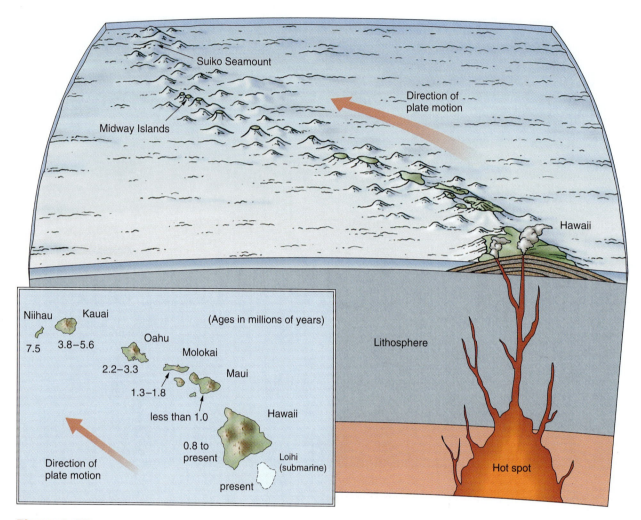

Figure 3–25

Hawaiian Islands—Emperor Seamount Chain. The chain of islands and seamounts that extends from Hawaii to the Aleutian trench results from the movement of the Pacific Plate over an apparently stationary hot spot. Radiometric dating of the Hawaiian Islands shows that the volcanic activity decreases in age toward the island of Hawaii. (Reprinted with permission from Tarbuck, E. J., and Lutgens, F. K., *The Earth: An introduction to physical geology,* 3d ed., 1990, fig. 18.28.)

It was not until 60 Ma that a noticeable separation between North America and Europe began, and the island of Greenland appeared to be moving on the same plate as Europe. By 30 Ma, the northern end of the Mid-Atlantic Ridge shifted to the east of Greenland, putting it back on the North American Plate. North America and South America were not fully connected by the Isthmus of Panama until quite recently, about 5 Ma. This event had a marked effect on the patterns of ocean circulation and the distribution of marine life.

In the Southern Hemisphere (Figure 3–26B), the early stages of Pangaea's breakup show South America and a continent composed of India, Australia, and Antarctica separating from Africa. By 120 Ma, there is a clear separation between South America and Africa, and

India had moved northward, away from the Australia-Antarctica mass, which began moving toward the South Pole. As the Atlantic Ocean continued to open, India moved rapidly northward and collided with Asia about 35 Ma. Subsequent to this collision, the spreading center that ran between India and Australia was replaced by one that began to separate Australia from Antarctica. The scars from the northward movement of the Indian Plate are clearly visible in Figure 2–18.

The overall effect of global plate tectonic events over the past 180 million years is the creation of the new Atlantic Ocean, which continues to grow, and the division of the ancient ocean of Panthalassa into the Pacific Ocean and Indian Ocean. The fate of the Indian Ocean is uncertain. But it is clear that the Pacific Ocean

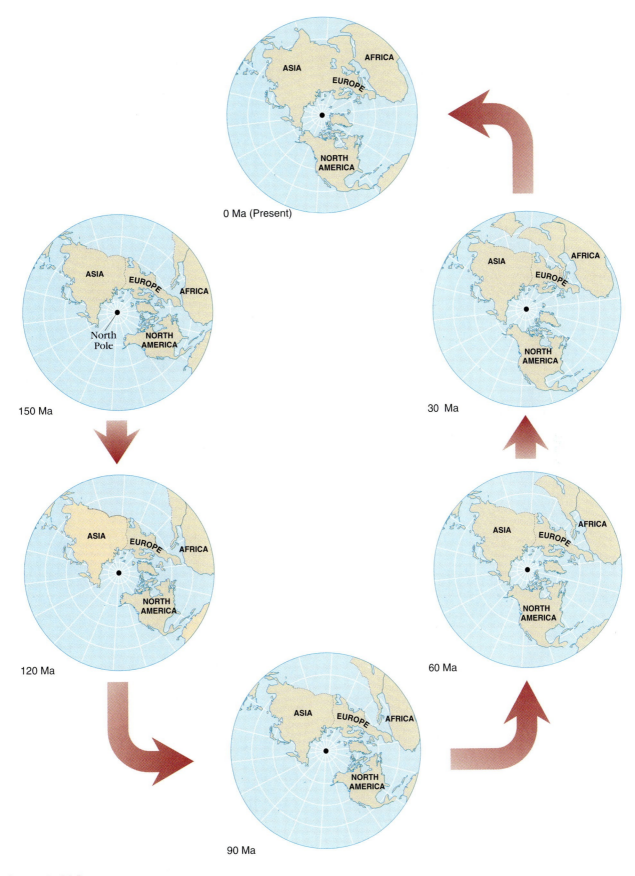

0 Ma (Present)

150 Ma

North Pole

120 Ma

90 Ma

60 Ma

30 Ma

Figure 3–26A

Paleogeographic Reconstruction—Northern Hemisphere. These North Polar views show the positions of the continents over the last 150 million years. Ages are in Ma (millions of years ago) at intervals of 30 million years.

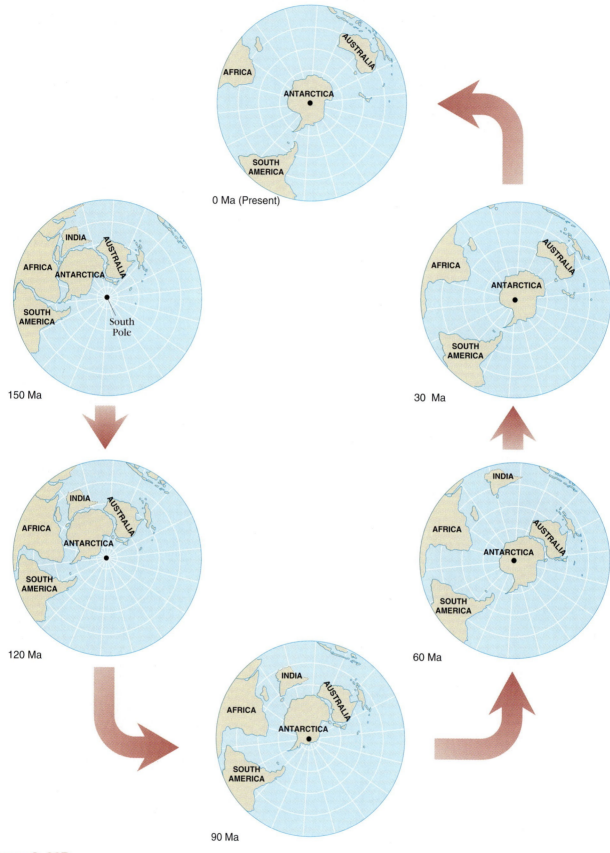

Figure 3–26B

Paleogeographic Reconstruction—Southern Hemisphere. These South Polar views are based on data from the first 60 legs of the Deep Sea Drilling Project, conducted during a decade of deep-ocean drilling, 1968 to 1978. (A and B courtesy of P. L. Firstbrook, B. M. Funnell, A. M. Hurley, and A. G. Smith in association with Scripps Institutions of Oceanography, University of California, San Diego.)

continues to shrink as oceanic plates produced within it subduct into the many trenches that surround it and continent-bearing plates bear in from east and west.

Age of the Ocean Floor

Much can be learned of the history of ocean basins by mapping the *pattern of age distribution* for the ocean crust. Observing Figure 3–27, you can see that the Atlantic Ocean has the simplest and most symmetrical pattern of age distribution. As Pangaea was rifted by the new Mid-Atlantic Ridge, the first separation involved North America and Africa over 175 Ma.

The Pacific Ocean has the least symmetrical pattern because of the large amount of subduction that surrounds it. The ocean floor east of the East Pacific Rise that is older than 40 Ma has already been subducted, whereas ocean floor in the northwestern Pacific is over 170 Ma. The Rise itself has even disappeared under North America. The broader age bands on the Pacific Ocean floor are clear evidence that its spreading rate has been greater than in the Atlantic or Indian oceans.

Metallic Ores—A Gift from Plate Tectonics

Some continental deposits of copper, lead, zinc, and silver appear to have arrived via converging plates (Figure 3–28). Copper sulfide ores from oceanic crust have been pushed up onto continental plates. A good example of this kind of ore deposit is the Troodos Massif along the southern coast of Cyprus, which has been mined for copper since the early days of Mediterranean civilization.

Spreading centers at oceanic ridges may be the source of this metallic enrichment. The metals found

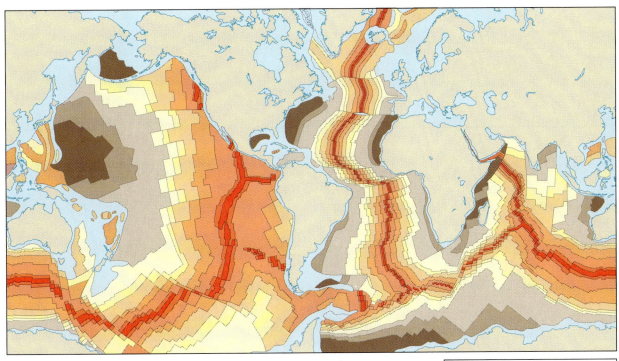

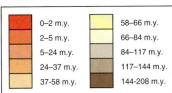

0–2 m.y.	58–66 m.y.
2–5 m.y.	66–84 m.y.
5–24 m.y.	84–117 m.y.
24–37 m.y.	117–144 m.y.
37–58 m.y.	144–208 m.y.

Figure 3–27

Relative Age of Oceanic Crust beneath Deep-Sea Deposits. Notice that the youngest rocks (bright red areas) are found along the oceanic ridge crests and the oldest oceanic crust (brown areas) is located adjacent to the continents. When you observe the Atlantic basin, a symmetrical pattern centered on the Mid-Atlantic ridge crest becomes apparent. This pattern verifies the fact that seafloor spreading generates new oceanic crust equally on both sides of a spreading center. Further, compare the widths of the ageing stripes in the Pacific basin with those in the South Atlantic. Because these stripes were produced during the same time period, this comparison verifies that the rate of seafloor spreading was faster in the Pacific than in the Atlantic. (Reprinted from Tarbuck, E. J., and Lutgens, F. K., 1990, *The Earth: An introduction to physical geology*, 3d ed., fig. 19.16, after *The Bedrock Geology of the World*, by R. L. Larson et al. Copyright © 1985 by W. H. Freeman.)

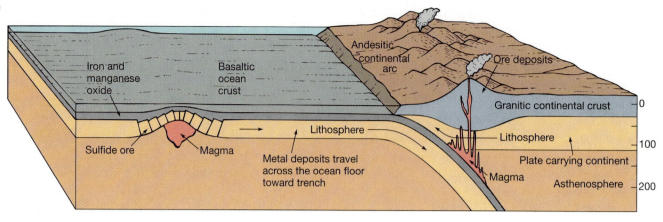

Figure 3–28

Theory of Metallic Ore Production.

here are predominantly iron and manganese, with significant amounts of copper, nickel, cobalt, zinc, and barium. Analysis of such sediments in a 200-square-kilometer (77-square-mile) area of the Atlantis Deep in the Red Sea have shown that they contain more than 3

million tons of zinc, 1 million tons of copper, almost 1 million tons of lead, and 5000 tons of silver. Similar iron- and manganese-rich deposits were recovered by submersibles diving along the Mid-Atlantic Ridge during Project FAMOUS in 1974.

SUMMARY

Scientific investigation of continental drift began with Alfred Wegener at the start of the 20th century. Many geologists and geophysicists who had resisted this hypothesis throughout the first half of the 20th century acknowledged that the continents were indeed moving when evidence gathered by the late 1960s indicated that all the present continents were combined into one large continent, Pangaea, about 200 million years ago.

The discovery that Earth's magnetic field has changed polarity throughout history and that the record of these changes is permanently recorded in the ocean crust made it possible to develop a hypothesis of seafloor spreading to explain how the continents could move. According to the hypothesis, magma chambers feed lava to fractures along oceanic ridges and rises, creating new seafloor, adding it to lithospheric plates. These plates are dense mantle rocks with a thin crust of basaltic lava beneath the ocean basins.

As material is added, it moves away down the slopes of the ridges and rises to make room for new material surfacing at the axes. Thus, crustal rocks beneath the oceans increase in age from the axes of the mountain ranges into the deep-ocean basins. Continents located on a lithospheric plate are carried passively along as the plate moves away from the oceanic ridge or rise. Lithospheric movement appears to be made possible by a plastic region of mantle material, the asthenosphere, immediately be-

neath the lithosphere. The seafloor spreading hypothesis was confirmed when the *Glomar Challenger* began the Deep Sea Drilling Project to sample ocean sediments and crust in 1968.

As new mass is added to the lithosphere at the oceanic rises and ridges (divergent boundaries), the opposite ends of the plates are subducted into the mantle at ocean trenches or beneath continental mountain ranges such as the Himalayas (convergent boundaries). Additionally, oceanic ridges and rises are offset and plates slide past one another along transform faults.

Although Earth is over 4.6 billion years old, all rocks found on the ocean floor are younger than 180 million years old because they are recycled back into the mantle by subduction.

As the lithospheric plates move away from the high-heat-flow oceanic ridges and rises toward low-heat-flow trenches, they cool and thicken. The age and depth of the ocean floor increase with increasing distance from the spreading centers.

Along the axes of oceanic ridges and rises, seawater percolates down through fractures to be heated and returned to the surface through hydrothermal vents with temperatures from 10° to 350°C (50° to 662°F). This process has produced most of Earth's metal sulfide ore deposits, and new biological communities were discovered in 1977. These vent biocommunities depend on chemosynthesis by bacteria that oxidize hydrogen sulfide rather than the photosynthetic plankton that support most marine life.

Seamounts and guyots are intraplate features that form as volcanic peaks and islands along the spreading centers. As they move away from the spreading centers on the cooling lithospheric plates, the volcanoes become inactive within 30 million years and submerge. Where average monthly temperatures exceed 18°C (64.4°F) and deposition of sediment is minimal, coral reefs develop around volcanic peaks. With continued subsidence, they can change from fringing reefs to barrier reefs and atolls.

A hot spot is a region of local upwelling of molten material. A hot spot is producing the island of Hawaii and has produced all the volcanic peaks of the Hawaiian Island chain and the Emperor Seamounts during the past 70 million years.

Paleogeography, the study of the changing shape and location of ocean basins and continents, has discovered changes in the physical and biological character of the oceans. The study of these changes is called paleoceanography. Since the breakup of Pangaea, which began 180 Ma, the Atlantic Ocean has come into existence and Panthalassa has been reshaped into the Pacific Ocean and Indian Ocean. There is evidence that metallic ore deposits associated with continental mountain ranges originated as part of the global plate tectonics process.

KEY TERMS

Absolute dating (p. 54)
Asthenosphere (p. 60)
Atoll (p. 74)
Barrier reef (p. 74)
Bullard, Edward (p. 53)
Continental drift (p. 52)
Convection cell (p. 60)
Convergent boundary (p. 64)
Divergent boundary (p. 64)
East Pacific Rise (p. 64)
Fossil (p. 53)
Fringing reef (p. 74)
Galápagos Rift (p. 65)
Global plate tectonics (p. 52)
Gondwanaland (p. 55)

Half-life (p. 53)
Hot spots (p. 63)
Hydrothermal vent (p. 65)
Isotope (p. 53)
Juan de Fuca Ridge (p. 59)
Laurasia (p. 55)
Lithosphere (p. 59)
Lithospheric plates (p. 59)
Magnetic anomaly (p. 59)
Magnetic dip (p. 55)
Mid-Atlantic Ridge (p. 64)
Paleoceanography (p. 75)
Paleogeography (p. 75)
Paleomagnetism (p. 55)
Pangaea (p. 52)

Panthalassa (p. 55)
Plume (p. 63)
Radioactive dating (p. 53)
Relative dating (p. 53)
San Andreas Fault (p. 66)
Seafloor spreading (p. 52)
Spreading center (p. 59)
Spreading rate (p. 64)
Subduction (p. 60)
Tethys Sea (p. 55)
Transform fault (p. 66)
Transform boundary (p. 64)
Wegener, Alfred (p. 52)

QUESTIONS AND EXERCISES

1. Why couldn't Alfred Wegener convince most scientists that the continents were indeed moving across Earth's surface?

2. What evidence was developed in the mid-20th century that supported the possibility of continental drift?

3. How do the magnetic properties of the continents and ocean basins differ? Why?

4. Describe how fossil evidence is used in relative dating of rock units found on Earth.

5. What discovery made absolute dating of rock units possible? Why was this technique a significant improvement over relative dating?

6. What is the age of a rock that has a ratio of U^{235} to Pb^{207} of 1:3?

7. Describe the evidence found on the present continents that suggests that Laurasia and Gondwanaland existed between 250 and 350 million years ago.

8. How does the magnetic dip of magnetite particles found in igneous rocks tell us the latitude at which they formed?

9. How does continental movement account for the two apparent wandering curves of the north magnetic pole as determined from Europe and North America?

10. What property of the magnetic record found in oceanic crustal rocks gave rise to the idea of seafloor spreading?

11. Describe a possible mechanism responsible for moving the continents over Earth's surface, including a discussion of the lithosphere and asthenosphere.

12. Describe the general relationships that exist among distance from the spreading centers, heat flow, age of the ocean crustal rock, and ocean depth.

13. Discuss the three types of plate boundaries—spreading centers of the oceanic ridges, trenches, and transform faults. Explain why earthquake activity is usually confined to depths less than 10 kilometers (6 miles) along the spreading centers and transform faults but may occur as deep as 670 kilometers (415 miles) near the trenches. Construct a plan view and cross section showing each of the three boundary types and direction of plate movement.

14. What evidence indicates that the plates are more likely being pulled down in the trench areas than being pushed apart from the oceanic ridge spreading centers?

15. Describe the general process of hydrothermal circulation associated with spreading centers and its relation-

ship to mining ore deposits and chemosynthetically supported biocommunities.

16. Explain why seamounts and guyots increase in age and depth with increased distance from the oceanic ridges and rises. Discuss the length of time they probably were active volcanoes.

17. Discuss the possible relationship of coral reef development to seafloor spreading. Include the effects of both vertical and horizontal motions.

18. How are the alignments and age distribution patterns of the Emperor Seamount and Hawaiian Island chains explained by the hot spot theory?

19. When does evidence for the following events first show up on the paleogeographic reconstructions of Figure 3–26?
    a. Greenland separates from Europe
    b. South American begins movement away from Africa
    c. India separates from Antarctica
    d. Australia separates from Antarctica

20. Describe the relationship of metallic ore deposits in the lithosphere to the plate tectonics process.

REFERENCES

Apperson, K. D. 1991. Stress fields of the overriding plate at convergent margins and beneath active volcanic arcs. *Science* 254:5032, 670–678.

Bercovici, D.; Schubert, G.; and Glatzmaier, G. A. 1989. Three-dimensional spherical models of convection in the earth's mantle. *Science* 244:4907, 950–954.

Bullard, Sir Edward. 1969. The origin of the oceans. *Scientific American* 221:66–75.

Burnett, M. S.; Caress, D. W.; and Orcutt, J. A. 1989. Tomographic image of the magma chamber at 12°50′N on the East Pacific Rise. *Nature* 339:206–208.

Carrigan, C. R., and Gubbins, D. 1979. The source of the earth's magnetic field. *Scientific American* 240–2:118–133.

Davies, P. J.; Symonds, P. A.; Feary, D. A.; and Pigram, C. J. 1987. Horizontal plate motion: A key allocyclic factor in the evolution of the Great Barrier Reef. *Science* 238:1697–1700.

Dietz, R. S., and Holden, J. C. 1970. The breakup of Pangaea. *Scientific American* 223:30–41.

Hamilton, W. B. 1988. Plate tectonics and island arcs. *Bulletin of the Geological Society of America* 100:1503–1526.

Hess, H. H. 1962. *History of ocean basins. Petrologic studies: A volume to honor A. F. Buddington,* Engel, A. E. J.; Lames, H. L.; and Leonard, B. F., eds., 559–620. New York: Geological Society of America.

Liu, M.; Yuen, D. A.; Zhao, W.; and Honda, S. 1991. Development of diapiric structures in the upper mantle due to phase transitions. *Science* 252:1836–1839.

Menard, H. W. 1986. *The ocean of truth: A personal history of global tectonics.* Princeton, N.J.: Princeton University Press.

Mid-ocean ridges. 1992. *Oceanus* 34:4, 1–111.

Olson, P.; Silver, P. G.; and Carlson, R. W. 1990. The large-scale structure of convection in the Earth's mantle. *Nature* 344:209–214.

Sclater, J. G.; Parsons, B.; and Jaupart, C. 1981. Oceans and continents: Similarities and differences in the mechanisms of heat loss. *Journal of Geophysical Research* 86:11, 535–552.

Shepard, F. P. 1977. *Geological oceanography.* New York: Crane, Russak.

Smith, D. K., and Cann, J. R. 1993. Building the crust at the Mid-Atlantic Ridge. *Nature* 365:6448, 707–715.

Tarbuck, E. J., and Lutgens, F. K. 1990. *The earth: An introduction to physical geology,* 3d. ed. New York: Macmillan.

_____. 1991. *Earth science.* 6th ed. New York: Macmillan.

van der Hilst, R.; Engdahl, R.; Spakman, W.; and Nolet, G. 1991. Tomographic imaging of subducted lithosphere below northwest Pacific island arcs. *Nature* 353:37–42.

SUGGESTED READING

Sea Frontiers

Burton, R. 1974. Instant islands. 19:3, 130–136. A description of the 1973 eruption on the island of Heimaey south of Iceland and its relationship to plate tectonics processes.

Dietz, R. S. 1976. Iceland: Where the mid-ocean ridge bares its back. 22:1, 9–15. A description of the rift zone of Iceland and its relationship to the Mid-Atlantic Ridge and seafloor spreading.

_____. 1977. San Andreas: An oceanic fault that came ashore. 23:5, 258–266. A discussion of the San Andreas Fault and its relationship to global plate tectonics.

_____. 1971. Those shifty continents. 17:4, 204–212. A readable and informative presentation of the crustal features and the possible mechanism of plate tectonics.

Emiliani, C. 1972. A magnificent revolution. 18:6, 357–372. A discussion of advances in studying Earth science from the sea, including climate cycles, plate tectonics, and the economic potential of resources lying within marine sediments.

Mark, K. 1974. Earthquakes in Alaska. 20:5, 274–283. A discussion of the origin, nature, and effects of earthquakes in Alaska.

_____. 1972. Ocean fossils on land. 18:2, 95–106. A discussion of the significance of marine fossils found in rocks now many miles inland is centered on the work of an early American paleontologist, James Hall.

Rona, P. 1984. Perpetual seafloor metal factory. 30:3, 132–141. A discussion of how metallic mineral deposits form in association with hydrothermal vents located on oceanic ridges and rises.

Scientific American

Bloxham, I., and Bubbins, D. 1989. The evolution of the earth's magnetic field. 261:6, 68–75. The theory of how Earth's magnetic field is associated with motions in Earth's liquid outer core.

Bonatti, E., and Crane, K. 1984. Oceanic fracture zones. 250:5, 40–51. The role of oceanic fracture zones in the plate tectonics process is related to the spherical shape of Earth.

Brimhall, G. 1991. The genesis of ores. 264:5, 84–91. After being emplaced into Earth's crust at oceanic ridges, metal deposits undergo a number of processes before they become mineable ores embedded in continental rocks.

Jeanloz, R., and Lay, T. 1993. The core-mantle boundary. 261:5, 48–55. The boundary between the liquid core that generates Earth's magnetic field and the overlying solid mantle is surprisingly active. Core fluids may rise into the mantle hundreds of meters above the average position of the core-mantle boundary.

Macdonald, K. C., and Fox, P. J. 1990. The mid-ocean ridge. 262:6, 72–95. An interesting and comprehensive overview of present knowledge of the oceanic ridges and rises is presented.

CHAPTER 4
Marine Sediments

More than half of the rocks exposed on the continents above the present shoreline are made of sediment laid down in past ocean environments. The examples are familiar: sandstone, limestone, shale, and others. This relationship between the continents and ocean basins had puzzled geologists until global plate tectonics theory explained the mechanism by which ocean sediments are incorporated into the continents. Simply, folded mountain ranges form when two plates converge and compress the sediments against the margin of an existing continent.

Geologists once believed that sediments in the deep-ocean basins existed in unchanged locations since Earth's first oceans formed. They felt that within the sediments lay an undisturbed record of Earth's history. Knowledge of the plate tectonics process, and the discovery that the oldest seafloor dates back only 170 million years, has evaporated that dream. But much of the ancient record is still available in marine sediments that became lithified—turned to rock—and that today are readily accessible in continental mountain ranges.

The accumulation of sediments on the ocean floor may seem like the most boring process one could imagine. However, increased understanding of the intricacies of this process has opened a most revealing window into the history of our planet. Most of what we know of Earth's past geology, climate, and biology has been learned through study of these ancient marine sediments—they are *our information highway into the past.*

Although it seems unlikely that sediment representing more than 200 million years of Earth's 4.6-billion-year history will be found in the deep-ocean basins, this does not dim the marine geologist's interest in the study of these sediments. They are helping us determine the details of the more recent history of the present ocean basins. Clues to past climates, movements of the ocean floor, ocean circulation patterns, and nutrient supply for marine plants and animals are embedded in the sedimentary deposits throughout the ocean basins.

As stated, most of our knowledge of Earth's ancient history has been obtained by studying sedimentary rocks, some of which date back 3.8 billion years. The farther back we go, the more difficult it is to obtain information from these deposits. Yet, with the expanding technology that can be used to extract information from these rocks, we hope to develop an increasingly precise understanding of the processes that have shaped Earth throughout its history.

In the following pages, we will consider important properties of sediments and then some specific sedimentary deposits found in today's oceans.

Sediment Texture

Sediment texture is determined primarily by grain size. The abbreviated Wentworth scale presented in Table 4–1 classifies particles in sizes ranging from boulders down through cobbles, pebbles, granules, sand, silt, and clay to colloids. (Colloid-size particles are so tiny that they stay suspended in a fluid indefinitely.) Sediment size indicates the energy condition under which a deposit is laid down. Deposits that are laid down in areas where wave action is strong (areas of high energy) may be composed primarily of larger particles—cobbles and boulders. The deposition of clay-sized particles occurs in areas where the energy level is low and the current speed is minimal.

Sediments composed of particles that are primarily within the same size classification are *well sorted.* A beach sand is usually a well-sorted deposit. *Poorly sorted* deposits may contain particles ranging in size from the colloid to boulders. An example is sediment

Table 4–1
Wentworth Scale of Grain Size for Sediments

Size Range (mm)	Particle	
256	Boulder	
	Cobble	
64	Pebble	
4		
	Granule	
2	Very coarse sand	
1	Coarse sand	SAND
½	Medium sand	
¼	Fine sand	
⅛	Very fine sand	
1/16	Coarse silt	
1/32	Medium silt	SILT
1/64	Fine silt	
1/128	Very fine silt	
1/256	Coarse clay	
1/640	Medium clay	CLAY
1/1024	Fine clay	
1/2360	Very fine clay	
1/4096	Colloid	

Source: Wentworth, 1922, after Udden, 1898.

carried by glaciers that dropped out as the glacier melted.

As particles are carried from the source to their point of deposition, they increase in *maturity.* This results from their association with moving water, which has the capacity to carry particles of a certain size away from a deposit and leave behind other, larger particles. As the time in the transporting medium increases, particles of sand size or larger become more rounded through chemical weathering and abrasion. Increasing **sediment maturity** is indicated by (1) decreasing clay content, (2) increased sorting, and (3) increased rounding of the grains within the deposit.

The poorly sorted glacial deposits, which contain relatively large quantities of clay-sized and larger particles that have not been well rounded, is an example of an *immature* sedimentary deposit. The beach sand we used as an example of a well-sorted sediment contains very little clay-sized material and is usually composed of well-rounded particles that have undergone considerable transportation. The beach sand is an example of a *mature* sedimentary deposit. Figure 4–1 illustrates the nature of sediments of various degrees of maturity based on these criteria.

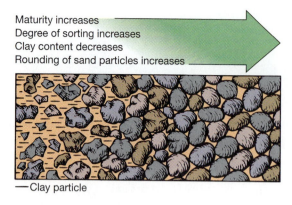

Maturity increases
Degree of sorting increases
Clay content decreases
Rounding of sand particles increases

—Clay particle

Figure 4–1

Sediment Maturity. As sediment maturity increases—left to right—the degree of sorting and rounding of particles increases, whereas clay content decreases.

Sediment on the Move

The sediment eroded from the continents and carried to the margins of the ocean by rivers and glaciers either settles out to fill bays, becomes a part of a delta, is spread across the continental shelf by currents, or is carried beyond the shelf to the deep-ocean basin. The greatest volume of sediment transport in the ocean is achieved by currents near the margins of the continent called *longshore currents.* Lower-energy currents distribute the finer components of the sediment along the margin of the continental shelf and even into the deep-ocean basin.

Figure 4–2 shows the relationship between current velocity and the erosion, transportation, and deposition of particles ranging in size from 0.001 to 100

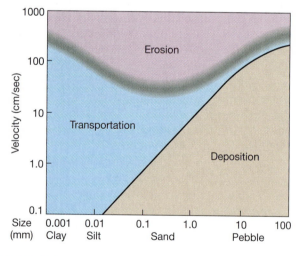

Figure 4–2

Hjulstrom's Curve. Horizontal current velocity vs. erosion—deposition of sediment by particle size. (Reprinted with permission from Tarbuck, E. J., and Lutgens, F. K., *The earth: An introduction to physical geology,* 3d ed. (New York: Macmillan, 1990), fig. 10.15.)

millimeters (0.00004 to 4 inches). As might be expected, the curve that separates transportation of particles from their deposition shows that larger particles will settle out at higher current velocities than smaller ones.

Observing the curve that separates the process of erosion from transportation, we can see that it takes higher-velocity currents to erode (pick up and carry away) larger particles, from the sand-sized particles up. However, clay-sized particles require a higher-velocity current than that needed to erode the larger sand-sized particles. This surprising fact results from the fact that clay-sized particles are flat rather than round, like typical sand-sized particles. Thus they have a greater surface area in contact with one another when they are laid down in a deposit. This causes a great cohesive force in clay sediments. Overcoming this cohesive attraction and picking up these particles requires a high-velocity current.

Particles of a diameter less than 20 micrometers (1 micrometer = 1 millionth of a meter) can be carried far out over the open ocean by prevailing winds. Such particles either settle out as the velocity of the wind decreases or find their way into the ocean during precipitation. They commonly serve as nuclei around which raindrops and snowflakes form, and thus fall to Earth. Some particles that reach high altitudes can be carried very rapidly by the jet stream that exists in the mid-latitudes.

Sediment from Rocks, Organisms, Water, and Space

We think of sediment as coming from eroding rocks, but sediment also derives from organisms, from minerals dissolved in water, and even from space. Clues to the origin of sediments are found in their mineral composition. Certain minerals, such as quartz and feldspar, are common rock-forming minerals. Their presence in a marine sediment indicates that the sediment was derived from rocks. Sediments derived from organisms often contain silica (SiO_2) and calcite ($CaCO_3$). Sediments that precipitate from ocean water have a characteristic composition, as do sediments from space. We will now describe the various origins of marine sedimentary particles.

Lithogenous Sediment

Lithogenous means "derived from rocks." Most lithogenous sediment comes from the mass of rock that makes up the continents. Volcanic islands in the open ocean are also important sources of lithogenous sediment.

All parts of Earth's crust originally formed from the solidification of molten material into *igneous* rocks (*ignis* is Latin for "fire"). Igneous rocks solidified at temperatures and pressures well above those of today's surface conditions and in an environment where free oxygen and water were scarce. These conditions caused the formation of distinctive minerals and rocks.

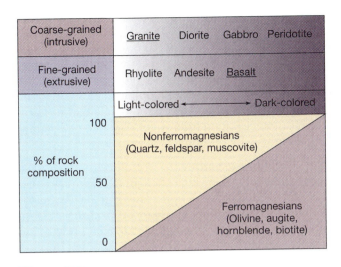

Figure 4–3

Igneous Rock Classification. This figure shows the most common types of igneous rocks. The coarse–grained ones cool slowly deep beneath Earth's surface, and granite, the typical intrusive rock of the continents, is the most common variety. Peridotite is found in Earth's mantle. The fine-grained extrusive rocks, which are associated with volcanic activity, cool rapidly at or near Earth's surface. The most common variety is basalt, which is characteristic of the oceanic crust. The rocks on the left of the table are light colored and contain mostly nonferromagnesian minerals. Moving to the right, ferromagnesian minerals replace nonferromagnesian minerals, and the rocks become darker in color.

Igneous rocks are composed of discrete crystals of naturally occurring compounds called *minerals.* These minerals may be grouped into two basic categories: *ferromagnesian* (iron-magnesium) and *nonferromagnesian.* Ferromagnesian minerals are relatively dense and dark and contain significant amounts of the elements iron and magnesium. Common varieties are olivine, augite, hornblende, and biotite mica. The nonferromagnesian minerals contain no iron or magnesium and have a lower density and lighter color. Examples are quartz, feldspar, and muscovite mica.

All these minerals are—to various degrees—unstable under the low temperature conditions that prevail at Earth's surface. Consequently, when exposed, they immediately begin to undergo chemical and physical breakdown known as **weathering.** As they break down, the particles are carried to the ocean, where they are deposited. By far the greatest quantity of lithogenous material is found around the margins of the continents. However, there are no parts of the ocean basins where traces of this sediment are completely absent.

Figure 4–3 shows the major types of igneous rocks and their general mineral composition and texture. Intrusive rocks, such as granite, cool and solidify slowly beneath the surface of Earth. The individual crystals grow to large size and give the rocks a coarse-grained texture. Extrusive rocks, such as basalt, cool more rapidly at or near Earth's surface, so the mineral crystals don't have time to grow large. They have a fine-grained texture. Either type weathers and its particles eventually end up on the seafloor. Many of these particles are *clays* (clay-size particles, Table 4–1). The most notable are chlorite, montmorillonite, illite, kaolinite, and a group called zeolites.

A large percentage of the lithogenous particles that find their way into the deep-ocean sediments far from the continents are transported by prevailing winds that remove small particles from the subtropical desert regions of the continents. Figure 4–4 shows where small shards of quartz are present in the surface sedi-

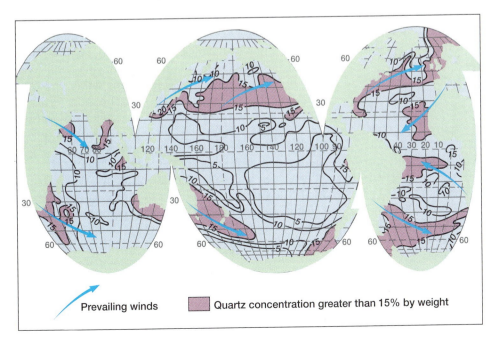

Figure 4–4

Lithogenous Quartz in Surface Sediments of the World's Oceans. Occurring as chips and shards of mostly 5 to 10 micrometers (0.0002 to 0.0004 inches), most quartz in deep-sea sediments is believed to have been transported by wind. Desert areas of Africa, Asia, and Australia are important source areas. Turbidity flows have provided significant quartz in the equatorial, northwestern, and northern Atlantic Ocean and the Bay of Bengal, where deep-bottom currents have modified sediment distribution. The distribution of lithogenous clays transported by wind may have a similar pattern. (After Leinen et al., 1986.)

Prevailing winds   Quartz concentration greater than 15% by weight

ments of the ocean floor. It reveals a close relationship between the location of strong prevailing wind systems and high concentrations of quartz shards in ocean sediment. The pattern of clay particle distribution in these sediments would likely be similar.

Biogenous Sediment

Biogenous means "derived from organisms." The hard remains of organisms, such as bones and teeth of animals and the protective coverings of minute algae and protozoans, are deposited on the ocean bottom by the billions. The most common chemical compounds in biogenous sediment are **calcium carbonate** ($CaCO_3$), which is similar to limestone, and **silica** (SiO_2), which is similar to quartz or glass.

Contributing most of the silica are microscopic algae called **diatoms** and protozoans called **radiolarians. Foraminifers,** close relatives of radiolarians, are the source for most of the calcium carbonate, but other significant contributors are pteropods (relatives of snails) and algae that leave behind hard **coccoliths** (Figure 4–5).

Coral reefs are made of the calcium carbonate skeletons secreted by corals and algae, so they are biogenous sediments. In fact, coral reefs are massive sedimentary rock structures, discussed in Chapter 15.

The Biological Carbon Pump. Please take a moment to study Figure 4–6, which shows where carbon is stored on Earth. Without carbon, life would not exist. Its storage and movement among rocks, oceans, air, and living things is crucial to maintaining life as we know it.

The **biological carbon pump** is the process by which carbon from carbon dioxide (CO_2) is incorporated into organisms through photosynthesis and through their secretion of carbonate shells. Over geologic time, more than 99 percent of the carbon dioxide added to the atmosphere by volcanic activity has been removed by the biological pump action and deposited in marine sediments as biogenous calcium carbonate and fossil fuels (oil and natural gas).

As a result of human activities, the atmospheric concentration of CO_2 has increased by 25 percent over the past 125 years, to a level of 350 parts per million. The concentration increases by 1.2 parts per million each year. This doesn't sound like much, but when each million molecules of air in the vast atmosphere acquires another 1.2 molecules of CO_2, the overall increase is a massive 2.1 billion tons (gigatons, or Gt).

Human activities add some 5.3 Gt/yr to the atmosphere, but only 2.1 Gt/yr *stay* in the atmosphere. Where does the missing 3.2 Gt go? It very likely goes into the ocean. Because nearly all the CO_2 ever put into the atmosphere has ended up in marine sediments, it is vital that we learn much more precisely the role of biological pumping in the cycling of CO_2 throughout Earth's ecosystem.

Hydrogenous Sediment

Hydrogenous means "derived from water." Chemical reactions within seawater cause certain minerals to **precipitate,** which means that they drop out of the water as solids and fall to the seafloor. Manganese deposits, phosphates, and glauconite are minerals that form by chemical precipitation from water. The rate of accumulation of this type of sediment is very slow.

Manganese Nodules. The element *manganese* is important for making high-strength steel alloys. Among the sediments on the deep-ocean floor having major economic potential are **manganese nodules** (Figure 4–7). The *Challenger* discovered that such nodules are relatively abundant in all of the major ocean basins.

The major components of these nodules are *manganese dioxide* (around 30 percent by weight) and *iron oxide* (around 20 percent). Also occurring in the nodules are copper, cobalt, and nickel. Copper is valuable in electrical wiring, pipe, and making brass and bronze. Cobalt is alloyed with iron to make strong magnets and steel tools. Nickel is used in making stainless steel. Although their concentrations are usually less than 1 percent, they can exceed 2 percent by weight, and are economically interesting.

Phosphates. These phosphorus-bearing compounds occur abundantly as a precipitate in nodules, as a thin crust on the continental shelf, and on banks at depths above 1000 meters (3300 feet). Concentrations of phosphates in such deposits commonly reach 30 percent by weight. Phosphates are valuable as fertilizers, and old marine deposits on land are extensively mined to supply agriculture.

Glauconite. Another hydrogenous sediment is *glauconite,* a greenish mineral that lends color to some marine "green sands" and "green muds."

Carbonates. The two most important carbonates in marine sediment are *aragonite* and *calcite.* Both are calcium carbonate ($CaCO_3$), but they have different crystalline structures. Aragonite is less stable and over time changes to calcite. The most common forms of precipitated carbonate are short aragonite crystals (less than 2 millimeters, or 0.08 inches) and **oolites,** onion-like spheres that precipitate around a nucleus and reach diameters of less than 2 millimeters (0.08 inches).

Ocean water is essentially saturated with calcium ions. The only factor preventing precipitation of more calcium carbonate is the absence of carbonate ions. Carbon dioxide (CO_2) is important in carbonate chemistry, for it is the source of "carbonate" in calcium carbonate.

The presence of carbon dioxide and the carbonate ion in ocean water is mutually exclusive. A high concentration of CO_2 means there will be few or no carbonate ions in the water. Where photosynthesis removes signif-

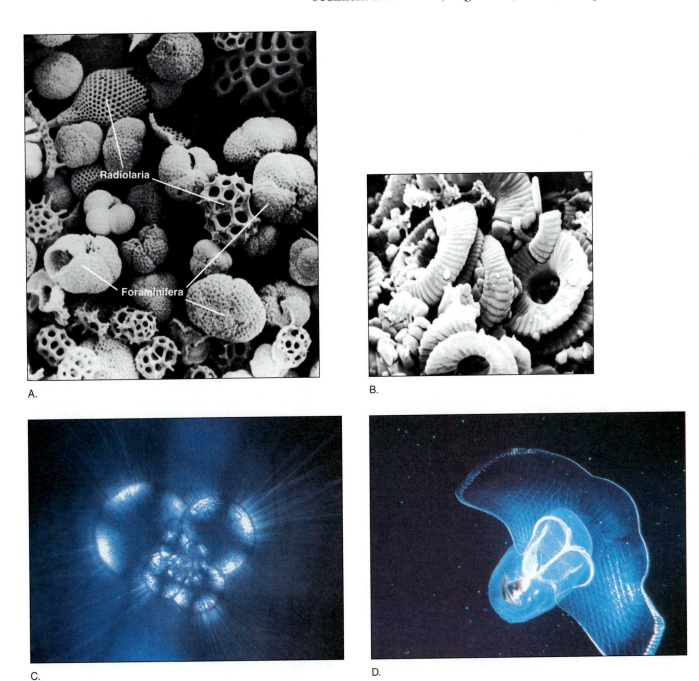

A.

B.

C.

D.

Figure 4–5

Microscopic Skeletons from the Deep Sea. *A:* Magnified 160 times by scanning electron microscopy, the skeletons are foraminifers and radiolarians from sediment cores taken in the west Pacific Ocean by the Deep Sea Drilling Project. They were recovered in water about 2800 meters (9300 feet) deep, with further penetration of 130 meters (425 feet) into the ocean bottom. Foraminifer skeletons are calcareous. Radiolarian skeletons are siliceous. These tiny organisms live near the ocean surface, but upon death their skeletons fall to the seabed to become entombed in sediment. By carefully studying them, scientists can determine how long ago they lived and can learn much about the history of the ocean. (Courtesy of the Deep Sea Drilling Project, Scripps Institution of Oceanography, University of California, San Diego.) *B:* Coccoliths magnified about 10,000 times. (Photo courtesy of Deep Sea Drilling Project, Scripps Institute of Oceanography.) *C:* Foraminifer, *Orbulina universa,* magnified 300 times. (Photo courtesy of Dr. Howard Spero.) *D:* Snail-like pteropod, *Corolla spectabilis.* (Photo courtesy of James M. King, Graphic Impressions, P. O. Box 21626, Santa Barbara, CA 93121.)

Figure 4–6

Flows and Reservoirs of Carbon. Each year, 0.04 billion tons (gigatons, or Gt) of carbon, in the form of carbon dioxide, is emitted into the atmosphere by volcanic activity. Human activities release 5.3 Gt/y of carbon dioxide into the atmosphere. What happens to the 3.2 Gt that is removed from the atmosphere is an area of intensive investigation. A significant amount of it may enter the ocean. The present reservoirs of carbon are estimated as shown.

icant carbon dioxide from the water, more carbonate ions are available for carbonate precipitation. Also, warm water contains less carbon dioxide. Both factors are at work in places such as the Bahama Banks, where water is heated and algal photosynthesis is high. The

factors reduce the carbon dioxide content of the water and enhance precipitation of carbonates.

Ancient marine carbonate or limestone deposits constitute 2 percent of Earth's crust and 25 percent of all sedimentary rocks on Earth. They form the bedrock un-

A.

B.

Figure 4–7

Hydrogenous Sediment. *A:* Surface of a manganese nodule, magnified 156 times by a scanning electron microscope. The small tube-and-dome structures were built by organisms for shelter. Nodules may owe their existence to organisms that participate in their formation. (Photo courtesy of Scripps Institution of Oceanography, University of California, San Diego.) *B:* Manganese nodules on the Indian Ocean seafloor are formed as chemical precipitates, 5 to 10 centimeters in diameter. They contain quantities of manganese, iron, copper, nickel, and cobalt. Although the manganese is inferior to manganese ores mined on land, it is still a valuable commodity, making harvesting of the lumps a probability in the future. (Photo by Bruce Heezen, courtesy of Lamont-Doherty Earth Observatory, Columbia University.)

derlying Florida and many midwestern states from Kentucky to Michigan and from Pennsylvania to Colorado. Percolation of groundwater through these deposits has dissolved the limestone to produce dramatic phenomena, such as formation of sudden sinkholes and spectacular cavern features.

Cosmogenous Sediment

Cosmogenous means "derived from space." Typical particles of cosmic origin found in marine sediment are microscopic spherules. Some are rocky and others are rich in iron. They had long been thought to have been shed from meteors entering Earth's atmosphere. But investigators now believe they form in the asteroid belt between the orbits of Mars and Jupiter. They are believed to be sparks produced when asteroids collide. They then rain down on Earth as a general component of space dust.

A crater 170 kilometers (106 miles) in diameter along the northern coast of Mexico's Yucatán peninsula may be the site of a meteor impact 65 Ma. This impact could have kicked up so much dust that it blocked sunlight, chilled Earth's surface, and brought about the extinction of dinosaurs and other species.

Analysis of cores taken in the Antarctic Basin about 1400 kilometers (870 miles) west of Cape Horn indicates that a meteorite struck that area 2.3 million years ago. This is the only known impact of a meteor into a deep-ocean basin (Figure 4–8). The evidence is

in the form of three types of particles—a nonterrestrial basalt that shows signs of impact stress, vesicular particles that formed from melted meteoric material, and a few pieces of metal.

The metallic element iridium (Ir) occurs in greater concentrations in meteoric material than on Earth. Therefore, layers of sediment that contain unusually high concentrations of iridium are evidence of meteoric impact if there is no evidence of another source. Based on iridium concentrations shown in Figure 4–8, the pattern suggests that the meteorite that hit the Pacific Ocean was between 0.5 and 1 kilometers (0.3 and 0.6 miles) in diameter. It hit the ocean without producing a crater in the ocean floor, no doubt because the deep water absorbed all of the impact energy. One can only imagine the splash and the giant waves that traveled worldwide.

This event occurred near the start of the Pleistocene ice age. Although it surely didn't *cause* the climatic change, it could have *contributed* to a more rapid decrease in global temperature resulting from the meteoric dust and water vapor that the event could have sent into the stratosphere.

Neritic and Oceanic Sediments

The relative amounts of lithogenous, biogenous, hydrogenous, and cosmogenous particles found in marine sedimentary deposits vary considerably. The two largest

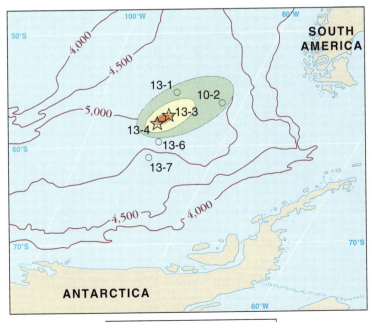

Figure 4–8

Evidence of Meteorite Impact in Antarctic Basin. An asteroid between 0.5 and 1.0 kilometers (0.3 and 0.6 miles) in diameter approached from the southwest and struck the Antarctic basin, according to the distribution of iridium-rich particles recovered in piston cores labeled on the map. The greatest concentrations of Ir-rich particles occur in cores 13–3 and 13–4, which are starred. (Data courtesy of Frank T. Kyte, University of California, Los Angeles.)

IRIDIUM CONCENTRATION IN
2.3 MILLION YEAR-OLD SEDIMENTS
(billionth of a gram per square centimeter)

10–15	50–200
15–50	200

categories into which marine sedimentary deposits can be divided are neritic deposits, found near the continents, and oceanic deposits, characteristic of the deep-ocean basins.

Neritic sediment refers to material with a wide range of particle sizes that is composed primarily of lithogenous particles derived from the continents. It accumulates rapidly on the continental shelf, slope, and rise. Neritic sediment also contains biogenous, hydrogenous, and cosmogenous particles, but these constitute only a minor percentage of the total sediment mass because of the rapid deposition rate of lithogenous particles near the continents. Neritic sediment is distributed along the continental shelf by surface currents and carried down the submarine canyons by turbidity currents. Here the sediment is further distributed by deep boundary currents.

Oceanic sediment is the fine-grained material that accumulates at a slower rate on the deep-ocean-basin floor. A greater variety of oceanic sediment accumulates because many fewer lithogenous particles arrive from the continental margins. The lithogenous component does dominate the sediment found in most of the deeper basins of the ocean floor, but biogenous and hydrogenous components are abundant in many areas.

Continental Margin Sediments (Neritic)

Neritic sediments include relict sediments, turbidites, glacial deposits, and carbonates.

At the end of the last Ice Age, around 18,000 years ago, glaciers melted and sea level rose. As a result, many rivers of the world today deposit their sediment in estuaries (drowned river mouths) rather than carry it onto the continental shelf as they did during the geologic past. In many areas the sediments that cover the continental shelf, **relict sediments,** were deposited from 3000 to 7000 years ago and have not yet been covered by more recent deposits. Sea level has not risen much over the last 3000 years. Such sediments presently cover about 70 percent of the world's continental shelf.

Turbidites. Although it seems unlikely that wave action and ocean current systems could carry coarse material beyond the continental shelf into the deep-ocean basin, there is evidence that much neritic sediment has been deposited at the base of the continental slope forming the continental rise. These accumulations thin gradually toward the abyssal plains. Such deposits are called **turbidites** and are thought to have been deposited by turbidity currents that periodically move down the continental slopes through the submarine canyons, carrying loads of neritic material that spreads out across the continental rise (Figure 4–9).

Turbidites have *graded bedding* (Figure 4–9B). This means that each deposit that is laid down has coarser material at its base and decreasing particle size toward the top of the deposit. This results from particles of different sizes settling out of a moving water mass as the velocity decreases. The coarser material settles out first, whereas the finer material stays in suspension longer. The seaward end of a turbidite deposit may be composed mostly of fine clay-sized particles.

During the millions of years that sediments accumulated at the margins of continents, they built a *sedimentary wedge* (Figure 2–10). The continental shelf, slope, and rise developed as the surface features of this wedge. The wedge is more than 10 kilometers (6 miles) thick and contains over 75 percent of the sediment on the entire ocean floor. Less than 25 percent of marine sediment occurs in the deep-ocean basins that cover in excess of 80 percent of the ocean floor.

Glacial Deposits. Poorly sorted deposits containing particles of all sizes, from boulders to clay, may be found in the high-latitude portions of the continental shelf. These glacial deposits were laid down by melting glaciers that covered the continental shelf area during the Pleistocene epoch, when glaciers were more widespread than today and sea level was lower. Glacial deposits are still forming around the continent of Antarctica and around the island of Greenland by *ice-rafting.* In this process, rock particles trapped in glaciers are carried out to sea by *icebergs* that break away from glaciers as they push into the coastal ocean. As the icebergs melt, the lithogenous particles of many sizes are released and settle to the ocean floor.

Carbonate Deposits. During the geologic past, deposits of limestone ($CaCO_3$) in the marine environment appear to have been widespread. Some limestones contain evidence of a biogenous origin (fossil seashells). Others appear to have formed as chemical precipitates. There appear to be very few places where nonbiogenous carbonates are presently forming. All these deposits are found in low-latitude shallow waters of continental margins or adjacent to islands.

The Bahama Banks is the largest region where significant carbonate deposition is presently occurring, although deposits are also forming on Australia's Great Barrier Reef, in the Persian Gulf, and in other local areas in the low latitudes. Coral reefs are another example of biogenous deposits of calcium carbonate found extensively on the continental margins of low-latitude continents and around oceanic islands.

Deep-Ocean Sediments (Oceanic)

The continental rise is a transitional feature between the continental margin and the deep-ocean basin. Consequently, some neritic material is deposited in significant quantities in the deep-ocean basin as part of the deep-sea fan deposits making up the continental rise. However, in the total area of the deep-ocean basin, more

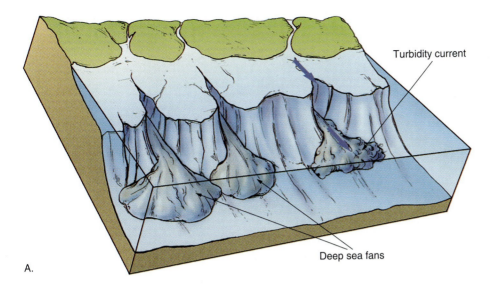

A.

Turbidity current

Deep sea fans

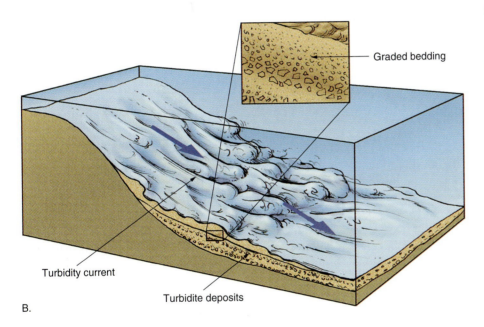

Graded bedding

Turbidity current

Turbidite deposits

B.

Figure 4–9

Turbidites and Graded Bedding. *A:* Turbidity currents flow down submarine canyons and deposit sediment as deep-sea fans. *B:* Examination of the deep-sea fans shows that they are composed of a series of deposits called turbidites. Each turbidite represents the sediment deposited by one turbidity current event. Each turbidite displays graded bedding in which the sediment particles grade from coarse at the bottom to fine at the top. (Reprinted with the permission of Macmillan Publishing Company from *The Earth,* Fourth Edition, figs. 19.6 and 19.7, by Edward J. Tarbuck and Frederick K. Lutgens. Copyright © 1993 by Macmillan Publishing Company.)

commonly no neritic material exists. Covering most of the deep-ocean floor are two types of oceanic sediments, abyssal clay and oozes.

Abyssal Clay. Accumulating at a rate of approximately 1 millimeter (0.04 inches) per 1000 years, **abyssal clays** cover most of the deeper ocean floor. They are commonly red-brown or buff in color because they contain oxidized iron. Abyssal clay is predominantly derived from the continents and carried by winds or ocean currents. These particles, along with some cosmic and volcanic dust, settle slowly to the ocean floor.

Oozes. At somewhat shallower depths, biogenous material is a significant portion of the sediment. It consists primarily of the minute protective hard coverings of microscopic organisms. If sediment contains 30

percent or more skeletal material by weight, oceanographers call that deposit an **ooze.** Depending on chemical composition, it may be a **calcareous ooze** ($CaCO_3$) or a **siliceous ooze** (SiO_2).

Oozes are not present on the continental margin. This is not for lack of skeletal remains of plants and animals; rather, the rate of accumulation of lithogenous sediment is so great that organically derived material never makes up 30 percent or more of the sediment. Oozes can form only where the deposition of material other than the remains of plants and animals occurs at a very low rate.

More specific names may be given to oozes that indicate the types of organic remains most abundant in the deposits. Thus we have *diatom ooze, radiolarian ooze, foraminifera ooze,* and *pteropod ooze.* Most com-

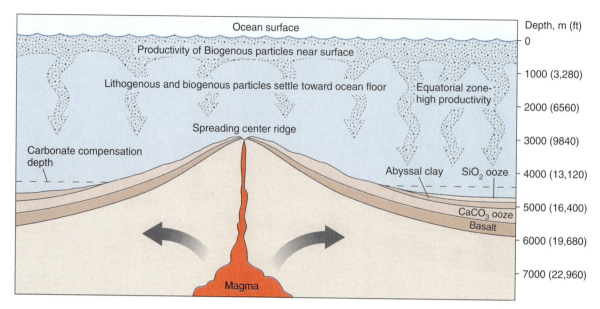

Figure 4–10

Seafloor Spreading and Sediment Accumulation. New basaltic oceanic crust is forming at ridges that run across the ocean floor. The ridges rise to depths of about 3000 meters (10,000 feet). As the crust moves away from the ridges and sinks to greater depths, the type of sediment that accumulates on it changes. Until it sinks below the carbonate compensation depth, the typical sediment that forms is $CaCO_3$ ooze. Below this depth $CaCO_3$ is dissolved, and abyssal clays dominate. If the crust passes beneath a region of high biological productivity, such as the Pacific equatorial region, oozes could again accumulate.

mon is *Globigerina ooze,* named for a foraminifer, which is especially widespread in the Atlantic and South Pacific oceans.

The rate of accumulation of biogenous material on the ocean floor depends upon three fundamental processes—productivity, destruction, and dilution. *Productivity* is the reproduction of plankton (floating minute organisms) that contribute most of the biogenous material in oceanic sediments.

Destruction of skeletal remains occurs by dissolving in the seawater. The ocean is undersaturated with silica at all depths, so seawater continually dissolves silica. Siliceous shells must be thick to last until they can be incorporated in the sediment.

In the case of calcium carbonate, destruction (solubility) varies with depth. At the warmer surface, the water is generally saturated with calcium carbonate. At greater depths, the colder water contains more carbon dioxide, allowing the calcareous fragments to be readily dissolved. Below 4500 meters (15,000 feet), carbon dioxide increases until calcium carbonate dissolves quite readily. Carbonate oozes are generally rare below 5000 meters (16,400 feet).

The depth at which calcium carbonate is dissolved as fast as it falls from above is the **carbonate compensation depth (CCD).** This may be as deep as 6000 meters (20,000 feet) in portions of the Atlantic Ocean, but

may be above 3500 meters (11,500 feet) in regions of low biological productivity in the Pacific Ocean.

Dilution refers to the inability of oozes to form where other sediments keep biologically derived silica or calcium carbonate below 30 percent of the sediment. Biogenous sediment deposition rates range between 1 and 15 millimeters (0.04 and 0.6 inches) per 1000 years, depending on the combined effect of productivity and destruction. Figure 4–10 shows the relations among carbonate compensation depth, seafloor spreading, productivity, and destruction in determining what type of sediment will accumulate on the ocean floor.

Figure 4–11 shows the percentage by weight of calcium carbonate accumulation in the surface sediments of ocean basins. It shows that high concentrations (exceeding 80 percent) are found high on the oceanic ridges, whereas little is found in deep-ocean basins such as the North Pacific, where the bottom lies beneath the carbonate compensation depth.

Distribution of Neritic and Oceanic Sediments

Figure 4–12 illustrates the distribution of neritic and oceanic sediments. It is interesting to compare this map

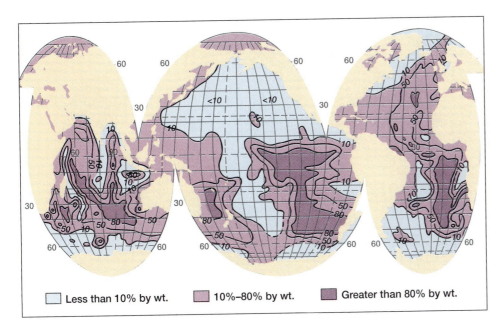

| Less than 10% by wt. | 10%–80% by wt. | Greater than 80% by wt. |

Figure 4–11

Calcium Carbonate in Surface Sediments of the World's Oceans. The deeper the ocean floor, the less calcium carbonate occurs in the sediment. It is also low in sediments accumulating beneath cold, high-latitude waters. Extensive concentrations of 80 percent or greater calcium carbonate in the sediment are associated with relatively shallow ocean floor on the Carlsberg Ridge in the Indian Ocean, the East Pacific Rise, and the Mid-Atlantic Ridge. Note that the deepest ocean basin, the North Pacific, lies for the most part beneath the carbonate compensation depth and has very little calcium carbonate in its accumulating sediment. (After Biscaye et al., 1976; Berger et al., 1976; and Kolla and Biscaye, 1976.)

with the seafloor painting in Figure 2–18; you can see the result of all the principles we have just discussed.

Oceanic sediments cover about 75 percent of the ocean bottom. Figure 4–13 shows the portion of each ocean floor that is covered by each sediment type. The bottom of the graph shows the collective total area for the entire world ocean. Note that calcareous oozes cover almost 48 percent of the entire deep-ocean floor. Abyssal clay covers 38 percent and siliceous oozes 14 percent of the total area. Calcareous oozes dominate the Indian and Atlantic oceans, whereas abyssal clay dominates the Pacific (also see Figure 4–12). This is undoubtedly related to the fact that the Pacific Ocean is deeper, so more of its bottom lies beneath the carbonate compensation depth.

Siliceous oozes cover a smaller percentage of the ocean bottom in all the oceans because regions of high productivity of diatoms and radiolarians, the major components of these deposits, are restricted to areas of high biological productivity along the equator and in the Antarctic.

One major problem that puzzled marine geologists is that sediments on the deep-ocean floor very closely reflect the particle composition of the surface water directly above. This was difficult to understand

because it would typically take these very tiny particles from 10 to 50 years to sink from the ocean surface to abyssal depths. During this interval, a horizontal ocean current of only 1 centimeter/second (about 0.02 miles/hour) would carry them from 3000 to 15,000 kilometers (1800 to 9300 miles) laterally before they reached the deep-ocean floor.

However, study of GEOSECS samples shows that 99 percent of particles that fall to the ocean floor do so as part of **fecal pellets** produced by tiny animals living in the water column above. These pellets (Figure 4–15), though still small, are large enough to allow the particulate matter to reach the abyssal ocean floor in 10 to 15 days. This could explain the similarity in particle composition of the surface waters and the sediment immediately below.

Ocean Sediments as a Resource

The seafloor is rich in potential mineral and organic resources. Because of inaccessibility, and therefore high cost, few are likely to be exploited in the near future. Some resources are being exploited, however, and they are presented here.

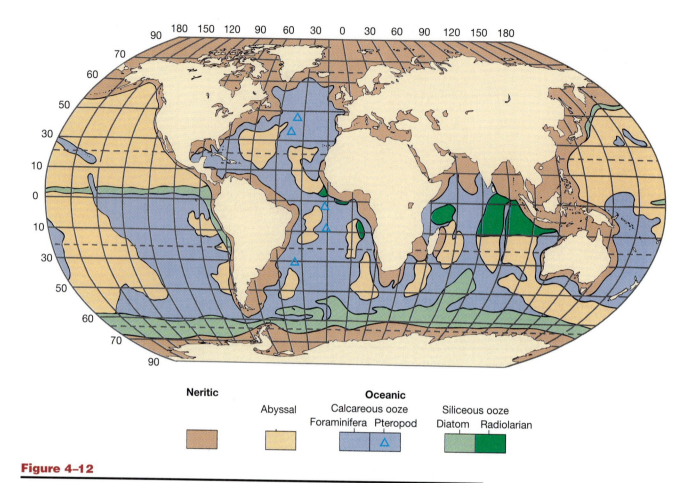

Figure 4–12

World Distribution of Neritic and Oceanic Sediments. Beyond the neritic deposits of the continental margins lie the oceanic marine deposits. Abyssal clays are found in the deeper ocean basins, where the bottom lies beneath the carbonate compensation depth. Calcareous oozes are found best developed on the relatively shallow deep-ocean environments of the oceanic ridges and rises. Siliceous oozes are found beneath areas of unusually high biological productivity such as the Antarctic and equatorial Pacific Ocean.

Petroleum

The remains of microscopic organisms, buried within marine sediments before they could decompose, are the source of our most valuable marine resource—petroleum. Of the nonliving resources extracted from the oceans, more than 95 percent of the value is in oil and gas.

The term **petroleum** encompasses both oil and natural gas. It provides most of the energy on which the world economy depends. There is believed to be enough oil available with present technology until about the year 2020, which will occur before your 50th birthday if you are a typical student. Beyond that date, major changes in the energy supply will be required. We will then need to convert to unconventional sources such as extra-heavy oil, tar sands, and other high-cost sources. Converting to natural gas would extend the fuel supply only a few years.

The percentage of world oil production coming from offshore has increased from a trace in 1930 to over 30 percent in 1990 (Figure 4–16). Major offshore reserves exist in the Persian Gulf, in the Gulf of Mexico, off southern California, in the North Sea, and in the East Indies. Additional reserves are expected to be found off the north coast of Alaska and in the Canadian Arctic, Asian seas, Africa, and Brazil, but there is little likelihood that they will provide enough to reverse the existing trend. With almost no hope of finding major undiscovered reserves on land, and only a slightly increased likelihood of finding them offshore, future oil exploration will still be intense in offshore waters.

Sand and Gravel

The offshore sand and gravel industry is second in value only to that of petroleum. This resource, which

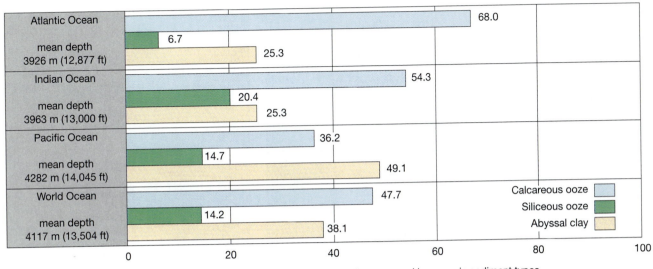

Figure 4–13

Distribution of Oceanic Sediment. This graph shows that less ocean basin floor is covered by calcareous ooze in deeper basins. This is probably because deeper ocean basins have more floor that lies beneath the carbonate compensation depth. The depths shown exclude shallow adjacent seas, where little oceanic sediment accumulates. Notice that the dominant oceanic sediment in the deepest basin, the Pacific, is abyssal clay, whereas calcareous ooze is the most widely deposited oceanic sediment in the shallower Atlantic and Indian oceans. (After Sverdrup, Johnson, and Fleming, 1942.)

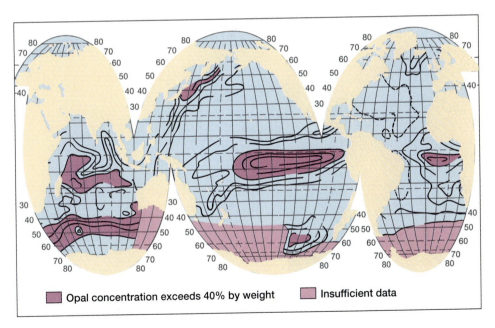

Figure 4–14

Biogenous Silica (Opal) in Surface Sediments of the World's Oceans. Opal in surface sediments of the ocean is concentrated in areas of greatest biological productivity. The particles are produced by diatoms and radiolarians in the surface waters. In the equatorial and northwest Pacific, opal is produced predominantly by radiolarians. Elsewhere, diatom remains are more abundant. The highly productive waters south of the Antarctic Convergence show up well in the south Indian Ocean and southeast Pacific Ocean. Data for the southern Atlantic Ocean and southwest Pacific Ocean are insufficient to determine the relationship between highly productive surface waters and opal concentrations in the sediment. (After Leinen et al., 1986.)

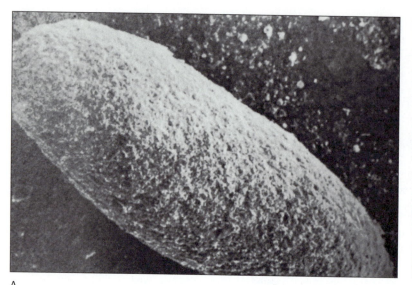

A.

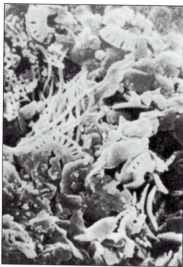

B.

Figure 4–15

Fecal Pellets. *A:* Fecal pellet produced by zooplankton. The pellet is 200 micrometers (0.008 inches) long, large enough to fall fairly rapidly from the surface to the seafloor. *B:* Surface of the fecal pellet showing the remains of small phytoplankton and other debris. (Photos courtesy of Susumu Honjo, Woods Hole Oceanographic Institution.)

Figure 4–16

Offshore Drilling Rig. This rig in the Gulf of Mexico off the Texas coast is drilling for gas and oil. (© Walter Frerck/Odyssey/Chicago.)

includes rock fragments and shells of marine organisms, is mined by technology similar to that used on land. The resource is used primarily for beach fill, land fill, and aggregate for mixing concrete.

Offshore deposits are a major source of sand and gravel in New England, in New York, and throughout the Gulf Coast. Many European countries, Iceland, Israel, and Lebanon also depend heavily on such deposits.

Some gravel deposits are rich in valuable minerals. Diamonds are recovered from gravel deposits in South Africa and Australia. Sediments rich in tin have been mined for years from Thailand to Indonesia. Platinum and gold have been found in deposits in gold mining areas throughout the world.

A major concern in sand and gravel mining is that it not be conducted too near shore. Nearshore mining operations may damage beaches because beach material is washed by wave action and currents offshore into the mined areas.

Salt Domes and Sulfur

When the Atlantic Ocean began to form, isolated basins of seawater evaporated. This left a thick bed of *halite* (ordinary table salt—sodium chloride, NaCl) mixed with the mineral gypsum, which contains sulfur. Denser sediments subsequently accumulated over the salt and gypsum in areas such as the present Gulf Coast. The salt, being less dense, has risen through the sediments to produce *salt domes* (Figure 4–17). In time, the salt on the top of the dome is dissolved by water, and the less soluble gypsum remains. Bacteria then convert the gypsum to free **sulfur.**

Using the Frasch process, the sulfur is mined by pumping hot water and air into it, melting the sulfur and forcing it back to the surface through the space between the hot water pipe and another larger pipe that surrounds it. However, with the recovery of large amounts of sulfur by today's pollution control equipment, most sulfur mining operations have been shut down.

Phosphorite (Phosphate Minerals)

Phosphorite is a sedimentary rock made of various phosphate minerals, which contain the element phosphorus, an important plant nutrient. Although no commercial phosphorite mining from the oceans is presently occurring, the marine reserve is estimated to exceed 50 billion tons. Phosphorite occurs at depths less than 300 meters (1000 feet) on the continental shelf and slope. Deposits can be used to produce phosphate fertilizer if economic conditions make them recoverable.

Some shallow sand and mud deposits contain up to 18 percent phosphate. Some phosphorite deposits

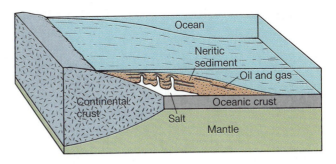

Figure 4-17

Sulfur Deposits. Evaporation of ocean water produced deposits of salt and gypsum on the floor of the newly formed Atlantic Ocean 170 to 180 million years ago when the ocean began to form. These deposits were subsequently covered by denser lithogenous sediments. The less dense salt flowed upward through the denser sediments to form salt domes. Free sulfur forms as a cap on the domes. Oil and gas deposits form in porous sand formations that are domed up or pierced by the rising salt. The oil and gas, being less dense than the salt water trapped in the sediments, rises until trapped by an impermeable layer of shale or salt.

occur as nodules, with a hard crust formed around a nucleus. The nodules may be as small as a sand grain or up to a meter in diameter and may contain over 25 percent phosphate. Most land sources have had their phosphate concentration enriched to more than 31 percent by groundwater leaching.

Polymetallic Crusts and Nodules

Manganese nodules have been known in the deep ocean since the voyage of the HMS *Challenger* (1872–1876). As explained earlier, the nodules contain significant concentrations of manganese and iron and smaller concentrations of copper, nickel, and cobalt. Mining companies began in the 1960s to assess the feasibility of mining them. Exploration found the richest metallic content in the eastern Pacific Ocean between Hawaii and Mexico (Figure 4–18).

Of the four metals, cobalt is a "strategic" metal that is not mined in the United States. It is required to produce dense, strong alloys with other metals for use in high-speed cutting tools, powerful permanent magnets, and jet engine parts. Major enrichments of cobalt in Earth's crust are confined to central southern Africa and deep-ocean nodules and crusts. Because of the unstable political situation in Africa, the United States has been looking to the ocean floor as a more dependable source.

However, interest has faded in the face of a depressed metals market. A second concern is the uncer-

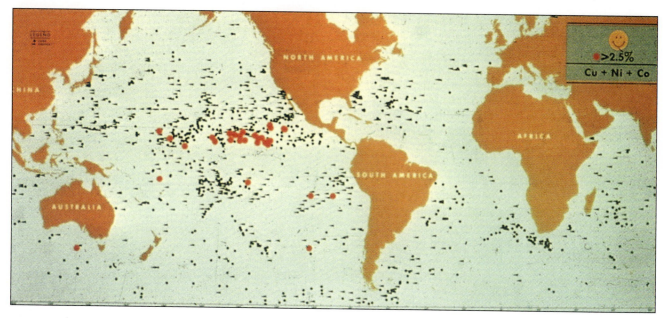

Figure 4–18

Distribution of Deep-Water Manganese Nodules Containing More Than 2.5 Percent Combined Copper, Nickel, and Cobalt. The manganese nodule deposits with the greatest economic potential have the highest concentrations of copper, nickel, and cobalt, and occur between the Clarion and Clipperton fracture zones in the eastern Pacific Ocean (see Figure 2–18). (Courtesy Scripps Institution of Oceanography, University of California, San Diego.)

tainty of ownership of the nodules, due to disagreement over provisions of the United Nations Law of the Sea Convention. This convention has still not been signed by most of the developed nations that have mining interests.

In 1981 **cobalt-rich manganese crusts** were found on the upper slopes of islands and seamounts that lie within the exclusive economic zones (EEZs) of the United States and its allies. The cobalt concentrations in these crusts are half again as rich as in the best African ores and at least twice as rich as in the deep-sea manganese nodules (Figure 4–19).

The greatest cobalt concentrations seem to be at depths between 1000 and 2500 meters (3300 and 8200 feet) on the flanks of islands and seamounts in the central Pacific Ocean. Additional sampling is needed to ensure that these crust deposits are extensive enough to provide a dependable source of cobalt. If they are, the first deep-sea mining operations may be for cobalt and other metals that lie within the EEZs of the United States or one of its allies. There are at least 100 of these seamounts in the EEZs of the Line Islands and Hawaiian Islands, and each could yield up to 4 million metric tons of ore.

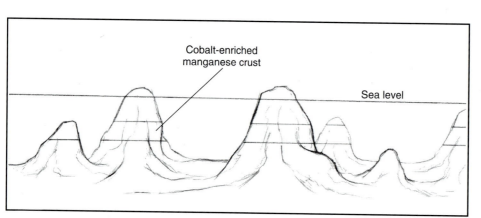

Figure 4–19

Cobalt-Rich Ocean Crusts. Cobalt-rich manganese crusts, composed primarily of iron and manganese oxides, contain up to 2.5 percent cobalt. This high cobalt concentration in the crusts, which are up to 2 centimeters (0.8 inches) thick, has aroused interest in its economic potential.

High-Level Nuclear Waste: Burial at Sea?

For nearly five decades, high-level nuclear waste has been accumulating from the production of nuclear weapons and from commercial power generation. The United States alone has more than 75 million gallons of waste from weapons production and 12,000 metric tons of "spent" reactor fuel rods that must be safely disposed of until they no longer are dangerous.

This material will be radioactive for more than one million years, but a "safe" disposal system would likely require a shorter time of highly secure confinement. The wastes are composed of over 50 isotopes of various elements. Each has different chemical, half-life, abundance, and radioactive emission characteristics. Investigators consider safe confinement to allow no more release of radiation into the atmosphere than that of the natural uranium ore from which the waste was generated (Figure 4–20).

The major emphasis is on disposal at land sites because the materials remain accessible. However, since 1974, research has been conducted into disposal of high-level nuclear waste in the deep sea. The United States stopped research in 1986 because of budget constraints. International research ceased in 1987 when international agreement was reached to pursue land-based disposal. However, the seabed disposal research did produce encouraging results. Given the political realities that may be faced as work proceeds toward land-based disposal, deep-sea disposal may be considered again. The major features of the seabed disposal program are described here.

Given the requirements that the waste must be kept away from human activities, kept safe from exposure by natural erosion, and kept away from seismically active regions, the *centers of oceanic lithospheric plates* might be ideal disposal sites. Such regions are beneath at least 5000 meters (16,000 feet) of water, far from the marine activities of humans. Fishing, petroleum production, and mining activities are now conducted primarily on the continental shelves. Only a potential area of manganese nodule mining between Mexico and the Hawaiian Islands would need to be avoided.

The lithospheric plates are covered with up to 1000 meters (3300 feet) of fine sediment that has accumulated uninterrupted for over 100 million years in some regions. This pattern of sediment accumulation indicates long periods of stability as the plates move across Earth's surface, and the pattern can be expected to continue for millions of years. In addition, the sediment could serve as an ideal medium in which to place the high-level nuclear waste. Many of the radioactive isotopes, if released from their storage canisters (which may fail within 1000 years), would naturally adhere to the clay particles in the sediment. This would slow diffusion of radioactivity away from the burial site.

A number of mid-plate, mid-gyre sites (MPGs) that are distant from ocean currents have been identified in the North Atlantic and North Pacific oceans (Figure 4–21). The initial plan is to place the canisters under 30 to 100 meters (100 to 330 feet) of sediment at intervals of at least 100 meters. How to emplace the canisters has not been decided yet, but it may be possible to simply drop them from an appropriate distance above the ocean floor (Figure 4–22).

Models indicate that heat release from the waste (radioactive waste gives off considerable heat) will be transferred almost entirely by conduction. Therefore, upward convection of radiation by water may be negli-

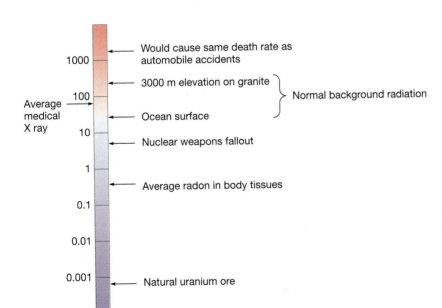

Figure 4–20

Radiation Doses from Various Sources in Millirems per Year. The U.S. Environmental Protection Agency sets minimum performance requirements. The levels of radiation emitted from high-level disposal sites should be no greater than that of natural uranium ore.

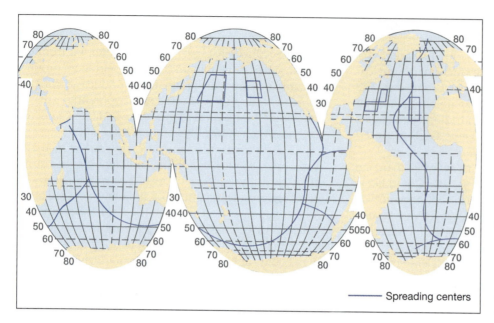

Figure 4–21

MPG Regions. Mid-plate, mid-gyre (MPG) regions in northern oceans where the environment can be expected to be stable for millions of years and thus safely entomb high-level nuclear waste until it no longer is dangerous.

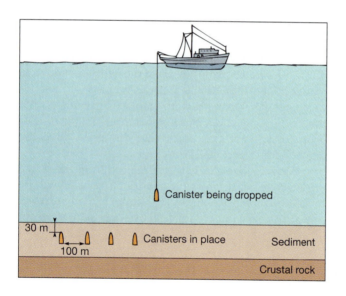

Figure 4–22

Nuclear Waste Canister Placement. Plan shows canisters buried at least 30 meters (100 feet) beneath the ocean floor, spaced 100 meters (330 feet) apart. Because of the softness of the sediment, it may be possible to implant the canisters by simply dropping them from an appropriate distance above the sediment surface. (Figure not to scale.)

gible. Although these results are promising, we await models showing how well the burial holes will reseal themselves and the means by which radiation will be transported through the water column once it reaches the sediment surface.

Should the sea-disposal option ever be used, the present thinking is that it will not be for at least another 50 years. Much additional testing must be conducted before such a form of disposal can be put into operation.

SUMMARY

Sediments that accumulate on the ocean floor are classified by origin as lithogenous (derived from rock), biogenous (derived from organisms), hydrogenous (derived from water), or cosmogenous (derived from space). Sediment texture, determined in part by the size and sorting of particles, is affected greatly by the type of transportation that brought it to the deposit (water, wind, or ice) and the energy conditions under which it was deposited.

Lithogenous sediment is rock fragments. *Biogenous* sediment is calcium carbonate ($CaCO_3$) from the remains of foraminifers, pteropods, and coccoliths from algae, and silica (SiO_2) from diatoms and radiolarians. *Hydrogenous* sediment includes a wide variety of materials that precipitate from the water or form by interaction of substances dissolved in the water with materials on the ocean floor. Manganese nodules, phosphorite, carbonates, and zeolites are examples. *Cosmogenous* sediment is composed of nickel-iron and silicate spherules that may result from asteroid collision.

Neritic sediment accumulates rapidly along the margins of continents. It is dominated by sediment of lithogenous origin. Due to the recent rise in sea level from melting glaciers, many rivers throughout the world, including those flowing into the Atlantic along the east coast of the United States, are now depositing sediment not on the continental shelves but rather in their estuaries. About 70 percent of the sediment covering the continental shelves is relict sediment that is 3000 to 7000 years old.

Turbidity currents are thought to transport shelf sediment down submarine canyons and deposit it as turbidites on the deep-ocean floor. More than 75 percent of the sediment mass on the ocean floor is part of the thick sediment wedge underlying the continental shelves, slopes, and rises. In high latitudes, this accumulation includes poorly sorted glacial deposits.

Oceanic sediment accumulates at low rates on the floor of the deep-ocean basins, far from the continents. In the deeper basins, where most biogenous sediment is dissolved before reaching bottom, abyssal clay deposits predominate. Oozes composed of over 30 percent by weight of biogenous sediment are found in shallower basins on ridges and rises. The rate of biological productivity measured against the rates of destruction and dilution of biogenous sediment determines whether abyssal clay or oozes will form on the ocean floor. Much of the destruction of calcium carbonate occurs because the carbon dioxide content of the ocean water increases with depth. At the carbonate compensation depth, ocean water dissolves calcium carbonate at a rate equal to the rate it settles from above, thus preventing accumulation on the ocean floor below that depth.

Although it would take from 10 to 50 years for individual sediment particles in the ocean surface waters to settle to the bottom, most appear to be combined into larger aggregates as fecal pellets of small marine animals and reach the ocean floor in 10 to 15 days. This process of particle transport explains the similarity in particle composition of the surface waters and sediment on the ocean floor immediately below them.

The most valuable resource from the ocean today is petroleum, which accounts for more than 95 percent of the value of nonliving resources presently being exploited. Sand and gravel are removed from the coastal ocean for beach fill, land fill, and concrete aggregate. Sulfur was once a significant mineral recovered from the caps of salt domes but is now being replaced by sulfur recovered with pollution control equipment. Phosphorite is deposited in shallow coastal waters and has the potential to be mined for phosphate fertilizer. Manganese nodules and cobalt-rich manganese crusts have created interest but will not likely be mined in the near future.

Although present plans call for the disposal of high-level nuclear waste on land, a deep-sea disposal program has been studied. It involves burying the waste in deep-sea sediments.

KEY TERMS

Abyssal clay (p. 93)
Biogenous (p. 88)
Biological carbon pump (p. 88)
Calcareous ooze (p. 93)
Calcium carbonate (p. 88)
Carbonate compensation depth (CCD) (p. 94)
Cobalt-rich manganese crust (p. 100)
Coccoliths (p. 88)
Cosmogenous (p. 91)
Diatom (p. 88)

Fecal pellet (p. 95)
Foraminifers (p. 88)
Hydrogenous (p. 88)
Lithogenous (p. 86)
Manganese nodules (p. 88)
Neritic sediment (p. 92)
Oceanic sediment (p. 92)
Oolites (p. 88)
Ooze (p. 93)
Petroleum (p. 96)
Phosphorite (p. 99)

Precipitate (p. 88)
Radiolarians (p. 88)
Relict sediment (p. 92)
Sediment maturity (p. 85)
Silica (p. 88)
Siliceous ooze (p. 93)
Sulfur (p. 99)
Turbidites (p. 92)
Weathering (p. 87)

QUESTIONS AND EXERCISES

1. What characteristics of marine sediment indicate increasing maturity?
2. Why is a higher-velocity current required to erode clay-sized particles than larger, sand-sized particles?
3. List the four basic sources of marine sediment.
4. Refer to Figure 4–3; describe how the rocks granite, andesite, and basalt differ in texture, color, mineral composition, and density.
5. List the two major chemical compounds of which most biogenous sediment is composed and the organisms that produce them.
6. Describe the ocean environment in which calcium car-

bonate precipitation is most likely to occur. Discuss biological processes and water temperature conditions that enhance precipitation.

7. Describe the most common types of cosmogenous sediment and give the probable source of these particles.

8. Describe the basic differences between neritic sediment and oceanic sediment.

9. Explain why many rivers are not now carrying sediment to the continental shelf but depositing it in their estuaries.

10. Discuss the processes by which sediments are carried to and distributed across the continental margin.

11. How do oozes differ from abyssal clay? Discuss how productivity, destruction, and dilution combine to determine whether an ooze or abyssal clay will form on the deep-ocean floor.

12. Refer to Figure 4–10; explain why abyssal clay deposits on the floor of deep-ocean basins are commonly underlain by calcareous ooze.

13. How do fecal pellets help explain why the particles found in the ocean surface waters are closely reflected in the particle composition of the sediment directly beneath? Why would one not expect this?

14. Discuss the present importance and the future prospects for the production of petroleum, sand and gravel, sulfur, phosphorite, and polymetallic crusts and nodules.

15. What areas of concern must be addressed before high-level nuclear waste could be deposited in deep-sea sediments?

REFERENCES

Berger, W. H.; Adelseck, C. G., Jr.; and Mayer, L. A. 1976. Distribution of carbonate in surface sediments of the Pacific Ocean. *Journal of Geophysical Research* 81:15, 2617–2629.

Biscaye, P. E.; Kolla, V.; and Turedian, K. K. 1976. Distribution of calcium carbonate in surface sediments of the Atlantic Ocean. *Journal of Geophysical Research* 81:15, 2592–2602.

Broadus, J. M. 1987. Seabed minerals. *Science* 235:853–860.

Emery, K. O., and Uchupi, E. 1984. *The geology of the Atlantic Ocean.* New York: Springer-Verlag.

Hildebrand, A. R., and Boynton, W. V. 1990. Proximal Cretaceous-Tertiary boundary impact deposits in the Caribbean. *Science* 248:4957, 843–846.

Honjo, S. 1992. From the Gobi to the bottom of the North Pacific. *Oceanus* 35:4, 45–53.

Howell, D. G., and Murray, R. W. 1986. A budget for continental growth and denudation. *Science* 233:4762, 446–449.

Inter-University Program of Research on Ferromanganese Deposits of the Ocean Floor. Phase 1 Report. Unpublished.

Kolla, V., and Biscaye, P. E. 1976. Distribution of calcium carbonate in surface sediments of the Atlantic Ocean. *Journal of Geophysical Research* 81:15, 2602–2616.

Kyte, F. T.; Zhou, L.; and Wasson, J. T. 1988. New evidence on the size and possible effects of a late Pliocene oceanic asteroid impact. *Science* 241:4861, 63–65.

Leinen, M.; Cwienk, D.; Heath, G. R.; Biscaye, P. E.; Kolla, V.; Thiede, J.; and Dauphin, J. P. 1986. Distribution of biogenic silica and quartz in recent deep-sea sediments. *Geology* 14:3, 199–203.

Masters, C. D.; Root, D. H.; and Attanasi, E. D. 1991. Resource constraints in petroleum production potential. *Science* 253:146–152.

Power to the people: A survey of energy. 1994. *The Economist* 331:7868 (18-page insert following p. 60).

Sverdrup, H. U.; Johnson, M. W.; and Fleming, R. H. 1942. Renewal 1970. *The oceans: Their physics, chemistry, and general biology.* Englewood Cliffs, N.J.: Prentice Hall.

Udden, J. A. 1898. Mechanical composition of wind deposits. *Augustana Library Pub.* 1.

Weaver, P. P. E., and Thomson, J., eds. 1987. *Geology and geochemistry of abyssal plains.* Palo Alto: Blackwell Scientific.

Wentworth, C. K. 1922. A scale of grade and class terms for clastic sediments. *Journal of Geology* 30, 377–392.

SUGGESTED READING

Sea Frontiers

Dietz, R. S. 1978. IFOs (Identified Flying Objects). 24:6, 341–346. The source of Australasian tektites and microtektites is discussed.

Dudley, W. 1982. The secret of the chalk. 28:6, 344–349. An informative discussion of the formation of marine chalk deposits.

Dugolinsky, B. K. 1979. Mystery of manganese nodules. 25:6, 364–369. The problems of origin, growth, and the environmental implications of manganese nodule mining.

Feazel, C. T. 1986. Asteroid impacts, seafloor sediments, and extinction of the dinosaurs. 32:3, 169–178. A discussion of the possibility that high concentrations of iridium and osmium in a marine clay deposited at the time dinosaurs and many other species died out 65 million years ago may have resulted from the collision of Earth with a meteor 6 miles in diameter.

Feazel, C. T. 1989. Inner space: Porosity of seafloor sediments. 35:1, 49–52. A petroleum geologist discusses how porosity forms in marine sediments and the significance of the pores.

Mark, K. 1988. Ancient ocean rocks: High in the Sangre de Cristo Mountains. 34:1, 22–29. The history of an ancient ocean is revealed in the sediments of the mountains of Colorado and New Mexico.

Prager, E. J. 1988. Curious nodules on Florida's outer shelf. 34:6, 354–358. The origin of calcareous nodules found near Molasses Reef south of Key Largo is discussed.

Shinn, E. A. 1987. Sand castles from the past: Bahamian stromatolites discovered. 33:5, 334–343. The formation of stromatolites, sand domes trapped by algal growth, in the Bahamas is discussed.

Victory, J. J. 1973. Metals from the deep sea. 19:1, 28–33. The formation and economic potential of mining manganese nodules are discussed.

Wood, J. 1987. Shark Bay. 33:5, 324–333. The history of Shark Bay, Australia, since its discovery by Dirk Hartog in 1616 is summarized. The stromatolites and "tame" dolphins that attract tourists are also discussed.

Scientific American

Nelson H., and Johnson, K. R. 1987. Whales and walruses as tillers of the sea floor. 256:2, 112–118. The 200,000 walruses and 16,000 gray whales that feed by scooping sediment from the Bering Sea continental shelf suspend large amounts of sediment that is transported by bottom currents.

C H A P T E R 5
Properties of Water

Before we further investigate the ocean, we must understand the remarkable properties of the substance that makes up 96.5 percent of the ocean's mass—water. At first you might not consider water to be an unusual substance; you probably think of it as the most common substance on Earth. Water most certainly is abundant, but it has unique properties that make it a very uncommon substance. Generally, these properties include how its molecules stick together, its remarkable heat-storage capacity, and its solvent ability.

The importance of water cannot be overstated. Water controls the distribution of heat over Earth's surface, which is not surprising considering that 71 percent of the planet's surface is water covered. To a large extent, water controls our daily weather. Most living things, including humans, are roughly 70 percent water, so without it we would not exist. The very presence of water on our planet makes life possible.

The Electrically Charged Water Molecule

Two hydrogen atoms combine with one oxygen atom to form a single water molecule. They join in such a manner that the hydrogen atoms are not on opposite sides of the oxygen atom (180° apart), as you might expect, but are separated by an angle of 105° (Figure 5–1A). As the electrons move around within this structure, the arrangement of atoms produces a greater concentration of electrons around the nucleus of the oxygen atom than around the hydrogen nuclei, as you can see in the figure. As a result, a positive charge from the unshielded proton in each hydrogen atom nucleus is concentrated at the locations shown in Figure 5–1A.

This situation gives the entire molecule an electrical polarity, a bit like a very weak battery, in which the oxygen-atom end of the molecule is slightly more negatively charged and the end with the hydrogen atoms is slightly more positively charged. Thus, the water molecule has two electrical poles, or is **dipolar,** and water is known as a *polar substance*. Although these electrical charges are weak, they help determine water's remarkable properties.

How Water Dissolves Salt and Other Substances

We think of water as a good conductor of electricity, but *pure* water is not. Pure water is a very poor conductor because the molecules will not *move* toward either the negatively charged pole or the positively charged pole in an electrical system. Instead, if a battery is connected to a container of water, the water molecules simply rotate in place until they become oriented with their positively charged hydrogen nuclei toward the negative pole of the battery and their negatively charged oxygen end toward the positive pole of the battery (remember, opposites attract). This orientation of water molecules in an electrical field tends to neutralize the field.

Owing to their polar nature, water molecules form bonds between one another. These are referred to as **hydrogen bonds.** The positively charged hydrogen areas of each water molecule attract the negatively charged oxygen end of the neighboring water molecules. The molecules become bonded together by *electrostatic forces*.

When polar water molecules are bound together in groups (as in a cup of water), their ability to reduce the intensity of an electrical field acting on the water is greatly increased. Therefore, the electrostatic attraction between ions (electrically charged atoms or molecules) of opposite charges introduced into water is greatly reduced. In fact, this electrostatic attraction can be reduced to $\frac{1}{80}$ its value out of water.

Consider ordinary table salt, or sodium chloride, a compound held together by **ionic bonds.** An ionic bond is the electrostatic attraction that exists between ions that have opposite charges (Figure 5–1B). Simply by placing this solid substance in water, we reduce the electrostatic attraction (ionic bonding) between the sodium and chloride ions by 80 times. This sharply reduced attraction makes it much easier for the sodium

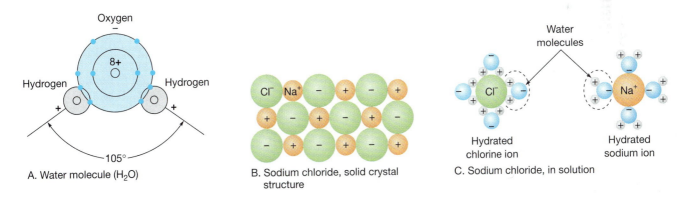

A. Water molecule (H₂O)

B. Sodium chloride, solid crystal structure

C. Sodium chloride, in solution

Figure 5–1

Water as a Solvent. *A:* A water molecule has an angle of 105° between its two hydrogen atoms, forming a dipolar (two-pole) molecule. The oxygen end of the molecule is negatively charged, and the hydrogen regions exhibit a positive charge. This single fact is the key to much of water's behavior. *B:* When ordinary salt, sodium chloride, is in its solid state, ionic bonds hold together its sodium ions (Na^+) and chloride ions (Cl^-). This forms the compound sodium chloride (NaCl). *C:* When sodium chloride is in solution, the positively charged hydrogen end of the water molecule is attracted to the negatively charged Cl^- ion. The negatively charged oxygen end is attracted to the positively charged Na^+ ion.

ions and chloride ions to dissociate, or spread apart, which is the reason why salt dissolves in water so easily. Once these ions dissociate, the positively charged sodium ions become attracted to the negative ends of the water molecules. The negatively charged chloride ions become attracted to the positive ends of the water molecules (Figure 5–1C).

As more and more ions of sodium and chloride are freed by the weakening of the electrostatic attraction that binds them together, they become surrounded by the polar molecules of water. This condition of being surrounded by water molecules is called *hydration*. The sodium ions are surrounded by water molecules whose negative ends are toward the positive sodium ion. The chloride ions are surrounded by water molecules whose positively charged ends are toward the negative chloride ion.

Water's Thermal Properties

Water is unusual because it exists on Earth in three states (solid, liquid, gas), and it has the capacity to store great amounts of heat energy.

Importance of Water's Freezing and Boiling Points

The Celsius (centigrade) temperature scale was constructed on the basis of water's characteristics at 1 atmosphere of pressure (standard sea level pressure). The scale's designers set the **freezing point** of water—the temperature at which it changes state from a liquid to a solid—at 0°. They set the **boiling point** of water, the temperature at which it changes state from a liquid to a gas, at 100°. Doing so made the liquid range of water—0° to 100°—an even 100 units, giving the scale its original name, *centigrade* (cent = 100, grade = graduations). (In contrast, our everyday Fahrenheit scale sets water's freezing and boiling points at 32° and 212°, for a spread of 180 units, which is awkward to work with.)

The freezing and boiling points of water are uniquely high for a compound of its type (H_2O—two hydrogen atoms plus one atom of another element). Similar compounds occur at Earth's surface temperatures as gases, with freezing and boiling points well below 0°. Water is unique because it occurs in all three states of matter—gaseous, liquid, and solid—all within the narrow range of atmospheric conditions.

Heat and Changes of State

To change the state of a compound such as water, what must happen? Among all molecules of any compound, there exists a relatively weak attraction called the **van der Waals force,** named for a Dutch scientist. It be-

comes significant only when molecules are very close together, as in the solid and liquid states (but not the gaseous state). Generally, the heavier the molecule, the greater the van der Waals attraction between individual molecules of the compound. Therefore, with increasing molecular weight, a greater amount of energy is needed to overcome the van der Waals force and allow a change of state to occur, from a solid to a liquid or a liquid to a gas. Consequently, the melting and boiling points of compounds generally increase as the molecular weight increases.

What form of energy changes the state of matter? Very simply, *heat,* as you know from experience with ice cubes and boiling water. But before proceeding, we must define heat, temperature, and calorie:

- **Heat** *is the energy of moving molecules.* It is proportional to the energy level of molecules, and thus is the total kinetic energy of a sample.
- **Kinetic energy** *is energy of motion.* Heat may be generated by combustion (a chemical reaction we commonly call "burning"), some other chemical reaction, friction, or radioactivity.
- *A* **calorie** *is the amount of heat required to raise the temperature of one gram of water by 1 Celsius degree.* To measure the amount of heat energy that is being added to or removed from molecules, we use the calorie.
- **Temperature** *is the direct measure of the average kinetic energy of the molecules that make up a substance.* Kinetic energy is energy of motion, so the higher the temperature, the faster the molecules of the substance are moving.

Let us return to the van der Waals force, which causes molecules of a substance such as water to be attracted to one another. It is this attraction that must be broken if the state of the substance is to be changed from solid to liquid or liquid to gas. To break this attraction, heat energy must be given to the molecules so that they can move faster to overcome this force. As heat energy is added to molecules, their motion increases. Molecules can have three types of motion, shown in Figure 5–2.

In the **solid state,** a substance has a rigid form and structure and does not flow. Bonds are constantly being broken and reformed. But the prevailing relation between molecules is one of rather firm attachment, produced by the nearness of the molecules, which causes the van der Waals force to be greater. In the solid state, the dominant type of **molecular motion** is *vibration,* as the molecules vibrate with energy but remain in relatively fixed positions (Figure 5–2, left).

In the **liquid state,** molecules are unstructured and can flow to take the shape of their container. The molecules have gained enough energy to overcome many of the van der Waals forces that bound them to-

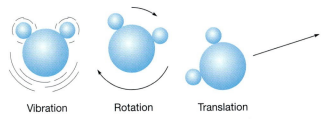

Figure 5–2

Motion of Molecules. *Vibration* is the typical movement of molecules in a crystalline solid (such as ice). *Rotation* occurs only in liquids and gases (such as liquid water and water vapor). All gas molecules move in random directions by *translation*.

gether in the solid state. The molecules have enough freedom to move relative to one another. In this state we see all forms of molecular movement—*vibration, rotation, and translation* (Figure 5–2). The molecules are free to move relative to one another, but are still attracted by one another. Bonds are being formed and broken at a much greater rate than in the solid state.

In the **gaseous state,** molecules are unstructured and flow very freely, filling whatever container they are placed in. *Translation* has become the dominant type of motion (Figure 5–2, right). Molecules now are moving at random, and there exists no significant attraction

among individual molecules. The only effect that one molecule may have on another results from collision during random movement.

The general relationship of water molecules to one another in the solid, liquid, and gaseous states is shown in Figure 5–3.

The fact that water freezes at 0°C (32°F) and boils at 100°C (212°F) demonstrates the great significance of the polarity of the water molecule and the hydrogen bond that this structure produces. The high freezing and boiling points of water are manifestations of the additional heat energy required to overcome two bonds to achieve a change of state: both the van der Waals bonds and the hydrogen bonds.

Heat Capacity of Water

Another direct result of the hydrogen bond between water molecules is the high *heat capacity* of water. Because of the great strength of the hydrogen bonds between water molecules, more heat energy must be added to accelerate the molecules and raise the water temperature than is required for substances in which the dominant intermolecular bond is the weaker van der Waals force. As noted, 1 calorie is the amount of heat required to raise the temperature of 1 gram of water 1°C (in English units, 0.035 ounce 1.8°F).

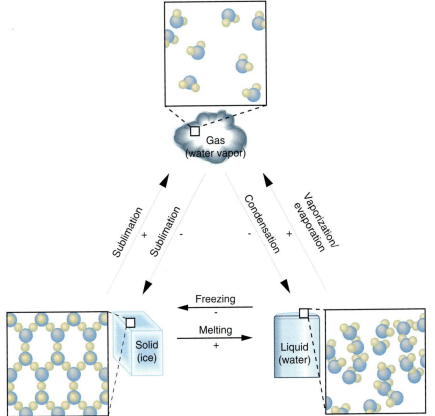

Figure 5–3

The Three Physical States of Water. Water molecules arrange themselves in response to heat, or absence of heat, in their environment. Note the molecular arrangement in each state—solid, liquid, or gas—and how polarity controls the arrangement. (Used with permission from R. W. Christopherson, *Geosystems,* 2d ed., Macmillan, 1994, Figure 7–7, p. 186.)

The heat capacity of water, compared with that of most other substances, is great. We use this exceptional capacity of water to absorb heat in cooking, home heating, car cooling systems, and industrial cooling. Water is the standard against which the heat capacities of other substances are compared. **Heat capacity** *is the amount of heat that is required to raise the temperature of 1 gram of any substance 1°C.* Heat capacity is 1 calorie for water and less than 1 calorie for most other substances. In other words, water gains or loses much more heat than other common substances while undergoing an equal temperature change.

The difference in the heat capacities of ocean water and the rocks that make up the continents is strikingly illustrated in Figure 5–4. It shows the difference in day and night temperatures averaged worldwide for the month of January 1979. Throughout most of the ocean area, day and night temperatures vary less than 1°C because the high heat capacity of water easily accommodates the daily heat gains and losses. By contrast, the much lower heat capacity of continental rocks, soil, and vegetation can cause day and night temperatures to fluctuate between 15°C and 30°C. It is the heat capacity of water that is responsible for the moderate climates of the coastal regions of the continents.

Water's Latent Heats of Melting and Vaporization

Closely related to water's unusually high heat capacity are its high *latent heat of melting* and *latent heat of vaporization.* The word *latent* means hidden; it refers to

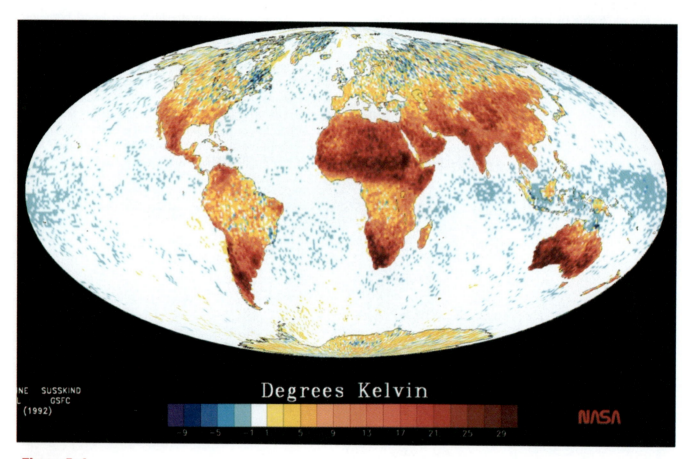

Figure 5–4

Day and Night Temperature Differences. This map shows the *difference* between day and night temperatures, averaged for the month of January 1979. For example, in white areas of the open ocean, there is only a 1°K difference in day and night temperatures (note degrees Kelvin scale). In land areas such as South America, Africa, and Australia, there is a difference of up to 30°K from day to night. This dramatically demonstrates the difference in heat capacity between water and land. The high heat capacity of water limits the day/night difference to about 1°K for the oceans. Continental regions, with their much lower heat capacity, have much greater differences. The map was made by subtracting 2 A.M. temperatures from 2 P.M. temperatures. (Courtesy of Moustafa T. Chahine, Jet Propulsion Laboratory, California Institute of Technology.)

the fact that heat is stored invisibly and dormantly in water. The only time the effect of this latent heat can be observed is when a change of state occurs—ice to water or vice versa, and water to water vapor or vice versa. During any of these changes of state, a startling amount of heat is either absorbed (notice how cool your skin feels when water evaporates from it) or released (if you ever have been scalded by water vapor—steam—you know how much latent heat it releases).

If enough heat energy is added continuously to a solid, it converts to a liquid. This happens at a temperature called the substance's *melting point.* Also, it will change from a liquid back to a solid at its *freezing point,* which is the same temperature. For water, this melting/freezing point is 0°C (32°F).

If enough heat energy is added continuously to a liquid, it converts to a gas. This happens at a temperature called its *boiling point.* Also, it will change from a gas back to a liquid at its *condensation point,* which is the same temperature. For water, this boiling/condensing point is 100°C (212°F).

As a substance changes state, there may be no temperature change at the point where a change of state occurs, even though heat is continuously being added. The reason is that the heat energy is being used entirely to break all the intermolecular bonds that must be broken to complete the change of state. The heat energy is doing this instead of increasing the molecular movement, which is indicated by temperature. While these bonds are being broken, a mixture of the substance in both states is in equilibrium. Only upon completion of the change of state will the temperature again rise.

The **latent heat of melting** is the amount of heat that must be added to 1 gram of a substance, at its melting point, to break the required bonds to complete the change of state from solid to liquid.

The **latent heat of vaporization** is the amount of heat that must be added to 1 gram of a substance, at its boiling point, to break the required bonds to complete the change of state from liquid to vapor (gas).

Again, the word *latent* describes the heat energy that is stored or "hidden" within a mass of water or water vapor. We have looked at what happens when heat energy is added and temperature increases. Now let us examine what happens when heat energy is removed, and temperature falls. When water vapor is cooled sufficiently, it returns to a liquid state. We say that it **condenses** and heat is released into the surrounding air. On a small scale, the heat released is enough to cook food, as in a "steamer." On a large scale, this amount of heat is tremendous, and powers thunderstorms and hurricanes.

Release of heat also occurs with the freezing of water, which is the change of state from liquid to solid. In Earth's polar regions, heat released by formation of sea ice is absorbed by the heat-deficient atmosphere.

Note that the *amount* of heat involved in the latent heat of vaporization and condensation is identical: vaporization calories = condensation calories. Also, the *amount* of heat involved in the latent heat of melting and freezing is identical: melting calories = freezing calories.

A practical application of these principles of heat transfer can be seen in the use of ice for refrigeration. Put a bag of ice in a picnic cooler and it will lower the temperature of food and drinks in the cooler. This happens because heat energy travels from the food and drinks to the ice, where it is absorbed to change the ice from its solid state to a liquid state.

The principle of cooling air with water is similar. In arid climates, the hot dry air that passes through a surface coated with liquid water quickly loses heat energy to the water, and the water is converted to a vapor. Thus, after passing across a water-covered surface, the hot air becomes considerably cooler.

An Example: Watching H₂O Change State

To help clarify this phenomenon, let us examine the change of state of H_2O from a solid at –40°C (–40°F), through liquid, through the vapor state, to 100°C. In Figure 5–5, the vertical scale is temperature, ranging from –40° to 120°C. The horizontal scale begins at 0 calories of heat energy and continues through 800 calories.

Let us begin with 1 gram of ice. The addition of 20 calories of heat raises its temperature by 40°, from –40°C to 0°C (you can see this at the lower-left corner of the figure). Once the temperature has been raised to 0°C (32°F), note the plateau on the graph, during which the continuous addition of heat does not increase the temperature further, until 80 more calories of heat energy have been added. We see no temperature increase during the addition of these 80 calories because all the heat energy goes to break the intermolecular bonds that structure the water molecules into ice crystals. The temperature must remain unchanged until most of the bonds are broken and the mixture of ice and water has changed completely to 1 gram of water.

This amount of heat required to convert 1 gram of ice to 1 gram of water, 80 calories, is termed the *latent heat of melting.* It is greater for water than for any other common substance. To give you a sense of how much heat 80 calories is, it is $\frac{1}{1000}$ the food energy in one egg.

The bonds that are broken in converting most substances from a solid to a liquid are van der Waals bonds. In water, however, some of the hydrogen bonds also must be overcome. It is necessary only to break the ice structure into numerous small clusters, surrounded by individual water molecules, so that these remaining ice clusters can move relative to one another and allow the mass to assume a liquid state. Liquid water, particularly at low temperature near the freezing point, could well be described as a "pseudocrystalline" liquid, for it

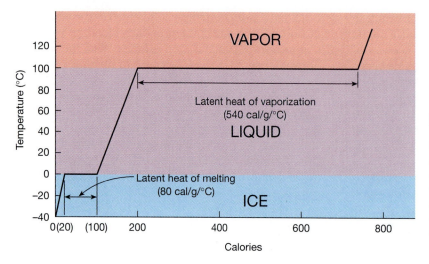

Figure 5–5

Latent Heats and Changes of State of Water. The latent heat of melting (80 calories) is much less than the latent heat of vaporization (540 calories). The reason is that only a few hydrogen bonds must be broken to convert 1 gram of ice to a liquid, but *all* remaining hydrogen bonds must be broken to convert 1 gram of liquid water to a gas.

still contains many small clusters of ice crystals. But once enough hydrogen bonds have been broken to allow freedom of movement among the clusters and individual water molecules, the temperature of the water will again rise with the addition of heat.

After the change from ice to liquid water has occurred at 0°C, the addition of heat to the water causes the temperature to rise again. As it does, note that it requires 1 calorie of heat to raise the temperature of water 1°C (or 1.8°F). We must therefore add another 100 calories before 1 gram of water reaches the boiling point of 100°C (212°F). (So far, we have added a total of 200 calories to make this happen.)

At this temperature, 100°C, we see another plateau (Figure 5–5). It is far more prominent on the graph than the plateau for the latent heat of melting, which was 80 calories/gram. This plateau, at 100°C, represents the addition of 540 calories, which is the latent heat of vaporization, to the gram of water. This 540 calories must be added before complete conversion to the vapor state oc-

curs. To give you a sense of how much heat 540 calories is, it is 1/1000 the food energy in a large chocolate malted.

Why is so much more heat energy required to convert 1 gram of water to water vapor (540 calories) than was required convert 1 gram of ice to water (80 calories)? Refer to Figure 5–6, and recall that in a gas the molecules move at random, free from the influence of other molecules. To make the conversion from ice to water, *not all the hydrogen bonds had to be broken,* just enough to allow movement among the ice clusters and individual molecules. But to convert water to water vapor, *every molecule must be freed* from the attraction of other water molecules. Therefore, every hydrogen bond must be broken (Figure 5–6). This requires a much greater amount of heat energy: 540 calories per gram of water instead of 80 calories.

Now you can see that large amounts of latent heat energy are stored in water, and that this heat energy is absorbed or released by water as it changes state—particularly if the change is from liquid to vapor or vice versa.

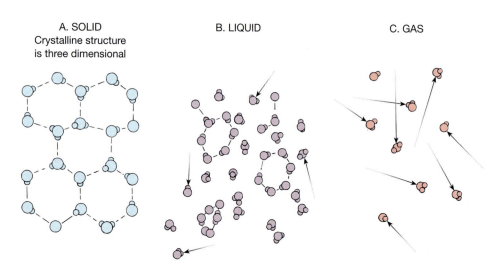

Figure 5–6

Hydrogen Bonds in H₂O. Below 0°C (A), there is too little heat energy to break the hydrogen bonds that bind water molecules to one another. Between 0° and 100°C (B), heat energy makes molecules active enough to break most bonds between them. Above 100°C (C), heat energy makes molecules so active that hydrogen bonds cannot hold any molecules together.

Of course, we do not have boiling temperatures of 100°C at the surface of the ocean, where conversion of water to vapor occurs in nature. Sea-surface temperatures average about 20°C (68°F) or less. The conversion of a liquid to a gas below the boiling point is called **evaporation.** At ocean-surface temperatures, individual molecules that are being converted from the liquid to the gaseous state have a lower amount of energy than do the molecules of water at 100°C. Therefore, to gain the additional energy necessary to break free of the surrounding ocean water molecules, an individual molecule must capture heat energy from its neighbors.

This phenomenon explains the cooling effect of evaporation. The molecules left behind have lost heat energy to those that escape. To produce 1 gram of water vapor from the ocean surface at temperatures less than 100°C requires more than the 540 calories of heat that are required to make this conversion at the boiling point. For instance, the **latent heat of evaporation** is 585 calories/gram at 20°C. This higher value is due to the fact that more hydrogen bonds must be broken at this lower temperature for a gram of water molecules to enter the gaseous state.

Now that you understand the evaporation-condensation cycle, and the great heat energy involved, here is why it is so important to us. The sun radiates energy to Earth, where some of it becomes stored in the oceans. Evaporation removes this heat energy from the oceans and carries it high into the atmosphere as the water vapor rises. In the cooler environment of the upper atmosphere, the water vapor condenses into clouds, releasing the latent heat energy. *It is water's latent heat of evaporation that accounts for the removal of great quantities of heat energy from the low latitude oceans by evaporation.* It is later released in the heat-deficient higher latitudes after the vapor is transported through the atmosphere and condenses there as rain and snow (Figure 5–7).

Water's Surface Tension

You can fill a container with water, not only to the brim, but even higher (Figure 5–8A). Water can be piled a short distance above the container's rim, forming a convex (pushed-up) surface. This demonstrates water's

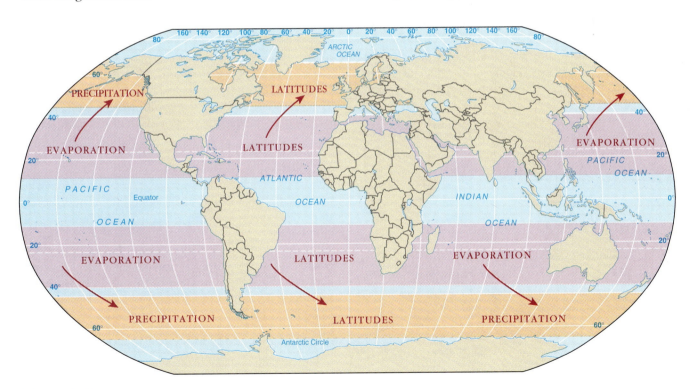

Figure 5–7

Atmospheric Transport of Surplus Heat from Low Latitudes into Heat-Deficient High Latitudes. A vast quantity of heat is removed from the ocean at lower latitudes by evaporation ("evaporation latitudes"). As this occurs, we see how water's high *latent heat of evaporation* plays a major role in regulating Earth's climate. The heat removed from the tropical ocean is carried poleward and is released at higher latitudes through precipitation ("precipitation latitudes"). For every gram of water that condenses in cooler latitudes, the same amount of heat is released to warm these regions as was removed from the tropical ocean when that gram of water was evaporated to become water vapor.

surface tension, the phenomenon in which its surface molecules cling together, behaving like a weak membrane. Without surface tension, water poured into a container would overflow as soon as the container's top was reached. Water droplets also clearly reveal water's surface tension. Again, without surface tension, water would spread out over a surface in a very thin film instead of beading into droplets.

In freshwater streams and lakes, as well as in the oceans, insects and spiders can be seen skittering across this molecule-thin "skin" that exists at the surface of all water bodies. Some creatures even are adapted to hang, bat-like, from the underside of the layer, down into the water.

These phenomena result from the tendency of water molecules to attract one another, or to cohere at the surface of any accumulation of water. Because of this cohesive tendency, it is possible to float on water objects that are much denser. A razor blade carefully laid on the surface will float, as will a metal paper clip or steel needle, although they are typically five times as dense as water. Many insects use the surface of water as if it were a solid surface, scurrying and skating across it at will; this gives the common "water strider" its name.

Surface tension is a manifestation of water's hydrogen bonds. Molecules on the surface of water are strongly attracted to the water molecules below them by their hydrogen bonds (Figure 5–8B). The air above the surface has a very low density of molecules compared with the water, so the attraction between surface water molecules and atmospheric molecules is slight.

Water also clings to the surface of many substances; we refer to this everyday phenomenon as *wetting*. Water will adhere strongly to glass, all organic substances, and inorganic material such as rocks and soil. When water is poured into a container made of a substance to which it strongly adheres, the adhesive force, or the force of attraction between water molecules and molecules of the other substance, will cause the surface tension layer to take on a form unlike the convex one described previously. In this case the surface tension layer is concave (pushed-in), because the water is drawn up on the container's sides (Figure 5–8C).

If the container is glass, the positively charged portions of the water molecules (the hydrogen ends) are attracted to unbonded electrons (negative charge) in the oxygen atoms that are part of the makeup of the glass container. Being strongly attracted by these oxygen atoms, the water molecules will literally "climb" up the side of the container. They are held back only by the hydrogen bond attraction that exists between individual water molecules.

In fact, if the diameter of the container is decreased to a very fine bore, such as a very thin, hollow glass tube, the water will climb to great heights. The reason is a combination of cohesion, which holds the water molecules together, and the adhesive attraction between the water molecules and the glass container. This phenomenon is called **capillarity,** in which water is drawn up in small tubes as a result of surface tension.

Salinity of Ocean Water

It is obvious that ocean water is salty; you can taste it. But what exactly does the salt consist of? Where does it come from?

Most of the physical properties that we observe in ocean water are simply properties of water, because water averages 96.53 percent of the ocean's mass. The remaining 3.47 percent is dissolved solids that we call *salts*. From place to place, the proportion of water to dissolved salts varies within the ocean, although only by tenths and hundredths of a percent (Figure 5–9).

The ratio of major elements that make up the salts is remarkably constant worldwide. This **rule of constancy of composition** was firmly established by English chemist William Dittmar in his analysis of the *Challenger* expedition water samples discussed in Chapter 1.

Chemically, a salt is a compound formed by the replacement of the hydrogen ion (H^+) of an acid with some positively charged chemical unit, usually a metal, such as sodium. For example, most of the "salt" in the ocean results from replacing the hydrogen ion in hydrochloric acid (H^+Cl^-) with sodium ions (Na^+, to form

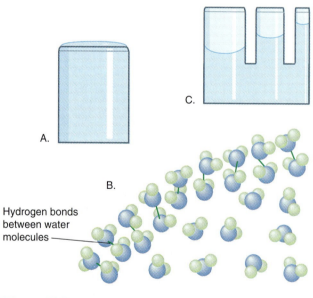

Figure 5–8

Hydrogen bonds between water molecules

Surface Tension. A surface tension "skin" or "membrane" forms as hydrogen bonds create a strong attraction between the outermost layer of water molecules and the underlying molecules. Water has the greatest surface tension of all common liquids except the element mercury, which is the only metal that is a liquid at normal surface temperatures.

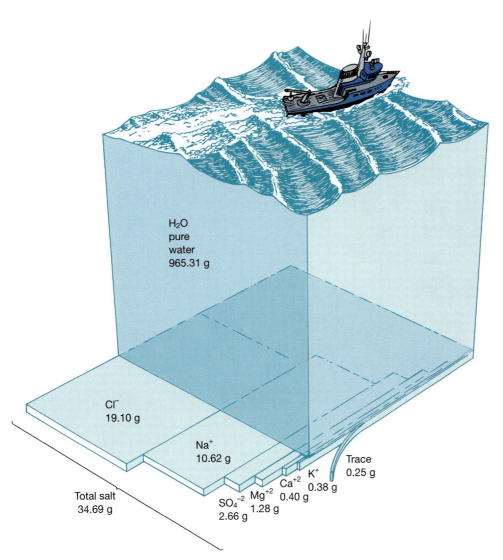

Figure 5–9

Constituents of Ocean Salinity. The composition of 1 kilogram of ocean water in grams (g). Note that the 965.31 grams of pure water + the 34.69 grams of salts = 1000.00 grams. In other words, average seawater contains about 965 parts per thousand (ppt) water and 35 ppt salt.

NaCl, or sodium chloride), or replacing the hydrogen ion with potassium ions (K^+, to form KCl, or potassium chloride). Most of the remaining "salts" are formed by replacing the hydrogen ions in sulfuric acid ($H_2^{++}SO_4^-$) with magnesium ions (Mg^{++}, to form $MgSO_4$) or calcium ions (Ca^{++}, to form $CaSO_4$).

It seems probable that every known element will be found dissolved in ocean water, when we eventually develop techniques for making such minute measurements. However, only six elements account for over 99 percent of the dissolved solids in ocean water. These are chlorine, sodium, sulfur (in the form of the sulfate ion, SO_4), magnesium, calcium, and potassium (Table 5–1 and Figure 5–9).

Salinity is the amount of dissolved solids in seawater. Because the oceans are very well mixed and the relative abundance of the major constituents is essentially constant, we have a condition that makes the chemical measurement of salinity relatively simple.

Because of seawater's constancy of composition, we need only measure the concentration of one major constituent to determine the salinity of a given water sample. The constituent that occurs in the greatest abundance and is the easiest to measure accurately is the chloride ion, Cl. The weight of this ion in a water sample is called *chlorinity,* and it usually is expressed as *grams of chloride per kilogram of ocean water* (grams/kilogram). It also is expressed as *parts per thousand.* (The familiar symbol for parts per *hundred* is %; the corresponding symbol for parts per *thousand* is ‰.)

In any sample of ocean water worldwide, the chloride ion always accounts for a 55.04 percent proportion of the dissolved solids. Therefore, by measuring the chloride ion concentration, we can determine the total salinity of a seawater sample by the following relation:

$$\text{Salinity (‰)} = 1.80655 \times \text{chlorinity (‰)}$$

For example, given the average chlorinity of the ocean, which is 19.2 ‰, the ocean's average salinity is 1.80655 × 19.2 ‰, which rounds to 34.7 ‰. In other words, on average there are 34.7 parts salt in every 1000 parts of seawater.

Table 5–1

Ocean Salinity. Considering the average salinity of the world ocean, a 1000-g sample would contain 24.7 g of dissolved solids (34.7 parts per thousand by weight). Only six elements account for 99.28 percent of these dissolved solids.

Major Constituents (over 100 parts per million)

Ion	Percentage
Chloride, Cl^-	55.04
Sodium, Na^+	30.61
Sulfate, SO_4^{2-}	7.68
Magnesium, Mg^{2+}	3.69
Calcium, Ca^{2+}	1.16
Potassium, K^+	1.10
	99.28

Minor Constituents (1–100 parts per million)

	ppm
Bromine, Br	65.0
Carbon, C	28.0
Strontium, Sr	8.0
Boron, B	4.6
Silicon, Si	3.0
Fluorine, F	1.0

Trace Elements (less than 1 part per million)

Nitrogen, N	Iodine, I
Lithium, Li	Iron, Fe
Rubidium, Rb	Zinc, Zn
Phosphorus, P	Molybdenum, Mo

(Technical note: the number 1.80655 comes from dividing 1 by the .5044—the chloride ion's proportion of 55.04 percent. However, if you actually divide this, you will get 1.81686, not 1.80655. But oceanographers have agreed to use 1.80655 because the constancy of composition has been found to be an approximation.)

Standard seawater consists of ocean water analyzed for chloride ion content to the nearest ten-thousandth of a part per thousand by the Institute of Oceanographic Services in Wormly, England. It is then sealed in ampules and sent to laboratories throughout the world for use as a reference standard in analysis.

The chloride content in seawater can be measured very accurately, but this requires much time and care. Fortunately, with advanced oceanographic instruments, this task has been greatly simplified. Oceanographers know that the electrical conductivity of ocean water increases with salinity, so salinity now is most commonly determined by measuring the conductivity of ocean water with a *salinometer*. Salinity can be determined to better than 0.003 parts per thousand with this method.

Ocean Salinity Over Time

Ions of salinity are continually added to the ocean from two basic sources: rivers that dissolve the ions from continental rocks and carry them to the sea (Table 5–2) and volcanic eruptions, both on the land and on the seafloor. Figure 5–10 illustrates these sources.

Salts do not remain in the ocean forever but are removed by several processes (Figure 5–10). When waves break at sea, a salt spray releases many tiny salt particles into the atmosphere, where they may be blown over land before being washed back to Earth by precipitation. At oceanic ridges, infiltration of ocean water causes magnesium and sulfate ions to become incorporated into seafloor mineral deposits. Calcium, sulfate, and sodium are deposited in ocean sediments in the shells of dead microscopic organisms and animal feces. The most important means of removal is the adsorption (physical attachment) of essentially all ions dissolved in ocean water to the surfaces of sinking clay and biological particles.

Residence Time

The **residence time** is the average length of time that an atom of an element resides in the ocean. It can be calculated as follows:

$$\text{Residence time} = \frac{\text{Amount of element in the oceans}}{\substack{\text{Rate at which element is added to} \\ \text{or removed from the oceans}}}$$

Oceanographers consider the oceans to be well mixed and in a steady state because of the constancy of com-

Table 5–2

Major Constituents of Dissolved Solids in Streams. In figure 5–9, you can see the concentrations of ions that comprise the ocean's salinity. This table shows the ion content of freshwater streams, which is far lower, measured in parts per million rather than the parts per thousand in figure 5–9. This low concentration is why "fresh" water does not taste salty. For example, although calcium (Ca^{+2}) is a major constituent in streams, its concentration there is slight compared with seawater, where it is more than 26 times greater.

Constituent	Parts per million
Carbonate ion, HCO_3^-	58.4
Calcium ion, Ca^{2+}	15.0
Silicate, SiO_2	13.1
Sulfate ion, SO_4^{2-}	11.2
Chloride ion, Cl^-	7.8
Sodium ion, Na^+	6.3
Magnesium ion, Mg^{++}	4.1
Potassium ion, K^+	2.3
Total	119.2

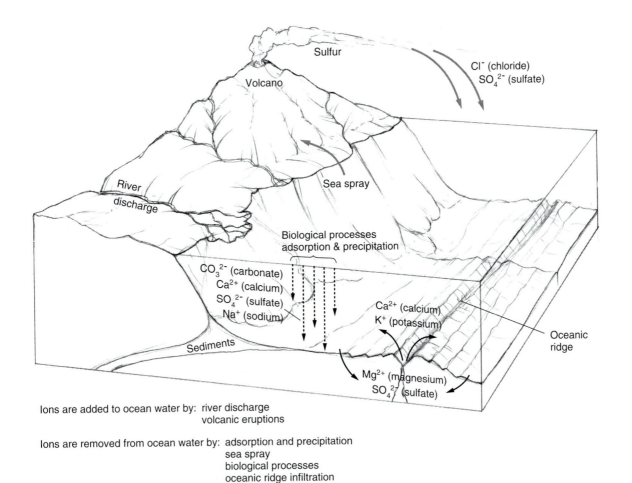

Ions are added to ocean water by: river discharge
volcanic eruptions

Ions are removed from ocean water by: adsorption and precipitation
sea spray
biological processes
oceanic ridge infiltration

Figure 5–10

How the Constituents in Seawater Are Controlled. Salt ions are added to the oceans by stream runoff and volcanic eruptions. They are removed at an equal rate by a number of processes. Salt spray releases ions into the atmosphere. Ions are incorporated into ocean sediments by adsorption and incorporation into shells of tiny organisms. Ions are also removed from ocean water and incorporated into mineral deposits associated with submarine volcanoes.

position. This implies that the rate at which an element is removed from the oceans must equal the rate at which it is being added. For this reason, we can use either the amount added or the amount removed as the denominator in the equation.

The residence time of an element in the ocean depends on how reactive it is with the marine environment. More reactive elements have a shorter residence time. For example, the reactive elements aluminum and iron have residence times of 100 and 140 years, respectively, whereas less reactive sodium has a residence time of 260 million years (see Table 5–3).

Elements occur in seawater in different forms, and the different forms have different residence times. Aluminum, for example, occurs in particulate form and dissolved form. The short residence time for aluminum shown in Table 5–3 (100 years) is the *average* for particulate and dissolved aluminum. Recent investigation has shown that dissolved aluminum has a longer resi-

dence time of 1400 years, which means that particulate aluminum must be removed from ocean water at a very rapid rate to achieve the 100-year average.

Water itself has a residence time in the ocean. Each year, the upper 1 meter (3.3 feet) of the ocean's

Table 5–3
Residence Time of Some Elements in the Oceans

Element	Amount in Ocean (g)		Residence Time (in years)
Sodium, Na	14,700,000	$\times 10^{15}$	260,000,000
Potassium, K	530,000	$\times 10^{15}$	11,000,000
Calcium, Ca	560,000	$\times 10^{15}$	8,000,000
Silicon, Si	5,200	$\times 10^{15}$	10,000
Iron, Fe	14	$\times 10^{15}$	140
Aluminum, Al	14	$\times 10^{15}$	100

Source: Data from Goldberg and Arrhenius, 1958.

surface evaporates. This water is returned directly by precipitation over the ocean, or indirectly by precipitation over the continents, which runs off into rivers and returns to the ocean. The ocean has an average depth of 4000 meters (over 13,000 feet). Thus, if the ocean had a completely uniform depth of 4000 meters, it would completely evaporate in 4000 years. Because the water cycles through the atmosphere and is returned, 4000 years would be the time required to completely recirculate all ocean water. Thus, the residence time of water in the oceans would be 4000 years. In actuality, because ocean depth varies widely, residence time varies widely from place to place.

The process by which water is recycled among the ocean, atmosphere, and continents is called the water cycle, or the **hydrologic cycle** (Figure 5–11). Of the total yearly evaporation of water from Earth's surface, 83 percent evaporates from the oceans and 17 percent evaporates from the surface of the continents. It returns to Earth by precipitation from the atmosphere: 76 percent falls directly back into the oceans and 24 percent falls onto the continents.

During a year, the atmosphere transports to the continents about 8.5 percent of the water vapor that evaporates from the oceans. This water is returned to the oceans by stream runoff and as groundwater. At any given time, 97 percent of Earth's water is contained in the ocean basins, 2 percent is in glaciers, 0.6 percent is groundwater, 0.001 percent is in the atmosphere, and about 0.02 percent is in rivers and lakes.

Because evaporation from the ocean exceeds precipitation directly back into it, 35 percent of the precipitation on continents is evaporated ocean water. On the continents, precipitation exceeds evaporation, so the excess returns to the oceans as stream runoff. The residence time of water in the atmosphere is about 10 days.

Water in the atmosphere has another important consequence: The greenhouse effect of atmospheric

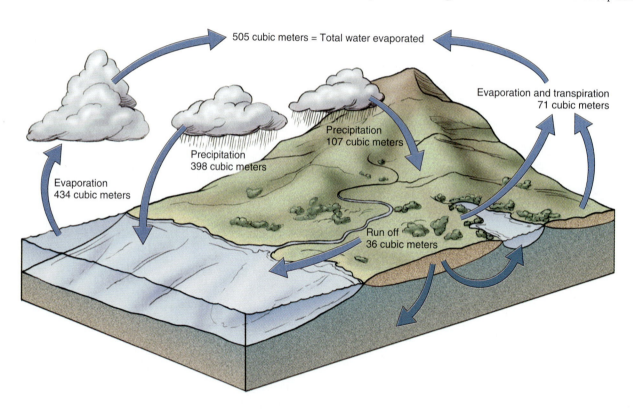

505 cubic meters = Total water evaporated

Evaporation and transpiration
71 cubic meters

Precipitation
107 cubic meters

Precipitation
398 cubic meters

Evaporation
434 cubic meters

Run off
36 cubic meters

Figure 5–11

Hydrological Cycle. Earth's water supply exists in these proportions: 97 percent in the world ocean, 2 percent frozen into glaciers, 0.6 percent groundwater, 0.001 percent water vapor in the atmosphere, and about 0.02 percent in streams, rivers, and lakes. Although some evidence indicates that glaciers are entering a stage of melting, this process is very slow, so we will consider these ice masses to be in a steady state. All of this water is in motion. Water is removed from the liquid reservoirs (ocean, streams, lakes) by evaporation, and it returns to them by precipitation. The atmosphere transports water vapor from the oceans to the continents, where it is deposited as precipitation and returns to the oceans by stream flow or migrating through the soil and rocks as ground water. (Reprinted with the permission of Macmillan Publishing Company from *The Earth*, 4th ed. (fig. 10.2), by Edward J. Tarbuck and Frederick K. Lutgens. Copyright © 1993 by Macmillan Publishing Company.)

water is nearly double the effect of all other greenhouse gases combined. You will see the significance of this in Chapter 6.

Water Density

Density simply refers to how tightly the molecules of a substance are packed together. The density of water varies in its different states and at different temperatures. Density is of great importance in considering the movement of water in the ocean because water of greater density sinks, whereas water of less density rises. Further, anything that is more dense than water will sink into it, and a substance that is less dense will float on the surface. We define **density** as *mass per unit of volume, usually in grams per cubic centimeter (g/cm³)*.

Density is affected by temperature, salinity, and pressure. With most substances, we observe that "colder is denser." This density increase happens because the same number of molecules occupy less space as they cool down and lose energy. This condition, thermal contraction, also occurs in water. As water cools, it gets denser—as long as the temperature decrease occurs above 4°C (39.2°F). But as the temperature of water is lowered from 4° to 0°C (32°F), we observe that its density *decreases*. In other words, the water has stopped contracting and is actually expanding. This is unusual among Earth's many substances. The result is that ice actually takes up *more* volume than liquid water, just the opposite of most other substances. Why?

This anomalous behavior of freezing water can be explained only by considering water's molecular structure and the hydrogen bond. Please refer to Figure 5–12

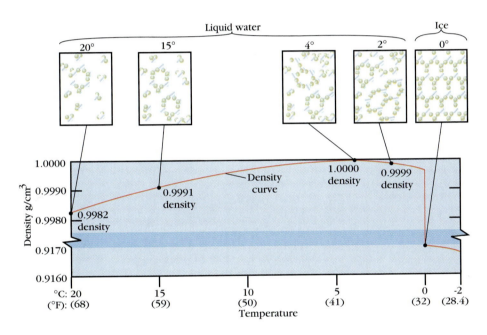

(a)

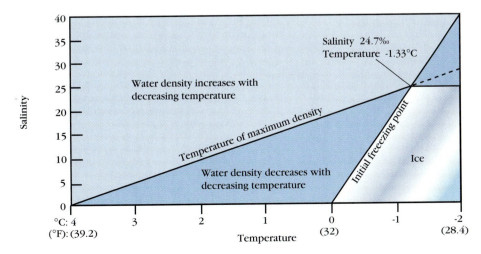

(b)

Figure 5–12

The Freezing of Water. *A:* Formation of ice clusters in freshwater. The density of freshwater ranges from 0.9982 g/cm³ at 20°C to a maximum of 1.000 g/cm³ at 4°C. But the density of ice (solid water) is only 0.9170 g/cm³, so it floats on water. *B:* Salinity, freezing point, and temperature of maximum density.

during this explanation. Starting at the left in the diagram, as we lower the temperature of water from 20°C (68°F) and reduce the amount of thermal agitation of the water molecules, the unbonded water molecules occupy less volume because of their decreased energy. The water contracts. But as we approach the freezing point below 4°C, this reduction in volume is not sufficient to compensate for another phenomenon: Ice crystals are becoming more abundant. Ice crystals are bulky, open, six-sided structures in which water molecules are widely spaced. The greater rate of increase of ice crystals as the temperature approaches the freezing point accounts for the decreased density of water below 4°C. By the time water fully freezes, the density of the ice is only about 0.9 that of water at 4°C.

As you can see on the curve, water attains its maximum density at 4°C (39.2°F). This is very important, for water at this temperature is denser than either warmer or colder water and thus will sink, creating vertical currents. Conversely, water at higher or lower temperature, whether liquid or frozen, will float, also creating vertical currents.

The **temperature of maximum density** for freshwater, 4°C (39.2°F), can be lowered by two means. Increasing the pressure on water will lower the temperature of maximum density. Adding solid particles, such as salt, also lowers the temperature of maximum density. The reason is that both inhibit the formation of bulky ice clusters. Thus, to produce crystals equal in volume to those that could be produced at 4°C in freshwater, more energy must be removed, causing a reduction in the temperature of maximum density.

This interference with the formation of ice crystals also produces a progressively lower freezing point for water as more solids are added. This is why most seawater never freezes, except near Earth's frigid poles, and

Table 5–4
Summary of the Properties and Significance of Water

Properties	Physical and Biological Significance
Physical states. Water is the only substance that occurs as a gas, liquid, and solid within the range of surface temperatures on Earth.	*Gaseous water* (water vapor) is an important component of the atmosphere, because water vapor transfers great quantities of heat from warm, low latitudes to cold, high latitudes. Liquid water also contributes to this process through ocean currents. *Liquid water* runs across land, dissolving minerals from the rocks and carrying them to the oceans. Liquid water makes up over 85 percent of the mass of most marine organisms. Liquid water is the medium in which occur the chemical reactions that support life. *Ice* formation at temperatures of about –1.8°C (28.7°F) during winter at higher latitudes increases surface water salinity, increasing the water's density and thus making it sink. This sinking water, with its dissolved oxygen, is the only source of oxygen for the deep ocean.
Solvent property. Water can dissolve more substances than any other common liquid, hence its nickname, "the universal solvent."	*Ocean water* carries dissolved within it the nutrients required by marine algae and the oxygen needed by animals. It is this property that has produced the "saltiness" of the oceans.
Heat capacity. The quantity of heat required to change the temperature of 1 g of a substance by 1°C. Water has the highest heat capacity of all common liquids. The heat capacity of water is used as the unit of heat quantity, *calorie.*	*Heat capacity* is a major factor in making water the most important moderator of climate. It explains the narrow range of temperature change occurring at any ocean location. Water gains or loses much more heat than other common substances while undergoing an equal temperature change.
Latent heat of melting. The quantity of heat gained or lost per gram by a substance changing from a solid to a liquid, or from a liquid to a solid, without a temperature change. For water, it is 80 cal at 0°C (32°F), the highest of any common substance.	*Heat energy lost when ice forms* is mostly absorbed by the heat-deficient atmosphere. *Heat energy gained by water when ice melts* is manifested as molecular energy of the liquid water. This prevents the high-latitude ocean from becoming much warmer or colder than the –1.8°C (28.7°F) freezing temperature of ocean water.

why people in mid-latitudes spread salt on roads and sidewalks during winter (to lower the freezing point).

In Figure 5–12, you can see that the *freezing point* and *temperature of maximum density* converge as they drop. They coincide when the salinity reaches 24.7 parts per thousand (‰), at a temperature of –1.33°C (29.61°F). When salinity is greater than 24.7 ‰, the density of water increases as temperature drops, until the water freezes. Thus, unlike freshwater with its density anomaly at 4°C, average seawater (34.7 ‰) has no such maximum density anomaly.

Because the components of salt (sodium and chlorine) have a density greater than that of water, increasing water salinity also increases its density. Physical oceanographers are greatly concerned with the relations among water density, salinity, and temperature, because density distribution is an important force in ocean circulation. Unfortunately, it is difficult to directly measure ocean water density. To determine density distribution within a region of study, oceanographers measure temperature and salinity and then use these values to calculate the density.

Table 5–4 summarizes the physical and biological significance of the properties of water.

Light Transmission in Ocean Water

As Figure 5–13 reveals, most solar energy is in the group of wavelengths we call *light*. It is essential to understand this radiant energy from the sun, because it powerfully affects the oceans with three major consequences:

- Ocean currents and wind-driven ocean waves ultimately derive their energy from solar radiation.
- A thin layer of warm water at the ocean surface overlies the great mass of cold water that fills most of the

Properties	Physical and Biological Significance
Latent heat of vaporization. The quantity of heat gained or lost per gram by a substance changing from a liquid to a gas, or from a gas to a liquid, without a temperature change. It is greater for water, 540 cal at 100°C (212°F), than for any other common substance. It is 585 cal at 20°C (68°F), at which much of the evaporation from the ocean surface occurs.	*Global heat transfers*—a great amount of excess heat energy is removed from the low-latitude ocean by evaporation. It is released through precipitation into the atmosphere at heat-deficient higher latitudes. This greatly moderates temperatures at the poles and equator, which otherwise would be far more extreme.
Surface tension. Cohesive attraction of hydrogen bonds causes a molecule-thick "skin" to form on water surfaces. Water has the greatest of all common liquids.	*Surface tension*—some organisms use this "skin" as a walking surface. Others hang from its undersurface.
Light transmission. No sunlight penetrates below a depth of 1000 m (3300 f). Red wavelengths are absorbed by a depth of 20 m (65 f) and yellow by 100 m (330 f). Only blue-green wavelengths penetrate below 250 m (800 f).	*Wind-driven ocean waves and currents* derive their energy from solar radiation. The sun heats the ocean from the surface, producing a warm surface layer and cold deep water masses. The energy for life in the oceans, most of which is supported by photosynthesis, is dependent on solar radiation. Without this radiation to illuminate objects, vision would be useless.
Sound transmission. The velocity of sound increases with greater temperature, salinity, and pressure. A zone of rapid temperature decrease beneath the warm surface layer of ocean water is a low-velocity sound channel called the SOFAR channel. This channel traps sound and can conduct it over great distances.	*SOFAR channel* may be used by marine mammals to communicate over long distances. Human science has made great use of it in studying the nature of the oceans.
Density. Mass per unit volume (g/cm³). Ocean water density increases at lower temperatures and as salinity and pressure increase. For pure water, the temperature of maximum density is 4°C (39.2°F), but ocean water with a salinity greater than 24.7°/∞ will get denser as the temperature is lowered to the freezing point.	*Plankton* that stay near the surface through buoyancy and frictional resistance to sinking are greatly influenced by the effect of temperature on density. In low-density warm water, plankton must be smaller or more ornate to obtain the increased ratio of surface area to body mass necessary to remain afloat.

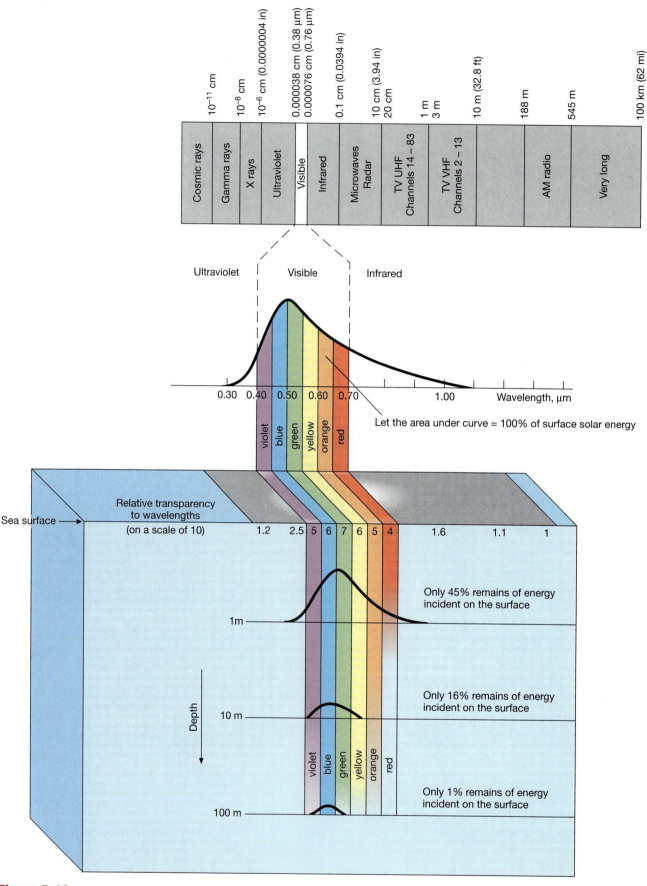

Figure 5–13

The Electromagnetic Spectrum and Transmission of Visible Light in Water. The spectrum (top) runs from extremely short cosmic rays (left) with increasing wavelength toward the right, through ultraviolet, visible light, infrared, microwaves, and broadcasting, to extremely long waves. The narrow portion of the spectrum that we see as light is shown at bottom, passing through seawater. As light penetrates seawater, the water filters longer wavelengths (red, orange, and yellow).

ocean basins. This is the "life layer" where most sea life exists, and the warm surface waters also strongly affect world climates. This thin layer of heated water exists at the surface because the oceans are heated by solar radiation.

- Phytoplankton (floating photosynthetic cells) are the basis of the ocean food chain. They produce their own food through photosynthesis. Photosynthesis can occur only where sunlight penetrates the ocean water, so phytoplankton and most animals that eat them must live where the light is, in the thin layer of sunlit surface water.

We will now take a look at the phenomenon of light from the sun, and how it interacts with ocean water.

Why are some areas of the ocean blue, whereas others are greenish? And when you go down 100 meters (330 feet) or so, why does everything look blue? Ocean water ranges from a deep indigo blue to a yellow-green. The indigo blue is typical of tropical and equatorial regions, where there is little primary biological productivity (plant growth that provides food for animals). The yellow-green occurs in the coastal waters of high-latitude areas, where primary biological productivity occurs seasonally at a very high rate.

The lack of particulate matter in tropical waters minimizes the molecular scattering of solar radiation, which causes the water to appear blue. The same process is also responsible for the blue color of the sky. Greater concentrations of particulate matter, especially where phytoplankton is abundant, result in greater scattering and absorption of light. This decreases the transmission of solar radiation, producing the greenish color characteristic of such waters.

Before proceeding, you need to know something about the electromagnetic spectrum and the narrow portion of it we see as light. The many wavelengths of electromagnetic energy radiated by the sun are displayed in Figure 5–13 (upper portion), the **electromagnetic spectrum.** The very narrow segment of this spectrum designated as *visible light* can be divided by wavelength into violet, blue, green, yellow, orange, and red energy levels. Combined, these different wavelengths of light produce white light. The shorter wavelengths of energy to the left of visible light are highly dangerous and include X rays and gamma rays. To the right of the visible segment of the electromagnetic spectrum are longer wavelengths of energy. Our technology uses these wavelengths for heat transfer and communication.

We will further consider the electromagnetic spectrum when we examine the role of solar energy in setting the ocean masses in motion. For now, we will consider only that portion of the spectrum that includes visible light. We call it "visible" light because the electromagnetic sensors we call eyes, and those of most animals, are adapted to sense light wavelengths.

Why do we see things in color? Our basic source of light is the sun, and its radiation includes all the visible colors. Most of the light we see is reflected from objects. Objects are selective reflectors, reflecting different wavelengths of light, and each wavelength represents a color in the visible spectrum. For example, vegetation absorbs most wavelengths except green and yellow, which they reflect, so plants look green.

If a certain wavelength of light is missing from the light that falls upon an object, that color cannot be seen. The ocean is a selective absorber of visible light, and its absorption is greater for the longer-wavelength colors (red, orange). Thus, the shorter-wavelength portion of the visible spectrum is all that can be transmitted to greater depths (Figure 5–13, lower portion).

Consequently, red wavelengths are absorbed within the upper 10 meters (30 feet), and yellow disappears before a depth of 100 meters (330 feet) has been reached, but green light can still be perceived down to 250 meters (over 800 feet). Only the blue and some green wavelengths extend beyond these depths, and their intensity becomes low. It is because of this absorption pattern that objects in the ocean usually appear blue-green. Only in the surface waters can the true colors of objects be observed in natural light, since only in the surface waters can all wavelengths of the visible spectrum be found. No sunlight penetrates below a depth of 1000 meters (3300 feet).

To measure the transmission of visible light in the ocean, a *Secchi disc* is attached to a rope that is marked off in meters. After the disc (about 30 centimeters, or 12 inches, in diameter) is lowered into the ocean, the depth at which it can last be seen indicates the water's *turbidity*—the amount of suspended material in it. Increased turbidity increases the degree of light absorption, decreasing the transmission of visible light.

Sound Transmission in Ocean Water

Sound is transmitted much more efficiently through water than through air. This has made it possible to develop sound systems for determining the position and distance of objects in the ocean. The basic technique is called **sonar,** an acronym for *sound navigation and ranging.* Sonar relies on echoes—determining the time that it takes an echo to return to the source of the sound.

Average sound velocity in the ocean is 1450 meters/second (4750 feet/second). This is over four times faster than sound velocity in the air, which is 334 meters/second (1100 feet/second) in a dry atmosphere at 20°C. Further, sound travels faster in the ocean with increases in temperature, salinity, and pressure. Because these factors vary with the seasons, the velocity of sound through water in a particular area varies seasonally. To determine the exact velocity of sound for any

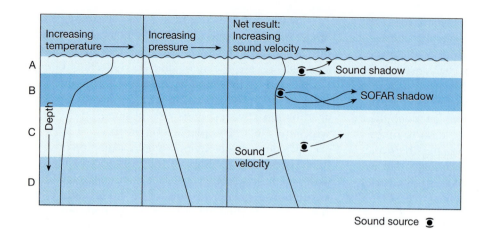

Figure 5–14

Transmission of Sound in the Ocean. Sound travels much faster in the ocean than in the air. In the water, sound travels faster with increasing temperature and pressure, shown in these graphs. Water pressure increases steadily with depth, and sound travels faster through the compacted water molecules. However, the rapid decline in temperature with depth offsets the effect of the pressure increase. The net result is the lower-velocity sound channel shown. Waves bend (refract) into low-velocity areas. This produces *shadow zones* where velocity maxima exist and a channel traps sound energy in the low-velocity zone. The sound source beneath the SOFAR channel (SOund Fixing And Ranging) illustrates how refraction bends sound. If we consider sound waves that are traveling to the right from the sound source in band C, the sound at the bottom of the band will travel faster than the sound at the top of it due to increasing water pressure. If this sound band moves as a coherent unit, it will be bent upward into the SO-FAR channel. As it passes through the SOFAR channel, the top of the band will be moving faster than the bottom and will bend it back down into the channel—trapping the sound in the SOFAR channel.

point in the ocean, these three variables—temperature, salinity, and pressure—must first be determined, or a *velocimeter* can measure it directly.

At a depth of around 1000 meters (3300 feet), there exists a layer of ocean water where these three variables combine to create a relatively low velocity of sound transmission. Sound originating above and below this low-velocity layer travels to the layer and there becomes retracted, or bent, into the layer. In effect, it becomes trapped in the layer. The sound then travels in this channel and can be transmitted across entire ocean basins (Figure 5–14). This channel is called the **SOFAR channel** (an acronym for SOund Fixing And Ranging) because of its practical application in distance determination.

Because sound travels so efficiently in the SOFAR channel, it may be used by marine mammals, especially whales, in communicating over long distances. A study that is planned to investigate whether the oceans are experiencing any warming due to the greenhouse effect will use sound transmissions over thousands of kilometers within the SOFAR channel. (See Table 5–4.)

Geosecs (Geochemical Ocean Sections)

The GEOSECS project mentioned in Chapter 1 is a huge systematic program designed to improve understanding of circulation patterns and mixing in the oceans. All major ocean basins have been sampled throughout their depths. The sampling system included electronic sensors that telemeter temperature, salinity, dissolved oxygen, pressure, and particulate content to a console aboard ship. As rosettes of 30-liter (approximately 8-gallon) sampling bottles are lowered into the ocean, oceanographers can decide to sample at any depth where an interesting value for any of the properties appears on the console (Figure 5–15).

The collected water is analyzed for 23 chemicals, 15 isotopes, and the amount and type of particulate matter. From results so far, oceanographers think the GEOSECS program not only will provide a much improved understanding of physical processes but also will reveal much about biological cycles in the ocean. Figure 5–16 shows the location of one GEOSECS station in the Pacific and graphs of data acquired there.

Figure 5–15

GEOSECS Sampling Bottles. One of the two rosettes carried aboard the Scripps Institution of Oceanography's research vessel *Melville* during GEOSECS Pacific operations. The nonmetallic bottles are used in collecting seawater samples. The end of each bottle can be triggered to close by remote control by the scientist operating the shipboard console. (Photo from Scripps Institution of Oceanography, University of California, San Diego.)

Desalination

Earth's growing population is consuming freshwater supplies in greater volumes each year. As freshwater becomes a scarcer commodity, scientists and governments look to the seas for water. Saltwater is not drinkable, so **desalination,** or salt removal from seawater, must be employed to take advantage of the greatest water supply on Earth—the oceans. Desalination can provide freshwater for domestic, home, and agricultural use. Some desalination methods are described here.

Distillation is boiling saltwater and capturing the water vapor, leaving the salt behind. The water vapor is passed through a cooling condenser, where it condenses and is collected as freshwater. This simple procedure is expensive, but it requires large amounts of heat energy to boil the water (remember, 540 calories per gram at 100°C). Increased efficiency is required to make distillation practical on a large scale.

Another process demanding large amounts of energy is *electrolysis*. In this method, two volumes of freshwater—one containing a positive electrode and the other a negative electrode—are placed on either side of a volume of seawater. The seawater is separated from each of the freshwater reservoirs by semipermeable membranes. These are permeable to salt ions but not to water molecules. When an electrical current is applied, positive ions such as sodium ions are attracted to the negative electrode, and negative ions such as chloride ions are attracted to the positive electrode. In time, enough ions are removed through the membranes to convert the seawater to freshwater.

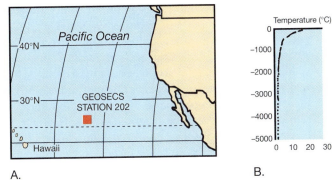

A.

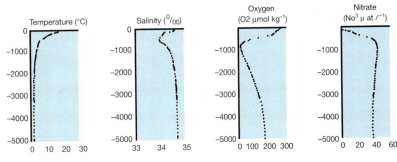

B.

Figure 5–16

GEOSECS Data. *A:* Map showing the location of GEOSECS Station 20 at latitude 26°N, longitude 139°W. (The dashed line is the Tropic of Cancer.) *B:* Four curves show temperature, salinity, dissolved oxygen, and nitrate measured from the surface down to 5000 meters depth (16,400 feet) at Station 202. The curves show that ocean water properties may change significantly from the surface to a depth of 1000 meters (about 3300 feet). The rate of change below that depth is much less. (Plots courtesy of Woods Hole Oceanographic Institution.)

Another application that is limited to small-scale use is *freezing*. The process of freezing requires plentiful energy, so it may not be practical on a large scale. Another way is to use natural ice. Imaginative thinkers have proposed towing icebergs to coastal waters off countries that need freshwater. Here, the freshwater that is produced as the icebergs melt could be captured and pumped ashore.

Solar humidification does not require supplemental heating and has been successfully used in large-scale agricultural experiments in Israel, West Africa, and Peru. It involves the evaporation of saltwater in a covered container that is heated by direct sunlight. Saltwater in the container evaporates, and the water vapor that condenses on the cover runs into collection trays.

A method that has much potential for large-scale projects is **reverse osmosis.** In natural osmosis, water molecules pass through a water-permeable membrane from a freshwater solution into a saltwater solution. But in reverse osmosis, pressure is applied to the saltwater solution, forcing it to flow "backwards" through the water-permeable membrane into the freshwater solution (Figure 5–17). This is the reverse of natural osmosis, giving the method its name. A significant problem is that the membranes must be replaced frequently. At least 30 countries located in arid climates, including Saudi Arabia and most Middle Eastern states, have built reverse osmosis units.

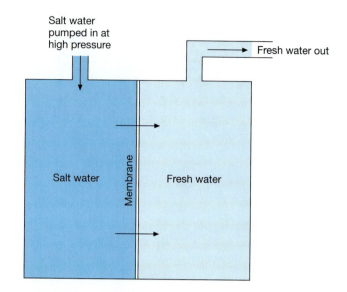

Figure 5–17

Reverse Osmosis Process Used to Produce Freshwater from Saltwater. This very simple process requires that pressure be applied to a reservoir of saltwater to force it through a water-permeable membrane into a reservoir of freshwater, from which it is collected.

S U M M A R Y

Water is unusual because it exists on Earth in three states (solid, liquid, gas), and it has the capacity to store great amounts of heat energy.

The hydrogen bond (the bond between water molecules resulting from the bipolar nature of the molecule) plays the major role in giving water its unusual properties.

Water is a great solvent because its bipolar molecules can attach themselves to charged particles—ions—that make up many substances. This hydrates them and places them into solution.

The hydrogen bond also accounts for the unusual thermal properties of water, such as its high freezing point (0°C) and boiling point (100°C), high heat capacity (1 calorie, the heat required to raise the temperature of 1 gram of water 1°C), high latent heat of melting (80 calories/gram), and high latent heat of vaporization (540 calories/gram).

The surface tension phenomenon that makes water form droplets and allows it to be poured into a container until it stands well above the sides results from the water molecules being strongly bound by hydrogen bonds to other water molecules beside and beneath them. This cohesion, combined with the adhesive attraction of water to glass, organic substances, rock, and soil, can pull water to great heights in small tubular structures.

Salinity is the amount of dissolved solids in ocean water, averaging about 34.7 grams of dissolved solids per kilogram of ocean water. Salinity is usually expressed in parts per thousand (‰). Over 99 percent of the dissolved solids in ocean water are accounted for by six ions—chloride, sodium, sulfate, magnesium, calcium, and potassium. In any seawater sample, these ions are in the same proportion, so salinity can be determined by measuring the concentration of only one, usually the chloride ion.

The density of pure water increases as temperature drops, as does the density of most substances, but only to 4°C (39.2°F), its temperature of maximum density. Below 4°C, water density decreases with temperature, due to formation of bulky ice crystals. By the time water fully freezes, the density of the ice is only about 0.9 that of water at 4°C.

Density increases with salinity and pressure.

The residence time of elements in seawater ranges from 100 years for aluminum to 260 million years for sodium. The time required to fully recirculate the water in the oceans, or the water's residence time, is 4000 years.

In ocean water of low biological productivity, the molecule-sized particles scatter the short wavelengths of visible light, producing a blue ocean. Greater amounts of dissolved organic matter in more productive ocean water scatter more green light, which produces a green ocean. With increasing depth in the ocean, the colors of the visible spectrum are absorbed by ocean water in a way that removes red and yellow at relatively shallow depths. The blue and green wavelengths of light are the last to be removed. Turbidity refers to the amount of suspended material in water; high turbidity greatly reduces the depth to which light will penetrate.

Velocity of sound transmission in the ocean increases with temperature, salinity, and pressure. A low-velocity SOFAR sound channel can conduct sound over great distances in the oceans.

The ocean contains about 97 percent of Earth's water. Water is recycled among the ocean, atmosphere, and continents by the hydrologic cycle, which involves evaporation, precipitation, and surface and groundwater runoff from the continents.

The GEOSECS program has improved oceanographers' understanding of physical and biological ocean processes. It features an advanced system of sampling and chemically analyzing the properties and content of ocean water.

Desalination of ocean water to provide freshwater for domestic, home, and agricultural use may be achieved by distillation, electrolysis, freezing, solar humidification, and reverse osmosis of seawater. The last two methods show the greatest potential for practical applications.

KEY TERMS

Boiling point (p. 108)
Calorie (p. 108)
Capillarity (p. 114)
Condenses (p. 111)
Constancy of composition, rule of (p. 114)
Density (p. 119)
Desalination (p. 125)
Dipolar (p. 107)
Electromagnetic spectrum (p. 123)
Evaporation (p. 113)
Freezing point (p. 108)

Gaseous state (p. 109)
Heat (p. 108)
Heat capacity (p. 110)
Hydrogen bonds (p. 107)
Hydrologic cycle (p. 118)
Ionic bonds (p. 107)
Kinetic energy (p. 108)
Latent heat of evaporation (p. 113)
Latent heat of melting (p. 111)
Latent heat of vaporization (p. 111)
Liquid state (p. 108)
Molecular motion (p. 108)

Residence time (p. 116)
Reverse osmosis (p. 126)
Salinity (p. 115)
SOFAR channel (p. 124)
Sonar (p. 123)
Solid state (p. 108)
Standard seawater (p. 116)
Surface tension (p. 114)
Temperature (p. 108)
Temperature of maximum density (p. 120)
Van der Waals force (p. 108)

QUESTIONS AND EXERCISES

1. Describe what condition exists in water molecules to make them dipolar.
2. Discuss how the dipolar nature of the water molecule makes it such an effective solvent for ionic compounds.
3. Define freezing point and boiling point.
4. There is a fundamental difference between the intermolecular bonds that result from the dipolar nature of the water molecule (hydrogen bond) and the van der Waals force as compared with the chemical bonds. What is it?
5. Why are the freezing and boiling points of water higher than would be expected for a compound of its molecular makeup?
6. How does the heat capacity of water compare with that of other substances? Describe the effect this characteristic of water produces on climate.
7. The heat energy added as latent heat of melting and latent heat of vaporization do not increase water temperature. Why? Why is the latent heat of vaporization so much greater than the latent heat of melting?
8. Describe how excess heat energy absorbed by Earth's low-latitude regions is transferred to heat-deficient higher latitudes through a process that uses water's latent heat of evaporation.
9. How does hydrogen bonding produce the surface tension phenomenon of water?
10. As water cools, two distinct changes in the behavior of molecules take place: Their slower movement tends to increase density, whereas the formation of bulky ice crystals decreases density. Describe how the relative rates of their occurrence cause pure water to have a temperature of maximum density at 4°C (39.2°F) and make ice less dense than liquid water.
11. As water becomes more saline, the temperature of maximum density and the freezing temperature of water decrease and converge. At what level of salinity

does water cease to have a temperature of maximum density above its freezing temperature?

12. What condition of salinity makes it possible to chemically determine the total salinity of ocean water by measuring the concentration of only one constituent, the chloride ion?

13. If there is an estimated 18×10^{20} grams of magnesium (Mg) in the ocean, and it is being added (or removed) at the rate of 56.5×10^{12} grams per year, what is the residence time of magnesium in the ocean? (See appendix I.)

14. List the reservoirs of water on Earth and the percentage of Earth's water that each holds. Explain how the hydrologic cycle moves water among these reservoirs.

15. How is the color of the ocean surface water related to biological productivity? Why does everything in the ocean at depths below the shallowest surface water take on a blue-green appearance?

16. How does the decrease in temperature at the base of the warm surface water produce a SOFAR, or sound channel, below the ocean's surface?

17. Briefly describe how the following methods of desalination would produce freshwater: distillation and reverse osmosis.

REFERENCES

Davis, K. S., and Day, J. S. 1961. *Water: The mirror of science*. Garden City, N.Y.: Doubleday.

Hammond, A. L. 1977. Oceanography: Geochemical tracers offer new insight. *Science* 195:164–166.

Parvey, H. W. 1960. *The chemistry and fertility of sea waters*. New York: Cambridge University Press.

Kuenen, P. H. 1963. *Realms of water*. New York: Science Editions.

MacIntyre, F. 1970. Why the sea is salt. *Scientific American* 223:5, 104–115.

Packard, G. L. 1975. *Descriptive physical oceanography*. 2nd ed. New York: Pergamon Press.

Kevelle, R. 1963. Water. *Scientific American* 209:93–108.

Yokohama, Y.; Guichard, F.; Reyss, J-L.; and Van, N. H. 1978. Oceanic residence times of dissolved beryllium and aluminum deduced from cosmogenic tracers [10]Be and [26]Al. *Science* 201:1016–1017.

SUGGESTED READING

Sea Frontiers

Friedman, R. 1990. Salt-free water from the sea. 36:3, 48–54. The economics and nature of the various processes used in desalination are discussed.

Gabianelli, V. J. 1970. Water: The fluid of life. 16:5, 258–270. The unique properties of water are lucidly described and explained. Topics covered include hydrogen bond, capillarity, heat capacity, ice, and solvent properties.

Smith, F., and Charlier, R. 1981. Saltwater fuel. 27:6, 342–349. The potential for using the salinity difference between river water and coastal marine water to generate electricity is considered.

Scientific American

Baker, J. A., and Henderson, D. 1981. The fluid phase of matter. 245:5, 130–139. The structure of gases and liquids is modeled using hard spheres.

MacIntyre, F. 1970. Why the sea is salt. 223:5, 104–115. A summary of what is known of the processes that add and remove the elements dissolved in the ocean.

CHAPTER 6
Air-Sea Interaction

Physical processes called *weather* occur in the atmosphere. They are characteristic of climate conditions found at each latitude. Similarly, "weather" also occurs in the waters of the oceans—flowing, turbulence, storms, heat and cold, even "clouds." The reason is simple: All *fluids,* which include gases and liquids, behave according to the same natural laws. For instance, the winds of the atmosphere correspond to currents in the oceans. The biggest difference is that, because the ocean is 1000

times denser than the atmosphere, oceanic "weather" systems develop more slowly and last longer than they do in the atmosphere.

Most of the currents and waves we observe in the surface ocean are created directly by atmospheric winds. The winds themselves are created by heating, provided by solar energy. In other words, the ultimate source of energy for all of these forms of fluid motion, in the atmosphere and in the ocean, is radiant energy from the sun. The atmosphere and ocean constantly exchange this energy by reflection, absorption, evaporation, condensation, and precipitation. Together, the atmosphere and ocean are a powerful team that shape the weather worldwide.

This chapter is the first of three that will explain the relations between the weather of Earth's atmosphere and the "weather" of its oceans.

In addition to the day-to-day influence of the oceans on our weather, periodic extremes of atmospheric weather, such as droughts and profuse precipitation, probably are tied to periodic changes in oceanic conditions. The well-publicized "El Niño" events of the equatorial Pacific Ocean appear to have such a relation.

In fact, El Niño was one of the first major phenomena to spur study of air-sea interaction, because in the 1920s it was recognized that El Niño, an ocean event, was tied to worldwide climate changes. (The *El Niño-Southern Oscillation* is discussed in Chapter 7.)

Then, in 1988, studies confirmed that Earth's average temperature had risen over the past century. This led to much investigation of the *greenhouse effect.* This phenomenon of Earth's atmosphere was seen as possibly warming our planet's entire life zone, due to human-caused increases in carbon dioxide and other gases that absorb heat.

To what extent are the oceans involved in all of this?

- Do changes in the ocean produce changes in the atmosphere that lead to the El Niño phenomenon—or vice versa?
- Is there a complex feedback system that involves both air and sea to a generally equal degree?
- Atmospheric carbon dioxide has increased only half as much as predicted from human activities; where did the rest of the carbon dioxide go?
- How does carbon dioxide enter the oceans?

Answers probably will be long in coming. But in this chapter we will examine the issues so you will understand new research as it unfolds during your lifetime.

Seawater Density

Scientists assign to pure water a density of 1.0000 grams per cubic centimeter (g/cm³). This serves as a standard against which the density of all other substances can be measured. Seawater is, of course, not pure water; it has various salts dissolved in it. Thus, seawater density ranges from 1.02200 to 1.03000 g/cm³ in the open ocean, which means it is 2 percent to 3 percent heavier than pure water.

Density is an important property of ocean water because density differences make water masses sink or float, thus determining the vertical position and movement of ocean water masses. For example, if saltwater

(say of density 1.0027) is added to freshwater (density 1.0000), the denser saltwater would sink to the bottom.

In stable ocean water, gravity rules: Denser water sinks, so low-density water rests at the top of any column of water one chooses to study, and the density increases with depth. Oceanographers study the density of water *in place,* or wherever it actually occurs in the ocean environment. Density in deep ocean trenches is about 5 percent greater than at the ocean surface because of the compressive effect of pressure from the great mass of water overhead.

As noted in the preceding chapter, oceanographers are greatly concerned with the relations among water density, salinity, and temperature, because density distribution is an important factor in ocean circulation. Figures 6–1, 6–2, and 6–3 present the relations among seawater density, salinity, and temperature in relation to depth and latitude.

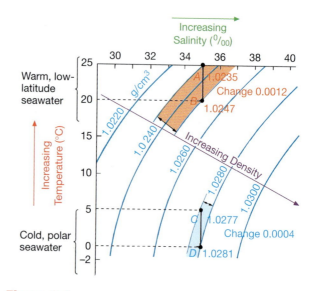

Figure 6–1

Seawater Density Varies with Temperature and Salinity. Seawater density ranges from 1.0220 to 1.0300 grams per cubic centimeter (g/cm³) in the open ocean, depending on water saltiness and temperature (blue curves). This range is only 0.0080 g/cm³, but very significant. In high-latitude (polar) areas of cold water, temperature has less effect on density than in high-temperature, low-latitude (equatorial) areas. Points A, B, C, and D show density measured at a salinity of 35‰. Note that the density change is much greater over a 5° temperature span at a higher temperature than at a lower temperature. The density change across a temperature range of 20 to 25°C (68 to 77°F) is 0.0012, compared with a change of only 0.0004 across an equal range at the lower temperatures of 0 to 5°C (32 to 41°F). Thus, a change in the temperature of warm, low-latitude water has about three times the effect on density that an equal change in temperature occurring in colder, high-latitude waters. (After G. L. Pickard, *Descriptive physical oceanography,* © 1963.)

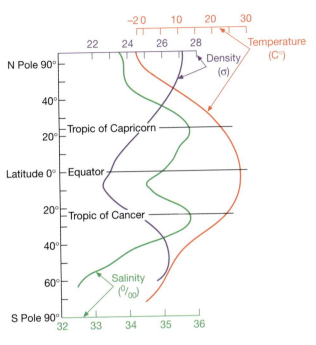

Figure 6–2

Average Ocean Surface Temperature, Salinity, and Density Vary with Latitude. As you would expect, seawater temperature is lowest at the poles and highest at the equator. Salinity is lowest at the poles and equator and peaks at the Tropics of Cancer and Capricorn. In response to these variations in temperature and salinity, density increases from about 1.022 g/cm³ near the equator to maxima of 1.026 to 1.027 near 50° to 60° north and south latitude in the ocean's surface water. At higher latitudes, it decreases slightly. Salinity maxima in the tropics seem not to affect density, pointing out the greater importance of temperature in controlling density in low latitudes. (After G. L. Pickard, *Descriptive physical oceanography,* © 1963. By permission of Pergamon Press Ltd., Oxford, England.)

Figure 6–1 shows that temperature change affects density (curving lines) much more in warm, equatorial areas than in polar regions. It can be noted that the lines of constant density more nearly parallel the temperature axis at the bottom (low temperatures) than top (high temperatures). This means that for each unit of temperature change greater density change occurs in the high-temperature range.

Figure 6–2 reveals much about the ocean's characteristics *by latitude*. Note the red temperature curve, shaped as you would expect: coldest toward the poles and warmest at the equator. The green salinity curve is interesting, with lowest salinity toward the poles and around the equator, but highest salinities around the Tropic of Cancer and the Tropic of Capricorn, due to higher evaporation rates in these regions.

Figure 6–3 shows the change with depth of density, temperature, and salinity. In ocean water, most

properties change more vertically than horizontally. This is certainly true of density (Figure 6–3A). In the case of temperature, surface water density in the open ocean is affected primarily by temperature changes (Figure 6–3B). The gain or loss of heat energy at the ocean surface is important in controlling water density. But in the extreme polar areas of the ocean, where temperatures remain relatively constant, salinity changes can significantly effect density (Figure 6–3C). Figure 6–3 shows the vertical distribution of density in various regions of the ocean.

In the equatorial region, there is a shallow zone of low-density water near the surface. Below this lies a zone where density increases very rapidly, called the **pycnocline** (density slope). The pycnocline is a *stable barrier to mixing* between low-density water above and high-density water below. The pycnocline has a high gravitational stability because it would require greater energy to move a mass of water up or down within the pycnocline than it would to move an equal mass the same distance where density changes very slowly with increasing depth.

The pycnocline (density slope) results from the combined effect of rapid vertical temperature change (the **thermocline,** or temperature slope), and salinity change (the **halocline,** or salinity slope). The interrelation of these three zones determines the degree of separation between the **upper-water** and **deep-water** masses (Figure 6–3).

Blue lines in Figure 6–3A show that the pycnocline is lacking in high latitudes. You can see that little difference exists between surface water density and bottom water density in these polar areas. Thus, these water columns are more inclined to mix than water columns possessing a pycnocline. In the next chapter, we will consider the importance of low-stability water columns in polar latitudes to mixing ocean waters worldwide.

Solar Energy Received by Earth

Essentially all of the energy available to Earth comes from the sun. Solar energy strikes Earth at an average rate of 2 calories per square centimeter each minute. Considering that the side of Earth facing the sun at any moment exposes about 255 trillion square centimeters to solar radiation, this is a tremendous amount of incoming energy. It is this energy that drives the global ocean-atmosphere engine, creating pressure and density differences that stir currents and waves in both air and sea.

The worldwide average temperature of Earth and the lower atmosphere is about 15°C (59°F). However, if the atmosphere contained no water vapor, carbon dioxide, methane, or other trace gases, the worldwide average temperature would be far colder, about −18°C

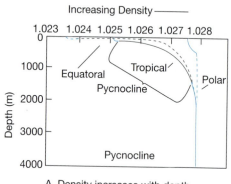

A. Density increases with depth

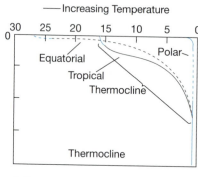

B. Temperature (°C) decreases with depth

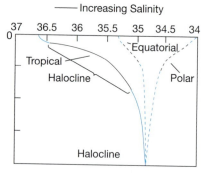

C. Salinity (⁰/₀₀) varies with depth

Figure 6-3

Seawater Density Varies with Depth and Latitude. In parts A, B, and C, black indicates the pycnocline, thermocline, or halocline zones. *A: Pycnocline*—in response to gravity, density generally increases with depth. However, the *rate* of this increase varies with latitude. In equatorial waters, a thin layer of low-density surface water is separated from high-density deep water by a zone of rapid density change, the *pycnocline*. The pycnocline is poorly developed or missing in polar waters. It is the absence of the stable pycnocline that facilitates vertical mixing in some high-latitude areas, due to the density difference. *B: Thermocline*—the pycnocline primarily results from, and coincides with, a rapid vertical temperature decline, the *thermocline* (compare the generally similar curves in A and B). The thermocline is better developed in equatorial latitudes. A zone of rapid vertical change in salinity, the *halocline,* also may affect density. (The temperature scale is shown increasing to the left; this is unconventional, but was done to make the pycnocline and thermocline curves easier to compare.) *C: Halocline*—in equatorial and tropical latitudes, the halocline usually represents decreasing salinity with depth. However, in polar latitudes, the halocline may show increasing salinity with depth. *D: The ocean temperature/pressure profile*—the mixed surface layer represents water of uniform temperature resulting from wave mixing. The zone of the thermocline/pycnocline is a relatively light zone called *upper water.* It is well developed throughout the equatorial and mid-latitudes. It is underlain by the denser, cold *deep-water* mass that extends to the deep-ocean floor. (*A* and *B* after G. L. Pickard, *Descriptive physical oceanography,* © 1963.)

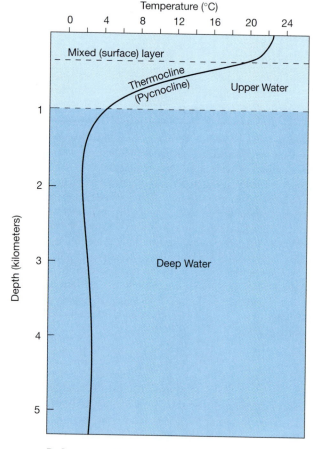

D. Ocean profile of temperature and pressure

(−4°F). All of the water on Earth would be frozen. The existing, more pleasant temperatures are made possible by the greenhouse effect of these gases.

The Atmosphere's Greenhouse Effect

The **greenhouse effect** is analogous to the operation of a greenhouse for raising plants. Energy radiated by the sun covers the full electromagnetic spectrum, but most of the energy that reaches Earth's surface is in short wavelengths, in and near the visible portion of the

spectrum. In a greenhouse, shortwave sunlight passes through the transparent glass or plastic. Striking plants, floor, and other objects inside, it is converted to heat. Heat has longer wavelengths, and when heat energy tries to escape the greenhouse, it finds the glass or plastic to be opaque. The heat energy becomes trapped inside, and temperature climbs.

In the upper atmosphere, most solar radiation within the visible spectrum (0.38 to 0.76 micrometers) penetrates the atmosphere to Earth's surface, like sunlight coming through greenhouse glass. After scattering

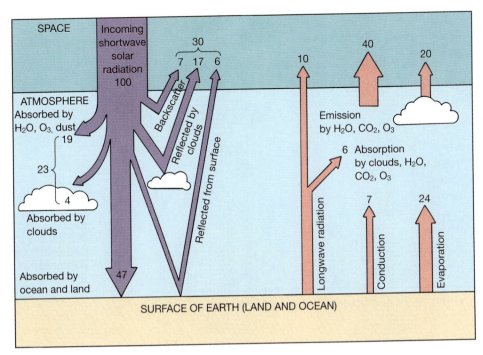

Figure 6–4

Earth's Heat Budget. Short-wave solar radiation from the sun (mostly visible light) is reflected, scattered, and absorbed by various components of the Earth-atmosphere system. The diagram shows 100 units of incoming solar radiation and how they are distributed. All of the absorbed energy is radiated back into space by Earth as longwave infrared radiation at varying rates, so, over time, solar input equals Earth's radiation output. (Values shown are estimates.)

by atmospheric molecules and reflection off clouds, about 47 percent of the solar radiation that is directed toward Earth is absorbed by the oceans and continents (Figure 6–4). About 23 percent is absorbed by the atmosphere and clouds, and about 30 percent is reflected into space by atmospheric backscatter, clouds, and Earth's surface. Figure 6–4 diagrams the roles of the various components of Earth's **heat budget.** The values are estimates because they are not known with any high degree of accuracy.

If it is true that Earth has maintained a constant average temperature over long periods, Earth must be reradiating energy back to space at the same rate at which it absorbs solar energy. Just like the heat energy radiated from objects inside a greenhouse, the energy reradiating from Earth falls within the infrared range (0.76 micrometers to 0.1 centimeters). Figure 6–5 shows that the intensity of energy radiated by the sun peaks at 0.48 micrometers in the visible spectrum, but that reradiation from Earth peaks at 10 micrometers in the infrared range. *It is this change of wavelengths that is the key to the greenhouse effect.*

Although Earth's atmosphere and clouds absorb only 23 percent of the shorter-wavelength incoming solar radiation, the outgoing longer-wavelength infrared radiation from Earth is absorbed in much greater quantities by water vapor, carbon dioxide, and other gases in the atmosphere. This is the same as air in a greenhouse being warmed by trapped heat energy.

Some of the infrared energy absorbed in the atmosphere becomes reabsorbed by Earth to continue the process (the rest of the energy is lost to space). Therefore, the solar radiation received is retained for a time within our atmosphere. It moderates temperature fluctuations between night and day and between seasons.

Which Gases Contribute to the Greenhouse Effect? Research has disclosed which greenhouse

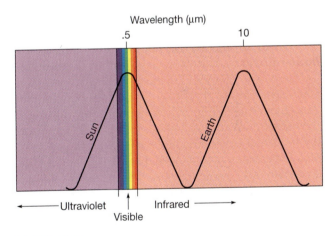

Figure 6–5

Energy Spectra Radiated by the Sun and Earth. Most of the energy coming to Earth from the sun is within the visible spectrum and peaks at green (0.48 micrometers—μm—or 0.0002 inches). The atmosphere is transparent to most of this radiation, and it is absorbed at Earth's surface. When Earth reradiates this energy back toward space, it does so as longer-wavelength infrared or "heat" radiation, with a peak at a wavelength of 10 μm (0.004 inches). The carbon dioxide and water vapor in the atmosphere absorb some of this infrared radiation and reradiate it to produce the greenhouse effect. This keeps the atmosphere's temperature warmer than it otherwise would be.

Gas	Concentration (ppbv)	Rate of Increase (% per year)	Relative Contribution (%; totals 100%)
Carbon dioxide CO_2	353,000	0.5	60
Methane CH_4	1,700	1	15
Nitrous oxide N_2O	310	0.2	5
Ozone (tropospheric) O_3	10–50	0.5	8
Chlorofluorocarbon CFC-11	0.28	4	4
Chlorofluorocarbon CFC-12	0.48	4	8

Table 6–1
Estimated Contribution of "Greenhouse Gases" to Increasing the Greenhouse Effect, Based on Present Concentration and Observed Rate of Increase.

ppbv = parts per billion by volume (not by weight). Source: After H. Rodhe, 1990.

gases are increasing in the atmosphere due to human activity. They are listed in Table 6–1 in order of their potential contribution to increasing the greenhouse effect.

Clearly, carbon dioxide dominates. The other trace gases—methane, nitrous oxide, tropospheric ozone, and chlorofluorocarbons—are present in far lower concentrations. However, they are important because, per molecule, they absorb far more infrared radiation than does carbon dioxide (Table 6–2). Thus, none of these gases can be ignored when considering the greenhouse effect.

What Should We Do about Increasing Greenhouse Gases? Earth's average surface temperature has risen by 0.5°C (0.9°F) in the last 100 years. But there is no clear, simple proof that this has resulted from the 25 percent increase in atmospheric carbon dioxide or the increase in other greenhouse gases that began with the Industrial Revolution in the late 1700s (Figure 6–6). Many scientists are convinced that it has; others are not so certain. Also, some scientists who develop computer models of climate believe the greenhouse warming will evaporate more seawater and thus produce more cloud cover, which will block the sun's rays and significantly reduce the warming effect.

What *is* very clear, however, is that our agricultural and industrial activities are producing changes in the environment. Although we cannot eliminate modifications of the environment caused by human activity, we must seriously reduce our combustion of fossil fuels and reduce the widespread deforestation that accompanies the spread of agriculture. We must preserve Earth's plant communities and replace much of what is removed.

We cannot wait until we have proof that our activities are harming Earth's ecosystem. We must modify activities that change the environment most severely,

while research improves our understanding of how the system works.

Distribution of Solar Energy

Pretend for a moment that Earth is not spherical, but a flat plate in space, with the flat side directly to the sun

Table 6–2
Effect of Each Greenhouse Gas on the Greenhouse Effect. The ability of one molecule of each of these gases to absorb infrared radiation is compared with one molecule of carbon dioxide. It is clear that all of these gases absorb infrared radiation much more efficiently than does carbon dioxide. Table 6–1 shows that their smaller overall contribution to increased greenhouse effect in the atmosphere is due to their low concentrations in the atmosphere compared with carbon dioxide.

Gas	Relative Radiation Absorption per Molecule
Carbon dioxide CO_2	1
Methane CH_4	25
Nitrous oxide N_2O	200
Ozone (tropospheric) O_3	2000
Chlorofluorocarbon CFC-11	12,000
Chlorofluorocarbon CFC-12	15,000

Source: After H. Rodhe, 1990.

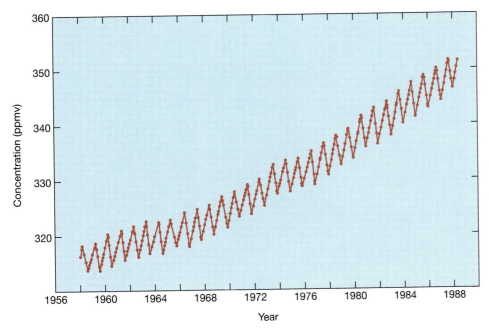

Figure 6–6

Atmospheric Carbon Dioxide Increase since 1958. This graph shows the concentration of atmospheric carbon dioxide at Mauna Loa Observatory in Hawaii, expressed in parts per million. The annual cycle reflects the seasonal cycles of photosynthesis (which consumes CO_2) and respiration (which releases CO_2) by biota in the Northern Hemisphere. The overall increase, however, is due primarily to the burning of fossil fuels. Analysis of centuries-old air trapped in continental ice sheets shows that the atmospheric CO_2 concentration has increased 25 percent since 1885.

(perpendicular). If this were the case, sunlight would fall equally on all parts of Earth. But Earth is spherical, so solar radiation strikes its surface at 90° (perpendicular) only at the center of the sphere (Figure 6–7, point A). Anywhere else on the sphere, sunlight strikes at angles less than 90° (point B), down to 0° at the edge of the sphere. The significance of this is that solar radiation is most intense at the perpendicular point but becomes progressively less intense toward the poles because it is spread out more.

Consider A in the figure, where a sample of solar radiation is striking Earth near the center of the illuminated portion. The sample covers exactly 1 square kilometer (0.4 square miles). By contrast, consider the same cross-sectional area of sunlight falling on a polar region of Earth's surface (B), where it does not strike at a right angle. Here the square kilometer of radiation spreads out over considerably more than a square kilometer of Earth's surface. The point is that *the intensity of radiation is greatly decreased compared with that available in the equatorial region.*

The angle at which direct sunlight strikes the ocean surface is important in determining how much of the solar energy is absorbed and how much is reflected. If the sun shines down on a smooth sea from directly overhead, only 2 percent of the radiation is reflected, but if the sun is at a very low angle, only 5° above the horizon, 40 percent of the radiation is reflected back into the atmosphere (Table 6–3).

Calculating the amount of radiation intercepted by any point on Earth is complicated by three key Earth movements: its daily rotation, its tilt, and its annual revolution around the sun.

- The rotation ensures that the sunlight received varies daily, simply due to the daily rotation through daylight and darkness.
- Earth's tilt is more complicated. Our planet is not "straight up-and-down" but is tilted 23.5° with respect to the sun. (Picture Earth as a spinning top, but one that is leaning a good bit—see Figure 6–7, top right.) Some areas "lean into" the sunlight, whereas others "lean away" from it.
- As Earth revolves on its annual trip around the sun, the Northern Hemisphere take turns "leaning toward" the sun every six months. Consequently, solar radiation received varies, causing the change of seasons. During the Northern Hemisphere winter, a significant portion of Earth's surface north of the **Arctic Circle** (66.5°N latitude) spends up to six months in darkness (dark area in figure). Half a year later, during the Southern Hemisphere winter, the same situation prevails south of the **Antarctic Circle** (66.5°S latitude; dark area in lower-right figure).

Elevation of sun above horizon	90°	60°	30°	15°	5°
Reflected radiation (%)	2	3	6	20	40
Absorbed radiation (%)	98	97	94	80	60

Table 6–3
Reflection and Absorption of Solar Energy Varies with the Angle of Incidence on a Flat Sea.

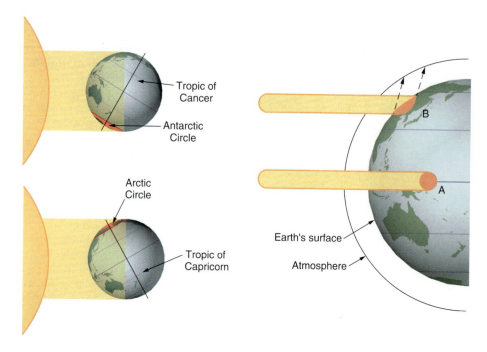

Figure 6–7

Importance of Angle at which Solar Radiation Is Received by Earth. The amount of solar energy received at higher latitudes is much less than that at lower latitudes for three reasons: (1) An area of solar radiation that strikes Earth at a high latitude (A) is spread over more surface than the same area of radiation that strikes Earth perpendicular to the surface at a lower latitude (B). (2) Radiation that strikes Earth at high polar latitudes also passes through a greater thickness of atmosphere. (3) More of the energy is reflected due to the low angle at which it strikes Earth's surface. *Northern Hemisphere winter*—areas north of the Arctic Circle receive no direct solar radiation at all, and the Northern Hemisphere receives reduced radiation because of the low angle. However, areas south of the Antarctic Circle receive continuous radiation ("midnight sun") for months. *Northern Hemisphere summer*—the situation is reversed. Note that, throughout the year, the sun is directly over some latitude between the tropics, affording intense, closer-to-vertical radiation to the hot equatorial area.

As Earth orbits the sun, its perpendicular radiation slowly migrates from the **Tropic of Cancer** (23.5°N), across the equator, to the **Tropic of Capricorn** (23.5°S), and slowly back. The belt between these two tropics receives much greater annual radiation per square kilometer than polar areas.

Oceanic Heat Flow and Atmospheric Circulation

Because of the low angle at which solar radiation strikes Earth's surface toward the poles, and the very reflective ice cover in polar latitudes, more energy is reflected back into space than is absorbed. In contrast, because of the high angle at which sunlight strikes Earth between 35°N and 40°S, more energy is absorbed than is radiated back into space. Figure 6–8 shows how this phenomenon is manifested in the average daily heat flow in oceans of the Northern Hemisphere.

You might expect that, as a direct result of this condition, the equatorial zone would grow warmer as the years pass and the polar regions would become progressively cooler. This is not the case; the polar regions always are considerably colder than the equatorial zone, but the temperature difference is not increasing. Clearly, excess heat at equatorial latitudes is being transferred toward the poles by some mechanism. This mechanism involves both the oceans and the atmosphere.

Figure 6–9 shows how heat is transferred in the atmosphere. The greater heating of the atmosphere over the equator causes air to expand, to decrease in density, and therefore to rise. As it rises, it cools by expansion, and the water vapor contained in the rising air mass condenses and falls as rain in the equatorial zone. After losing its moisture, this dry air mass travels some distance north and south of the equator. It then descends in the subtropical regions, around 30° latitude in both Northern and Southern Hemispheres.

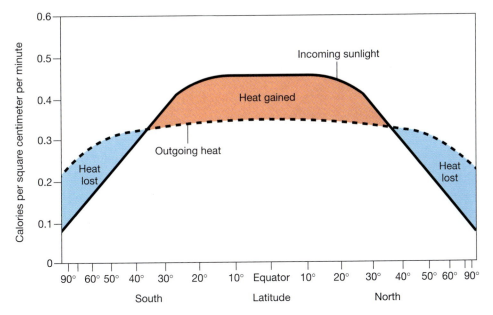

Figure 6–8

Heat Gain and Heat Loss from Oceans. Heat gained by the oceans at equatorial latitudes (red) equals heat lost at polar latitudes (blue). This creates a balanced heat budget for Earth's oceans.

As the descending air approaches Earth's surface, it is warmed by compression. Upon reaching the surface it spreads away from the subtropics, moving both toward the equator and toward higher latitudes. This descending, denser air creates a high-pressure belt in the subtropics, as shown. Conversely, over the equator, the low density of the rising air creates a ring of low pressure that girdles the globe.

The masses of air that move across Earth's surface from the subtropical high-pressure belts toward the equatorial low-pressure belt constitute the **trade winds.** Some of the air that descends in the subtropical regions moves along Earth's surface to higher latitudes as the **westerly wind belts.** These masses rise over the dense, cold air moving away from the polar high-pressure caps at the subpolar low-pressure belts located near 60°N and 60°S latitude. The air moving away from the poles produces the **polar easterly wind belts.** The air that rises at the 60° latitudes cools, releases precipitation in these regions, and ultimately descends in the polar regions or in the subtropics.

These latitudinal patterns exist, but are idealized here to help you understand atmospheric circulation. In reality, the patterns are significantly altered by the uneven distribution of land and ocean over Earth's surface. The general effects of this idealized system are, however, clearly visible on a broad scale. In the following paragraphs, we will examine how these air masses are affected by Earth's rotation as they move across Earth's surface.

Coriolis Effect

Imagine that you have three rockets, each of which will travel upward and then fall back to Earth in exactly one hour. You launch the first one at the North Pole (or

South Pole), straight up. Where does it land an hour later? Earth is rotating beneath it, but you are on the axis of rotation, so it lands right where you launched it, on the pole. No surprises here.

Now from the North Pole, let's launch the second rocket. This time, aim it toward a point along the 30°N latitude line, say the Canary Islands off the western coast of Africa. Where does the rocket land an hour later? Along the 30° latitude line, Earth is rotating eastward at about 1400 kilometers/hour (870 miles/hour). Thus, the rocket will land in the mid-Atlantic, 1400 kilometers west of its target (see Figure 6–10).

Now let's fire the third rocket at the Canary Islands from a point on the equator directly south of the Canary Islands. The point on the equator from which the rocket is fired is moving east at 1600 kilometers (1000 miles) per hour, 200 kilometers (124 miles) per hour faster than are the Canary Islands. This means that the rocket is also moving east 200 kilometers per hour faster than the Canary Islands. Thus, when the rocket returns to Earth one hour later at the latitude of the islands, it will land near the African Coast 200 kilometers east of the Canary Islands.

The point is that any object traveling horizontally above Earth's surface over a significant distance for significant time *will veer to the right in the Northern Hemisphere due to Earth's rotation.* (In the Southern Hemisphere, it will veer to the left.) This phenomenon is called the **Coriolis effect,** named for the French engineer who first calculated it.

Coriolis Effect and the Trade Winds

The Coriolis effect is important to airlines, pilots, anyone who launches rockets, meteorologists, and oceanographers. The reason it is important to weather people

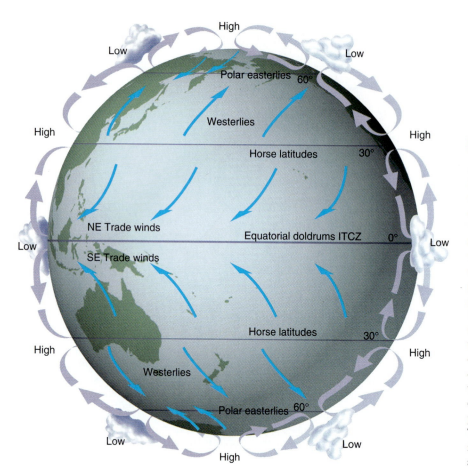

High
Low
Polar easterlies 60°
Westerlies
Horse latitudes 30°
NE Trade winds
Equatorial doldrums ITCZ 0°
SE Trade winds
Horse latitudes 30°
Westerlies
Polar easterlies 60°
Low
High
High
Low
Low
High
High
Low
High
Low

Figure 6–9

Air Mass Circulation around Earth. Intense, year-round heating of the equatorial zone creates rising columns of warm air, developing a "permanent" low-pressure belt along the equator. In the 60° latitude regions, winds riding up over cold polar air rise, creating two more zones of low-pressure air. Descending cool, dry air produces high-pressure belts in the 30° latitude regions (subtropics). Air movements are primarily vertical in these belts. Between the belts, however, strong lateral air movements occur, producing the westerlies and trade winds. Due to cold, dense air masses overlying the polar regions, high-pressure conditions also exist there. The pattern shown here is general, but is extensively modified by the seasons and distribution of the continents.

and to ocean scientists is that the Coriolis effect *helps to steer winds and ocean currents*. In the following paragraphs, we will see how this occurs. We will consider only the Northern Hemisphere, but the same phenomenon occurs in the Southern Hemisphere with the directions reversed.

Recall from Figure 6–9 that air masses are moving away from descending air in the subtropics, both southward toward the equator and northward toward high latitudes. If Earth did not rotate, these would be simple north and south movements. But the "Northeast Trade Winds" do not blow directly from the north but from the northeast, and the westerlies blow from the southwest. This is the result of Coriolis effect.

Figure 6–10A shows that as Earth rotates on its axis, points at different latitudes rotate at different velocities. The velocity depends on the distance from the latitude to the pole (axis of rotation). It ranges from 0 kilometers/hour at the poles to over 1600 kilometers/hour (nearly 1000 miles/hour) at the equator.

The trade winds blow from the northeast because, as the air mass moves southward from the subtropical region near 30°N latitude, it is also moving with the rotating Earth in an easterly direction at about 1400 kilometers/hour (870 miles/hour). The air mass starts moving southward toward a point on the equator at the same longitude. But this point on the equator is moving eastward faster, at 1600 kilometers/hour (about 1000 miles/hour). The distance that the air mass must cover to reach the equator is about 3200 kilometers (2000 miles). If it moves to the south at 32 kilometers/hour (20 miles/hour), it will require 100 hours to arrive at the equator.

The air mass is moving south at 32 kilometers/hour and east at 1400 kilometers/hour. For every hour that the air mass moves in the southerly direction, the point on the equator toward which it started moves farther east by 200 kilometers (124 miles) than does the air mass (1600 kilometers/hour −1400 kilometers/hour = 200 kilometers). So, during the 100 hours it takes for the air mass to make the trip, the point on the equator toward which it started will be 20,000 kilometers east of the air mass when it reaches the equator (calculation: 100 h × 200 kilometers/hour = 20,000 kilometers).

While the air mass was moving the 3200 kilometers in a southerly direction, it *appears* as a result of Earth's rotation also to have moved 20,000 kilometers in a westerly direction. Stated another way, as the air mass covered 30° of latitude from north to south, it appeared to move across 180° of longitude in a westerly direction. Certainly, if we were to encounter this air mass aboard a ship at some point between the equator and 30°N latitude, it would be coming out of the east-northeast.

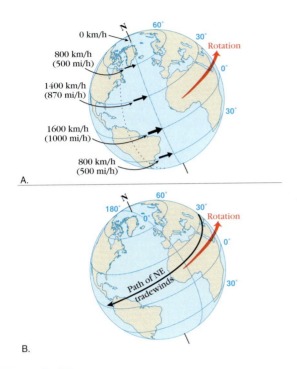

Figure 6-10

Coriolis Effect. *A:* Earth spins at a steady rate, but the actual velocity of any point varies with latitude. Because Earth measures about 40,000 kilometers (about 25,000 miles) around the equator, dividing that distance by 24 (hours in a day) means that Earth's equator is traveling at more than 1600 kilometers/hour. Moving poleward, the velocity diminishes to 0 kilometers/hour at the poles. *B:* To see the Coriolis effect, assume that an air mass is rotating with Earth at 30° latitude, which means it is traveling along with Earth's surface at 1400 kilometers/hour (870 miles/hour). If it heads southward, toward the equator, at 32 kilometers/hour, it will reach the equator 100 hours later. But because the equator is rotating eastward faster than the air mass—some 200 kilometers/hour faster—the air will arrive at the equator at a point that is 20,000 kilometers west of the point on the equator that was directly south of the air mass when it started moving (200 kilometers/hour × 100 h = 20,000 kilometers). This air mass could represent the northeast trade winds, which follow a similar path.

Coriolis Effect and the Westerlies

A similar but different situation prevails for the winds that move northward from 30° latitude toward 60° latitude. These masses also are moving, along with Earth's surface, eastward at 1400 kilometers/hour. But the point due north at 60° latitude, toward which the air mass is headed, is moving in an easterly direction at only 800 kilometers/hour (500 miles/hour).

Unlike the situation for the trade winds, this westerly air mass is moving eastward at a velocity *greater* than that of the point at 60°N latitude toward which it started. For every hour that this mass moves northward,

it also moves 600 kilometers (about 375 miles) farther east than the point at 60° latitude toward which it started. If we were to encounter this air mass at some point between 30° and 60°N latitude, it would appear to come from the west-southwest.

If the air mass moves northward at the same velocity as a trade wind mass moves southward, the amount of deviation to the right will be greater for the westerlies than for the trade winds. This happens because the degree of Coriolis effect depends on the different rate at which points at different latitudes rotate about Earth's axis.

Intensity of the Coriolis Effect

The rotational velocity of these points ranges from 0 kilometers/hour at the poles to more than 1600 kilometers/hour at the equator, but the rate of change with each degree of latitude is not constant. As we approach the pole from the equator, the rate of change of rotational velocity per degree of latitude change increases.

As a comparison, we can note that there is a difference of 200 kilometers/hour in velocity from the equator to 30°N latitude, but there is a difference of 600 kilometers/hour in the velocity of rotation over the same distance, from 30°N to 60°N latitude. If we carry this consideration over the next 30° of latitude, from 60°N latitude to the pole, where the velocity is zero, the difference is over 800 kilometers/hour. This explains why the Coriolis effect increases with latitude, toward the poles.

However, the magnitude of the effect depends mainly on the length of time that an air mass is in motion. Thus, even at low latitudes, a large Coriolis deflection is possible if an air mass is in motion for a long time.

Heat Budget of the World Ocean

As explained, variations in latitude and absorption of solar energy make ocean temperatures vary from place to place and time to time (Figures 6–7 and 6–8). Also as explained, the temperature difference between polar and equatorial regions remains constant due to the transfer of heat energy from the equator to the higher latitudes by moving air and ocean masses (Figure 6–9).

When wind blows over the ocean surface, friction drags the water, creating waves and currents. Consequently, driven by the moving air masses shown in Figure 6–9, large surface current systems exist in the world's oceans. They play an important role in the transfer of heat energy.

Temperatures at various places in the ocean depend upon the rate at which heat flows in and out of these areas. Each ocean location has a heat budget, which describes the heat flow into and out of the area. The heat budget is expressed as follows:

Rate of heat gain – Rate of heat loss = Net rate of heat loss or gain

In greater detail (Q is heat):

$$(Q_{sun} + Q_{current}) - (Q_{radiation} + Q_{evaporation} + Q_{conduction}) = Q_{net}$$

gain   gain/loss   loss   loss   loss   gain/loss

If Q_{net} is positive, it means the temperature of the ocean in that locality is rising. If it is negative, local temperature is falling. If it is zero, the temperature of the mass of water is not changing.

$Q_{current}$ and Q_{net} should be zero if we consider the world ocean over a long period of time. Seasonal increases and decreases in temperature should average, so that Q_{net} becomes zero. Therefore, the rate at which heat is gained in the world ocean through solar radiation should equal the rate at which heat is lost through radiation, evaporation, and conduction into the atmosphere (Figure 6–11).

The heat budget for the entire *world ocean* is expressed with these average values:

$$Q_{sun} = Q_{radiation} + Q_{evaporation} + Q_{conduction}$$

100% = 41% + 53% + 6%

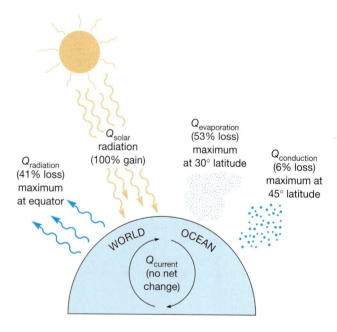

Figure 6–11

Avenues of Heat Flow between the Ocean and the Atmosphere. Solar radiation (Q_{solar}) provides heat to the oceans. It is circulated within the ocean by currents ($Q_{current}$). Heat is lost from the ocean through evaporation ($Q_{evaporation}$), radiation ($Q_{radiation}$), and conduction ($Q_{conduction}$). Note that the heat loss totals 100 percent, equaling the heat gain from the sun.

The amount of heat lost by radiation, primarily of infrared wavelengths, is greatest near the equator, where the greatest amount of solar energy is absorbed. The greatest heat loss by evaporation is in dry subtropics. The heat lost through conduction reaches a maximum along the western margins of ocean basins where warm water currents carry water into regions where the atmosphere is much colder than the water.

The Oceans, Weather, and Climate

The idealized pressure belts and resulting wind systems that we just studied are significantly modified by two factors:

1. Earth's tilt on its axis of rotation, which produces seasons.
2. The air over continents gets colder in winter and warmer in summer than the air over adjacent oceans.

As a result, during winter the continents usually develop atmospheric high-pressure cells due to the weight of cold air centered over them, and during the summer they develop low-pressure cells. Figure 6–12 shows this situation during January.

As air moves away from high-pressure cells and toward low-pressure cells, the Coriolis effect comes into play. It produces a counterclockwise flow of air around low-pressure cells (cyclonic) and a clockwise flow of air around high-pressure cells (anticyclonic), shown in Figure 6–13. Thus, wind patterns associated with continents may reverse themselves seasonally, as winter high-pressure cells are replaced by summer low-pressure cells. (These directions are for the Northern Hemisphere; they are reversed in the Southern Hemisphere.)

At high and low latitudes, day-to-day weather may change little. Polar regions are usually cold and dry regardless of the season. Near the equator, day-to-day weather also is the same year-round. The air is warm, damp, and still as the dominant direction of air movement in the doldrums belt is upward. Midday rains are common.

It is at mid-latitudes that weather becomes interesting. Due to the seasonal change of pressure systems over continents, air masses from the high and low latitudes may move into the mid-latitudes, meet, and produce severe storms. The United States is likely to be invaded by major polar air masses and tropical air masses that have definite areas of origin and distinctive characteristics (Figure 6–14). Note that some air masses originate over land (c = continental) and therefore are dryer, but most originate over the sea (m = maritime) and are moist. Some are colder (P = polar) and some are warm (T = tropical).

As polar and tropical air masses move into the mid-latitudes, they also move gradually in an easterly direction. A **warm front** is the contact between a warm

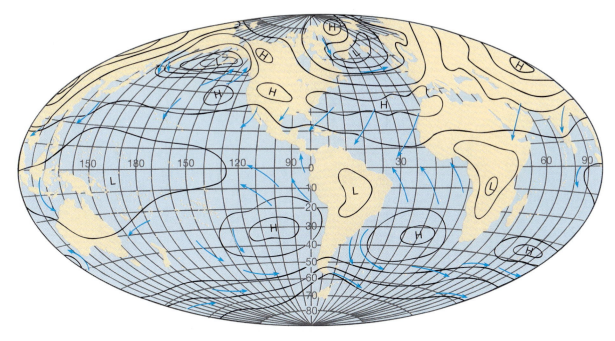

Figure 6–12

January Sea-Level Atmospheric Pressures and Winds. The generalized high-pressure and low-pressure belts shown in Figure 6–9 are modified by the seasons and distribution of continents, as shown here. These pressure patterns are January averages. Because continental rocks have a lower heat capacity than seawater, continents become hotter in the summer and colder in the winter than the adjacent oceans. The cold air over the continents of the Northern Hemisphere produces the high-pressure cells centered over the continent. Conversely, the warm air over continents in the Southern Hemisphere produces low-pressure cells.

air mass moving eastward into an area that is occupied by cold air. A **cold front** is the contact between a cold air mass moving eastward into an area that is occupied by warm air (Figure 6–15).

These confrontations are brought about by the movement of the **jet stream**, a fast-moving, easterly flowing air mass. It exists well above ground level, cen-

tered at about 10 kilometers (6 miles) above the midlatitudes. It usually follows a wavy path and may cause unusual weather by steering a polar air mass far to the south or a tropical air mass far to the north.

Regardless of whether the direction of colliding air masses makes the boundary between them a cold front or a warm front, the less-dense, warmer air always rises

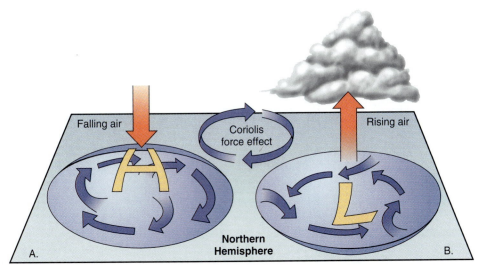

Figure 6–13

High-Pressure and Low-Pressure Cells and Air Flow. As air moves away from high-pressure cells (A) and toward low-pressure cells (B), the Coriolis effect causes the air masses (winds) to veer to the right in the Northern Hemisphere. This results in *clockwise winds around high-pressure cells (anticyclonic)* and *counterclockwise winds around low-pressure cells (cyclonic)*.

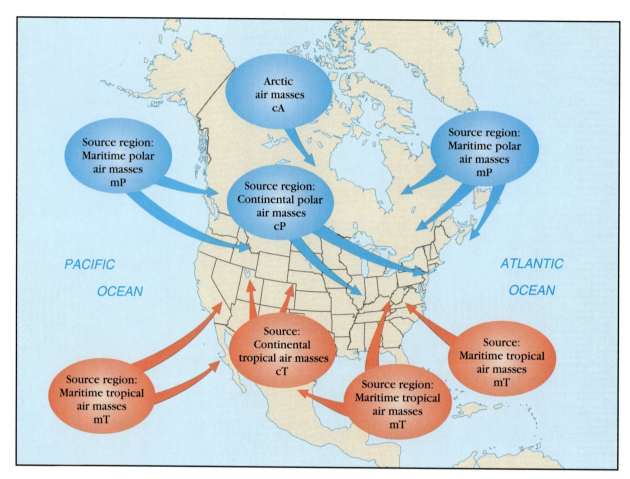

Figure 6–14

Air Masses that Affect U.S. Weather. Polar air masses (blue) are more likely to invade the United States during winter, and the tropical air masses (red) tend to move in from the south during summer. Air masses are classified on the basis of their source region. The designation continental (c) or maritime (m) indicates moisture content, whereas polar (P) and tropical (T) indicate temperature conditions. (Reprinted by permission from Lutgens, F. K., and Tarbuck, E. J., *The Atmosphere,* 6th ed., New York: Englewood Cliffs, NJ: Prentice-Hall, Inc., 1995, fig. 8.2.)

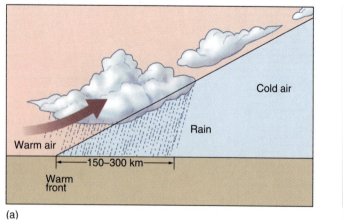

(a)

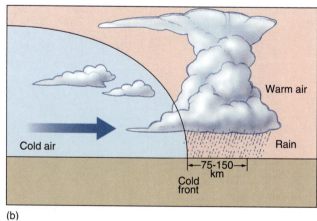

(b)

Figure 6–15

Warm and Cold Fronts. Cross sections through a warm front (A) and cold front (B).

above the denser cold air, where it cools, and the water vapor in it condenses as precipitation. A cold front is usually steeper, and the temperature differences across it is greater. Therefore, rainfall along a cold front usually is heavier and briefer than rainfall along a warm front.

Climate Patterns in the Oceans

We think of *climate* as applying to land areas where people live, but the term applies to regions of the ocean as well. We divide the open ocean into climatic regions that run generally east-west. They have relatively stable boundaries (Figure 6–16).

Figure 6–17 is a satellite view of water vapor over land and sea. Its patterns are revealing, for areas of greater water vapor indicate higher temperatures. In the *equatorial* region, the major air movement is vertical as the heated air rises, so surface winds are weak. Surface waters are warm, and the air is saturated with water vapor; this is quite noticeable in the figure. Heavy precipitation keeps salinity relatively low. Sailors once referred to this region as the *doldrums* because their sailing ships were becalmed by the lack of winds. Meteorologists refer to it as the *intertropical convergence zone (ITCZ)* because it is the region between the tropics where the trade winds converge (Figure 6–9).

Figure 6–18 is a satellite view of wind speeds. *Tropical* regions, those near the Tropic of Cancer or Tropic of Capricorn (Figure 6–16), are characterized by strong northeasterly trade winds in the Northern Hemisphere and southeasterly trade winds in the Southern Hemisphere (Figure 6–18). These winds push the equatorial currents and create moderately rough seas. Relatively little precipitation falls at higher latitudes within tropical regions, but precipitation increases toward the equator. Hurricanes and typhoons initiate in the doldrums and pass through the tropics as tropical storms. They carry large quantities of heat into higher latitudes.

The belts of high pressure previously described are centered in the *subtropical* regions (Figure 6–16). The dry air descending on the subtropics results in little precipitation and a high rate of evaporation, producing the highest surface salinities in the open ocean (Figure 6–19). Winds are weak in the open ocean, as are currents. How-

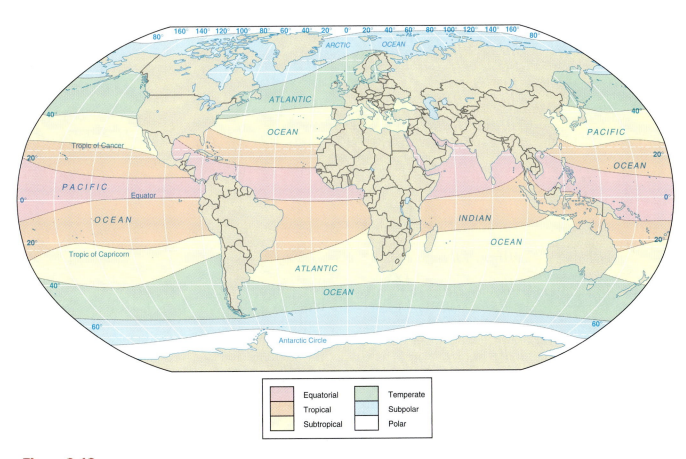

Figure 6–16

Climatic Patterns of the Open Ocean. The open ocean can be divided into climatic regions with relatively stable boundaries that run generally east-west.

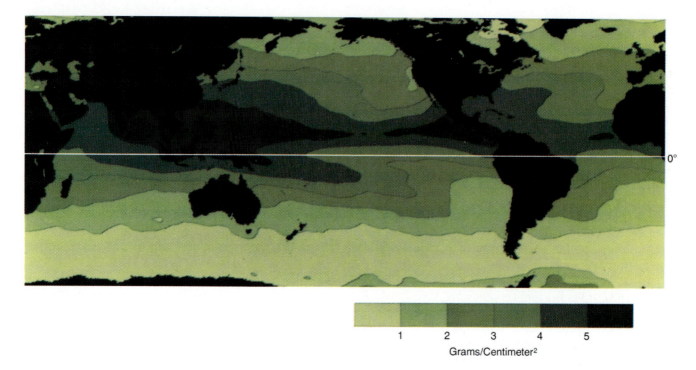

1 2 3 4 5
Grams/Centimeter²

Figure 6–17

***SEASAT* SMMR Water Vapor, July 7–October 10, 1978.** This image shows how much water vapor was in the air over land and sea during the few days when these data were recorded by a satellite's water vapor sensor. The atmosphere's ability to hold water vapor increases with temperature, so the greatest atmospheric water vapor occurs near the hot equator. This image reflects the great amount of heat energy being removed from the ocean by evaporation in the low latitudes. This belt of maximum atmospheric water vapor content is called the *doldrums*. It is a region where warm air slowly rises and minimal horizontal air movement (wind) exists to propel sailing vessels or to cool their crews. These data were obtained in late summer-early fall in the Northern Hemisphere, so the doldrums are shifted significantly northward of the equator. The western tropical Pacific Ocean contains the greatest amount of atmospheric water vapor. This produces a super-greenhouse effect that heats the underlying surface waters to the highest temperatures found in the world ocean (see Figure 7–4).

ever, strong boundary currents (along the boundaries of continents) flow north and south, particularly along the western margins of the subtropical oceans.

The *temperate* regions are characterized by strong westerly winds blowing from the southwest in the Northern Hemisphere and from the northwest in the Southern Hemisphere (Figures 6–16 and 6–18). Severe storms are common, especially during winter, and precipitation is heavy. In fact, the North Atlantic is noted for fierce storms in this zone, storms that have claimed many ships and lives over the centuries.

The *subpolar* ocean is covered in winter by sea ice (Figure 6–16). It melts away, for the most part, in summer. Icebergs are common, and the surface temperature seldom exceeds 5°C (41°F) in the summer months.

Surface temperatures remain at or near freezing in the *polar* areas, which are covered with ice throughout most of the year (Figure 6–16). In these areas, which in-

clude the Arctic Ocean and the ocean adjacent to Antarctica, there is no sunlight during the winter and no night during the summer.

Fog

There are two major sources of **fog** over ocean waters. The first forms over land on clear nights, which allow extensive heat loss by radiation. The fog usually forms in winter over marshy areas where the air has a high relative humidity during the day. The fog that forms at night may be carried by light winds out over coastal waters.

The second variety forms over water when warm, moist air moves from a region with warmer surface-water temperatures to an area having colder water temperatures. Such fog usually occurs in spring or summer when the air temperature may be significantly warmer than the

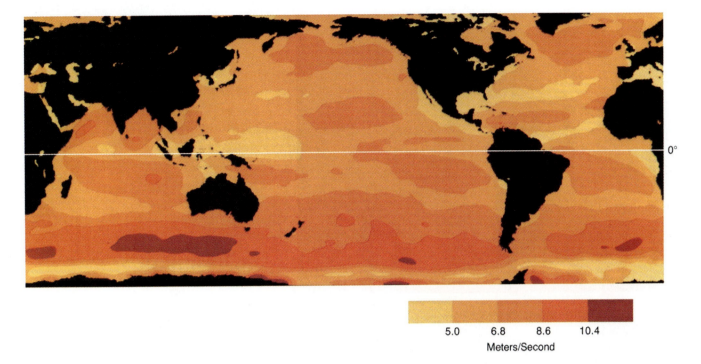

5.0   6.8   8.6   10.4
Meters/Second

Figure 6–18

***SEASAT* Altimeter Wind Speed, July 7–October 10, 1978.** Two of Earth's prevailing wind systems are the trade winds (which blow from 30° latitudes toward the equator) and the westerlies (which blow from 30° toward 60° latitudes). Because there are fewer continents in the Southern Hemisphere, the westerly wind belt there contains the strongest year-round winds. The lowest wind speeds occur at the doldrums, where the air is rising, and the horse latitudes (30°N and S latitudes), where air descends. Wind speeds gradually increase away from these belts of minimum wind speed.

water temperature. An example is the fog that forms when warm, moist air moves from over the Gulf Stream onto the Grand Banks off Newfoundland (Figure 6–20A).

Less common is **sea smoke,** or steam fog. It forms when cold air moves over ocean water that is about 10C° (18F°) warmer than the air. The lower layers of this cold air are warmed by water evaporated from the ocean surface and rise into the overlying cold air. Condensation occurs in a pattern that looks like smoke rising from the ocean surface (Figure 6–20B).

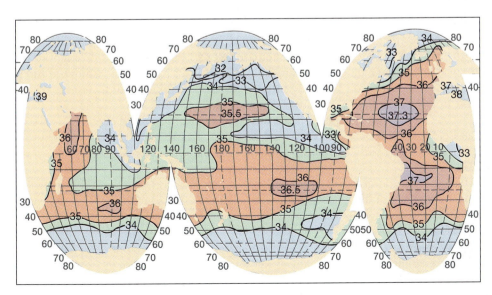

Figure 6–19

Surface Salinity of the Oceans in August (‰). The dry air descending on the subtropics results in little precipitation and a high rate of evaporation, producing the highest surface salinities in the open ocean. (After Sverdrup et al., 1942.)

A.

B.

Figure 6–20

Sea Fog and Sea Smoke. *A:* Water vapor and air are cooled by cold surface water and condense to produce sea fog. *B:* Arctic sea smoke, formed when water evaporated from the warm ocean surface rises into overlying cold air. (Photo *A* by Martin Bond, Science Photo Library; photo *B* by Howard Bluestein, Science Source.)

Sea Ice

Polar cold develops a permanent to nearly permanent ice cover on the sea surface. The term **sea ice** is used to distinguish such masses of frozen seawater ice from icebergs. **Icebergs** also are found at sea, but they are ice masses that originate by breaking off from glaciers that form on continental land.

The freezing point of pure water is 0°C (32°F). But the freezing point of seawater with a salinity of 35 parts per thousand (‰) is nearly 2 degrees lower, −1.91°C (28.6°F). As water freezes at the ocean surface, the dissolved solids cannot fit into the crystalline structure of the ice. Thus, the salts are largely left behind, to be dissolved in surrounding seawater, so that the salinity and density of the surrounding water increases. This greater salinity tends to lower the freezing point of the remaining water.

However, the low-temperature, high-density water that is excluded from the ice tends to sink and be re-

placed by warmer, less-dense water at the surface. This establishes a circulation that enhances the formation of sea ice, because freezing is aided by low salinity and calm water conditions. Sea ice is found throughout the year around the margin of Antarctica, within the Arctic Sea, and in the extreme high-latitude region of the North Atlantic Ocean.

Sea ice begins to form as small needlelike crystals of hexagonal shape (six-sided). They eventually become so numerous that a slush develops. As the slush begins to form into a thin sheet, it is broken by wind stress and wave action into disc-shaped pieces called **pancake ice** (Figure 6–21A). As further freezing occurs, the pancakes coalesce to form **ice floes.**

The rate at which sea ice forms is closely tied to temperature conditions. Large quantities of ice form in relatively short periods during which the temperature is very low (for example, −30°C or −22°F). Even at low temperatures, the rate of ice formation slows as ice thick-

A.

B.

C.

Figure 6–21

Sea Ice. *A:* Pancake ice, freezing slush that is broken by wind stress and wave action into disc-shaped pieces. *B:* Strong offshore winds blow across the Antarctic ice shelf near the Antarctic Peninsula. Ribbons of sea ice remain seaward of the shelf ice as the spring season begins (September in Antarctica). *C:* Rafted ice. As ice floes expand, they slide over and beneath one another to become "rafted." (Photo *A* by Stephen J. Krasemann; photo *B* courtesy of NASA.)

ness increases due to the poor heat conduction of ice, which is a good insulator. Although newly formed sea ice contains little salt, it does trap a significant quantity of brine (salty water) during the freezing process.

Depending upon the rate of freezing, newly formed ice may have a total salinity from 4 to 15‰. The more rapidly it forms, the more brine will be captured and the higher will be the salinity. Over time, the brine will trickle down through the coarse structure of the sea ice, and its salinity will decrease. By the time it is a year old, sea ice normally has become relatively pure.

Pack Ice, Polar Ice, and Fast Ice

In the Arctic Sea, ice that forms at the sea surface can be classified into one of three categories—*pack ice, polar ice,* or *fast ice.*

Pack ice forms around the margin of the Arctic Sea (called the Arctic Ocean in some atlases). The ice extends through the Bering Strait into the Bering Sea on the Pacific side, and as far south as Newfoundland and Nova Scotia in the North Atlantic. Pack ice reaches its maximum extent during May and breaks up to cover its least area in September (Figure 6–22). This ice can be

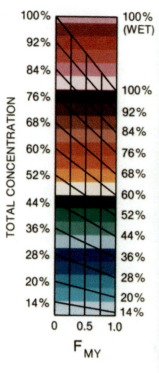

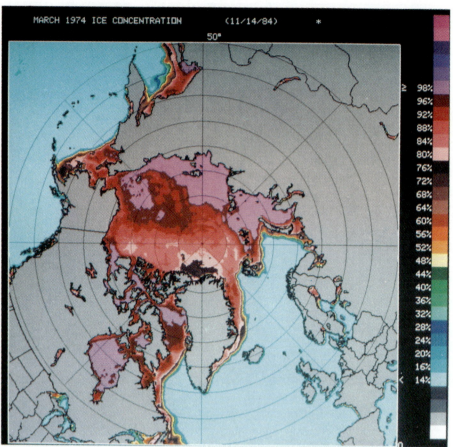

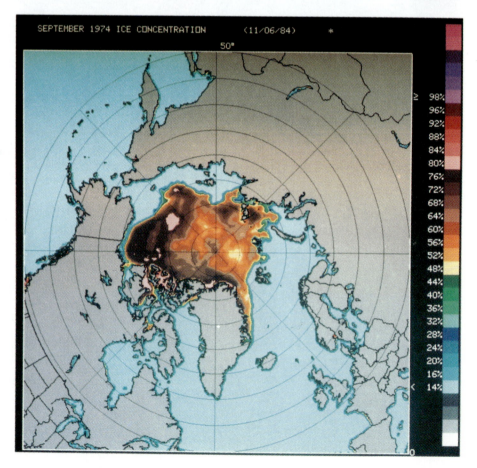

Figure 6–22

**Extent of Ice in Arctic
Sea.** These 1974 images show
the percent of sea ice concentration in March (top) and September (bottom). They are
accurate to within 15 percent
for first-year ice (pack ice) and
25 percent for multiyear ice
(polar ice cap). The scales of
ice concentration along the
right margins are for first-year
ice. The other scale can be used
for first-year ice (left side of
scale) and multiyear-ice (right
side of scale). The images were
developed from the ESMR
(Electrically Scanning Microwave Radiometer) aboard
the *Nimbus 5* satellite. (Courtesy
of NOAA.)

penetrated by ships called *icebreakers,* which have re-inforced hulls. The ice reaches a maximum thickness of about 2 meters (6.5 feet) in winter.

Pack ice is driven primarily by wind, although it also responds to surface currents, which produce stresses that continually break and reform its structure. Pack ice formation is achieved by the expansion of floes that begin to raft onto one another as they expand to cover the sea's surface (Figure 6–21C).

The *polar ice* that covers the greatest portion of the Arctic Sea, including the polar region, attains a maximum thickness in excess of 50 meters (over 160 feet). During summer, melting may produce enclosed water bodies called *polynyas,* although the polar ice never totally disappears. Its average thickness during summer is over 2 meters (6.5 feet).

Polar ice is constantly being exchanged, because floes from the adjoining pack ice are carried into the polar region during winter, and floes break out of the polar ice and reenter the pack ice during summer. Circling in a clockwise direction around the Arctic Ocean, about one-third of the pack and polar ice is carried into the North Atlantic by the East Greenland Current each year.

Developing in the winter from the shore out to the pack ice is the **fast ice,** which melts completely during the summer. The fast ice, so called because it is "held fast" or firmly attached to the shore, attains a winter thickness exceeding 2 meters (6.5 feet).

In the Southern Hemisphere, where the polar region is not an ocean but a continental mass, we might characterize all the sea ice that forms around the Antarctic continental margin as pack ice and fast ice that have a temporary existence. The pack ice rarely extends north of 55°S latitude and breaks up completely from October to January except in the very quiet bays. Winds that are frequently very strong help prevent the formation of a greater pack ice accumulation around the Antarctic continent.

Icebergs

As noted, icebergs are distinct from sea ice because they come from glaciers. In the Antarctic, the source glaciers cover the Antarctic continent. In the Arctic, they cover most of the island of Greenland and part of Ellesmere Island (Figure 6–23D). These vast ice sheets grow from snow accumulation on the landmass and spread outward until their edges push into the sea. There the ice sheets, which are less dense than seawater, are buoyed by the water and break up under the stress of current, wind, and wave action. This breakup that produces icebergs is called *calving.*

In the Arctic, icebergs originate primarily from the ice that follows narrow valleys into the sea along the

western coast of Greenland (Figure 6–23D). Icebergs also are produced along the eastern coasts of Greenland and Ellesmere Island. The East Greenland Current and the West Greenland Current (arrows on map) carry the icebergs up to 20 kilometers (12.4 miles) per day into the North Atlantic, where the Labrador Current may move them into North Atlantic shipping channels. They seldom are carried south of 45°N latitude, but during some seasons icebergs will move as far south as 40°N latitude and become shipping hazards.

Such an accumulation of icebergs had developed where the *Titanic* sank in April 1912. After receiving repeated warning of the iceberg hazard, the supposedly unsinkable 46,000-ton luxury ship proceeded with 2224 passengers at an excessive speed of 41 kilometers/hour (25.5 miles/hour) until it came to a grinding halt after hitting an iceberg with its starboard (right) bow. Within hours, the *Titanic* sank at 41°46′N latitude, 50°14′W longitude near the Grand Banks, claiming 1517 lives.

The tragedy brought about the formation of the ice patrol that since has prevented further loss of life from such accidents. The U.S. Navy began this patrol immediately following the *Titanic* disaster; it became international in 1914. The patrol, maintained today by the U.S. Coast Guard, is concentrated between 40°30′ and 48°N latitude and 43° and 54°W longitude.

The most recent major iceberg incident occurred in 1989. The Soviet cruise liner *Maxim Gorky* rammed an iceberg well north of the Arctic Circle between Greenland and Spitzbergen Island (north of the area shown in the map). Quick action by the crew and the Norwegian Coast Guard prevented any loss of life among the more than 1300 people on board.

In Antarctica, the edges of glaciers form **shelf ice.** They push into the marginal seas (seas around the margin of the continent), producing icebergs that have received less attention than those of the North Atlantic because they interfere less with shipping. Vast tabular bergs break from the edges of the shelf ice, with lengths exceeding 100 kilometers (60 miles) (Figure 6–23C). They may stand up to 200 meters (650 feet) above the ocean surface, although most probably rise less than 100 meters (330 feet) above sea level. In August 1991 an iceberg the size of Connecticut (13,000 square kilometers or 5000 square miles) broke loose from the ice shelf of the Weddell Sea.

Most calving occurs, as in the Arctic, during the summer months. When the sea ice breaks up, ocean swells driven by strong winds can reach the edge of the shelf ice. Large icebergs are calved and move to the north into the warmer, rougher water, in which they disintegrate. Carried by the strong West Wind Drift around Antarctica, these icebergs move easterly around the continent, rarely migrating farther north than 40°S latitude.

A.

B.

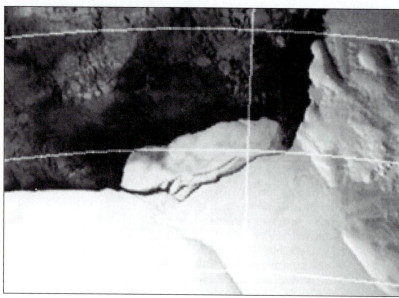

C.

Figure 6–23

Icebergs. *A:* Small North Atlantic iceberg. *B:* Part of a large tabular Antarctic iceberg with sea ice in the foreground. *C:* B-1 Iceberg. This infrared image shows the huge iceberg that broke from the Ross Ice Shelf in Antarctica in October 1987. The longitude and latitude lines intersecting near the east end of B-1 are 160°W and 78°S. This monster is 136 kilometers (84 miles) long, but two larger icebergs formed the year before in the Weddell Sea. *D:* Map showing North Atlantic currents and typical iceberg distribution (Δ). (Photo *A* by Rapho; photo *B* by Joyce Photographs; infrared image *C* courtesy of Robert Whritner, Manager, Antarctic Research Center, Scripps Institution of Oceanography, University of California, San Diego.)

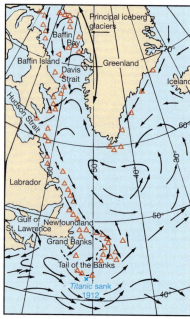

D.

Renewable Sources of Energy

As explained, the sun imparts great radiant energy to the sea and the atmosphere. Can this energy be harnessed to perform work? The potential for extracting energy from the movement and heat distribution patterns of the atmosphere and ocean is attractive for several reasons:

1. Work can be achieved without significant pollution of air or water.
2. The amount of energy available at any time is far greater than that in fossil fuels (coal, oil, natural gas) and nuclear fuel (uranium).
3. The energy is renewable as long as the sun continues to radiate energy to Earth, and will not be depleted.

In order of decreasing energy potential, the sources of renewable energy are (1) heat stored in the oceans, (2) kinetic energy of the winds, (3) potential and kinetic energy of waves, and (4) potential and kinetic energy of tides and currents. In later chapters we will consider the use and potential of winds, waves, tides, and currents. Here we will consider only the renewable source with the greatest store of potential energy: the warm surface layers of the tropical oceans.

Of Earth's surface between the Tropic of Cancer and Tropic of Capricorn, 90 percent is ocean. What makes this warm tropical surface water such an important source of energy is the presence of much colder water beneath the thermocline. With a temperature difference as small as 17°C (30.6°F), useful work can be done by **ocean thermal energy conversion (OTEC)** systems (Figure 6–24A).

In an OTEC unit, warm surface water heats a fluid, such as liquified propane gas or ammonia, which is un-

A.

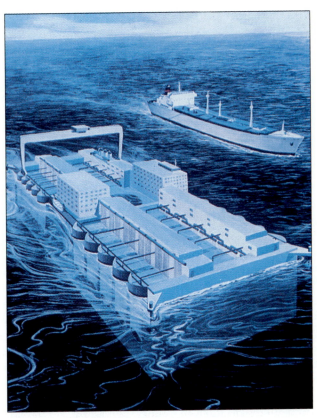

B.

Figure 6–24

Proposed Uses of Ocean Thermal Power. *A:* Ocean thermal energy conversion (OTEC) system with crew quarters and maintenance facilities. Attached around the outside are turbine-generators and pumps. It is over 75 meters (about 250 feet) in diameter and 485 meters (about 1600 feet) long, and it weighs about 300,000 tons. This unit is designed to generate 160 million watts of electrical power, enough to support a city of 100,000. *B:* OTEC plant for ammonia production. Moving slowly through tropical waters, this plant could produce 1.4 percent of the U.S. annual ammonia requirement and save 22.6 billion cubic feet of natural gas. (Photo *A* courtesy of Lockheed Missiles and Space Co., Inc.; photo *B* courtesy of U.S. Department of Energy.)

der pressure in evaporating tubes. Heating vaporizes the fluid, and the vapor pushes against the blades of a turbine, which turns an electrical generator. After passing through the turbine, the fluid is condensed by cold water that has been pumped up from the deep ocean. It is again ready for heating by warm surface water that will cause it to vaporize and pass through the turbine. Thus, heat energy stored in the ocean is converted to useful electrical energy. The system works in the opposite way from a typical refrigeration system and on a much larger scale.

Along the continental U.S. coast, the only region with good potential for ocean thermal energy conversion is a strip about 30 kilometers (19 miles) wide and 1000 kilometers (600 miles) long. It extends northward from southern Florida to St. Augustine, along the western margin of the Gulf Stream.

Testing is presently underway on the Seacoast Test Facility, a shore-based OTEC installation on the island of Hawaii. Although OTEC requires a large initial investment, the Hawaii test has demonstrated that OTEC power generation can be commercially successful.

Another proposed use of OTEC is for producing ammonia. A floating factory that could produce 586,000 tons of ammonia per year has been described in a feasibility report from the Johns Hopkins Applied Physics Laboratory to the U.S. Maritime Administration. The proposed demonstration ship would weight 68,000 tons and have a width of almost 60 meters (200 feet) and a length of over 144 meters (470 feet). It would move at less than 2 kilometers/hour through tropical waters and use energy derived from the temperature difference between the warm surface waters and cold deep waters to produce ammonia. Each such plant could produce about 1.4 percent of our nation's ammonia requirements—75 percent of which is used for fertilizer—and save 22.6 billion cubic feet of natural gas each year (Figure 6–24B).

SUMMARY

Seawater density increases with increased salinity and decreased temperature. Density change per unit of temperature change is greater in warmer water. Zones of rapid change in the density, temperature, or salinity in the water column are called a pycnocline, thermocline, and halocline, respectively.

Radiant energy reaching Earth from the sun is mostly in the ultraviolet and visible light range, whereas that radiated back to space from Earth is primarily in the infrared (heat) part of the spectrum. Water vapor, carbon dioxide, and other trace gases absorb the infrared radiation and heat the atmosphere. This phenomenon is called the greenhouse effect. Because human activities are increasing the concentrations of trace gases that enhance the greenhouse effect, there is concern that Earth's life zone may be warming.

The atmosphere is unevenly heated and set in motion because more energy is received than is radiated back into space at low latitudes, and because water vapor, which absorbs infrared radiation well, is unevenly distributed in the atmosphere. Belts of low pressure, where air rises, are generally found at the equator and at about 60° latitude. High-pressure regions, where dense air descends, are located at the poles and at about 30° latitude. The air at Earth's surface that is moving away from the subtropical highs produces trade winds moving toward the equator and westerlies moving toward higher latitudes.

Because Earth's surface rotates at different velocities at different latitudes, increasing from 0 kilometers/hour at the poles to over 1600 kilometers/hour (1000 miles/hour)

at the equator, objects in motion tend to veer to the right in the Northern Hemisphere and to the left in the Southern Hemisphere. This is called the Coriolis effect.

In the heat budget of the world ocean, the only heat source is the sun, whereas heat loss occurs through radiation, evaporation, and conduction into the atmosphere. The amount of heat lost by radiation, primarily of long infrared wavelengths, is greatest near the equator, where the greatest amount of solar energy is absorbed. The greatest heat loss by evaporation is in the dry subtropics. Heat lost through conduction peaks along the western margins of oceans in temperate latitudes where warm water currents carry water into regions where the atmosphere is much colder than the water.

Earth's tilt on its axis of rotation and the distribution of continents both modify the idealized pressure belts. High-pressure cells form over continents in winter and are replaced by low-pressure cells in summer. There is a counterclockwise cyclonic movement of air around the low-pressure cells and a clockwise anticyclonic movement around high-pressure cells. At mid-latitudes, cold air masses from higher latitudes meet warm air masses from lower latitudes and create cold and warm fronts that move from west to east across Earth's surface. Despite the modification of the idealized pressure belts, ocean climate patterns are closely related to them.

Sea ice forms when seawater is frozen in high latitudes. The process usually forms a slush, which breaks into pancakes that ultimately grow into floes. Sea ice develops as pack ice that forms each winter and melts almost entirely each summer, polar ice that is a permanent accumulation in polar regions of the Arctic Ocean, and fast ice that forms frozen to shore during the winter. Icebergs form when large chunks of ice break off the large continental glaciers that form on Ellesmere Island, Greenland, and Antarctica.

From the motions and patterns of heat distribution in the atmosphere and oceans, sources of energy can be exploited that are renewable and nonpolluting. These include heat stored in the ocean and potential and kinetic energy stored in the winds, waves, tides, and currents. Ocean thermal energy conversion (OTEC) is a process developed to use the difference in temperature between warm tropical surface water and the cold water below the thermocline to produce electricity and ammonia.

KEY TERMS

Antarctic Circle (p. 135)
Arctic Circle (p. 135)
Cold front (p. 140)
Coriolis effect (p. 137)
Deep water (p. 131)
Fast ice (p. 149)
Fog (p. 144)
Greenhouse effect (p. 132)
Halocline (p. 131)
Heat budget (p. 133)

Ice floe (p. 146)
Iceberg (p. 146)
Jet stream (p. 140)
Ocean thermal energy conversion (OTEC) (p. 151)
Pack ice (p. 147)
Pancake ice (p. 146)
Polar easterly wind belts (p. 137)
Pycnocline (p. 131)
Sea ice (p. 146)

Sea smoke (p. 145)
Shelf ice (p. 149)
Thermocline (p. 131)
Trade winds (p. 137)
Tropic of Cancer (p. 136)
Tropic of Capricorn (p. 136)
Upper water (p. 131)
Warm front (p. 140)
Westerly wind belts (p. 137)

QUESTIONS AND EXERCISES

1. A 5C° temperature change in ocean water from 0°C to 5°C does not cause the same amount of density change as a 5C° temperature change from 20°C to 25°C. See Figure 6–1 to describe the difference in density change.

2. The position of the pycnocline in the water column is determined by the combined effects of the thermocline and halocline. Describe these relationships in tropical waters (see Figures 6–2 and 6–3).

3. Describe the fundamental difference between solar radiation absorbed at Earth's surface and the back-radiation that is primarily responsible for heating Earth's atmosphere.

4. Discuss the greenhouse gases in terms of their relative concentrations and relative contributions to any increased greenhouse effect.

5. Describe the effect on Earth as a result of Earth's axis of rotation being angled 23.5° to the ecliptic. What would happen if Earth were not tilted on its axis?

6. Since there is a net annual heat loss at high latitudes and a net annual heat gain at low latitudes, why does the temperature difference between these regions not increase?

7. Why are there high-pressure caps at each pole and a low-pressure belt in the equatorial region?

8. Describe the Coriolis effect in the Northern and Southern Hemispheres and include a discussion of why the effect increases with increased latitude.

9. In considering the heat budget of the world over a long period, why should $Q_{current}$ (rate of heat gain or loss through ocean currents) and Q_{net} (net rate of heat gain or loss) be considered to be zero?

10. Describe the ocean regions where the maximum rates of heat loss by radiation, evaporation, and conduction occur, and explain why.

11. Discuss why the idealized belts of high and low atmospheric pressure shown in Figure 6–9 are modified (see Figure 6–12).

12. Name the polar and tropical air masses that affect U.S. weather. Describe the pattern of movement across the continent and patterns of precipitation associated with warm and cold fronts.

13. How are the ocean's climate belts (Figure 6–16) related to the broad patterns of air circulation described in Figure 6–9?

14. Under what conditions do fog and sea smoke develop?

15. Describe the formation of sea ice from the initial freezing of the ocean surface water through the development of polar ice in the Arctic Ocean.

16. What is the difference in the average size and shape of icebergs in the Arctic and Antarctic? Why do these differences exist?

17. Construct your own diagram of how an ocean thermal energy conversion unit might generate electricity, or make a flow diagram presenting the steps of the process.

REFERENCES

Changing climate and the oceans. 1987. *Oceanus* 29:4, 1–93.

Charlson, R.J.; Schwartz, S.E.; Hales, J.M.; Cess, R.D.; Coakley, J.A., Jr.; Hansen, J.E.; and Hofmann, D.J. 1992. Climate forcing by anthropogenic aerosols. *Science* 255:5043, 423–430.

Ocean energy. 1979. *Oceanus* 22:4, 1–68.

Oceans and climate. 1978. *Oceanus* 21:4, 1–70.

The oceans and global warming. 1989. *Oceanus* 32;2, 1–75.

Pickard, G.L. 1975. *Descriptive physical oceanography.* 2nd ed. New York: Pergamon Press.

Rodhe, H. 1990. A comparison of the contribution of various gases to the greenhouse effect. *Science* 248:4960, 1217–1219.

The enigma of weather—A collection of works exploring the dynamics of meteorological phenomena. 1994. *Scientific American.*

SUGGESTED READING

Sea Frontiers

Boling, G.R. 1971. Ice and the breakers. 17:6, 363–371. An interesting history of people in icy waters and the development and improvement of icebreakers.

Charlier, R. 1981. Ocean-fired power plants. 27:1, 36–43. The potential of ocean thermal energy conversion is discussed.

Houghton, R. A., and Woodwell, G. M. 1989. Global climate change. 260:1, 36–47. The history of global climate change.

Land, T. 1976. Europe to harness the power of the sea. 22:6, 346–349. An article emphasizing the clean, safe, permanent nature of ocean tides and waves as a source of power.

Mayor, A. 1988. Marine mirages. 34:1, 8–15. The nature of marine mirages and the research efforts that lead to our understanding them are considered.

Rush, B., and Lebelson, H. 1984. Hurricane! The enigma of a meteorological monster. 30:4, 233–239. An overview of the nature of hurricanes and the problem of predicting where they will go.

Scheina, R. L. 1987. The Titanic's legacy to safety. 33:3, 200–209. A brief summary of the sinking of the *Titanic* and an overview of the ice patrol and iceberg collision history subsequent to the sinking of the "unsinkable" luxury liner.

Smith, F. G. W. 1974. Planet's powerhouse. 20:4, 195–203. A description of how Earth's "heat engine" works, with an emphasis on the nature of tropical cyclonic storms.

———. 1974. Power from the oceans. 20:2, 87–99. A survey of the many tried and untried proposals for extracting energy from the oceans.

Sobey, E. 1979. Ocean ice. 25:2, 66–73. The formation of sea ice and icebergs as well as their climatic and economic effects are discussed.

———. 1980. The ocean-climate connection. 26:1, 25–30. Our increasing knowledge of the effect of the oceans on Earth's climate may be used to predict climatic trends of the future.

Scientific American

Gregg, M. 1973. Microstructure of the ocean. 228:2, 64–77. A discussion of the methods of studying the detailed movements of ocean water by observing temperature and salinity changes over distances of one centimeter and the motions they reveal.

Jones, P. D., and Wigley, T. M. 1990. Global warming trends. 263:2, 84–91. Data relating to evidence of global warming over the past 100 years are presented.

MacIntyre, F. 1974. The top millimeter of the ocean. 230:5, 62–77. Processes that are confined to a thin film at the surface of ocean water and their role in the overall nature of the oceans are discussed.

Penney, T. R., and Bharathan, D. 1987. Power from the sea. 256:1, 86–93. A prediction that generating electricity by ocean thermal energy conversion will be competitive with fossil fuel plants as the price of oil rises.

Pollack, H. N., and Chapman, D. S. 1993. Underground records of changing climate. 268:6, 44–53. Direct measurement of temperature shows that Earth has warmed over the past 150 years. Data from boreholes about ancient temperatures may give us a more complete picture of the history of global climate.

Revelle, R. 1982. Carbon dioxide and world climate. 247:2, 35–43. Some of the possible effects of increasing atmospheric temperature due to carbon dioxide accumulation are considered.

Stanley, S. M. 1984. Mass extinctions in the oceans. 250:6, 64–83. Geological evidence suggests that most major periods of species extinction over the last 700 million years occurred during the brief intervals of ocean cooling.

Stolarski, R. S. 1988. The Antarctic ozone holes. 258:1, 30–37. The discovery of the Antarctic ozone hole and its possible significance are discussed.

White, R. M. 1990. The great climate debate. 263:1, 36–45. The controversy over the degree of global warming we can expect in our future is discussed.

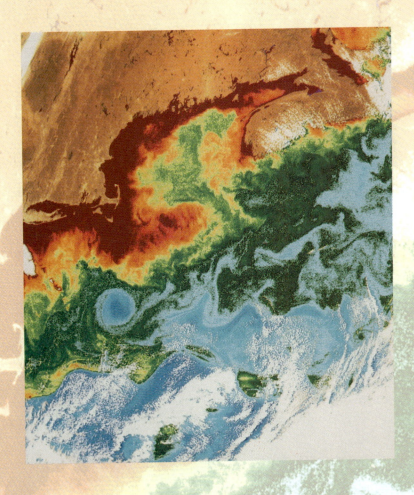

C H A P T E R 7
Ocean Circulation

The surfaces of the major oceans are dominated by huge current systems. They rotate clockwise in the Northern Hemisphere and counterclockwise in the Southern Hemisphere. These ocean currents are vast masses of water in motion. Ultimately, they are driven by radiant energy from the sun.

These currents are an important part of the heat engine that produces Earth's overall climatic patterns. More locally, they also affect the climates of coastal continental regions.

Cold currents flowing toward the equator on the western sides of continents produce arid conditions, whereas warm currents flowing poleward on the eastern sides of continents produce warm, humid conditions. It is ocean currents that make northern Europe and Iceland habitable, and make us shiver to think of living at similar latitudes along the Atlantic coast of North America (such as Labrador).

The effect of ocean currents on life is profound. Life in the deep sea is made possible only by a continuing supply of oxygen. This oxygen is carried there by cold, dense water currents that sink in polar regions and spread across the deep-ocean floor. Ocean currents greatly influence biological production (growth of algae, the basis for food chains) throughout the oceans. Ocean currents also powerfully affect the distribution of humans on land. They may even have enabled the travel of prehistoric peoples from Europe and Africa to the New World and on to the Pacific Ocean islands.

Ocean currents can be categorized as being either *wind driven* or *thermohaline*. Wind-driven currents are set in motion by moving air masses. This motion is primarily horizontal and occurs primarily in the upper waters of the world ocean. Thermohaline (temperature-salinity) circulation has a significant vertical component and accounts for the thorough mixing of the deep masses of ocean water. Thermohaline circulation is initiated at the ocean surface by temperature and salinity conditions that produce a high-density mass, which sinks and spreads slowly beneath the surface waters.

Even though we think of the ocean as having great depth, ocean basins are very shallow compared with their widths. If the ocean basins were scaled down to be the width of this page, they would be no deeper than this page is thick! Yet within this thin shell of water there is a rich dynamic structure.

Horizontal Circulation

Horizontal circulation in surface waters develops from friction between the ocean and the wind. On a tiny scale, you can simulate this simply by blowing gently and steadily across a vessel of water, not hard enough to cause waves, but enough to start a water current flowing.

Equatorial Currents, Boundary Currents, and Gyres

Trade winds blowing from the southeast in the Southern Hemisphere and from the northeast in the Northern Hemisphere are the principal drivers of the system of ocean surface currents. The trade winds thus set in motion the water masses between the tropics. They develop the **equatorial currents** that travel along the equator worldwide (Figure 7–1). They are called north and south equatorial currents, depending on their position north or south of the equator.

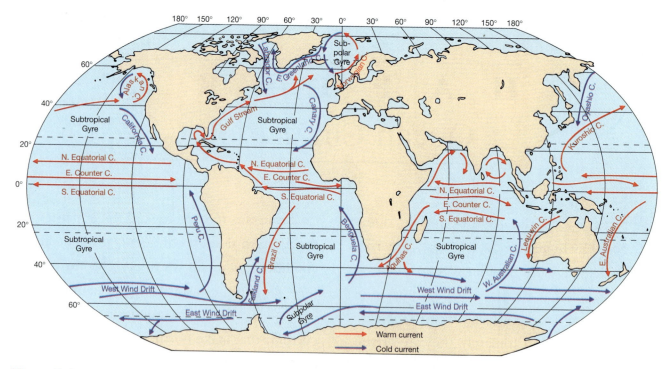

Figure 7–1

Wind-Driven Surface Currents in February and March. Major wind-driven surface currents of the world's oceans. The western and eastern boundary currents combine to form the subtropical gyres that dominate the five major ocean basins: the North and South Pacific oceans, the North and South Atlantic oceans, and the South Indian Ocean. They rotate clockwise in the Northern Hemisphere and counterclockwise in the Southern Hemisphere. The smaller subpolar gyres rotate in the reverse direction of the adjacent subtropical gyres. (After Sverdrup et al., 1942.)

These currents move westward parallel to the equator. Deflection by the Coriolis effect and deflection along continental margins partially directs these currents away from the equator, as warm **western boundary currents.** The name means that they are currents traveling along the western boundary of their ocean basin. Figure 7–1 shows these warm currents as red arrows; the Gulf Stream is an example.

At the same time, between 30° and 60° latitude in both hemispheres westerly winds blow surface water in an easterly direction. These winds blow from the northwest in the Southern Hemisphere and the southwest in the Northern Hemisphere. The Coriolis effect and continental barriers turn this water toward the equator as a cold **eastern boundary current** (eastern boundary of the ocean basin). These are shown in blue in Figure 7–1; the Peru Current is an example.

The combination of western boundary currents and eastern boundary currents produces the dominant feature of ocean basin surface circulation, the **subtropical gyre.** It rotates in a clockwise direction in the Northern Hemisphere (see the North Atlantic Ocean in Figure 7–1) and in a counterclockwise direction in the Southern Hemisphere (see the South Atlantic Ocean in Figure 7–1).

At subpolar latitudes, the polar easterlies drive the surface currents in a westerly direction and, in combination with the Coriolis effect, produce **subpolar gyres** that rotate in a pattern opposite that of the adjacent subtropical gyres. These subpolar gyres are best developed in the Atlantic waters between Greenland and Europe and the Weddell Sea off Antarctica (Figure 7–1).

Ekman Spiral and Transport

Why do water masses move as they do relative to wind direction? Recall the observations made by Fridtjof Nansen during the voyage of the *Fram* (Chapter 1). Nansen determined that the Arctic sea ice moved 20° to 40° to the right of the wind blowing across its surface. He passed this information on to V. Walfrid Ekman, a physicist who developed the mathematical relationships that explain Nansen's observations.

Ekman developed a circulation model that has been called the **Ekman spiral** (Figure 7–2). The model assumes that a uniform water column is being set in motion by wind blowing across its surface. Owing to the Coriolis effect, the surface current moves in a direction 45° to the right of the wind (in the Northern Hemisphere). This surface mass of water moves with a velocity no greater than 3 percent of the wind speed. The water moves as a thin "lamina," or layer. As it moves, it also sets in motion another "layer" beneath it.

It may be helpful to think of this in another way. As the wind energy put into the ocean surface is transmitted downward, it is consumed by water motion until none is left. Thus, with increasing depth, current speed decreases, and the Coriolis effect causes a gradual spiral to the right.

The energy of the wind is passed downward through the water column. Each successive "layer" of water is set in motion at a slower velocity, and in a direction to the right of the one above that set it in motion. At some depth, the momentum imparted by the wind to the moving water laminae is so slight that no

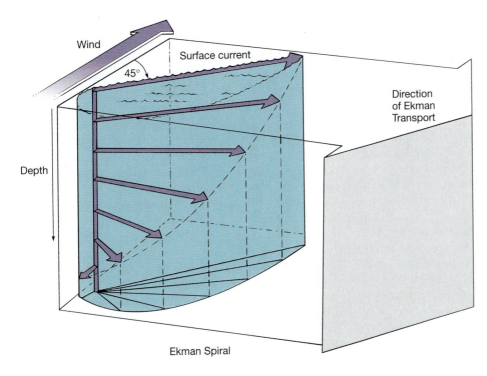

Figure 7–2

Ekman Spiral. Wind drives surface water in a direction 45° to the right of the wind in the Northern Hemisphere. Deeper water continues to deflect to the right and moves at a slower speed with increased depth. Ekman transport, which is the net water movement, thus is at right angles to the wind direction. This illustrates the principle, but in reality, the angles usually are somewhat smaller.

motion occurs as a result of wind at the surface. Although it depends on wind speed and latitude, this stillness normally occurs at a depth of about 100 meters (330 feet).

Figure 7–2 shows the spiral nature of this movement with increasing depth from the ocean's surface. The length of each arrow in the figure is proportional to the velocity of the individual lamina, and the direction of each arrow indicates its direction of movement.

Under these theoretical conditions, the surface layer should flow at an angle of 45° to the direction of the wind, as shown. But the overall average transport of all the laminae is a net water movement at 90° to the right of the wind direction. This is called the **Ekman transport.**

Of course, "ideal" conditions do not exist in the ocean, so movements actually resulting from wind friction on the ocean surface deviate from this idealized picture. Generally, we can say that the surface current will move at an angle of less than 45° to the direction of the wind and that the Ekman transport will be at an angle less than 90° to the direction of the wind. This is particularly true in shallow coastal waters, where all of the movement may be in a direction very nearly the same as that of the wind.

Geostrophic Currents

If we use as an example the gyre in the North Atlantic Ocean, and consider the Ekman transport's property of always turning water to the right in this hemisphere, you can see that a clockwise rotation in the Northern Hemisphere will tend to produce a **subtropical convergence** of water in the middle of the gyre. Water literally piles up in the center of the subtropical gyre. We find within all such ocean gyres a hill of water that rises as much as 2 meters (about 6.5 feet) above the water level at the margins of the gyres.

As Ekman transport continually pushes water into this "hill" structure, gravity also acts, but to move water down the surface slope. The Coriolis effect deflects the water flowing down the slope to the right in a curved path (Figure 7–3). The water piles up on these hills until the down-slope component of gravity acting on individual particles of water balances the Coriolis force. When these two forces balance, the net effect is a *geostrophic current* moving around the hill. (The term *geostrophic* means "Earth turning"—the currents behave as they do because of Earth's rotation.)

What was just described is accurate but idealized. Due to friction between water molecules, the water does converge and build up, but it gradually moves down the slope of the hill as it flows around it.

Westward Intensification

The apex of the hill formed within a rotating gyre is not in the gyre's center. The highest point of each hill is closer to the western boundary of the gyre. This is called **westward intensification.** Its causes are complex, but a good part of western intensification can be explained by the Coriolis effect. Remember that the Coriolis effect increases toward higher latitudes (poleward). Thus, eastward-flowing high-latitude water turns equatorward more strongly than westward-flowing equatorial water turns toward higher latitudes. This causes a broad equatorward flow of water across most subtropical gyres.

In Figure 7–3B, assume we have a steady state, with a constant volume of water rotating around the apex of the hill. The velocity of the water along the western margin will be much faster than the velocity around the eastern side. This can be seen clearly in the North Atlantic and North Pacific gyres, where western boundary currents move northward in excess of 5 kilometers/hour (3.1 miles/hour), but a flow along the eastern margin is more of a "drift," moving well below 0.9 kilometers/hour (0.6 miles/hour).

Western boundary currents commonly flow 10 times faster, and to greater depths, than eastern boundary currents. The warm western boundary currents are usually less than one-twentieth the width of the broad, cool drifts that flow equatorward on the east side of subtropical gyres.

Directly related to this difference in speed is the steepness of the hill's slope. The slope of the hill on the side of the slow-moving eastern boundary current is quite gentle; the western margin has a comparatively steep slope corresponding to the high velocity of the western boundary current.

Equatorial Countercurrents

The large volume of water that is driven westward in the north and south equatorial currents piles up against continents and very slowly "sloshes" back, creating **equatorial countercurrents.** Because the Coriolis effect is minimal near the equator, much of the water is not turned toward higher latitudes. Instead, it piles up at the western margin of an ocean basin.

This is particularly true in the western Pacific Ocean (Figure 7–1), where a dome of equatorial water is trapped in the island-filled embayment between Australia and Asia. This dome of water with very weak current flow displays the highest year-round surface temperature found anywhere in the world ocean, as you can see in Figure 7–4. Continued influx of water carried by the equatorial currents forces an eastward countercurrent of water that continues to South America.

Figure 7–5 is an image of average sea-surface elevation. The hills of water within the subtropical gyres of the Atlantic Ocean are clearly visible. In the Pacific Ocean, the hill in the North Pacific is easily seen, but the low equatorial elevations one might expect to see between the northern and southern subtropical gyres is

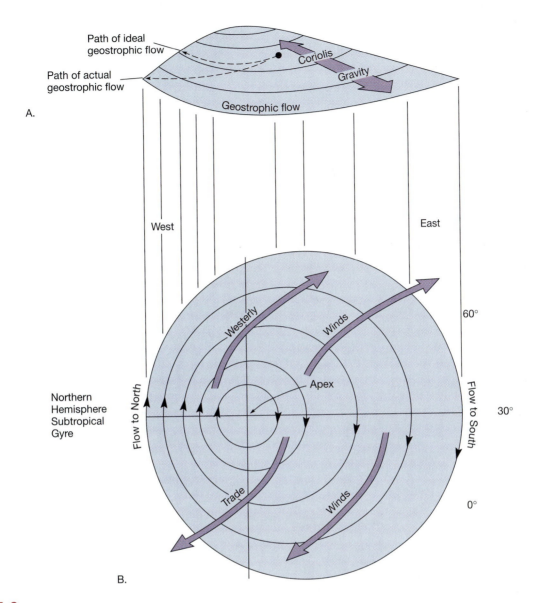

A.

B.

Figure 7–3

Geostrophic Current. *A:* As Earth's wind systems set ocean water in motion, circular gyres are produced. Water literally piles up inside the subtropical gyre, forming a "hill" up to 2 meters (6.6 feet) high. The apex of this hill is closer to the gyre's west margin because of Earth's rotation eastward. The theoretical geostrophic current would flow around the hill, in equilibrium between (1) the Coriolis force, which pushes water toward the apex through Ekman transport, and (2) gravity, which pulls the water downslope. However, friction between water molecules makes the current gradually run downslope. *B:* The Coriolis effect is stronger on water farther from the equator. Thus, eastward-flowing high-latitude water turns equatorward more strongly than westward-flowing equatorial water turns toward higher latitudes. This causes a broad, slow, equatorward flow of water across most of the subtropical gyre and forces the apex of the geostrophic hill toward the west. This phenomenon is referred to as *westward intensification.* Its main manifestation is a high-speed, warm, western boundary current that flows along the "hill's" steeper westward slope. The eastern slope has a slow drift of cold water toward the equator.

missing. These data were recorded during a moderate El Niño event (which will be explained shortly). Thus, the high stand of equatorial water is likely due to the fact the equatorial countercurrent is well developed.

Also contributing to the lack of definition between the two Pacific gyres is the fact that the South Pacific sub-tropical gyre is less intense than others. This is due to the great area it covers, plus its lack of confinement by continental barriers along its western margin. The South Indian Ocean hill is rather well developed, although its northeastern boundary stands high because of the influx of warm Pacific Ocean water through the East Indies islands.

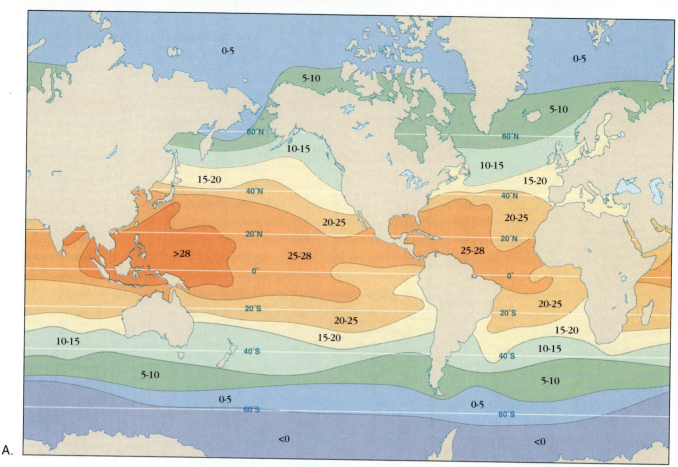

A.

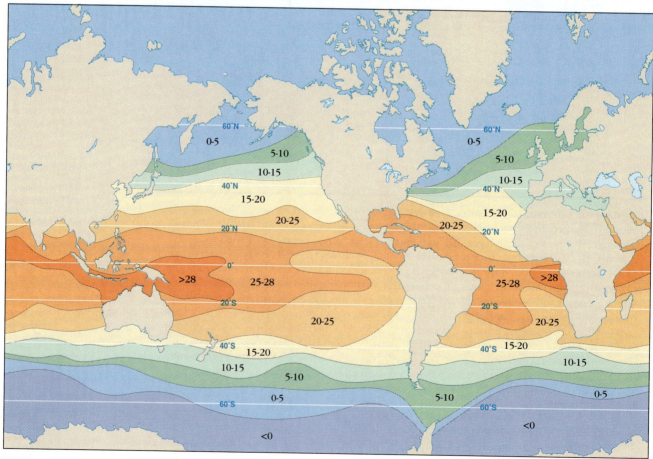

B.

Figure 7–4

Surface Temperature of the World Ocean (°C). Average surface temperature distribution for August (A) and February (B). Note the warmest area, in the Pacific between Asia and Australia, and how temperatures migrate north-south with the seasons. (After Sverdrup et al., 1942.)

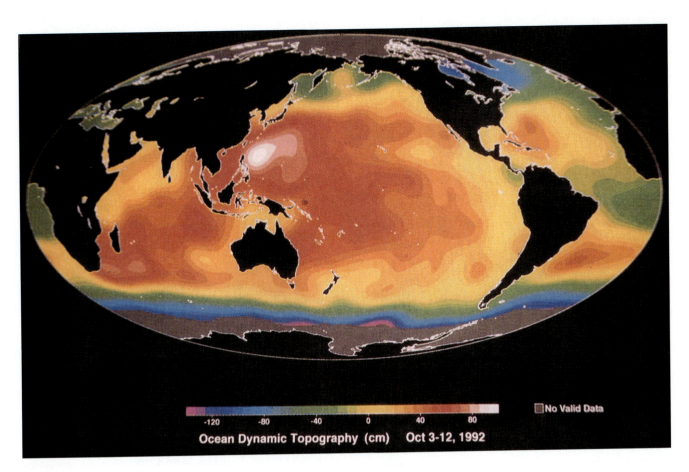

Figure 7–5

Dynamic Topography—A Satellite View. Dynamic topography of the ocean from data gathered by a satellite-borne radar altimeter during October 3–22, 1992. The elevations are measured in centimeters relative to an arbitrary datum. Westward intensification shows up well in all of the subtropical gyres. The total relief of the ocean surface due to current flow is about 2 meters (6.5 feet). The data are accurate to an astonishing degree of precision: to the nearest 2 centimeters (0.78 inches). (Courtesy of NASA.)

Antarctic Circulation

Now that you have an understanding of ocean currents in general, we will tour the world's oceans and examine the circulation in each. Although the oceanic mass surrounding the continent of Antarctica is not officially recognized as a distinct ocean, we will consider it first, because it features the mightiest of all ocean currents in terms of water volume transported. The *Antarctic Circumpolar Current* dominates the movement of water masses in the southern Atlantic, Indian, and Pacific oceans, south of 50° latitude. This latitude may be considered the northern boundary of what is unofficially referred to as the Southern Ocean.

A surface current called the *East Wind Drift,* so called because it is propelled by the polar easterlies, moves in a *westerly* direction around the margin of the Antarctic continent. The East Wind Drift is most extensively developed to the east of the Antarctic Peninsula

in the Weddell Sea region and in the area of the Ross Sea (Figure 7–6).

But the main circulation system in Antarctic waters is the **Antarctic Circumpolar Current**, extending northward to a position of approximately 40°S latitude. This mass is driven by the westerly winds, which are very strong throughout much of the year. The surface portion of this flow is named the **West Wind Drift.**

Thus there are two currents flowing around Antarctica in opposite directions. Recall that the Coriolis effect deflects moving masses to the left in the Southern Hemisphere, so the East Wind Drift is deflected toward the continent and the West Wind Drift is deflected away from it. This creates a zone of divergence, the *Antarctic Divergence,* between the two currents.

The Antarctic Circumpolar Current meets its greatest restriction as it passes through the Drake Passage (named for Sir Francis Drake) between the Antarctic Peninsula and the southern islands of South America.

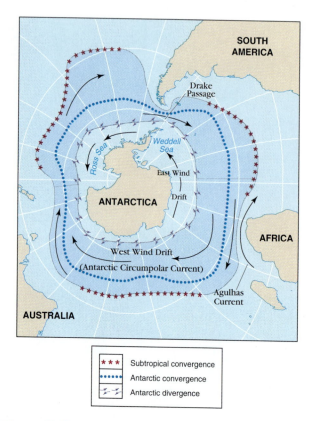

Figure 7–6

Antarctic Surface Circulation. At the right edge of Antarctica, note the East Wind Drift. This current flows around the entire continent westward, because it is driven by the polar easterly wind system. In the opposite direction, and farther out from the continent, the West Wind Drift circles Antarctica with an easterly flow. At the interface between these two currents, Ekman transport produces a zone of divergence, called the Antarctic Divergence. The cold West Wind Drift water converges with warmer water in each of the major ocean basins northward, at the Antarctic Convergence.

The passage is about 1000 kilometers (600 miles) long. Although the current is not speedy (its maximum surface velocity is about 2.75 kilometers/hour or 1.65 miles/hour), it does transport more water than any other ocean current, an average of about 130 million cubic meters per second.

A million cubic meters per second is a handy volume for describing ocean currents, so it has become a standard unit, named the *Sverdrup (sv)* after a noted oceanographer. We will use this unit in this chapter.

Atlantic Ocean Circulation

Atlantic Ocean surface currents are shown in Figure 7–7. It is helpful to study the currents on this map as they are described. In the Atlantic, the basic surface circulation pattern is that of two large gyres.

The North and South Atlantic Gyres

The *North Atlantic gyre* rotates clockwise and the South Atlantic gyre rotates counterclockwise, in response to the Coriolis effect. As stated, both rotations are driven by the trade winds and westerly winds. Each gyre consists of a poleward-moving warm current (red) and an equatorward-moving cold "return" current (blue). The two gyres are partially separated by the *Atlantic Equatorial Countercurrent*.

The *South Atlantic gyre* includes the *South Equatorial Current,* which reaches its greatest strength just below the equator and is split in two by topographic interference from the eastern prominence of Brazil. Part of the South Equatorial Current moves off along the northeastern coast of South America toward the Caribbean Sea and the North Atlantic. The rest is turned southward as the **Brazil Current,** which ultimately merges with the West Wind Drift and moves eastward across the South Atlantic. The Brazil Current is much smaller than its Northern Hemisphere counterpart, the Gulf Stream, due to the splitting of the South Equatorial Current.

The gyre is completed by a slow-drifting movement of cold water, the **Benguela Current,** that flows equatorward along Africa's western coast.

Outside the gyre, a significant northbound flow of cold water also moves along the western margin of the South Atlantic. This **Falkland Current** is an important cold current that moves along the coast of Argentina as far north as 25° to 30°S latitude, wedging its way between the continent and the southbound Brazil Current.

The Gulf Stream

The Gulf Stream moves northward along the U.S. East Coast, warming coastal states and moderating winters there (Figure 7–8). It is interesting to trace the origins and behavior of this important current.

The *North Equatorial Current* moves parallel to the equator in the Northern Hemisphere, where it is joined by that portion of the South Equatorial Current that is shunted northward along the South American coast. This flow then splits into two masses: the *Antilles Current,* which flows along the Atlantic side of the West Indies, and the **Caribbean Current,** which passes through the Yucatán Channel into the Gulf of Mexico. These masses reconverge as the Florida Current.

The **Florida Current** flows close to shore over the continental shelf, carrying a volume that at times exceeds 35 sverdrups. As it moves off North Carolina's Cape Hatteras and flows across the deep ocean in a northeasterly direction, it is called the **Gulf Stream.** It flows up to 9 kilometers/hour (5.6 miles/hour), the fastest current in the world ocean.

The western margin of the Gulf Stream can frequently be defined as a rather abrupt boundary that peri-

CONVERGENCES

Cold   →   ARC–Arctic

             STC–Subtropical

Warm →   ANC–Antarctic

CURRENTS

A–Antilles	G–Guinea
Bg–Benguela	GS–Gulf Stream
Br–Brazil	I–Irminger
C–Canary	L–Labrador
CC–Caribbean	NE–North Equatorial
EG–East Greenland	N–Norwegian
EW–East Wind Drift	SE–South Equatorial
EC–Equatorial Counter	WW–West Wind Drift
Fa–Falkland	
F–Florida	

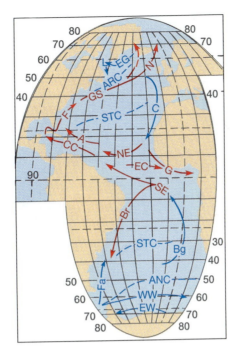

Figure 7–7

Atlantic Ocean Surface Currents. Fundamentally, Atlantic circulation consists of two gyres. (Base map courtesy of National Ocean Survey.)

odically migrates closer to and farther away from the shore. Its eastern boundary becomes very difficult to identify, as it is usually masked by filamentous meandering water masses that change their position continuously.

The Gulf Stream gradually merges with the water of the Sargasso Sea. The Sargasso Sea is the water that circulates around the rotation center for the North Atlantic gyre. Its name derives from a seaweed called *Sargassum* that floats on its surface.

A volume transport of over 90 sverdrups off Chesapeake Bay indicates that a large volume of Sargasso Sea water has been added to the flow provided by the Florida Current. However, by the time the Gulf Stream nears Newfoundland, the Gulf Stream's volume has been reduced to 40 sverdrups. This indicates that most of the Sargasso Sea water that joined the Florida Current to make up the Gulf Stream has returned to the diffuse flow of the Sargasso Sea.

Just how this loss of water occurs has yet to be determined. However, much of it may be caused by *meanders* (snakelike bends in the current) that pinch off to form large *eddies* (whirlpools). These are shown in Figure 7–8A, most notably near the center of the image. You can see meanders along the north boundary of the Gulf Stream pinching off and trapping warm Sargasso Sea water in eddies that rotate clockwise. These eddies move southwest at 3 to 7 kilometers a day (1.9 to 4.3 miles/day) toward Cape Hatteras, where they rejoin the Gulf Stream.

South of the Gulf Stream, counterclockwise-rotating eddies with cores of cold slope water move southward and westward. These eddies are large, ranging from 100 to 500 kilometers (60 to 300 miles) in diameter and extending to a depth of 1 kilometer (0.6

miles) in warm rings and 3.5 kilometers (2.2 miles) in the cold rings. They could remove large volumes of water from the Gulf Stream.

Southeast of Newfoundland, the Gulf Stream continues in a more easterly direction across the North Atlantic (Figure 7–8A). Here the Gulf Stream breaks into numerous branches. Two branches combine water through the mixing of the cold **Labrador Current** and the warm Gulf Stream. These are the **Irminger Current,** which flows along Iceland's west coast, and the **Norwegian Current,** which moves northward along Norway's coast. The other major branch crosses the North Atlantic and turns southward as the cool **Canary Current** passing between the Azores Islands and Spain. This broad, diffuse southward flow eventually joins the North Equatorial Current, completing the gyre.

Climatic Effects of North Atlantic Currents

The effects of the westward intensification of current flow in the North Atlantic Ocean can be seen from the surface temperatures shown in Figure 7–4B for February. From latitudes 20°N (Cuba) to 40°N (New York City) off the coast of North America, you can see a 20° temperature difference. By contrast, on the eastern side of the North Atlantic, only a 5° to 6° range in temperature can be observed between the same latitudes.

Figure 7–4A clearly shows how northwestern Europe is warmed by the Norwegian Current branch of the Gulf Stream, compared with the same latitudes along the North American coast. On the western side of the North Atlantic, the southward-flowing cold Labrador current keeps Canadian coastal waters much colder.

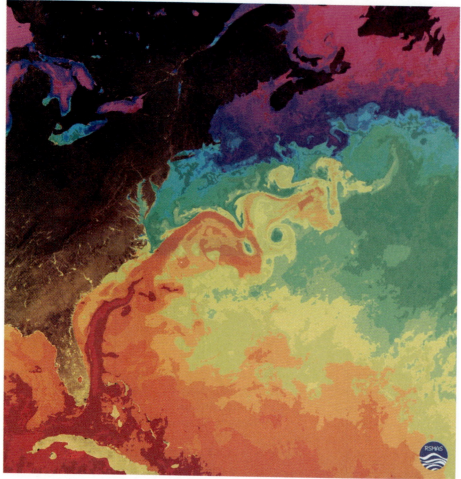

A.

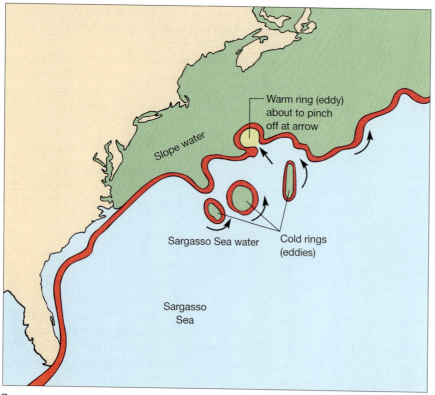

Warm ring (eddy)
about to pinch
off at arrow

Slope water

Sargasso Sea water

Cold rings
(eddies)

Sargasso
Sea

B.

Figure 7–8

The Gulf Stream and Sea-Surface Temperatures. *A:* The northwestern Atlantic ocean, off the U.S.-Canadian east coast, featuring the warm Gulf Stream. Sea-surface temperature data gathered by a NOAA satellite were processed to produce this false-color image. The warm waters of the Gulf Stream are shown in orange and red. Colder nearshore waters are shown in green, blue, and purple. Warm water from south of the Gulf Stream is transferred northward as warm core rings (yellow) surrounded by cooler (blue and green) water. Cold nearshore water spins off to the south of the Gulf Stream as cold core rings (green) surrounded by warmer (yellow and red-orange) water. As the Gulf Stream meanders northward, some of its meanders close, forming rings, which trap warm or cold water within them. The warm rings contain shallow, bowl-shaped masses of warm water about 1 kilometers (0.6 miles) deep, with diameters of about 100 kilometers (60 miles). The cold rings include cones of cold water that extend to the ocean floor. These rings may exceed 500 kilometers (310 miles) in diameter at the surface. The diameter of the cone increases with depth. *B:* This overlay of the satellite image in *A* clarifies the process of ring formation. The red band is the core of the Gulf Stream flow. Note the three cold rings that have formed by the pinching off (closure) of meanders. Also note the warm ring that is being created by pinching off the current. The cold rings have separated from the main Gulf Stream flow. The warm ring that is forming also will separate from the Gulf Stream. (*A:* Image courtesy of Dr. Charles McLain/Rosenstiel School of Marine and Atmospheric Science, University of Miami.)

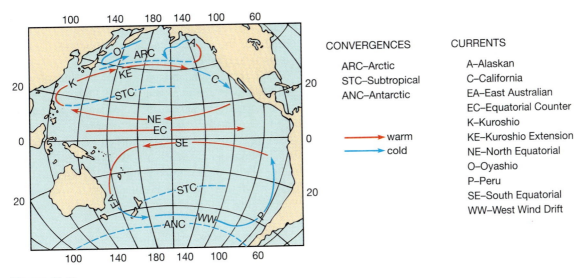

Figure 7–9

Pacific Ocean Surface Currents. Containing half of Earth's ocean water, the large gyres are somewhat less intense. However, the equatorial countercurrent is more strongly developed than in smaller ocean basins.

During the Northern Hemisphere winter (Figure 7–4B), North Africa's coastal waters, cooled by the southward-flowing Canary Current, are cooler than waters of Florida and the Gulf of Mexico.

Pacific Ocean Circulation

Surface circulation in the Pacific Ocean is generally similar to that described for the Atlantic, with two large gyres. The same general pattern of water movement and climatic affects exists. However, the *Equatorial Countercurrent* is much better developed in the Pacific Ocean than in the Atlantic (Figure 7–9).

El Niño–Southern Oscillation (ENSO) Events

For centuries, Peru's cool coastal waters have provided one of Earth's richest fishing grounds. But fishermen there know that an influx of warm water every few years reduces the population of anchovy, their main fishery and the food source for coastal birds. This warm-water phenomenon has usually occurred around Christmas and thus was given the name El Niño, meaning "the child."

A high-pressure system in the southeastern Pacific usually occurs in conjunction with a low-pressure condition over the Indo-Australian region (see Figures 7–10 and 6–12). Periodically, this pressure difference lessens

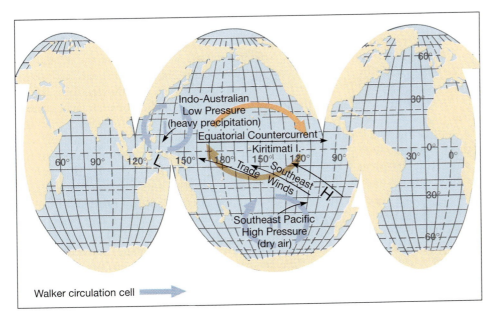

Walker circulation cell ➡️

Figure 7–10

Walker Circulation. Normal oceanic and atmospheric conditions. These are altered by El Niño–Southern Oscillation (ENSO) events.

when trade winds diminish, the temperature of surface waters in the eastern Pacific increases, and the eastward-flowing Equatorial Countercurrent increases. During the 1920s, G. T. Walker, a British meteorologist, studied this phenomenon and named it the Southern Oscillation (SO). The average period of this oscillation is 3 years, but it ranges from 2 to 10 years.

When it was discovered that the El Niño seemed to be related to the Southern Oscillation, the occurrences began to be referred to as **El Niño-Southern Oscillation (ENSO)** events. There were nine ENSO events between 1950 and 1993. Since 1993 the El Niño condition has only fluctuated without fully going away.

ENSO events are clearly related to other periodic climatic events observed worldwide, and the relations among these events are being studied intensely. Knowledge of these phenomena is important to our understanding of the world's climate; following is a brief overview of the present knowledge of ENSO and its associations with world climatic events.

The Southern Oscillation circulation cell is called the *Walker Circulation Cell* (Figure 7–10). Normally, this circulation cell affects the southeast trade winds, which converge on the Indo-Australian low-pressure cell, rise, and produce plentiful precipitation in the low-pressure area.

Recall that a large mass of the warmest water in the world ocean lies beneath this low-pressure cell (Figure 7–4), having been pushed there by the trade winds. The thermocline in this warm-water area doesn't begin to develop until a depth exceeding 100 meters (about 330 feet); in contrast, the thermocline develops within 30 meters (100 feet) of the surface in areas of the eastern equatorial Pacific.

Dry air descends within the southeastern Pacific high-pressure cell off the west coast of South America, and this coast is characterized by high rates of evaporation. These characteristics promote cool, nutrient-rich fishing waters off the Peruvian coast. All of these characteristics change during an ENSO event.

ENSO events have precursors. One is the movement of the Indo-Australian low-pressure cell to the east, beginning in October or November. In extreme cases, severe droughts can occur in Australia because the low-pressure cell moves so far east.

Concurrent with the eastward shift of the Indo-Australian low-pressure cell is another important shift. Earth's "meteorological equator" is the **Intertropical Convergence Zone** (ITCZ), where the northeast trade winds and southeast trade winds meet and rise (Figure 6–9). The ITCZ shifts southward as the Indo-Australian low-pressure cell moves eastward. The ITCZ's normal seasonal migration is from 10°N latitude in August to 3°N in February, but during ENSO events it may move south of the equator in the eastern Pacific.

Associated with this shift are weak trade winds, a decreased upwelling in the equatorial Pacific, and a deepening of the eastern Pacific thermocline. This results in a thickening layer of warm water (Figure 7–11A).

So far, these initial events are just amplifications of normal seasonal changes. But as the ENSO develops, the weakened trade winds and anomalous warmth of surface waters in the eastern Pacific spread westward. The coming of unusually warm surface waters to Kiritimati (Christmas Island; 2°N, 157°W) can be predicted by earlier observation of increased surface temperatures off the Peruvian coast. The event is fully developed by January (Figure 7–11B).

ENSO's Effects. Under these conditions, heavy rainfall from the southward shift of the ITCZ strikes the coast of Ecuador and Peru, which is usually arid, and spreads westward across the tropical Pacific. The intense eastward flow of the Equatorial Countercurrent causes a rise in sea level along the western coast of the Americas, and this rise progresses poleward in both hemispheres.

The event usually ends 12 to 18 months after it starts, with a gradual return to normal conditions that begins in the southeastern tropical Pacific and spreads westward (Figure 7–11C).

Although the below-average temperatures in the eastern Pacific (shown in Figure 7–11C) usually occur following an El Niño, such an event did not occur between 1975 and 1988. This cooling has been given the name *La Niña* (the "female child") and seems to be associated with weather phenomena opposite to those of El Niño at other locations on Earth. For instance, Indian Ocean monsoons are drier than usual in El Niño years but wetter than usual in La Niña years.

An intense 1982–1983 ENSO event caused severe drought in Australia and Indonesia. It was anomalous because it was initially confined to the central and western tropical Pacific but then spread eastward late in its development.

In November 1982 virtually all of the 17 million adult birds that normally inhabit Kiritimati had abandoned their nestlings. This event indicated the severity of the ENSO in the central Pacific, for such an abandonment was unknown before. It is assumed that the spread of a thick layer of warm water over the surrounding ocean prevented the rise of nutrients into the surface waters, causing the fish to leave in search of better feeding grounds. The birds, in turn, were forced to leave in search of the fish. If the birds did not find their necessary supply of fish, the nestlings and a large percentage of the adults may have died.

In Figure 7–12, the upper images show the warming that occurred off southern California due to the 1982–1983 ENSO. This warming was accompanied by a

Figure 7–11

Averaged ENSO Event Temperature Anomalies, 1950–1973.
A: After onset: average of March, April, and May temperature anomalies, °C. *B:* Maximum development: average of December, January, and February anomalies. *C:* Ending of event: average of May, June, and July anomalies. In a typical ENSO event, abnormally high surface-water temperatures first appear in the eastern Pacific, off the coast of Ecuador and Peru (*A*). This condition begins to be observable by December or January along the coast of South America and is well developed during the March-May period shown. In *B* it can be seen to develop in a westerly direction along the equator. It attains maximum development in the Central Pacific by the following February. By July (*C*), conditions return to near normal, with the exception of a significant negative temperature anomaly in the eastern Pacific. (Data courtesy of NOAA.)

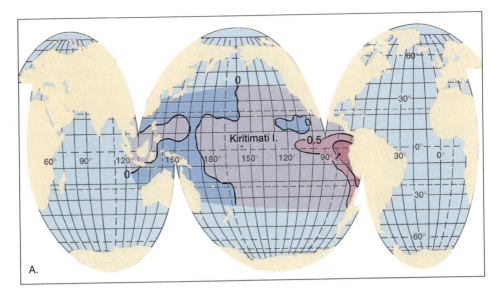

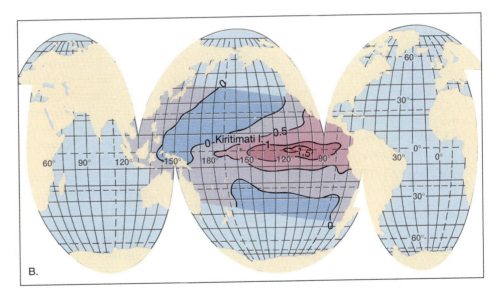

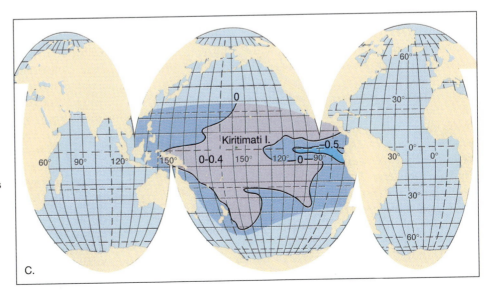

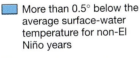 More than 0.5° below the average surface-water temperature for non-El Niño years

0 to 0.5°C below the average

0 to 0.5°C (0.9°F) above the average surface-water temperature for non-ENSO years

0.5° to 1.0°C (0.9° to 1.8°F) above average

1.0° to 1.5°C (1.8° to 2.7°F) above average

More than 1.5°C (2.7°F) above average

major reduction in biological productivity, as shown by the phytoplankton pigment images in the bottom images.

An ENSO event also may cause unusually voluminous rainfalls in California and the southeastern United States. It also can bring warm winters to Alaska and western Canada.

A 1991–1992 ENSO produced global weather modifications similar to those of the 1982–1983 event. Computer models were able to predict this ENSO event, even though the causes are not fully understood. The 1991–1992 ENSO was unusual because it maintained itself until the spring of 1995.

Indian Ocean Circulation

The Indian Ocean is mostly in the Southern Hemisphere, extending only to about 20°N latitude. Surface circulation varies considerably from that in the Atlantic and Pacific. From November to March, the equatorial circulation is similar to that in the other oceans, with two westward-flowing equatorial currents (North and South Equatorial Currents) separated by an eastward-flowing Equatorial Countercurrent. However, in contrast with the Atlantic and Pacific wind systems, which shift northward of the geographical equator, in the Indian Ocean the meteorological equator is shifted *southward*.

The Equatorial Countercurrent flows between 2° and 8°S latitude, bounded on the north by the North Equatorial Current (which extends as far as 10°N latitude) and on the south by the South Equatorial Current (which extends to 20°S latitude). The winds of the northern Indian Ocean have a seasonal pattern and are called **monsoon** winds (from the Arabic word meaning "season").

During winter, the typical northeast trade winds are called the *northeast monsoon*. They are reinforced because rapid cooling of air over the Asian mainland during winter creates a high-pressure cell, which forces atmospheric masses off the continent and out over the ocean, where air pressure is less (green arrows in Figure 7–13A).

During summer, the Asian mainland warms faster than the oceanic water. (Recall that this is due to the lower heat capacity of continental rocks and soil compared with water.) As a result, a summer low-pressure cell develops over the continent, allowing higher-pressure air to reverse direction and move from the Indian Ocean onto the Asian landmass. This gives rise to the *southwest monsoon* (green arrows in Figure 7–13B). (This may be thought of as a continuation of the southeast trade winds across the equator.)

During this season, the North Equatorial Current disappears and is replaced by the *Southwest Monsoon Current*. It flows from west to east across the North In-

dian Ocean. In September or October, the northeast trade winds are reestablished, and the North Equatorial Current reappears (Figure 7–13A).

Surface circulation in the southern Indian Ocean is similar to the counterclockwise circulation observed in other southern oceans. When the northeast trade winds blow, the South Equatorial Current provides water for the Equatorial Countercurrent and the **Agulhas Current,** which flows southward along Africa's eastern coast and joins the West Wind Drift. Turning northward out of the West Wind Drift is the **West Australian Current,** which completes the gyre by merging with the South Equatorial Current.

During the southwest monsoon, a northward flow from the equator along the coast of Africa, the **Somali Current,** develops with velocities approaching 4 kilometers/hour (2.5 miles/hour).

The eastern boundary current in the southern Indian ocean is unique. Other eastern boundary currents of subtropical gyres are cold drifts toward the equator that produce arid coastal climates, which receive less than 25 centimeters (10 inches) of rain per year. But in the southern Indian Ocean, the West Australian Current is displaced offshore by a southward-flowing current called the **Leeuwin Current.** This current is driven southward along the Australian coast from the warm-water dome piled up in the East Indies by the Pacific equatorial currents.

The Leeuwin Current produces a mild climate in southwestern Australia, which receives about 125 centimeters (50 inches) of rain per year. This current is weakened during ENSO events and contributes to Australian drought conditions associated with these events.

Vertical Circulation

So far we have discussed only horizontal currents in the ocean. But vertical movements are essential in stirring the ocean, delivering oxygen to its depths, bringing nutrients to the surface, and distributing heat energy.

Lateral movements of water masses also may bring about some vertical circulation within the upper-water mass. We refer to this shallow vertical circulation system as *wind-induced circulation*.

Of greater oceanwide importance in producing thorough mixing within the ocean is vertical circulation due to density changes at the ocean surface. Density changes cause water masses to sink. Changes that increase water density normally are changes in temperature and salinity, so this circulation is referred to as thermohaline (temperature-salt) circulation.

We will now look at both types of vertical ocean currents.

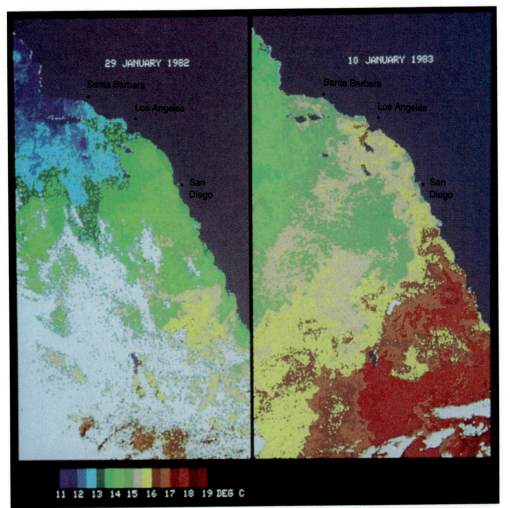

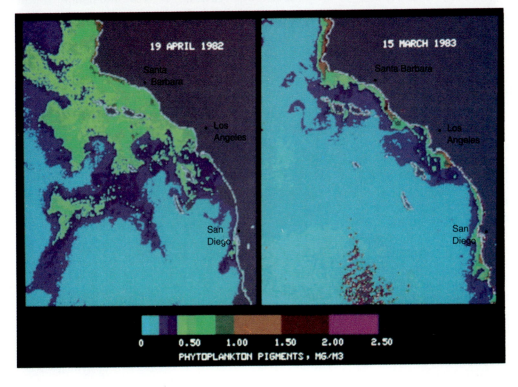

Figure 7–12

Water Temperatures and Phytoplankton off Southern California. *Top:* These sea-surface temperature images reveal that waters off southern California warmed from the January 1982 image (left) to the January 1983 image (right) by an average 1.72°C. (Images are from the Advanced Very High Resolution Radiometer, a satellite-mounted sensor.) *Bottom:* These phytoplankton-pigment images show a reduction in plant productivity over a similar time span, April 1982 to March 1983. (Images are from the Coastal Zone Color Scanner, another satellite-mounted sensor.) The 1982–1983 El Niño, the strongest ever recorded, pushed a layer of warm water northward from the equator. Coinciding with this event was a reduction of upwelling of nutrient-rich cold water along the coast. This reduction in the nutrient level of surface water reduced the level of biological productivity. Spectacular physical phenomena such as these and their biological consequences help highlight the close relation between the physical and biological phenomena of the oceans. (Courtesy Paul C. Fiedler/National Marine Fisheries Service.)

A.
WINTER: November–March,
Northeast monsoon
wind season

B.
SUMMER: May–September,
Southwest monsoon
wind season

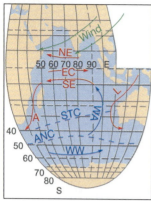
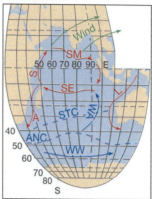

From November to March, Indian Ocean equatorial circulation is similar to other oceans, but Equatorial Countercurrent shifts southward.

Low pressure over mainland during summer draws Southwest Monsoon winds over the North Indian Ocean. This wind produces the Southwest Monsoon Current which flows easterly and replaces the North Equatorial Current.

Warm   Cold

CURRENTS A—Agulhas   EC—Equatorial Countercurrent
L—Leeuwin   NE—North Equatorial   S—Somali
SE—South Equatorial   SM—Southwest Monsoon
WA—West Australian   WW—West Wind Drift

CONVERGENCES STC—Subtropical   ANC—Antarctic

Figure 7–13

Indian Ocean Surface Currents.

Wind-Induced Circulation

There are regions throughout the world ocean where water masses rise or sink due to wind-driven surface currents that carry water away from these regions or toward them. Wherever surface flow is away from an area, **upwelling** occurs. If horizontal surface flow brings insufficient replacement water into the area, then water must come from beneath the surface—it must upwell—to replace water that was removed. This condition may occur in the open ocean or along the margins of continents.

Winds from the east drive three westerly currents on either side of the equator: the North Equatorial Current of the Indian Ocean and South Equatorial currents in the Atlantic and Pacific oceans. The Ekman transport of water on the north side of the equator moves it to the right, toward a higher northern latitude, while in the Southern Hemisphere it will move to the left toward a higher southern latitude (Figure 7–1). The net effect is a water deficiency at the surface between the two currents. Water from deeper within the upper-water mass comes to the surface to fill the void (Figure 7–14). This phenomenon is called *equatorial upwelling.*

Coastal upwelling is common along the margins of continents. It occurs where surface waters adjacent to the continents are carried out to the open ocean via wind-driven Ekman transport (Figure 7–15A). The replacement of this water comes from the lower portions of the upper water (an example is the Peru Current). Such areas are characterized by low surface temperatures and high concentrations of nutrients, both brought up from the bottom, making these areas high in biological productivity (plant growth).

Coastal downwelling is sinking of surface water caused by the reversal of the direction of coastal winds that cause upwelling. Wind blowing in the opposite direction pushes water toward shore instead, where it piles up and causes downwelling (Figure 7–15B).

Thermohaline Circulation

First we will consider something that does *not* cause vertical circulation. The intensity of solar radiation is much greater in the equatorial region. Therefore, you might expect heated equatorial surface water to expand and move away from the equator toward the poles, spreading over the colder, denser high-latitude waters (Figure 7–4). You might further expect some return flow from the high latitudes toward the equator. But this exchange does not occur. The reason is that energy imparted to surface waters by winds greatly exceeds the effect of density changes due to heating. Thus, wind overcomes any tendency for such a pattern of circulation to develop.

Instead, large-scale vertical circulation in the ocean results primarily from surface-density differences

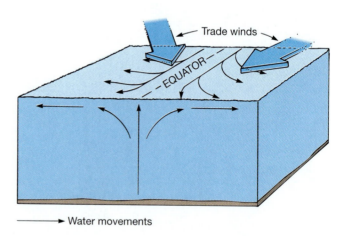

Trade winds

EQUATOR

Water movements

Figure 7–14

Equatorial Upwelling. Westward-flowing equatorial currents are driven by trade winds, and partly steered by the Coriolis effect, which pulls surface water away from the equatorial region. This water is replaced by upwelling subsurface water.

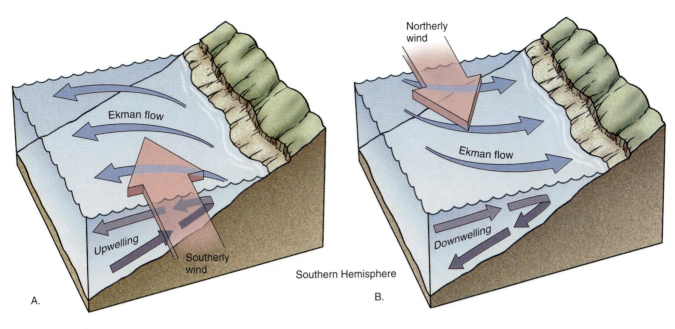

Figure 7–15

Coastal Upwelling and Downwelling. *A:* Where wind-driven coastal currents flow along the western margins of continents and toward the equator, Ekman transport carries surface water away from the continent. An upwelling of deeper water replaces the surface water that has moved away from the coast. *B:* A reversal of the direction of the winds that cause upwelling causes water to pile up against the shore and forces downwelling.

in oceanic water. The density differences result from temperature and salinity differences.

Density. Vertical mixing of ocean water is driven primarily through the sinking of cold water masses in high polar latitudes. Surface masses may sink to the ocean bottom and deep-water masses may rise to the surface.

Salinity. The other variable that affects surface water density is salinity (Figure 6–19). Salinity appears to have a minimal effect on the movement of water masses in equatorial latitudes. Density changes due to salinity are important only in very high latitudes, where low water temperature remains relatively constant.

For example, the highest-salinity water in the open ocean is found in the subtropical regions, but there is no sinking water in these areas because water temperatures are high enough to maintain a low density for the surface-water mass and prevent it from sinking. In such areas, a strong halocline, or salinity gradient, may develop. It features a relatively thin surface water layer with salinity exceeding 37‰. Salinity decreases rapidly with increasing depth to typical seawater salinities below 35‰.

Cold Deep Water. An important sinking of cold surface waters that become deep-water masses occurs in the subpolar regions of the Atlantic Ocean. In the North Atlantic, the major sinking of surface water is thought to occur in the Norwegian Sea. From there it

flows as a subsurface current into the North Atlantic. This flow becomes part of what is called **North Atlantic Deep Water** (Figure 7–16A). Additional surface water may sink at the margins of the Irminger Sea off southeastern Greenland and the Labrador Sea.

In the southern subpolar latitudes, the most significant area where deep-water masses form is the Weddell Sea, where rapid winter freezing produces very cold high-density water that sinks down the continental slope of Antarctica and becomes **Antarctic Bottom Water,** the densest water in the open ocean (Figure 7–16B).

On a broad scale, some surface-water masses converge, which may cause sinking (Figure 7–16). This convergence occurs within the subtropical gyres and in the Arctic and Antarctic. Subtropical convergences do not produce sinking because warm surface waters have relatively low densities. However, major sinking does occur along the *Arctic convergence* and **Antarctic Convergence (ANC)** (Figures 7–16 and 7–17). The major water mass formed from sinking at the Antarctic convergence is called the **Antarctic Intermediate Water** mass (Figure 7–16).

Worldwide Deep-Water Circulation. For every liter of water that sinks from the surface into the deep ocean, a liter of deep water is displaced and therefore must return to the surface. But it is difficult to identify specifically *where* this vertical flow to the surface is

A.

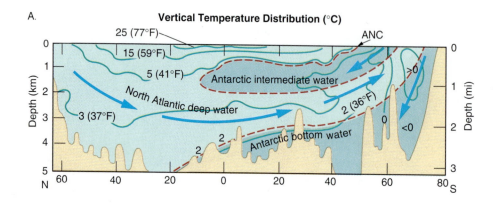

B.

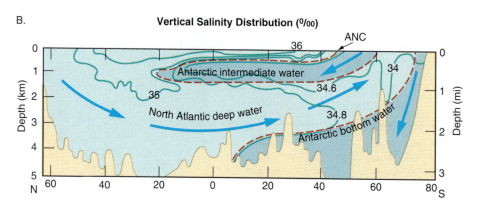

C.

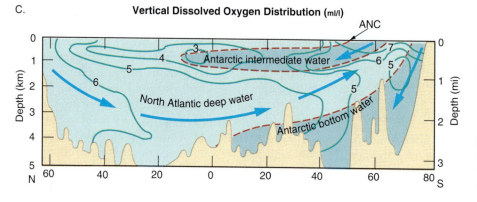

Figure 7–16

Atlantic Ocean Subsurface Circulation. Evidence for subsurface circulation comes from the physical properties of water. (Adapted from Sverdrup, Johnson, & Fleming, *The oceans: their physics, chemistry, and general biology,* © 1942.)

occurring. It probably occurs as a gradual, rather uniform return throughout the ocean basins. This return to the surface may be somewhat greater in low-latitude regions, where surface temperatures are higher.

Figure 7–18 shows what oceanographers believe to be the general deep-water circulation of the world ocean. The most intense deep-water flow is the western boundary current, because of Earth's rotation.

Figure 7–19 integrates deep thermohaline circulation and surface currents to show the probable overall pattern of global circulation. This circulation provides the deep ocean with oxygen-rich water by the sinking of dense surface water, primarily in the subpolar North Atlantic Ocean. Here, a volume of water equal to 100 Amazon rivers begins its long journey into the deep basins of all of the world's oceans (blue arrows).

Deep-water masses move southward through the Atlantic, across the South Indian Ocean, and into the North Pacific Ocean. As they do so, dissolved oxygen concentrations decrease because animal respiration and bacterial decomposition both consume oxygen. Bacterial decomposition also increases the concentration of nutrients used by plants, such as nitrogen and phosphorus compounds.

Dissolved oxygen decreases as it travels, from 4.5–6.5 milliliters per liter of water in the North Atlantic, where it has recently descended, to 3.5–4.5 in the North Pacific, nearing the end of its journey. Conversely, nutrient levels increase in deep-sea water from the North Atlantic Ocean to the North Pacific Ocean.

The return limb of this conveyor-belt-like system begins with surface water flowing out of the northeast-

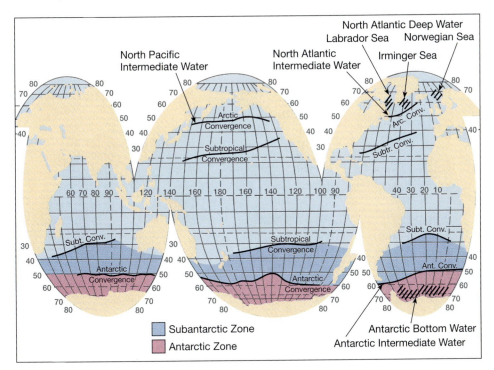

Figure 7-17

Regions of Sinking Intermediate and Deep-Water Masses. The only regions where deep and bottom water form are in the subpolar Atlantic Ocean. North Atlantic Deep Water forms north of the Arctic Convergence, and Antarctic Bottom Water forms near Antarctica. Best developed of all intermediate water masses, Antarctic Intermediate water forms at the Antarctic Convergence. The subtropical convergences all are centered within subtropical gyres.

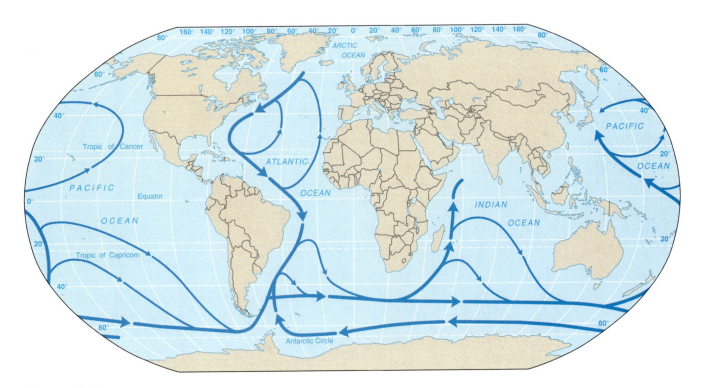

Figure 7-18

Stommel's Deep-Water Circulation Model. Based on information obtained to date, this highly schematic model of deep-water circulation developed by Henry Stommel in 1958 appears to be reasonably correct. Heavy lines mark the major western boundary currents. They result from the same forces that produce the more intense western boundary currents in the surface circulation. (After H. Stommel.)

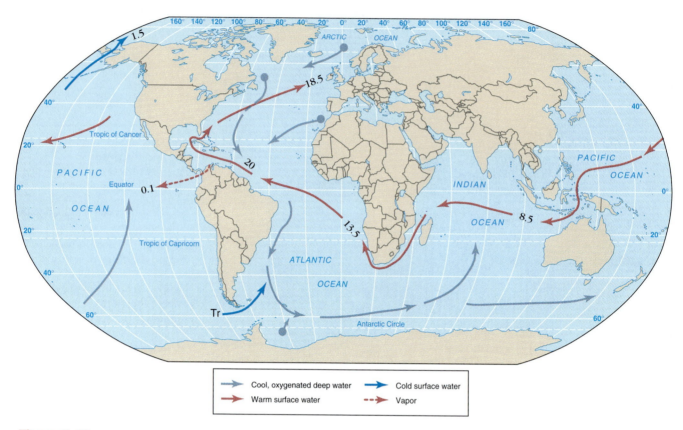

Figure 7–19

Global Cycle of Thermohaline Circulation. The deep ocean is replenished with oxygen by the sinking of North Atlantic Deep Water (NADW), shown with blue arrows. NADW is composed of water that sinks in the Norwegian Sea and is joined by water from Baffin Bay and the Mediterranean Sea. The total volume of this flow may be about 20 million cubic meters per second (Sverdrups, sv) as it enters the South Atlantic. This deep flow is recooled in the Antarctic and flows on into the deep Indian Ocean and Pacific Ocean. A broad pattern of upwelling returns the bottom water to the surface in all oceans. A surface flow of warmer, low-salinity water in the North Pacific (red arrows) starts the return journey to the North Atlantic. A flow of 8.5 sverdrups from the North Pacific increases to 13.5 sverdrups after crossing the Indian Ocean and reaches a total of 18.5 sverdrups by the time it is carried into the North Atlantic. A small amount of cold surface water enters the Atlantic from the Pacific through the Drake Passage (green arrow between Antarctica and South America) and 1.5 sverdrups flows into the Atlantic through the Bering Strait. At least 0.1 sverdrups of water is transferred from the Atlantic Ocean to the Pacific Ocean as water vapor across the Isthmus of Panama. (After Gordon, 1986.)

ern Pacific Ocean (red arrow off the U.S. West Coast). Water is added as the surface flow crosses the Indian Ocean and Atlantic Ocean on its return to the subpolar North Atlantic Ocean.

This pattern of return within the surface water may help explain the salinity pattern observed in the world ocean: The surface salinity of the North Pacific Ocean is the lowest and that of the North Atlantic is the highest. As the return flow begins in the North Pacific as low-salinity water, its salinity increases with time by evaporation. Cooling of this high-salinity water produces a density great enough to initiate sinking and for-

mation of the North Atlantic Deep Water by the time it returns to the subpolar North Atlantic Ocean.

The importance of this sinking of surface waters in the subpolar Atlantic Ocean to life on our planet cannot be overstated. Without this sinking and the return flow from the deep sea to the surface, distribution of life in the sea would be considerably different. Because of the lack of oxygen, there would be no life in the deep ocean. Life in surface waters would be significantly reduced and confined to the very margins of the oceans, where the only nutrient source would be runoff from streams on land.

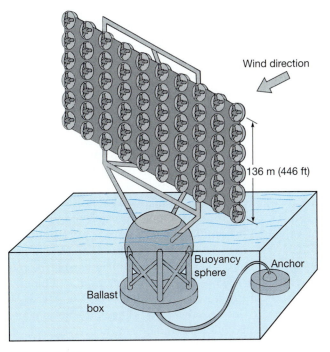

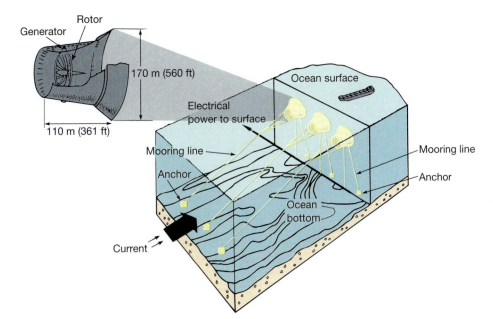

Figure 7–20

Offshore Windpower System (OWPS). There is some optimism that such a system could generate electricity to meet the needs of large areas of the U.S. North Atlantic coast. (Courtesy of Woods Hole Oceanographic Institution.)

The climatic effect of halting the sinking of surface waters is difficult to predict. Clearly, the oceans would become less efficient in absorbing and redistributing heat from solar radiation. This might cause much warmer surface water temperatures and much higher land temperatures than we now experience. One shud-

ders to think what this could mean to terrestrial plant and animal populations—our own included.

Power from Winds and Currents

Solar energy drives the winds. The winds, in turn, have driven ocean currents for billions of years and have powered society's machines for centuries. Westerly winds off the coast of New England are a nearshore wind resource that is reasonably sustained and contains a large amount of energy. There is some optimism that *offshore windpower systems (OWPS)* could generate electricity to meet the needs of large areas of the U.S. North Atlantic coast (Figure 7–20).

Many have considered the great amount of energy in the Florida-Gulf Stream Current System and dreamed of harnessing it to power society's work. Scientists and engineers met in 1974 to consider this possibility and concluded that some 2000 megawatts of electricity could be recovered along the east coast of southern Florida. Devices proposed for extraction range from underwater "windmills" to a water low-velocity energy converter (WLVEC), which is operated by parachutes attached to a continuous belt. Again, calculations indicate that such systems can be economically competitive.

The Coriolis Program was proposed in 1973 by William J. Moulton of Tulane University. Large hydroturbines are envisioned that could generate 43 megawatts of electricity from the movement of ocean currents (Figure 7–21). An array of 242 units covering an area 30 kilometers (18.6 miles) wide and 60 kilometers (37 miles) long could generate 10,000 megawatts—the equivalent of 130 million barrels of oil.

Figure 7–21

Coriolis Program. Buoyant Coriolis hydroturbines anchored in ocean current. It is estimated that an array of 242 such units placed in the Gulf Stream could provide 10 percent of the present electricity needs of Florida. (Courtesy of U.S. Department of Energy.)

Tests are now being conducted to establish an optimum size; identify suitable sites; and confirm engineering, economic, and environmental estimates. What remains to be seen is whether the designs can function as expected in the sometimes hostile marine environment.

Ocean Acoustic Tomography (OAT)

The motions of ocean water range across broad spectrums of size and time. Identifiable circulation patterns range from eddies a few tens of kilometers across, created by water flowing over and around seamounts, to slowly rotating subtropical gyres the size of ocean basins. Between these extremes are features such as the Gulf Stream rings, a few hundred kilometers in diameter, and coastal upwellings, such as that responsible for the highly productive waters along the coast of Peru.

Fully understanding these motions and their effects on climate and biological productivity is a longstanding goal of oceanographers. However, this goal has been stymied by the impossibility of continuously gathering data over a sufficient area for the necessary length of time. The principal problem is the high cost of sending multiple ships on long voyages.

But a new system may make such data gathering possible: *ocean acoustic tomography (OAT)*. "Tomography" is derived from the Greek word *tomos,* meaning a section or slice, in this case of the ocean. A *tomogram* is a picture of a slice of the ocean. Tomography is the "T" in medical CAT scans (computerized axial tomography). The CAT scan uses X rays to provide tomographic images (slices) of parts of the body, usually the brain. In oceanography, the leaders in OAT development have

been Walter Munk (Figure 7–22A) of the Scripps Institution of Oceanography and Carl Wunsch (Figure 7–22B) of Massachusetts Institute of Technology.

The use of tomography to map ocean circulation involves the transmission and reception of low-frequency sound. Increases in water temperature, salinity, and pressure increase the speed at which sound travels in seawater. As sound travels through cold rings or warm rings, for example, the sound speed will decrease or increase. If sound is transmitted across a slice of ocean from a series of transmitters on one side to a series of receivers on the other, features that cause a change in the speed of sound can be mapped.

A successful demonstration was performed for six months in 1981 southwest of Bermuda on an area of ocean that was 300 kilometers squared (186 miles squared) (Figure 7–23A). It used four transmitters and five receivers (Figure 7–23B). Two sensors of temperature and current flow were included, and survey ships periodically entered the area to record data against which the tomography could be checked.

In 1991 Walter Munk designed a test of OAT to see if it might aid in detecting global warming of the oceans. If so, OAT could become a valuable tool for monitoring effects from the increase in greenhouse gas concentrations in the atmosphere. His group set off acoustical signals for six days at a depth of 150 meters (about 500 feet) near Heard Island in the southern Indian Ocean. The signals were received after traveling as far as 18,000 kilometers (11,000 miles). Theoretically, if these signals are sent out periodically over a period of years and if the ocean warms up, these signals will be observed to travel the same distances, *but in less time.*

Based on the results of the Heard Island test, a 10-year study to investigate whether ocean warming is oc-

A.

B.

Figure 7–22

Munk and Wunsch. *A:* Walter H. Munk of the Scripps Institution of Oceanography. *B:* Carl Wunsch of Massachusetts Institute of Technology. Both are leaders in development of ocean acoustic tomography (OAT) techniques for the synoptic study of mid-scale to large-scale oceanic circulation patterns. (Walter Munk photo courtesy of University of California at San Diego, Office of Learning Resources, Scripps Institution of Oceanography, Photo Lab.)

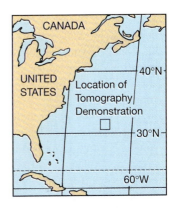

Figure 7-23

Ocean Acoustic Tomography (OAT). Ocean acoustic tomography was demonstrated in 1981 in the area shown on the map. *A* shows the test setup. Transmitters (T1 to T4) emitted low-frequency sound signals that were received by the receivers (R1 to R5). Temperatures and currents were measured by conventional means at moorings E1 and E2. B1, 2, 3 show the results of the demonstration. 1 and 3 are based on control data from the survey ships. 2 is the tomography. At the start of the survey, data from a depth of 700 meters (about 2300 feet) showed a large, cold eddy in the center of the survey area. It was separated from warm water to the northeast by a well-defined temperature front. The cold-water eddies observed moved off to the northwest, while the front progressed to the southwest during the demonstration. Cold eddies were indicated by the slowing of sound signals that passed through them. (Data courtesy of Woods Hole Oceanographic Institution.)

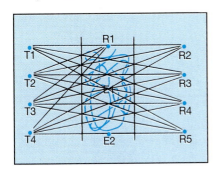

A. Tomography setup

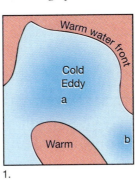

1.

B. Ship survey
(Start of survey)

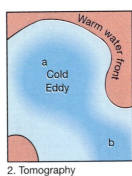

2. Tomography
(3 months into
demonstration)

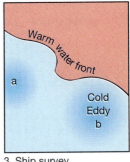

3. Ship survey
(end of survey)

curring was to begin in 1993. However, the test was delayed because some feared that the low-frequency test signals might harm or modify the behavior of marine mammals, particularly seals and whales. (The test signals would be at 75 hertz, a pitch near the lowest key on a piano, emitted for 20 minutes six times a day at 195 decibels sound pressure level, far louder than any imaginable band.)

It must be determined that the sound will not negatively affect mammals. The sounds that are to be emitted off the coasts of California and the island of Kauai, similar to the Heard Island test, would be detected at U.S. Navy listening posts as far as 16,000 kilometers (10,000 miles) away. This efficient travel of sound through the ocean is possible because of the SOFAR channel discussed in Chapter 5.

S U M M A R Y

Horizontal currents set in motion by wind systems are characteristic of the surface waters of the world ocean. According to the Ekman spiral concept, the friction between wind and water sets surface waters in motion, and the direction approaches 45° to the right of the wind in the Northern Hemisphere due to the Coriolis effect, and the net water movement approaches a right angle to the wind direction. Water thus is pushed toward the center of clockwise gyres in the Northern Hemisphere and counterclockwise gyres in the Southern Hemisphere, forming low "hills" in the gyres.

As water in the Northern Hemisphere runs down the slope of these hills, the Coriolis effect causes it to turn right and into the clockwise flow pattern. As a result, a geostrophic current flowing parallel to the contours of the hill

is maintained. The apex of the hill is west of the geographical center of the gyre.

The Circumpolar Current flows in a clockwise direction around Antarctica. It is the greatest-volume current in the world ocean, transporting over 130 sverdrups (million cubic meters of water per second) of water. The surface of this flow is called the West Wind Drift because it is driven by strong westerly winds. Two major deep-water masses form in Antarctic waters.

The highest-velocity ocean current is the Gulf Stream along the southeastern U.S. Atlantic coast.

Driven by the trade winds, north and south equatorial currents flow westward on either side of the equator. They may be separated by an equatorial countercurrent. The best-developed equatorial countercurrent system is found in the Pacific, where a large eastern water movement is achieved by the Pacific Countercurrent and the Equatorial Undercurrent, a ribbonlike subsurface flow. The formation of deep-water masses is not pronounced in the Pacific Ocean, as there appears to be no deep sinking of surface water.

A periodic phenomenon in the Pacific Ocean is the El Niño-Southern Oscillation (ENSO). It is associated with climatic changes worldwide.

The Indian Ocean circulation is dominated by the monsoon wind systems, which change direction with the seasons. They blow from the northeast in the winter and from the southwest in the summer.

The Antarctic Bottom Water, the densest deep-water mass in the oceans, forms primarily in the Weddell Sea and sinks along the continental shelf into the South Atlantic Ocean. Farther north, at the Antarctic Convergence, the low-salinity Antarctic Intermediate Water sinks to a depth of about 900 meters (2950 feet). Sandwiched between these two masses is the North Atlantic Deep Water, rich in plant nutrients after hundreds of years in the deep ocean.

Continued investigation of the energy potential of ocean winds and currents may lead to new renewable-energy technology that can be exploited by society.

Ocean acoustic tomography is a complex acoustical system designed to image a slice of the ocean and provide a picture of temperature distribution over large areas. Some time in the future, it may provide the first subsurface images of in situ temperature conditions within whole ocean basins.

KEY TERMS

Agulhas Current (p. 168)
Antarctic Bottom Water (p. 171)
Antarctic Circumpolar Current (p. 161)
Antarctic Convergence (p. 171)
Antarctic Intermediate Water (p. 171)
Benguela Current (p. 162)
Brazil Current (p. 162)
Canary Current (p. 163)
Caribbean Current (p. 162)
Coastal downwelling (p. 170)
Coastal upwelling (p. 170)
Eastern boundary currents (p. 157)
Ekman spiral (p. 157)

Ekman transport (p. 158)
El Niño-Southern Oscillation (ENSO) (p. 166)
Equatorial Countercurrents (p. 158)
Equatorial Currents (p. 156)
Falkland Current (p. 162)
Florida Current (p. 162)
Gulf Stream (p. 162)
Intertropical Convergence Zone (ITCZ) (p. 166)
Irminger Current (p. 163)
Labrador Current (p. 163)
Leeuwin Current (p. 168)

Monsoon (p. 168)
North Atlantic Deep Water (p. 171)
Norwegian Current (p. 163)
Somali Current (p. 168)
Subpolar gyre (p. 157)
Subtropical convergence (p. 158)
Subtropical gyre (p. 157)
Upwelling (p. 170)
West Australian Current (p. 168)
West Wind Drift (p. 161)
Western boundary current (p. 157)
Westward intensification (p. 158)

QUESTIONS AND EXERCISES

1. Compare the forces directly responsible for creating horizontal and deep vertical circulation in the oceans. What is the ultimate source of energy that drives both circulation systems?
2. On a base map of the world, plot the major surface circulation gyres of the oceans, the meteorological equators of each ocean, and the Subtropical, Arctic, and Antarctic Convergences. Superimpose the major wind belts of the world on the gyres. Label the currents using the symbols used in Figures 7–7, 7–9, and 7–13.
3. Diagram and discuss how the Ekman transport produces the "hill" of water within major ocean gyres that causes geostrophic current flow. As a starting place on the diagram, use the prevailing wind belts, the trade winds, and the westerlies.
4. What causes the apex of these geostrophic "hills" to be offset to the west of the center of the ocean gyre systems?
5. The largest current in the world ocean in terms of volume transport is the Antarctic Circumpolar Current.

Explain why its surface portion is referred to as the West Wind Drift. What is its maximum volume transport as compared with the maximum volume transport for the Gulf Stream?
6. Observing the flow of Atlantic Ocean currents in Figure 7–7, offer some explanations as to why the Brazil Current has a much lower velocity and volume transport than the Gulf Stream.
7. Explain why Gulf Stream eddies that develop northeast of the Gulf Stream rotate clockwise and have warm-water cores, whereas those that develop to the southwest rotate counterclockwise and have cold-water cores.
8. Describe the relations among El Niño–Southern Oscillation events, Walker circulation, trade winds, equatorial countercurrent flow, and the Intertropical Convergence Zone.
9. Describe the relations among the wind, the surface current it creates, and the development of equatorial and coastal upwellings.

10. Discuss why thermohaline vertical circulation is driven by sinking of surface water that occurs only in high latitudes.
11. Name the two major deep-water masses and give the locations of their formation at the ocean's surface.
12. The Antarctic Intermediate Water is identifiable throughout much of the South Atlantic on the basis of a temperature minimum, salinity minimum, and dissolved oxygen maximum. Why should it be colder and less salty and contain more oxygen than the surface-water mass above it and the North Atlantic Deep Water below it?

13. Surface salinities are high in the North Atlantic Ocean and low in the North Pacific Ocean. How can you explain this from the pattern of return of deep-water masses to the surface?
14. Where is the potential for using ocean windpower systems and ocean current power systems greatest, and why?
15. How can ocean acoustic tomography be used to check whether or not global warming is occurring?

REFERENCES

Behringer, D.; Birdsall, T.; Brown, M.; Cornuelle, B.; Heinmiller, R.; Knox, R.; Metzger, K.; Munk, W.; Spiesberger, J.; Spindel, R.; Webb, D.; Worcester, P.; and Wunsch, C. 1982. A demonstration of ocean acoustic tomography. *Nature* 299, 121–125.

Boyle, E., and Weaver, A. 1994. Conveying past climates. *Nature* 372: 41–42.

Gill, A. 1982. *Atmosphere-ocean dynamics.* Orlando, Florida: Academic Press.

Goldstein, R.; Barnett, T.; and Zebker, H. 1989. Remote sensing of ocean currents. *Science* 246:4935, 1282–1286.

Gordon, A. L. 1986. Interocean exchange of thermocine water. *Journal of Geophysical Research* 91:C4, 5037–5046.

Ledwell, J. R.; Watson, A. J.; and Law, C. S. 1993. Evidence for slow mixing across the pycnocline from an open-ocean tracer-release experiment. *Nature* 364:6439, 701–703.

Montgomery, R. 1940. The present evidence of the importance of lateral mixing processes in the ocean. *American Meteorological Society Bulletin* 21:87–94.

Pedlosky, J. 1990. The dynamics of the oceanic subtropical gyres. *Science* 248:4935, 316–322.

Philander, S. G. H. 1983. El Niño Southern Oscillation phenomena. *Nature* 302:5906, 295–301.

Pickard, G. L. 1975. *Descriptive physical oceanography.* 2nd ed. New York: Pergamon Press.

Stommel, H. 1987. *A view of the sea.* Princeton, N.J.: Princeton University Press.

___. 1958. The abyssal circulation. Letters to the editor. *Deep Sea Research* 5, New York: Pergamon Press.

Street-Perrott, A. F., and Perrott, R. 1990. Abrupt climate fluctuations in the tropics: The influence of Atlantic Ocean circulation. *Nature* 343:6259, 607–612.

Stuiver, M.; Quay, P. D.; and Ostlund, H. G. 1983. Abyssal water carbon-14 distribution and the age of the world oceans. *Science* 219:4586, 849–851.

Sverdrup, H. U.; Johnson, W.; and Fleming, R. H. 1942. Renewal 1970. *The oceans.* Englewood Cliffs, N.J.: Prentice Hall.

Wood, J. D. 1985. The world ocean circulation experiment. *Nature* 314, 501–511.

SUGGESTED READING

Sea Frontiers

Alper, J. 1991. Munk's hypothesis. 37:3, 38–43. Walter Munk's proposal to monitor global warming by transmitting sound signals across vast ocean reaches in the SOFAR channel is discussed.

Frye, J. 1982. The ring story. 28:5, 258–267. A discussion of the Gulf Stream, its rings, and how the rings are studied.

Miller, J. 1975. Barbados and the island-mass effect. 21:5, 268–272. A discussion of the phenomenon by which waters around tropical islands are much more productive than the surface waters of the open ocean.

Smith, F. G. W. 1972. Measuring ocean movements. 18:3, 166–174. Discussed are some practical problems related to current flow and some methods used to determine current direction, speed, and volume.

Smith, F. G. W., and Charlier, R. 1981. Turbines in the ocean. 27:5, 300–305. A discussion of the potential of extracting energy from ocean currents is presented.

Sobey, E. 1982. What is sea level? 28:3, 136–142. The factors that cause sea level to change are discussed.

Scientific American

Baker, D. J., Jr. 1970. Models of ocean circulation. 222:1, 114–121. This article discusses observations of a model depicting a segment of the surface of Earth over which fluids move and helps explain geostrophic flow within ocean gyres. Some knowledge of basic physics is necessary for full comprehension of the material presented.

Hollister, C. D., and Nowell, A. 1984. The dynamic abyss. 250:3, 42–53. Submarine "storms" are associated with deep, cold currents flowing away from polar regions toward the equator.

Munk, W. 1955. The circulation of the oceans. 191:96–108. A young physical oceanographer who has gone on to be highly honored by the scientific community presents his views of ocean circulation.

Spindel, R. C., and Worcester, P. F. 1990. Ocean acoustic tomography. 263:4, 94–99. The theoretical basis and practical application of OAT are clearly presented.

Stewart, R. W. 1969. The atmosphere and the ocean. 221:3, 76–105. The exchange of energy between the atmosphere and ocean and the resulting phenomena, currents, and waves are covered in this readable, comprehensive article.

Stommel, H. 1955. The anatomy of the Atlantic. 190:30–35. Another vintage article by a young oceanographer who went on to great achievements.

Webster, P. J. 1981. Monsoons. 245:5, 108–119. The mechanism of the monsoons and their role in bringing water to half Earth's population are discussed.

Welbe, P. H. 1982. Rings of the Gulf Stream. 246:3, 60–79. The biological implications of the large cold-water rings are considered.

C H A P T E R 8
Waves

The ocean surface seems restless, always in motion. The most obvious motion is that of waves traveling across its surface. The smaller, most common surface waves are driven by the wind and break in familiar fashion on the shore. Most of the time, they release their energy gently, although ocean storms sometimes push them ashore hard and fast, with devastating effect.

Undersea earthquakes can generate very large, low tsunami waves that extend to the surface. Fortunately, these potentially destructive waves are infrequent visitors to coastal regions, where they may inundate the shore up to 30 meters (100 feet) above normal sea level.

The tides are also a wave phenomenon. They are vast, low waves generated by the gravitational attraction of the sun and moon to every particle of ocean water.

Mariners in the days of sail always greeted imminent storms with a sense of challenge, because their humdrum daily shipboard routine would be replaced with excitement and—because of the waves—some moments of sheer terror.

Waves are among the best-known ocean phenomena. Yet it was not until well into the 1800s that some understanding of waves developed, including their causes and behavior. We will first look at the character of waves in general and then discuss oceanic waves.

How Waves Move

Waves are energy in motion. Wave phenomena involve transmission of energy by means of cyclic movement through matter. The medium itself (solid, liquid, or gas) does not actually travel in the direction of the energy that is passing through it. The particles in the medium simply oscillate, or cycle, in a back-and-forth or up-and-down or orbital pattern, transmitting energy from one particle to another. For example, if you thump your fist on a table, the energy waves travel through the table and can be felt by someone sitting at the other end, but the table itself does not move.

Waves move in different ways. Simple *progressive waves* (in which the waveform can be observed to progress or travel, as in water) are shown in Figure 8–1A. Progressive waves may be *longitudinal* or *transverse*.

In **longitudinal waves** (push-pull), such as sound waves, the particles that are in vibratory motion "push and pull" in the same direction that the energy is traveling, like a coil spring whose coils are alternately compressed and spread. A waveform travels through the medium by compressing and decompressing. Energy may be transmitted through all states of matter—gaseous, liquid, or solid—by this longitudinal movement of particles.

In **transverse waves** (side-to-side), energy travels at right angles to the direction of particle vibration. For example, if one end of a rope is fastened while the other end is moved up and down or side to side by hand, a waveform is set up in the rope and progresses along it, and energy is transmitted from the motion of the hand to the fastened end. If you watch any segment of the rope, you can see it move up and down (or side to side) at right angles to a line drawn from the fastened

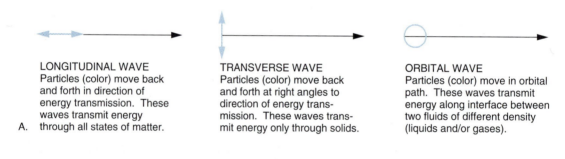

LONGITUDINAL WAVE
Particles (color) move back and forth in direction of energy transmission. These waves transmit energy
A. through all states of matter.

TRANSVERSE WAVE
Particles (color) move back and forth at right angles to direction of energy transmission. These waves transmit energy only through solids.

ORBITAL WAVE
Particles (color) move in orbital path. These waves transmit energy along interface between two fluids of different density (liquids and/or gases).

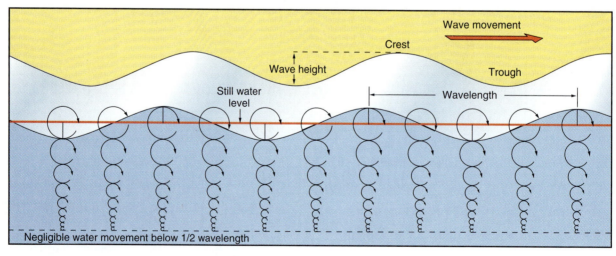

Figure 8–1

Types of Progressive Waves. *A:* Types of progressive waves. *B:* A simple sea wave and its components: height, wavelength, and depth. (*B:* From the Tasa Collection: Shorelines. Published by Macmillan Publishing Co., New York. Copyright © 1986 by Tasa Graphic Arts, Inc. All rights reserved.)

end to the hand that is putting energy into the rope. Generally, this type of wave transmits energy only through solids, because it is only in a solid that particles are strongly bound to one another.

Longitudinal and transverse waves are called *body waves,* as they transfer energy through a body of matter.

Waves that transmit energy along an interface between any two fluids of different density have particle movements that are neither longitudinal nor transverse. The best example of such an interface is that between air and water (atmosphere and ocean). We may say that the movement of particles along such an interface involves components of both longitudinal and transverse waves, because the particles move in circular orbits. Thus, waves at the ocean surface are **orbital waves,** or *interface waves.*

Wave Characteristics

When we observe the ocean surface, we see waves of various sizes moving in various directions. This produces a complex wave pattern that constantly changes. To introduce concepts needed to understand waves, we will look at a simple, idealized waveform (Figure 8–1B). It represents the transmission of energy from a single source and traveling along the ocean-atmosphere interface. Such a wave has uniform characteristics.

As our idealized progressive wave passes a permanent marker, such as a pier piling, you will notice a succession of high parts of the waves, **crests,** separated by low parts, **troughs.** If we mark the water level on the piling when the troughs pass and then do the same for the crests, the vertical distance between the marks is the **wave height, *H.***

The horizontal distance between any two corresponding points on successive waveforms, such as from crest to crest or from trough to trough, is the **wavelength, *L*.** The ratio of wave height to wavelength, ***H/L,*** is called **wave steepness.**

The time that elapses during the passing of one full wave, or wavelength, is the **wave period, *T*** (for time). Figure 8–2 shows the relations among wavelength, period, and speed. Because the period is the time required for the passing of one wavelength, if either the wavelength or period of a wave is known, the other can be calculated. Since $L(m) = 1.56$ (m/s)T^2,

$$\text{Speed }(S) = \text{Wavelength }(L) \text{ / Period }(T)$$

For example:

$$\text{Speed }(S) = L/T = 156 \text{ m/10 s} = 15.6 \text{ m/s}$$

The circular orbits followed by the water particles at the surface have a diameter equal to the wave height (Figure 8–1). While a particle is in the crest of a passing wave, it is moving in the direction of energy propagation. While it is in the trough, it is moving in the opposite direction, or backward. The half of the orbit accomplished in the trough is slower than the crest half of the orbit. Therefore, there is a small net transport of water in the direction the waveform is moving.

This condition results from the fact that particle speed slows with increasing depth below the still-water line. Also, the diameters of particle orbits decrease with increased depth until particle motion associated with our idealized wave ceases at a depth of one-half wavelength, *L/2.*

Deep-Water Waves

Ocean waves having the "ideal" characteristics just discussed belong to a category called **deep-water waves.** Such waves travel across the ocean where the water depth (*d*) is greater than one-half the wavelength (Figure 8–3A).

Included in deep-water waves are all wind-generated waves as they move across the open ocean. As you can see from Figure 8–2, the speed of deep-water waves is related to wavelength (*L*) and period (*T*). The easier of these characteristics to measure is period (the

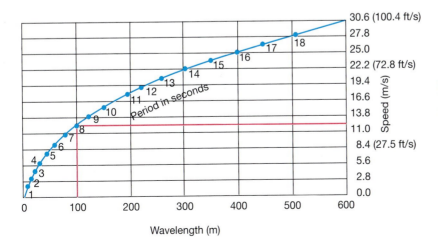

Figure 8–2

Speed of Deep-Water Waves. Ideal relations among wave speed, wavelength, and period for deep-water waves. Speed (meters per second) equals the wavelength (meters) divided by the period (seconds), or *S = L/T.* For example (shown with red lines), a wave with a period of 8 seconds has a wavelength of about 100 meters and a speed of about 12.5 meters per second.

A. DEEP-WATER WAVE

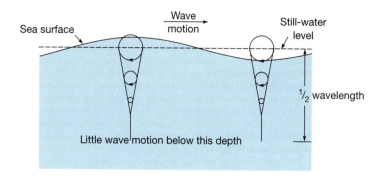

B. SHALLOW-WATER WAVE

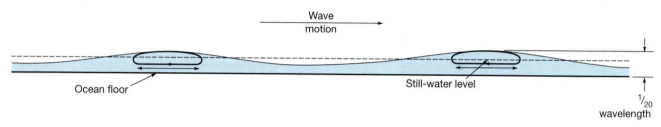

Figure 8–3

Deep-Water and Shallow-Water Waves. *A:* Wave profile and water-particle motions of a deep-water wave. Note the diminishing size of the orbits with increasing depth below the surface. *B:* Motions of water particles in shallow-water waves. Water motion extends to ocean floor.

time required for one wave to pass). Thus, the following equation is most commonly used for computing wave speed (*g* is acceleration due to gravity):

$$\text{Speed (meters/second)} = \text{gravity } (g) \times \text{period } (T) / 2\pi$$

Shallow-Water Waves

Waves in which depth (*d*) is less than $\frac{1}{20}$ of the wavelength (*L*/20) are classified as **shallow-water waves,** or *long waves* (Figure 8–3B). Included in this category are wind-generated waves that have moved into shallow nearshore areas; *tsunami* (seismic sea waves), generated by earthquakes in the ocean floor; and *tide waves,* generated by gravitational attraction of the sun and moon. In these waves, the wavelength is very great relative to water depth, and speed is determined by water depth.

Particle motion in shallow-water waves is in a very flat elliptical orbit that approaches horizontal oscillation. The vertical component of particle motion decreases with increasing depth. The presence of shallow-water waves can be detected to the ocean bottom and therefore can affect the bottom.

Transitional Waves

Transitional waves have wavelengths greater than twice the water depth, but less than 20 times. The speed

of transitional waves is controlled partially by wavelength and partially by water depth (Figure 8–4). Deep-water waves generated by winds at the ocean surface usually have periods of 10 to 12 seconds. They maintain their periods after encountering shallow coastal water.

Figure 8–5 shows the concentration of energy in ocean waves that have periods from less than 0.1 second to more than 1 day. The illustration also shows the principal causes of waves of different periods.

Wind-Generated Waves

Wind-generated waves have a life history that includes their origin in windy regions of the ocean, their movement across great expanses of open water without subsequent aid of wind, and their termination when they break and release their energy, either in the open ocean or against the shore.

"Sea"

As the wind blows over the ocean surface, it creates pressure and stress. These deform the ocean surface into small, rounded waves with extremely short wavelengths, less than 1.74 centimeters (0.7 inches). Commonly they are called *ripples,* but oceanographers call them **capillary waves.** The name comes from *capillar-*

Figure 8–4

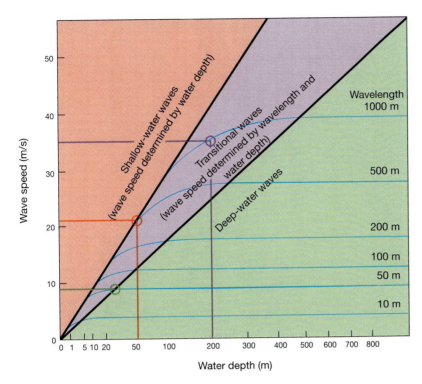

Wave Speed. This graph shows the relation of wave speed to wavelength and/or water depth. It is divided into three zones because water depth determines the speed of shallow-water waves (red), wavelength determines the speed of deep-water waves (green), and both affect the speed of transitional waves (purple). The blue lines are different wavelengths. *Shallow-water waves (red zone):* To determine wave speed, find the water depth on the bottom scale (red example: 50 meters). Then move up to the "shallow-water waves" line, and then over to the wave-speed scale. In the example, a wave in 50-meter-deep water travels about 21 meters per second. *Deep-water waves (green zone):* Here, water depth is not a factor. So to determine wave speed, simply extend the wavelength line (blue) to the left until it intersects the wave-speed scale. The green example is for a 50-meter wavelength, showing a wave speed of about 9 meters per second. *Transitional waves (purple zone):* Combine the shallow-water and deep-water procedures for transitional waves. The purple example is for a 1000-meter wavelength and 200-meter water depth. Note that the transitional waves have a slower wave speed than they would have had as deep-water waves.

ity (the surface tension of water). Capillarity is the dominant **restoring force** that works to destroy these tiny waves, restoring the smooth ocean surface once again. Capillary waves characteristically have rounded crests and V-shaped troughs (Figure 8–6).

As capillary wave development increases, the sea surface takes on a rougher character. This "catches" more of the wind, allowing the wind and ocean surface to interact more efficiently, transferring more energy to the water. As more energy is transferred to the ocean, **gravity waves** develop. They have wavelengths exceeding 1.74 centimeters (0.7 inches) and are shaped more like a sine curve (the familiar waveform shown in the middle of Figure 8–6). Because they reach greater height at this stage, gravity replaces capillarity as the dominant restoring force, giving these waves their name.

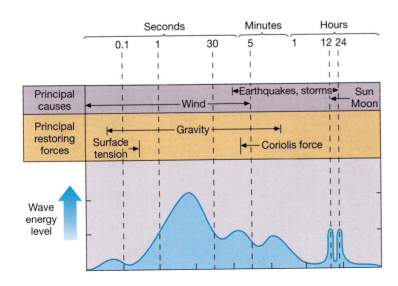

Figure 8–5

Distribution of Energy in Ocean Waves. This figure shows that most of the energy possessed by ocean waves is in wind-generated waves with a period of about 10 seconds. The two lower peaks on either side of the 5-minute period mark represent tsunami, whereas the two sharp peaks to the right represent ocean tides with their semidaily and daily periods. (After Blair Kinsman, *Wind waves: Their generation and propagation on the ocean surface.* © 1965.)

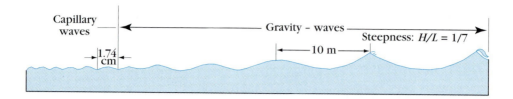

Figure 8–6

Capillary and Gravity Waves. As energy is transferred to the ocean surface by wind, small rounded waves with V-shaped troughs develop (*capillary waves*). As the water gains energy, the waves increase in height and length. When they exceed 1.74 centimeters (0.7 inches) in length, they take on the shape of sine waves and become *gravity waves*. Increased energy increases the steepness of the waves. The crests become pointed and the troughs rounded. As the steepness reaches the point where the wave height is 1/7 of its length, the waves become unstable and topple, forming *whitecaps* as they "break." (Not to scale.)

The length of these "young" waves is generally 15 to 35 times their height. As additional energy is gained, wave height increases more rapidly than wavelength. The crests become pointed and the troughs are rounded (Figure 8–6, right end). When the steepness reaches a ratio of 1/7, the speed with which the waves travel becomes 1.2 times that of typical deep-water waves.

Energy imparted by the wind increases height, length, and speed. When wave speed reaches that of the wind, neither height nor length can change because there is no net energy exchange, so the wave is at its maximum size.

Figure 8–7 shows important relations between wind and the area where waves are initiated. The area where wind-driven waves are generated is called, simply, **sea**. For clarity in this discussion, we will call it the *sea area*. It is characterized by choppiness and short wavelengths, with waves moving in many directions and having many different periods and lengths. The variety of wave periods and wavelengths is caused by frequently changing wind speed and direction.

Factors that are important in increasing the amount of energy that waves obtain are (1) wind speed, (2) the duration of time during which the wind blows in one direction, and (3) the *fetch*—the distance over which the wind blows in one direction.

Wave height is directly related to the energy in a wave. Wave heights in a sea area are usually less than 2 meters (6.6 feet), although it is not uncommon to observe waves with heights of 10 meters (33 feet) and periods of 12 seconds. As sea waves gain energy, their steepness increases. When steepness reaches a critical value of 1/7, open ocean breakers, called *whitecaps,* form. Figure 8–8 is a satellite image of average wave heights worldwide as they existed during the brief life of the *SEASAT* satellite in 1978.

The largest wind-generated wave authentically measured was 34 meters high (112 feet), with a period

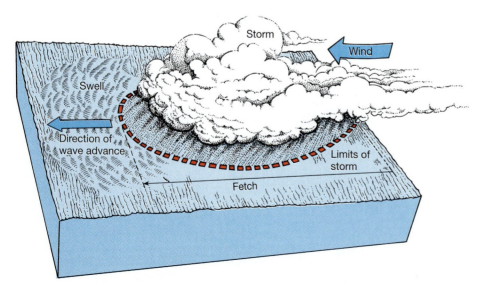

Figure 8–7

The "Sea" Area and Swell. As wind blows across the *sea area* (outlined in red), wave size increases with wind speed, fetch, and duration. As waves advance beyond the sea area, they continue to advance across the ocean surface as *swell*. Swell is free waves that are sustained by the energy they obtained in the sea area and not by new input of wind energy.

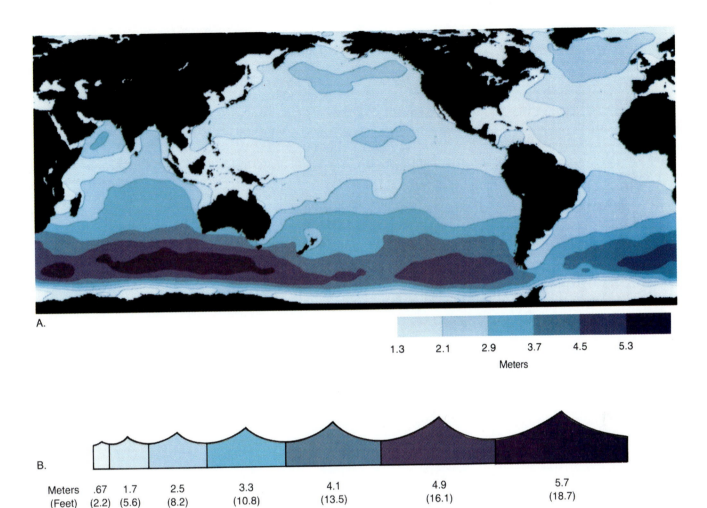

A.

	1.3	2.1	2.9	3.7	4.5	5.3

Meters

B.

Meters	.67	1.7	2.5	3.3	4.1	4.9	5.7
(Feet)	(2.2)	(5.6)	(8.2)	(10.8)	(13.5)	(16.1)	(18.7)

Figure 8–8

***SEASAT* Altimeter Wave Height, July 7–October 10, 1978.** Because the westerly wind belt in the Southern Hemisphere reaches the highest average wind speed on Earth, this wind system produces the largest ocean waves. The largest of the large occur in the southern Indian Ocean. The wave heights represented here are averages for the largest one-third of the waves occurring at any location.

of 14.8 seconds. It was seen in the North Pacific in 1935 by the crew of the U.S. Navy tanker U.S.S. *Ramapo*.

For a given wind speed, there is a maximum fetch and duration of wind beyond which the waves cannot grow. When both maximum fetch and duration are attained for a given wind velocity, a **fully developed sea** has been created. The reason it can grow no further is that waves are losing as much energy breaking as whitecaps under the force of gravity as they are receiving from the wind.

Swell

As waves generated in a sea area move toward its margins, where wind speeds diminish, they eventually move faster than the wind. When this occurs, the wave steepness decreases, and they become long-crested waves called **swell.** The swell moves with little loss of energy over large stretches of the ocean surface, transporting energy away from the wind-energy-input area of the sea area. Most of this energy eventually will be released along the continental margins as a major erosion agent.

Waves with longer length, traveling faster, leave the sea area first. They are followed by shorter-length, slower **wave trains,** or groups of waves.

This progression illustrates the principle of **wave dispersion,** which is a sorting of waves by their wavelength. In the generating area, waves of many wavelengths are present. In deep water, wave speed is a function of wavelength (Figure 8–2), so the longer waves "outrun" the shorter ones.

To illustrate the distance that can be traveled by swell without significant depletion of its energy, swell that originates from Antarctic storms has been recorded breaking along the Alaskan coast after traveling more than 10,000 kilometers (over 6000 miles).

As a group of waves leaves a sea area and becomes a swell wave train, the group moves across the ocean surface at only half the velocity of an individual wave in the group. Progressively, the leading wave disappears. However, the same number of waves always remains in the group. As the leading wave disappears, a new wave replaces it at the back of the group.

Interference Patterns. Because swells may move away from a number of storm areas in any ocean, it is inevitable that the swell from different storms will run together and the waves will clash, or interfere with one another. This gives rise to a special feature of wave motion: *interference patterns*. An interference pattern produced when two or more wave systems collide is the algebraic sum of the disturbance that each wave would have produced individually. The result may be a larger or smaller trough or crest, depending on conditions (Figure 8–9).

When swells from two storm areas collide, the interference pattern may be constructive or destructive, but it is more likely to be mixed. Ideally, **constructive interference** results when wave trains having the same wavelength come together *in phase,* meaning crest to crest and trough to trough. If we sum the displacements

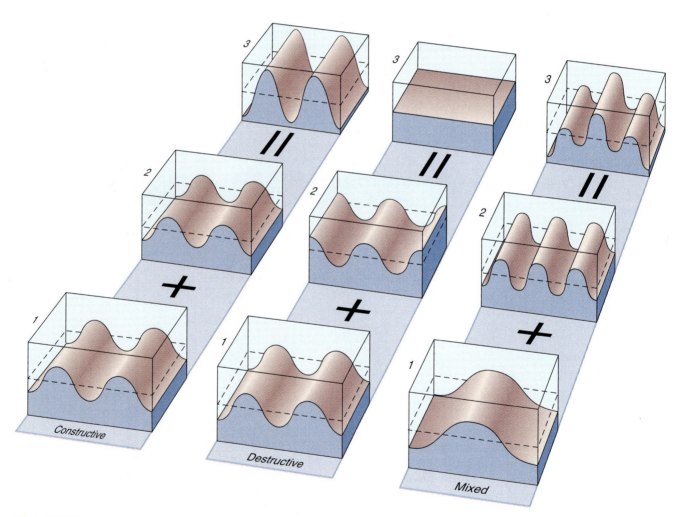

Figure 8–9

Wave Interference. As wave trains come together from different sea areas (1 and 2), three possible interference patterns may result (3). Very rarely, if the waves have the same length and come together *in phase* (crest to crest and trough to trough), totally constructive interference may occur. In this case, the amplitudes simply add to produce waves of the same length but of greater height. If two sets of waves have identical characteristics but come together exactly 180° *out of phase,* the result will be complete cancellation, or destructive interference, with no remaining wave at all. More commonly, waves of different lengths and heights encounter one another and produce a complex, mixed interference pattern.

that would result from each wave individually, we find that the resulting interference pattern is a wave having the same length as the two converging wave systems, but with a wave height equal to the sum of the individual wave heights (Figure 8–9, top).

Destructive interference results when wave crests from one swell coincide with the troughs from a second swell. If the waves have identical characteristics, the algebraic sum of the crest plus the trough would be zero, and the energy of these waves would cancel each other.

However, it is more likely that the two swell systems possess waves of various heights and lengths and would come together with a mixture of destructive and constructive interference. Thus, a more complex **mixed interference** pattern would develop. It is such interference that explains the occurrence of a varied sequence of high and lower waves and other irregular wave patterns we observe when swell approaches the seashore.

You can experience this effect in another fluid, the air, by listening to a steady, high-pitched sound in a room. An example is a steadily blown note on a brass instrument. As you move your head, the sound will grow loud (constructive interference) or virtually disappear (destructive interference), but mostly it will vary (mixed interference).

Rogue Waves

One of the ocean's mysteries is the cause of **rogue waves,** massive waves that can reach 10 stories in height. They result from rare coincidence in ordinary wave behavior. On average in the open ocean, one wave in 23 will be over twice the height of the wave average, one in 1175 will be three times as high, and only one in 300,000 will be four times as high. The chances of a truly monstrous wave are one in billions, but it does happen. Of course, this statistical information is useless in predicting specifically when and where a rogue wave will arise.

Lloyd's of London is the world's leading marine insurance underwriter. It reports that during the 1980s, storms claimed an annual average of 46 ships weighing over 500 tons (for example, a 15-meter-long (50-foot) fishing vessel). The total of vessels lost of all sizes may reach 1000 per year. Some are the victims of rogue waves.

Although it is impossible to predict their occurrence, a main cause of rogue waves is theorized to be extraordinary constructive wave interference. Also, it has been determined that rogue waves tend to occur more frequently in locations that are downwind from islands or shoals, and where storm-driven waves move against strong ocean currents, such as the Agulhas Current off the southeastern coast of Africa. This stretch of water, where Antarctic storm waves drive northeast into the Agulhas Current, is probably responsible for sinking more ships than any other place on Earth (Figure 8–10).

During the three-month life of the Seasat satellite, oceanographers learned more about ocean waves than from the centuries-long history of gathering data by ship, buoy, and aircraft. New satellites designed to go into service during the 1990s will provide great additional data, but they still will not make it possible to predict a rogue wave.

Surf

Most waves generated in the sea area by storm winds move across the ocean as swell. They release their energy along the margins of continents in the **surf zone,** where the swell forms breakers. As deep-water waves of the swell move toward continental margin over gradually *shoaling* water (water growing shallower), they eventually encounter water depths that are less than one-half of their wavelength (Figure 8–11).

These shoaling depths interfere with water particle movement at the base of the wave, so the wave slows. As one wave slows, the following waveform, which is still moving at unrestricted speed, overtakes the wave that is being slowed, thus reducing the wavelength. The energy in the wave, which remains the same, must go somewhere, so wave height increases. The crests become narrow and pointed and the troughs become wide curves, a waveform previously described for high-energy waves in the open sea. The increase in wave height accompanied by a decrease in wavelength increases the steepness (H/L) of the waves. As the wave steepness reaches 1/7, the waves break as surf (Figure 8–11).

If the surf is swell that has traveled from distant storms, breakers will develop relatively near shore in shallow water, the shoaling of which is primarily re-

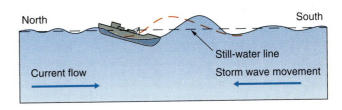

North   South

Still-water line

Current flow   Storm wave movement

Figure 8–10

Rogue Waves. This illustration shows an oil tanker encountering the rogue wave phenomenon typical off the southeastern coast of Africa. Riding the Agulhas Current southward to save fuel, it encounters a rogue wave that results from the northward travel of Antarctic storm waves. A deep trough, or "hole in the ocean" as it is called, drops 15 meters (50 feet) below the still-water line. The bow of the ship drops down into the trough, and the coming crest, which may be up to 15 meters above the still-water line, crashes onto the bow. This huge mass of water overcomes the structural capacity of the ship, so it sinks. Rogue waves have severely damaged a number of tankers and sunk others.

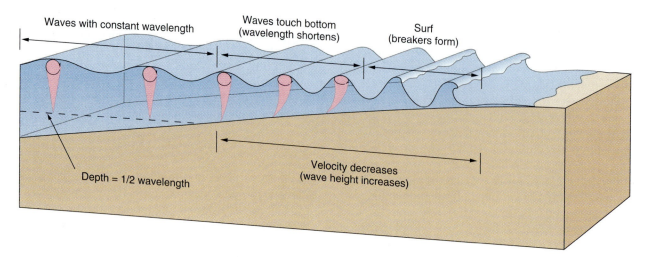

Figure 8–11

Surf Zone. As waves approach the shore and encounter water depths of less than one-half wavelength, friction removes energy from the waves. Each wave is traveling faster than the one in front and slower than the one behind it, so that wavelength decreases; wave height increases because the water must go somewhere, and the only way is up. When the water depth becomes 1/3 of wave height, the wave reaches a steepness of 1/7 and breaks on the shore. (The Tasa Collection: *Shorelines*. Published by Macmillan Publishing, New York. Copyright © 1986 by Tasa Graphic Arts, Inc. All rights reserved.)

sponsible for their breaking. By the time waves break, they have become shallow-water waves. The horizontal water motion associated with such waves moves water alternately toward and away from the shore in a sort of oscillation. The surf will be characterized by parallel lines of relatively uniform breakers.

However, if the surf is composed of waves that have been generated by local wind, the waves may not have been sorted into swell. The surf may be more nearly characterized by unstable, deep-water, high-energy waves with steepness already near 1/7. They will break shortly after feeling bottom some distance from shore, and the surf will be rough, choppy, and irregular.

Ideally, when the water depth is about 1.3 times the breaker height, the crest of the wave breaks, producing surf. When the water depth becomes less than ¹⁄₂₀ the wavelength, waves in the surf zone begin to behave as shallow-water waves (Figure 8–3). Particle motion is greatly hampered by the bottom, and a very significant transport of water toward the shoreline occurs (Figure 8–11).

Breaking in the surf results from severe restriction of particle motion near the wave's bottom. This slows the waveform, but at the surface, individual water particles that are orbiting have not been slowed, because they have no contact with the bottom. Thus, the top of the waveform leans shoreward as the bottom of the waveform "drags bottom." The entire waveform slows, and water moves faster in the circular orbits near the surface due to increasing wave height. This causes wa-

ter particles at the ocean/atmosphere interface to move faster toward shore than the waveform itself.

This motion can be seen particularly well in **plunging breakers** (Figure 8–12A), which have a curling crest that moves over an air pocket. This results because the curling particles literally outrun the wave, and there is nothing beneath them to support their motion. Plunging breakers form on moderately steep beach slopes.

The more commonly observed breaker is the **spilling breaker** (Figure 8–12B). These result from a relatively gentle slope of the ocean bottom, which more gradually extracts energy from the wave, producing a turbulent mass of air and water that runs down the front slope of the wave instead of producing a spectacular cresting curl. Because of the gradual extraction of energy of spilling breakers, they have a longer life span and give surfers a longer, if less exciting, ride than do the plunging breakers.

Recalling the particle motion of ocean waves in Figure 8–1, you can see that, in front of the crest, water particles are moving up into the crest. It is this force, along with buoyancy, that helps maintain a surfer's position in front of a cresting wave. When this upward motion of water particles is interrupted by the wave passing over water that is too shallow to allow this movement to continue, the ride is over. A skillful surfer, by positioning the board properly on the wave front, can regulate the degree to which the propelling gravitational forces exceed the buoyancy forces, and high speeds can be obtained while moving along the face of the breaking wave.

A.

B.

Figure 8–12

Breakers. *A:* Plunging breaker at Oahu, Hawaii. *B:* Spilling breaker. (Photo *A* © Tony Arruza/ Bruce Coleman, Inc.; photo *B* © Peter Arnold, Inc.)

Wave Refraction

As discussed, waves begin to bunch up and wavelengths become shorter as swell begins to "feel bottom" upon approaching the shore. However, it is seldom that swell will approach a shore at a perfect right angle (90°). Some segment of the wave can be expected to feel bottom first, and therefore it will slow before the rest of the wave. This bends the wave front, which is called **refraction,** as waves approach the shore. As shown in Figure 8–13, an irregular shoreline can result from an irregular bottom topography, because it slows portions of waves that approach the shore.

Wave refraction unevenly distributes wave energy along the shoreline. We can construct lines that are perpendicular to the wave fronts and space them so that the energy between lines is equal at all times. These are called *orthogonal lines* (Figure 8–13). They help us see

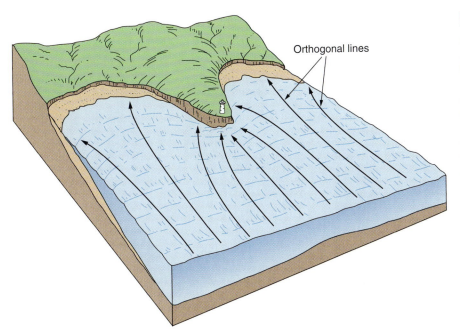

Orthogonal lines

Figure 8–13

Refraction—Bending Waves. As waves first "feel bottom" in the shallows off the headlands, they are slowed. But segments of the waves that move through deeper water leading into the bay are not slowed until they are well into the bay. As a result, the waves are *refracted* (bent). Their release of energy becomes concentrated on the headlands. Consequently, erosion is active on the headlands, whereas deposition occurs in the bay, where the energy level is low. Orthogonal lines, spaced so that equal amounts of energy are between any two adjacent lines, help to show this distribution of energy along the shore. (Adapted from Tarbuck, E. J., and Lutgens, F. K., *Earth science,* 6th ed. New York: Macmillan, 1991.)

how energy is distributed along the shoreline by breaking waves.

Orthogonals indicate the direction that waves travel. You can see that waves converge on headlands that jut into the ocean, but the same waves diverge in bays. Thus, a concentration of energy is released against the headlands, whereas energy is more dispersed in bays. The result is heavy erosion of headlands, whereas deposition may occur in bays. The greater energy of waves breaking on headlands is reflected in an increased wave height.

Wave Reflection

Not all of the energy of waves is expended as they rush onto the shore. A vertical barrier, such as a seawall, can reflect swell back into the ocean with little loss of energy. For this ideal reflection to occur without energy loss, the wave must strike the barrier at a right angle. Such a condition is rare in nature. Nonetheless, less ideal reflections will produce **standing waves,** also called *stationary waves.* Standing waves are the product of two waves of the same length moving in opposite directions, so there is no net movement.

The standing wave is a special interference pattern. In a standing wave, no net momentum is carried because the waves move in opposite directions. The water particles continue to move vertically and horizontally, but there is no more of the circular motion that we see in a progressive wave.

Standing waves are characterized by lines along which there is no vertical movement. There may be one or more such *nodes,* or nodal lines. *Antinodes,* crests that alternately become troughs, represent the points of the greatest vertical movement within a standing wave (Figure 8–14).

There is no particle motion when an antinode is at its greatest vertical displacement, and the maximum particle movement occurs when the water surface is level. At this time, the maximum movement of the water is in a horizontal direction directly beneath the nodal lines. The movement of water particles beneath the antinodes is entirely vertical.

We will consider standing waves further when we discuss the tides in the next chapter. Under certain conditions, the development of standing waves significantly affects the tidal character of coastal regions.

Reflection of wind-generated waves from coastal barriers occurs at an angle equal to the angle at which the wave approached the barrier, as shown in Figure 8–15. This type of reflection may produce a small-scale interference pattern similar to those previously discussed.

An outstanding example of reflection phenomena is the Wedge, a wave feature that develops west of the jetty at Newport Harbor, California (Figure 8–15A). As incoming waves strike the jetty and are reflected, a constructive interference pattern develops. When the crests of incoming waves merge with crests of reflected waves, plunging wedge-shaped breakers may exceed 8 meters (26 feet) height (Figure 8–15B). These waves present a fierce challenge to the most experienced body surfers. The Wedge has killed or crippled many who have come to try it.

Storm Surge

Large cyclonic storms have low pressure at Earth's surface. When such a storm develops over the ocean, the low pressure, compared with surrounding areas of the sea, produces a low "hill" of water. As the storm migrates across the open ocean, this hill moves with it. As the storm approaches shallow water near shore, the portion of the hill over which the wind is blowing shoreward will produce a **storm surge,** a mass of elevated, wind-driven water that produces an increase in sea level.

A storm surge can be extremely destructive to low-lying coastal areas, especially when it occurs at high tide. The coincidence of a storm surge with high tide in areas that normally have particularly high tides frequently produces major catastrophes, with great loss of life and property damage (Figure 8–16).

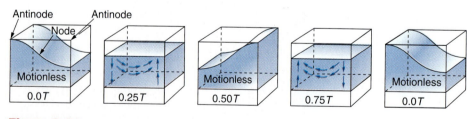

Figure 8–14

Reflection—Standing Waves. An example of water motion viewed at four points during a wave cycle. Water is motionless when antinodes reach maximum displacement. Water movement is maximum when the water is level. Movement is totally vertical beneath the antinodes, and maximum horizontal movement occurs beneath the node. The circular motion of particles in progressive waves does not exist in standing waves.

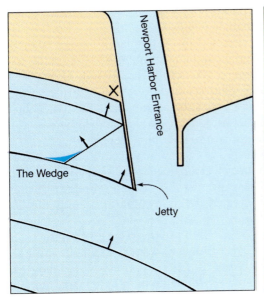

B.

Figure 8–15

Reflection—The Wedge. *A:* The Wedge, a wedge-shaped wave crest that may reach heights exceeding 8 meters (26 feet), develops from interference between incoming waves and reflected waves near the jetty that protects the entrance to Newport Harbor, California. The jetty is 400 meters long (1300 feet). *B:* View of The Wedge crest from the landward end of the jetty. The three dots in the water in front of the wave are the heads of body surfers waiting to catch the wave.

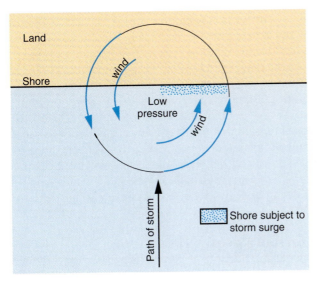

A.

B.

Figure 8–16

Storm Surge. *A:* As a cyclonic storm in the Northern Hemisphere moves ashore, the low-pressure cell around which the storm winds blow and the onshore winds of the storm produce a high-water *storm surge*. *B:* Low pressure and high onshore winds of Hurricane Kate produced a storm surge at Key West, Florida, in November 1985. (Photo © M. Laca/Weatherstock.)

One of the most remarkable examples of storm surge occurred at a lighthouse at Dunnet Head, Scotland. The lighthouse sits atop a cliff some 90 meters (about 300 feet) high. As storm surge waves broke against the cliff, they tossed stones that broke windows in the lighthouse.

Tsunami

The Japanese call the large, destructive waves that occasionally roll into their harbors **tsunami,** or "harbor waves." Tsunami are unusual waves that originate from earthquakes. The news media often mistakenly call them "tidal waves," which implies they are related to the tides, but no such relation exists.

Tsunami are usually caused by *fault movement,* or displacement in Earth's crust along a fracture. This causes not only an earthquake but also a sudden change in water level at the ocean surface above. Secondary events, such as underwater avalanches produced by the faulting, may also produce tsunami (Figure 8–17A).

Because the wavelength of a typical tsunami exceeds 200 kilometers (125 miles), it is obviously a shallow-water wave, the speed of which is determined by water depth. Moving at great speeds in the open ocean, well over 700 kilometers per hour (435 miles per hour), these waves have heights of only approximately 0.5 meters (1.6 feet) in the open ocean. Thus, tsunami are not readily observable until they reach shore. In shallow water, they slow, and the water begins to pile up. Tsunami may form crests that exceed 30 meters height (100 feet), rushing into unsuspecting harbors with destructive results (Figures 8–17B and 8–17C). A tsunami may consist of a single wave or multiple waves depending on how the earthquake releases energy.

The ocean most plagued by tsunami is the Pacific. It is ringed by a series of trenches that are unstable margins of lithospheric plates, along which large-magnitude earthquakes occur.

One of the most destructive tsunami ever generated came from the greatest release of energy from Earth's interior observed during historical times. On August 27, 1883, the volcanic island of Krakatau in what is now Indonesia exploded and essentially disappeared. The sound of the explosion was heard an incredible 4800 kilometers (2981 miles) away at Rodriguez Island in the western Indian Ocean. Dust from the explosion ascended into the atmosphere and circled Earth on high-altitude winds. This dust produced unusual and beautiful sunsets for nearly a year.

Not many were killed by the outright explosion of the volcano. But the displacement of water from the explosion was enormous. The resulting tsunami exceeded 30 meters (100 feet) height. It devastated the coastal region of the Sunda Strait between Sumatra and Java, taking more than 36,000 lives. The energy carried by this wave was detected in the English Channel, after the wave crossed the Indian Ocean, passed the southern tip of Africa, and moved north in the Atlantic Ocean.

Unlike a hurricane, whose high winds and waves threaten ships at sea and send them to the protection of a coastal harbor, a tsunami washes ships from their coastal moorings into the open ocean.

Before today's tsunami warning system existed, the first notice of a tsunami to most observers would be the rapid seaward recession of the shoreline. This is a recession created by a trough preceding the first tsunami wave crest. This recession would be followed in minutes by one or more destructive waves.

Such a recession was observed in the port of Hilo, Hawaii, on April 1, 1946. It was the beginning of a tsunami from an earthquake in the Aleutian Trench off the island of Unimak, Alaska, over 3000 kilometers (1850 miles) away. The recession was followed by a wave that elevated the water nearly 8 meters (26 feet) above normal high tide. The tsunami also struck Scotch Cap, Alaska, on Unimak Island. The lighthouse on the island, the base of which stood 14 meters (46 feet) above sea level, was destroyed by a wave that is estimated to have attained 36 meters (118 feet) height.

The wave produced by this disturbance in the Aleutian Trench was recorded throughout the coastal Pacific at tide-recording stations. It led to what is now the International Tsunami Warning System (ITWS).

Until 1948, it was impossible to warn of coming tsunami in time for people to avoid the destructive waves. But as a result of a wave that struck Hawaii in 1946, a tsunami warning system was established throughout the Pacific Ocean. When a seismic disturbance occurs beneath the ocean surface that has potential for producing a tsunami, observations are made at the closest tide-measuring stations for any indication of such a wave.

If one is detected, warnings are sent to all the coastal regions that might encounter a destructive wave, along with its estimated time of arrival. This warning allows for evacuation of people from low-lying areas and removal of ships from harbors before the waves arrive, if the disturbance has occurred at a great enough distance.

Internal Waves

We have to this point discussed waves that occur at the interface between the atmosphere and the ocean, where an obvious density discontinuity exists. But such density discontinuities also exist within the ocean water column itself. These are sites for generation of internal waves.

Just as energy can be transmitted along the interface between the atmosphere and the ocean, it also can

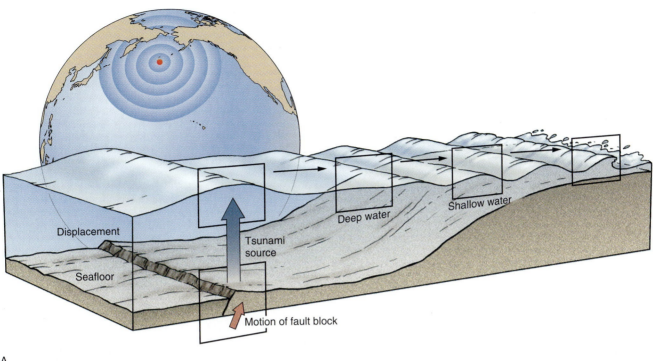

A.

B.

C.

Figure 8–17

Tsunami. *A:* Abrupt vertical movement along a fault in Earth's crust pushes up or drops the ocean water column above the fault. The energy released is then distributed laterally along the atmosphere-ocean interface in the form of massive long, low waves called tsunami. The energy is transmitted across the open ocean by these waves, which exceed 200 kilometers (125 miles) length, but are only about 0.5 meters (1.6 feet) high and thus are generally undetected. They release their energy upon reaching the shore, where they develop heights that can exceed 30 meters (100 feet). *B and C:* Tsunami at Sandy Beach, Oahu, Hawaii, just before and after arrival of the wave. (Photos courtesy of Y. Ishii. Permission of *Honolulu Advertiser.*)

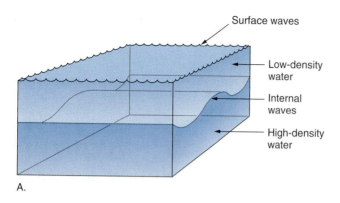

A.

Figure 8–18

Internal Wave. *A:* A simple internal wave moving along the density interface below the ocean surface. *B:* This view of internal waves in the Indian Ocean was taken by the crew of the Space Shuttle *Columbia* in 1990. The waves are coming in from the right, diffracting around shoals south of the Seychelles Islands and recombining to form interference patterns. The internal waves are the broad fronts running down the middle. (Photo courtesy of NASA.)

B.

be transmitted as an interface wave along density interfaces beneath the ocean surface in what are called **internal waves.** Internal waves may have heights exceeding 100 meters (330 feet), as shown in Figure 8–18. The greater the difference in density between the two fluids, the faster the waves will move.

Much remains to be learned about internal waves, but their existence is well documented. Many causes for them may be identified. Internal waves are known to have periods related to tidal forces, indicating that these forces may be a significant cause. Underwater avalanches in the form of turbidity currents, wind stress, and energy put into the water by moving vessels may also be causes of internal waves. It is thought that parallel slicks seen on surface waters may overlie troughs of internal waves. The slicks are caused by a film of surface debris that accumulates and dampens surface waves.

Internal waves reach greater heights from a smaller energy input than do the waves resulting from very large energy input observed at the ocean's surface. This is because they move along interfaces, such as the pycnocline, across which the density difference is considerably less than that which exists between the ocean surface and the atmosphere. They are thought to move as shallow-water waves at speeds considerably less than those of surface waves, with periods of 5 to 8 minutes and wavelengths of 0.6 to 0.9 kilometers (0.37 to 0.56 miles).

Power from Waves

There is great energy in moving water, which is why we have hydroelectric power plants on rivers to generate electricity. Even greater energy exists in ocean waves, if we can harness it efficiently. But significant problems must be overcome. Where waves refract and converge, as they do around the headland in Figure 8–13, energy is focused, creating a potential setting for power generation. Such a system might extract up to 10 megawatts of power per kilometer of shoreline. This would be comparable to the electricity consumed by 20,000 households for one month. It could produce significant power only when large storm waves broke against it.

Such a system would operate only as a power supplement, and a series of perhaps a hundred such structures along the shore would be required. Structures of this type could have a significant effect on natural processes, which could lead to serious coastal erosion problems in areas deprived of sediment.

Along shores endowed with suitable ocean floor topography, internal waves with their great heights may be effectively focused by refraction and induced to "break" against an energy-conversion device.

Two engineers at Lockheed Corporation have developed an interesting device to extract energy from ocean waves, called *Dam-Atoll* (Figure 8–19A). Waves

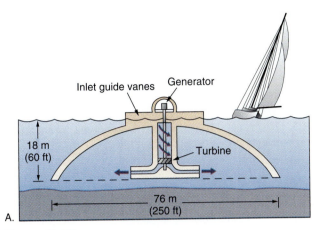

Figure 8–19

Dam-Atoll. Lockheed Corporation's Dam-Atoll designed to generate electricity from wave action. *A:* Water entering at the surface spirals down a central cylinder 18 meters (60 feet) across to turn a turbine located at the bottom of the cylinder. Each unit is designed to produce from 1 to 2 megawatts. The units are 76 meters in diameter (250 feet) and made of concrete. *B:* View from above. (Courtesy of Lockheed Corporation.)

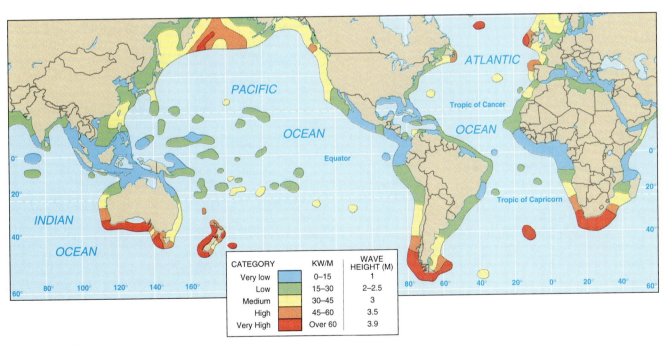

CATEGORY	KW/M	WAVE HEIGHT (M)
Very low	0–15	1
Low	15–30	2–2.5
Medium	30–45	3
High	45–60	3.5
Very High	Over 60	3.9

Figure 8–20

Global Coastal Wave Energy Resources. This map shows the effect of west-to-east movement of storm systems in the temperate latitudes. The western coasts of continents are struck by larger waves than eastern coasts, meaning that more wave energy is available along western shores. Further, larger waves appear to be associated with the westerly wind belts between 30° and 60° latitudes in both hemispheres. KW/M is kilowatts per meter; for example, every meter of "red" shoreline is a potential site for generating over 60 kilowatts of electricity. (Map constructed from data in U.S. Navy Summary of Synoptic Meteorological Observations, SSMO. Adapted from *Sea frontiers* [July-August 1987] "Sea Secrets," International Oceanographic Foundation [Vol. 33, No. 4] pp. 260–261; produced by The National Climatic Data Center with support from the U.S. Department of Energy.)

enter the top of the unit, just at the ocean surface. Water spirals into a whirlpool within a central core. The swirling water turns a turbine, the unit's only moving part. The turbine can provide a continuous electrical power output of 1 to 2 megawatts, according to inventors Leslie S. Wirt and Duane L. Morrow. In particularly good wave areas, such as the Pacific Northwest, they think 500 to 1000 units could provide power in quantities comparable to those provided by Hoover Dam, a major hydroelectric power-generating facility on the Colorado River near the Grand Canyon.

These units have many potential uses other than the generation of electricity. They could be used in cleaning up oil spills, protecting beaches from wave erosion, creating calm harbors in the open sea, and desalinating seawater through reverse osmosis.

Economic conditions in the future may enhance the appeal of converting wave energy to electrical energy. The coastal regions with the greatest potential for exploitation are shown in Figure 8–20.

S U M M A R Y

Wave phenomena transmit energy through various states of matter by setting up patterns of oscillatory motion in the particles that make up the matter. Progressive waves are longitudinal, transverse, or orbital depending on the pattern of particle oscillation. Particles in ocean waves move primarily in orbital paths.

Characteristics used to describe waves are wavelength (L), wave height (H), wave period (T), and wave speed (S). If water depth is greater than 1/2 wavelength, a progressive wave travels as a deep-water wave with a speed that is directly proportional to wavelength. If water depth is less than 1/20 wavelength, the wave will move as a shallow-water wave, the speed of which increases with increased water depth.

As wind-generated waves form in a sea area, capillary waves with rounded crests and wavelengths less than 1.74 centimeters (0.7 inches) form first. As the energy of the waves increases, gravity waves form, with increased wave speed, wavelength, and wave height. Energy is transmitted from the sea area across the ocean by low, rounded waves called swell.

Swell releases its energy in the surf as waves that break in the shoaling water near shore. If waves break on a relatively flat surface, the result is usually a spilling breaker, whereas breakers forming on steep slopes have spectacular curling crests and are called plunging breakers.

When swell approaches the shore, segments of the waves that first encounter shallow water are slowed, whereas other parts that have not been affected by shallow water move ahead, causing the wave to refract, or bend.

Refraction concentrates wave energy on headlands, whereas low-energy breakers are characteristically found in bays.

Reflection of waves off seawalls or other barriers can cause an interference pattern called a standing wave. In standing waves, crests do not move laterally as in progressive waves but form alternately with troughs at locations called antinodes. Separating the antinodes are nodes, where there is no vertical movement of the water.

During storms, the combination of low air pressure and onshore winds may produce a storm surge that raises the water level at the shore many meters above normal sea level. Such surges are particularly destructive if they coincide with high tide.

Tsunami, or seismic sea waves, are generated by seismic disturbances beneath the ocean floor. Such waves have lengths exceeding 200 kilometers (125 miles) and travel across the open ocean with undetectable heights of about 0.5 meters (1.6 feet) at speeds in excess of 700 kilometers per hour (435 miles per hour). On approaching shore, they may increase in height to over 30 meters (100 feet). Tsunami have been known to cause millions of dollars of damage and take tens of thousands of lives.

Internal waves are not well understood but are thought to form at density interfaces beneath the ocean surface, especially in connection with the pycnocline. They may be up to 100 meters (330 feet) high, with periods from 5 to 8 minutes.

Ocean waves could be harnessed to produce hydroelectric power, but significant problems must be solved.

K E Y   T E R M S

Capillary waves (p. 184)
Constructive interference (p. 188)
Crests (p. 183)
Deep-water waves (p. 183)
Destructive interference (p. 189)

Fully developed sea (p. 187)
Gravity waves (p. 185)
Internal waves (p. 196)
Longitudinal wave (p. 182)
Mixed interference (p. 189)

Orbital waves (p. 183)
Plunging breakers (p. 190)
Refraction (p. 191)
Restoring force (p. 185)
Rogue waves (p. 189)

QUESTIONS AND EXERCISES

1. Discuss longitudinal, transverse, and orbital wave phenomena, including the states of matter in which each can transmit energy.
2. Calculate the speed (*S*) for deep-water waves with the following characteristics:
   a. *L* = 351 meters, *T* = 15 seconds.
   b. *L* = 351 meters, *f* = 4 waves/minute. Express speed (*S*) in meters per second (m/s).
3. Describe the change in the shape of waves that occur as they progress from capillary waves to increasingly larger gravity waves until they reach a steepness ratio of 1/7. A change in which variable will make gravity the dominant restoring force, *H, L, S, T,* or *f?*
4. Waves from separate sea areas move away as swell and produce an interference pattern when they come together. If Sea A has wave heights of 1.5 meters (5 feet) and Sea B has wave heights of 3.5 meters (11.5 feet), what would be the height of waves resulting from constructive interference and destructive interference? Illustrate your answer (see Figure 8–9).
5. Describe changes, if any, in wave speed (*S*), length (*L*), height (*H*), and period (*T*) that occur as waves move across shoaling water to break on the shore.
6. Using orthogonal lines, illustrate how wave energy can be distributed along an uneven shore. Identify areas of high and low energy release.
7. Define the terms *node* and *antinode* as they relate to standing waves.
8. List three factors that may affect the height of storm surge. Make a diagram of a hurricane coming ashore from the south along an east-west shore and indicate the segment of shore along which you think the storm surge will reach maximum height. Explain why.
9. Why is it more likely that a tsunami will be generated by faults beneath the ocean along which vertical rather than horizontal movement has occurred?
10. What ocean depth would be required for a tsunami with a wavelength of 220 kilometers (136 miles) to travel as a deep-water wave? Is it possible that such a wave could become a deep-water wave any place in the world ocean?
11. Why is the development of internal waves likely within the thermocline?
12. Discuss some environmental problems that might result from developing facilities for conversion of wave energy to electrical energy.

REFERENCES

Bowditch, N. 1958. *American practical navigator.* Rev. ed. H.O. Pub. 9. Washington, D.C.: U.S. Naval Oceanographic Office.

Gill, A. E. 1982. *Atmosphere-ocean dynamics. International Geophysics Series,* Vol. 30. Orlando, Fla.: Academic Press, Inc.

Kinsman, B. 1965. *Wind waves: Their generation and propagation on the ocean surface.* Englewood Cliffs, N.J.: Prentice Hall.

Knamori, H., and Kikuchi, M. 1993. The 1992 Nicaragua earthquake: A slow tsunami earthquake associated with subducted sediments. *Nature* 361:6414, 714–716.

Melville, W., and Rapp, R. 1985. Momentum flux in breaking waves. *Nature* 317:6037, 514–516.

Pickard, G. L. 1975. *Descriptive physical oceanography: An introduction.* 2nd ed. New York: Pergamon Press.

Stewart, R. H. 1985. *Methods of satellite oceanography.* Berkeley, Calif.: University of California Press.

Sverdrup, H. U.; Johnson, M. W.; and Fleming, R. H. 1942. Renewal 1970. *The oceans: Their physics, chemistry, and general biology.* Englewood Cliffs, N.J.: Prentice Hall.

van Arx, W. S. 1962. *An introduction to physical oceanography.* Reading, Mass.: Addison-Wesley.

SUGGESTED READING

Sea Frontiers

Barnes-Svarney, P. 1988. Tsunami: Following the deadly wave. 34:5, 256–263. The origin of tsunami, a history of major occurrences, and the methods used to detect them are covered.

Changery, M. J. 1987. Coastal wave energy. 33:4, 259–262. The wave energy resources of the world's shores are considered.

Ferrell, N. 1987. The tombstone twins: Lights at the top of the world. 33:5, 344–351. A short history of the two lighthouses on Unimak Island, Alaska.

Land, T. 1975. Freak killer waves. 21:3, 139–141. The British design a buoy that will gather data in areas where 30 meter (98 feet) waves, which may be responsible for the loss of many ships, occur.

Mooney, M. J. 1975. Tragedy at Scotch Cap. 21:2, 84–90. A recounting of the events resulting from an earthquake off the Aleutians on April 1, 1946. The resulting tsunami destroyed the lighthouse at Scotch Cap, Alaska.

Pararas-Carayannis, G. 1977. The International Tsunami Warning System. 23:1, 20–27. A discussion of the history and operations of the International Tsunami Warning System.

Robinson, J. P., Jr. 1976. Newfoundland's disaster of '29. 22:1, 44–51. A description of the destruction caused by a tsunami that struck Newfoundland on November 18, 1929.

___. 1976. Superwaves of southeast Africa. 22:2, 106–116. A discussion of the formation and destruction caused by large waves that strike ships off the southeast coast of South Africa.

Smail, J. 1982. Internal waves: The wake of sea monsters. 28:1, 16–22. An informative discussion of the causes of internal waves and their effect on surface ships and submarines.

Smail, J. R. 1986. The topsy-turvy world of capillary waves. 32:5, 331–337. Capillary waves are clearly described, and their role in transmitting wind energy to the motion of waves and currents is discussed.

Smith, F. G. W. 1970. The simple wave. 16:4, 234–245. This is a very readable explanation of the nature of ocean waves. It deals primarily with the characteristics of deep-water waves.

___. 1971. The real sea. 17:5, 298–311. A comprehensive and readable discussion of wind-generated waves.

___. 1985. Bermuda mystery waves. 31:3, 160–163. A discussion of the possible source of large waves that struck Bermuda on November 12, 1984.

Truby, J. D. 1971. Krakatoa: The killer wave. 17:3, 130–139. The events leading up to the 1883 eruption of Krakatoa and the tsunami that followed are described.

Scientific American

Bascom, W. 1959. Ocean waves. 201:2, 89–97. An informative discussion of the nature of wind-generated waves, tsunami, and tides.

Koehl, M. A. R. 1982. The interaction of moving water and sessile organisms. 274:6, 124–135. A discussion of the adaptations of benthic shore-dwelling animals to the stresses of strong currents and breaking waves.

CHAPTER 9
Tides

If you have spent time at the seashore, you may be familiar with the rise and fall of tides. Even casual visitors notice that the edge of the sea slowly shifts landward and seaward daily, as sea level rises and falls. You may have watched sandcastles built at low tide disappear within hours as sea level rises to claim them. You probably also know that tides are caused by the moon's gravity and sun's gravity tugging at the blanket of water that covers 71 percent of Earth's surface. But the actual mechanism of the tides is complex.

People undoubtedly observed the rise and fall of the tides since they first inhabited the coastal regions of continents. However, we have no written record of tides prior to observations on the Mediterranean Sea by Herodotus (circa 450 B.C.). The first Greek observers concluded that the tides were related to the motion of the moon, for both followed a similar cyclic pattern. But it was not until **Sir Isaac Newton** (1642–1727) developed his universal law of gravitation that the tides could be explained adequately.

Tides are the ultimate manifestation of shallow-water wave phenomena. Tides are waves that possess lengths measured in thousands of kilometers and heights ranging to more than 15 meters (50 feet). Ocean tides are generated by the gravitational attraction of the sun and moon on the mass of the fluid ocean. The tides affect every particle of water from the surface to the deepest part of the ocean basin. Thus, the tide undoubtedly has a much more far-reaching effect on ocean phenomena than we can observe at the ocean surface.

Generating Tides

Tides are generated through a combination of gravity and motion among Earth, the moon, and the sun. We will now look at the forces that interact among these three celestial bodies to develop an understanding of the ocean's daily rhythms.

Newton's Law of Gravitation

Sir Isaac Newton published his *Philosophiae naturalis principia mathematica* (Philosophy of natural mathematical principles) in 1686; he stated the following in his preface:

> I derive from the celestial phenomena the forces of gravity with which bodies tend to the sun and several planets. Then from these forces, by other propositions which are also mathematical, I deduce the motions of the planets, the comets, the moon, and the sea.

What followed from his thinking was our first understanding of why tides behave as they do.

Newton's **law of gravitation** states that every particle of mass in the universe attracts every other particle of mass. This occurs with a force that is directly proportional to the product of their masses and inversely proportional to the square of the distance between the masses. The greater the mass of the objects and the closer they are together, the greater will be the gravitational attraction.

This gravitational attraction also causes tides in Earth's atmosphere and lithosphere. However, the atmospheric tide is not "sensible"—we cannot see or feel it with our unaided senses. And the lithospheric tides—a slight stretching of Earth's crust—are too slight to notice.

It is gravity that tethers the sun, its planets, and their moons together. And it is gravity that tugs every particle of water on Earth toward the moon and the sun, creating the tides. This part is not difficult to understand. What takes some effort to grasp is how tidal patterns are controlled by Earth's tilt on its axis, Earth's

rotation, and the celestial ballet of Earth and its moon moving as partners through space.

Gravitational Attraction: Mass Versus Distance

Tide-generating forces vary, which is why tides vary in height and frequency from place to place. Tide-generating forces vary inversely as the *cube* of the distance from the *center* of Earth to the *center* of the tide-generating object (moon or sun), instead of varying inversely to the *square* of the distance, as does gravitational attraction. Therefore, the tide-generating force, although it is basically gravity, is not linearly proportional to gravity.

Distance is a more highly weighted variable in the tide-generating forces than it is in gravitational attraction force. This is exemplified when we compare the Earth-moon relation to the Earth-sun relation.

Figure 9–1 illustrates the massiveness of the sun compared with the moon. Because the sun is 27 million times more massive than the moon, it should, solely on the basis of comparative mass, have a tide-generating force 27 million times greater than that of the moon. However, we must consider more significant factors: the Earth-moon *distance* and Earth-sun *distance*.

The sun is 390 times farther from Earth than the moon (see the scale in Figure 9–1). As was said, tide-generating forces vary inversely as the *cube* of the distance between objects. So the tide-generating force is reduced by 390^3, or about 59 million times compared with that of the moon. These conditions result in the sun's tide-generating force being $^{27}/_{59}$, or about 46 percent, that of the moon (Figure 9–1).

You probably were taught that "the moon orbits Earth," but it is not that simple. The two bodies actually orbit each other as they travel through space as a system. You can visualize this by imagining Earth and its moon as ends of a sledgehammer, flung into space, tumbling slowly end over end.

Thus, as Earth and its moon orbit the sun together as a system, they rotate around the center of mass for the Earth-moon system. This is called the *barycenter*. The barycenter is not located in the space between Earth and the moon, as you might think. Instead, the barycenter is within Earth's mantle, at a point about 4700 kilometers (2900 miles) from our planet's center. This is because Earth is much larger than the moon, and the two are close together in space.

It is the *barycenter* that follows a smooth orbit around the sun, whereas Earth and the moon themselves follow the wavy paths shown in Figure 9–2.

The tidal pattern we see on Earth primarily results from this "spinning sledgehammer" rotation of the Earth-moon system around its center of mass. Figure 9–3A shows the path followed by the center of Earth as it rotates around the barycenter for the Earth-moon system.

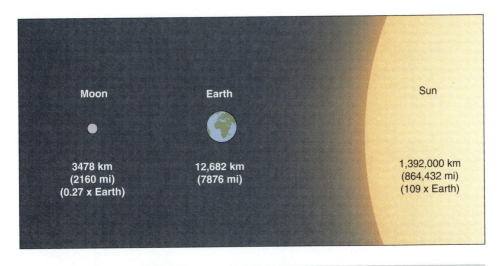

Tide-Generating Body	Distance from Earth (avg)	Mass (Metric tons)	Relative Tide-Generating Effects
Moon	384,835 km (234,483 mi)	7.3 x 10^{19}	Based on relative masses, the sun is 27 million times more massive than the moon and has 27 million times the tide-generating effect. However, since the sun is 390 times farther than the moon from Earth, its tide-generating effect is reduced by 390^3, or 59 million times.
Sun	149,758,000 km (93,016,845 mi)	2 x 10^{27}	

DETERMINATION OF TIDE-GENERATING FORCE OF SUN RELATIVE TO MOON

$$\text{Tide-generating force} \propto \frac{\text{Mass}}{(\text{Distance})^3} \propto \frac{\text{Sun–27 million times more mass}}{(\text{Sun–390 times farther away})^3}$$

$$(390)^3 = 59,000,000 \quad \text{Thus,} \quad \frac{27 \text{ million}}{59 \text{ million}} = 0.46 \text{ or } 46\%$$

The sun has 46% the tide-generating force of the moon.

Figure 9–1

Why the Moon Generates Bigger Tides than the Sun. Earth has a diameter of 12,682 kilometers (7876 miles). The diameter of the moon is 3478 kilometers (2160 miles), roughly one-fourth that of Earth. The diameter of the sun is 1,392,000 kilometers (864,432 miles), which is 109 times the diameter of Earth. Their relative sizes are shown to scale. The tabular data compare the masses of the sun and moon and their distances from Earth. These factors are important in determining their relative tide-generating effect.

Tide-Generating Force

To understand the tide-generating force, you need to understand centripetal force. **Centripetal force** "tethers" an orbiting body to its parent, pulling the object inward toward the parent, "seeking the center" of its orbit. As an example, if you tie a string to a rock and swing the tethered rock around your head (Figure 9–3B), the string pulls the rock toward your hand. The string provides a centripetal force to the rock, forcing the rock to seek the center of its orbit. If the string

breaks, the force is gone, the rock no longer can maintain its circular orbit, and it will fly off in a straight line, a tangent to the circle (Figure 9–3B).

The same is true of planets orbiting the sun. They are tethered not by string but by gravity. Thus, gravity operates as a centripetal force. If the gravity of the sun and its planets could abruptly be shut off, centripetal force would vanish, and all the planets would fly off into space along straight-line paths.

To understand how centripetal force helps generate the tides, it is helpful to imagine Earth divided into

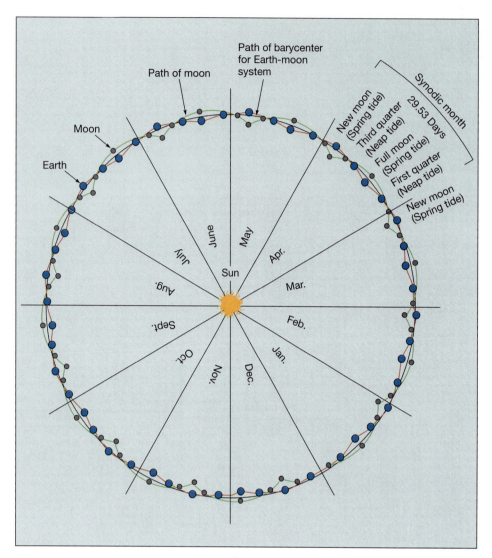

Figure 9–2

Earth-Moon-Sun Motion and Tides. As the Earth-moon system orbits the sun, the center of each body follows its own wavy path, because both also rotate about the barycenter (center of gravity) of the Earth-moon system (*see* Figure 9–3A). We can think of this system as a sledgehammer, with Earth being the heavy business end and the moon being at the opposite end of the handle. The "handle" that connects Earth and moon is gravity. Obviously, the center of mass is not in the middle of this system, because Earth is so much larger than the moon. The center of mass is heavily shifted toward Earth's end, so much so that the center of mass actually lies within Earth's mantle. If this "hammer" were thrown into space, the center of the moon on one end of the handle would orbit around the center of mass at a far greater distance than would the center of Earth. Therefore, the wobble of the moon is more pronounced than Earth's, and it is shown exaggerated in the figure. During a month, the moon moves from a position between the sun and Earth (new moon) to a position that puts Earth between the moon and the sun (full moon) and then back again.

millions of tiny particles of equal mass. Centripetal force varies a bit from particle to particle. Figure 9–4 shows this variation. Each particle of equal mass is located at a different distance and/or direction from the center of the moon. This causes the force of gravitational attraction between each particle and the moon to be different.

As a result of the differences between the required centripetal force and the gravitational force of attraction between each particle and the moon, small lateral forces ("sideways" forces) are produced, and these are the *tide-generating forces*. These push the ocean's water into bulges at points nearest the moon and sun.

For the two tide-generating bodies (moon and sun), the more distant they are from Earth, the shorter these tide-generating force arrows will be. Thus, the tide-generating force decreases inversely with the cube of the distance instead of the square of the distance, as

does gravitational force. This is why the moon controls tides far more than the sun.

Idealized Theory of Tides

It is easier to understand the tides if we assume an ideal Earth that has two tidal bulges, one toward the moon and one away from the moon, as Earth rotates on its axis, and an ideal ocean of uniform depth, with no friction between the seawater and the seafloor.

Although this simplification cannot give us an accurate means of predicting real tides that occur across Earth's surface, it will help you understand the gross characteristics of tides in the world ocean. Then we will consider the **dynamic theory** of tides that deals with reality: variable ocean depth, the presence of continents

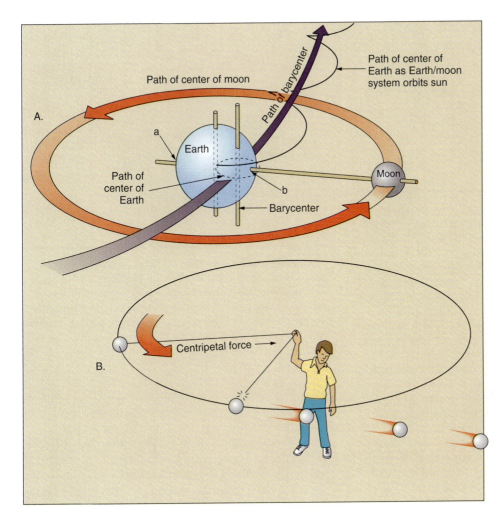

Figure 9–3

Earth-Moon-System Rotation. Earth and its moon move around a common center of gravity (barycenter) that is inside Earth. *A:* The dashed line through Earth's center is the path followed by its center as it moves around the common center of gravity of the Earth-moon system. The circular paths followed by point a and point b have the same radius as that followed by Earth's center. *B:* If you tie a string to a rock and swing it in a circle around your head, it stays in the circular orbit because the string exerts a centripetal (center-seeking) force on the rock. This force pulls the rock toward the center of the circle. If the string breaks with the rock at the position shown, the rock will fly off along a straight path, at a tangent to the circle.

that disrupt the ocean surface, and friction between the ocean water and ocean floor.

The Rotating Earth

Our ideal, uniformly deep ocean is modified only by the tide-generating force that causes bulges on opposite sides of Earth, as shown in Figure 9–5. Let us assume that a stationary moon is aligned with Earth's equator so that the maximum bulge will occur on the equator on opposite sides of Earth.

Earth requires 24 hours for one complete rotation, so on the equator you would experience two high tides each day. The time that would elapse between high tides, the *tidal period,* would be 12 hours. If you moved to any latitude north or south of the equator, you would experience a similar tidal period, but the high tides would be less high, because you would be at a lower point on the bulge rather than at its apex.

But high tides do not occur every 12 hours on Earth's surface. Instead, they occur every 12 hours and 50 minutes. This period is called the *lunar day.* The **lunar day** is the time that elapses between two succes-

sive passages of the moon over any longitude line on Earth. If you observe the time at which the moon rises on successive nights, you will see that it rise 50 minutes later each night (Figure 9–6).

Where do the additional 50 minutes come from? During the time that Earth is making a full rotation in 24 hours, the moon has continued moving another 12.2° to the east. Thus, Earth must rotate an additional 50 minutes to catch up and place the moon directly over the observer's longitude. (If we divide the lunar day of 24 hours and 50 minutes into 24 lunar hours, each *lunar hour* will be about 1 hour and 2 minutes of solar time in length.)

So far, we have considered the effects of Earth's rotation and the revolution of the Earth-moon system about its center of mass on the prediction of tides. We have ignored the effect of the sun. In the following discussion we will consider the combined effect of the moon and sun on Earth's tides.

Combined Effects of Sun and Moon

Figure 9–2 shows how the Earth-moon system revolves around the sun. The oscillating paths of Earth and the

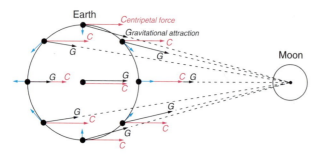

Figure 9–4

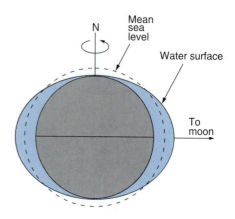

Figure 9–5

Tide-Generating Force. The length and direction of the arrows represent the magnitude and direction of the forces acting upon each of nine identical masses shown on the figure. *Red arrows (centripetal force)*—the centripetal force required to tether each particle in the identical circular orbit. *Black arrows (gravitational attraction)*—the gravitational attraction between particles and the moon. This force provides the required centripetal force, but gravity equals the required centripetal force only for the particle at the center of Earth. Also, it is in the required direction only for particles on a line connecting the centers of Earth and the moon. For all particles on the half of Earth facing the moon, the force is greater than required. For particles on the half of Earth facing away from the moon, the force is less. *Blue arrows (residual force)*—for all particles (except at the center of Earth), a residual force results, because the gravitational force varies from the required centripetal force. Residual force is determined by constructing a vector from the tip of the C arrow to the tip of the G arrow. This force is small, averaging about one-millionth of the magnitude of Earth's gravity. Therefore, where forces act perpendicular to Earth's surface, as does gravity, there is no tide-generating effect. However, where they have a significant horizontal component—tangent to Earth's surface—they aid in producing tidal bulges on Earth. Because there are no other large horizontal forces on Earth with which they must compete, these small, ever-present residual forces can push water across Earth's surface. (Consider all vector arrows to extend from the point they represent.)

Idealized Equilibrium Tide. Assuming an ocean of uniform depth covering Earth, and the moon aligned with the equator, tide-generating forces produce two bulges in the ocean surface. One extends toward the moon and the other away from the moon. As Earth rotates, all points on its surface (except the poles) experience two high tides daily because Earth rotates beneath the two bulges.

moon can be seen as they rotate together like twins around their barycenter. In the upper right quadrant, you can see that the moon cycles through its phases approximately every 29½ days. When the moon is between Earth and the sun, it cannot be seen for a few days, and it is then called the **new moon.** When the moon is on the side of Earth opposite the sun, its entire disk is brightly visible, and we call it the **full moon.** A **quarter moon** results when the moon is at right angles to the sun relative to Earth. (Viewed from Earth, it looks like half a moon.)

Refer to Figure 9–7. When the sun and moon are aligned, either with the moon between Earth and the sun (new moon) or with the moon on the side opposite the sun (full moon), the tide-generating forces of the sun and moon add together. You can see this in Figure 9–7A. At this time, we experience the maximum **tidal range,** the vertical difference between high and low

tide. This maximum tidal range is called the **spring tide,** because the tide surges or "springs forth." (The name has no connection with the spring season.)

When the moon is in either quarter phase (Figure 9–7B), the tide-generating force of the sun is working at right angles to the tide-generating force of the moon, and we experience minimum tidal range. This is called **neap tide.** (To help you remember this name, think of it as a "nipped" tide that is small.) The time that elapses between successive spring tides (full moon and new moon) or neap tides (first and third quarters) is about two weeks.

Declination of the Moon and Sun

Up to this point, for learning purposes, we have assumed that the moon and sun always remain aligned over the equator, but of course this is not the case. Most of the year, the sun and moon are north or south of the equator. This angular distance of the sun or moon above or below Earth's equatorial plane is called **declination.**

Earth revolves around the sun along an invisible ellipse in space. Imagine a plane in space that includes this ellipse. This plane is called the **ecliptic.** Earth is not perpendicular ("upright") on the ecliptic; it leans on its side by 23.5°.

The tilt of Earth's axis relative to its plane of revolution around the sun (ecliptic) is shown in Figure 9–8. You can see that Earth's tilted axis always points the same direction throughout the yearly cycle.

This tilt causes the seasons—spring, summer, fall, and winter (Figure 9–8).

- At the **spring equinox,** which occurs about March 21, the sun is directly overhead along the equator.

Figure 9-6

The Lunar Day. A lunar day is the time that elapses between successive appearances of the moon on the meridian directly over a stationary observer. As Earth rotates, the Earth-moon-system rotation moves the moon in the same direction (toward the east). During one complete rotation of Earth (the 24-hour solar day), the moon moves eastward 12.2°, and Earth must rotate an additional 50 minutes to place the observer in line with the moon again.

0 h	8 h	16 h	24 h	24 h 50 min
	8 h	8 h	8 h	50 min

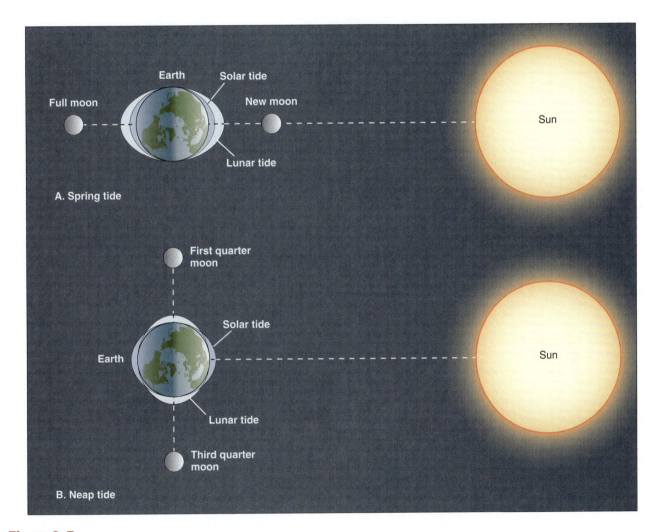

A. Spring tide

Full moon — Earth — Solar tide — New moon — Lunar tide — Sun

B. Neap tide

First quarter moon — Solar tide — Earth — Lunar tide — Third quarter moon — Sun

Figure 9-7

Earth-Moon-Sun Positions and the Tides. When the moon is in the new or full position, the tidal bulges created by the sun and moon are aligned, producing constructive interference and therefore larger bulges, which "spring forth" as *spring tides*. When the moon is positioned halfway between the new and full phases (called the first and third quarters), the tidal bulge produced by the moon is at right angles to the bulge created by the sun. The magnitudes of the bulges are reduced by destructive interference, and the resulting bulges are smaller, producing *neap tides*. New moon and full moon phases produce spring tides with maximum tidal ranges, whereas the first and third quarter phases of the moon produce neap tides with minimal tidal ranges. (From The Tasa Collection: *Shorelines*. Published by Macmillan Publishing Co., New York. Copyright © 1986, by Tasa Graphic Arts, Inc. All rights reserved.)

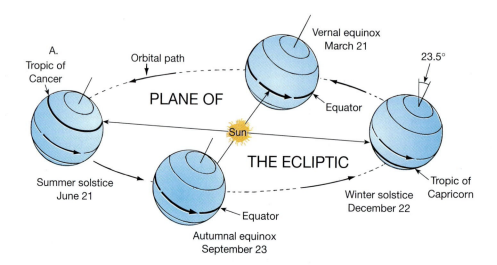

Figure 9–8

Orbital Planes of Earth and Moon. As Earth orbits the sun during one year, its axis of rotation constantly tilts 23.5° from perpendicular, relative to the plane of the ecliptic. The sun shines directly overhead along the Tropic of Cancer (23.5°N) on the summer solstice, about June 21. Three months later, the sun shines directly overhead along the equator (0°) during the autumnal equinox, about September 23. Three months later, the sun shines directly overhead along the Tropic of Capricorn (23.5°S) on the winter solstice, about December 22. Three months later, the sun shines directly overhead along the equator on the vernal equinox, about March 21. Three months later, the yearly orbit is completed, and the sun again shines directly overhead along the Tropic of Cancer.

- On about June 21, the **summer solstice** occurs. At this time the sun reaches its most northerly point in the sky, directly overhead along the Tropic of Cancer, which is at 23.5°N latitude.
- The sun then moves southward in the sky each day, and on about September 23 it is directly overhead along the equator again, producing the **autumnal equinox.**
- During the next three months the sun is more southerly in the sky until the **winter solstice** on about December 22, when the sun is directly overhead along the Tropic of Capricorn, at 23.5°S latitude.

Without Earth's 23.5° tilt, seasonal differences would disappear. Because of the tilt, the sun's declination varies between 23.5° north and 23.5° south of the equator on a yearly cycle.

To further complicate matters, the plane of the moon's orbit is at an angle of 5° to the ecliptic. This 5° adds to the 23.5° of Earth's tilt, so that the declination of the moon's orbit relative to Earth's equator can reach 28.5°. The declination will change from 28.5° south to 28.5° north and back to 28.5° south of the equator in a period of one month.

You now can see that we must expect tidal bulges rarely to be aligned with the equator. They occur mostly north and south of the equator. Because the moon is the dominant force that creates tides in Earth's oceans, tidal bulges follow the moon as it shifts position throughout

its monthly journey across the equator, ranging from a maximum of 28.5° north to a maximum of 28.5° south of the equator (Figure 9–9).

Effects of Distance

Additional considerations that affect the tide-generating force of the sun and moon on Earth are their shifting distances from Earth. Earth follows not a circular orbit around the sun but an elliptical orbit. Thus, the distance between Earth and the sun varies by 2.5 percent (be-

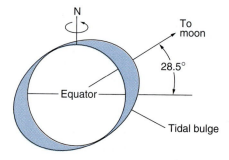

Figure 9–9

Maximum Declination of Tidal Bulges from Equator. The center of the tidal bulges may lie at any latitude from the equator to a maximum of 28.5° on either side of the equator, depending on the season of the year (solar angle) and the moon's position at the moment.

tween 148.5 million kilometers, or 92.2 million miles, during the Northern Hemisphere winter and 152.2 million kilometers, or 94.5 million miles, during summer). Simultaneously, the Earth-moon distance varies over 8 percent (between 375,000 kilometers, or 233,000 miles, and 405,800 kilometers, or 252,000 miles). One complete lunar cycle takes 29½ days (Figure 9–10).

Because of these movements, the greater "spring" tides have wider ranges during the Northern Hemisphere winter than in the summer.

Idealized Tide Prediction

To predict tidal patterns for our idealized water-covered Earth, let us return to the effect of declination. The declination of the moon will determine the position of the tidal bulges. In Figure 9–11, this is 28° north of the equator. If you stand at this latitude on a permanent point, your observation of the tides will be different from observations at the equator.

Begin your observations when the moon is directly overhead. At this time, you will observe high tide (Figure 9–11A). Six lunar hours later (6 hours and 12½ minutes solar time), you will see low tide (Figure 9–11B). Six lunar hours later, it will be followed by another high tide, but it will be much lower than the initial high tide (Figure 9–11C). Six hours later, at the end of a 24-lunar-hour period (24 hours and 50 minutes Earth time), you will have passed through a complete lunar-day cycle of high-low-high-low (two high tides and two low tides).

A representative curve of the type of tide you experience is in Figure 9–11E. Tide curves showing the heights of the same tides during one lunar day at the equator and at 28°S latitude are also provided.

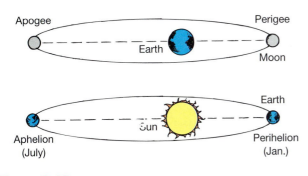

Figure 9–10

Effects of Elliptical Orbits. *Top:* The moon moves from its closest point to Earth (perigee) of 375,200 kilometers (233,000 miles) to its most distant point (apogee) of 405,800 kilometers (252,000 miles). Greater tidal ranges are experienced when the moon is closest to Earth. *Bottom:* At its closest, Earth is within 148.5 million kilometers (92.2 million miles) of the sun (perihelion). At its most distant, Earth is 152.2 million kilometers (94.5 million miles) from the sun (aphelion). Greater tidal ranges are experienced when Earth is nearest the sun.

Here is a summary of the characteristics of the tides on our idealized Earth:

1. Any location will have two high tides and two low tides per lunar day.
2. Neither the two high tides nor the two low tides are of the same height because of the changing declination of the moon and the sun (except for the rare occasions when the sun and moon are simultaneously above the equator).
3. Yearly and monthly cycles of tidal range are related to the changing distances of Earth from the sun and moon.
4. Each two weeks, half a lunar month, we would experience spring tides separated by neap tides (Figure 9–11F).

Considering the great number of variables that are involved in predicting tides, it is interesting to consider when the conditions might be right to produce the maximum tide-generating force. This occurs when the sun and moon are closest, and the moon either is between Earth and the sun (new moon) or opposite the sun (full moon), and when both the sun and moon have zero declination. This condition occurs only once every 1600 years. The next occurrence is predicted for A.D. 3300.

Is this cause for concern? Yes, because even a less-optimized coincidence of Earth-moon closeness and spring tides was made woefully clear during 1983 storms in the North Pacific. Slow-moving low-pressure cells developed over the Alaska's Aleutian Islands (Figure 9–12A). This caused strong northwest winds to blow across the ocean from Russia's Kamchatka Peninsula toward the U.S. coast. Averaging about 50 kilometers/hour (30 miles/hour), the winds produced a near fully developed 3 meters (10 feet) swell along the coast from Oregon to Baja California. The storm surge from this condition would have been trouble enough under average conditions, but the situation was made worse by the high spring tides of 2.25 meters (7.4 feet).

These unusually high spring tides occurred because Earth was still closest to the sun in its orbit (January 2) when the moon also was closest to Earth on January 28. Some of the largest waves came ashore January 26–28, causing over $100 million in damage (Figure 9–12B). At least a dozen lives were lost, 25 homes were destroyed, over 3500 homes were seriously damaged, and many commercial and municipal piers collapsed. With each such occurrence, we learn more of the cost of developing the shore.

Dynamic Theory of Tides

Tidal bulges are directed toward the moon and away from the moon on opposite sides of Earth. Thus, as Earth rotates, the bulges (or wave crests) are separated

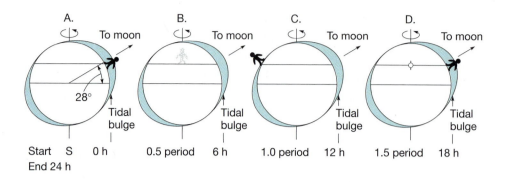

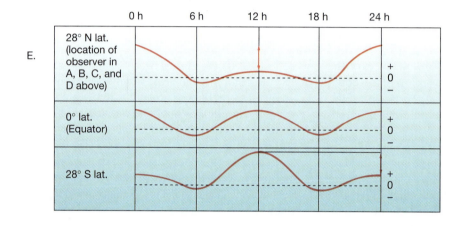

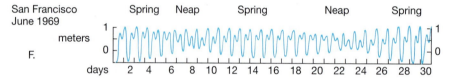

Figure 9–11

Predicted Idealized Tides. *A:* Follow the "tidal experience" you would have if you stood at 28°N latitude—for example, along a Florida beach near Cape Canaveral. At the start of the lunar day, you are at the center of the tidal bulge, and experience a *high tide*. *B:* Six lunar hours later (0.5 period), you experience *low tide* while located on the back side of Earth. *C:* After 12 lunar hours (1.0 period, or 0.5 lunar day), you again experience *high tide*. It is much lower than the first high tide, however, because you now are passing through the edge of the tidal bulge. *D:* You experience *low tide* again, 18 lunar hours (1.5 period) after start. At the end of one lunar day (24 hours and 50 minutes), you re-turn to *A* and experience a *high tide*. *E:* Tide curves for 28°N, 0°, and 28°S latitudes when the decli-nation of the moon is 28°N. (All curves for the same longitude). Note that tide curves for 28°N and 28°S have identical highs and lows but are out of phase by 12 hours. This results from the fact that the bulges in the two hemispheres occur on opposite sides of Earth. *F:* Along with the unequal heights for the two high and low tidal extremes occurring each lunar day, we expect to observe spring and neap tides, which are controlled by the sun-moon-Earth alignment. Thus, the tide curve for June in San Francisco demonstrates the general character of the predicted equilibrium tide. (*F* from Anikouchine and Sternberg. *The world ocean: An introduction to oceanography,* © 1973. Reprinted by permission of Prentice-Hall, Englewood Cliffs, New Jersey.)

by a distance of half Earth's circumference (about 20,000 kilometers, or 12,420 miles). We might expect the bulges to move across Earth at about 1600 kilome-ters/hour (1000 miles/hour).

However, the tides are an extreme example of shallow-water waves. As such, their speed is propor-tional to the water depth. For a tide wave to travel at 1600 kilometers/hour (1000 miles/hour), the ocean

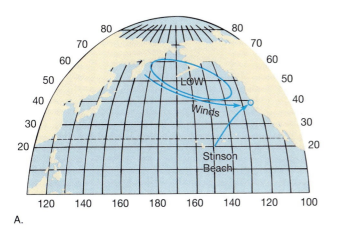

A.

B.

Figure 9–12

High Tides of January 1983. *A:* January 1983 storm winds blow uninterrupted across the North Pacific Ocean from the Kamchatka Peninsula of Russia to the U.S. West Coast. *B:* Homes threatened by storm waves and unusually high tides on January 27, 1983, at Stinson Beach north of San Francisco. (Wide World Photos.)

would have to be 22 kilometers (13.7 miles) deep! But the mean depth of the ocean is only 3.9 kilometers (2.4 miles). Thus, tidal bulges move as forced waves, with their velocity determined by ocean depth.

Based on the mean ocean depth, the mean speed at which tide waves can travel across the open oceans is about 700 kilometers/hour (435 miles/hour). With this limitation on the speed of tide waves, the idealized bulges that simply point toward and away from the tide-generating body cannot exist. Instead, they break up into a number of cells.

In the open ocean, the crests and troughs of the tide wave actually rotate around a point near the center of each cell. It is called an *amphidromic point* (which means "running around"). There is essentially no tidal range at this point. Radiating from this point, we can draw on a map **cotidal lines** that connect points along which high tide will simultaneously occur. Figure 9–13 shows cotidal lines, labeled to indicate the time of high tide in hours after the moon crosses the Greenwich Meridian.

The times indicate that the rotation of the tide wave is counterclockwise in the Northern Hemisphere and clockwise in the Southern Hemisphere. The wave makes one complete rotation during the tidal period. The size of the cells is limited by the fact that the tide wave must make one complete rotation during the period of the tide (usually 12 lunar hours).

Within an amphidromic cell, low tide is 6 hours behind high tide. For example, if high tide is occurring along the cotidal line labeled 10, low tide is simultaneously occurring along the cotidal line labeled 4.

We must also consider the effect of the continents, which interrupt the free movement of the tidal bulges

across the ideal unobstructed ocean surface considered earlier. The ocean basins between continents have set up within them free-standing waves. Their character modifies the forced astronomical tide waves that develop within the basin.

It is impossible to explain here all the tidal phenomena that occur throughout the world. For instance, high tide rarely occurs at the time the moon is at its highest point in the sky. The elapsed time between the passing of the moon and the occurrence of high tides varies from place to place. This is a result of the many factors that determine the characteristics of the tide at any given location.

Types of Tides

Ideally we expect two high tides and two low tides of unequal heights during a lunar day. But due to modification from various depths, sizes, and shapes of ocean basins, tides in many parts of the world exhibit different patterns: *diurnal tide* (daily), *semidiurnal tides* (twice daily), or *mixed tides*. These are shown in Figure 9–14.

The **diurnal tide** has a single high and low water each lunar day. These tides are common in the Gulf of Mexico and along the coast of Southeast Asia. Such tides have a tidal period of 24 hours and 50 minutes.

The **semidiurnal tide** has two high and two low waters each lunar day. The heights of successive high waters and successive low waters are approximately the same. Since tides are always growing higher or lower at any location, due to the spring-neap tide sequence, successive high tides and successive low tides can never be

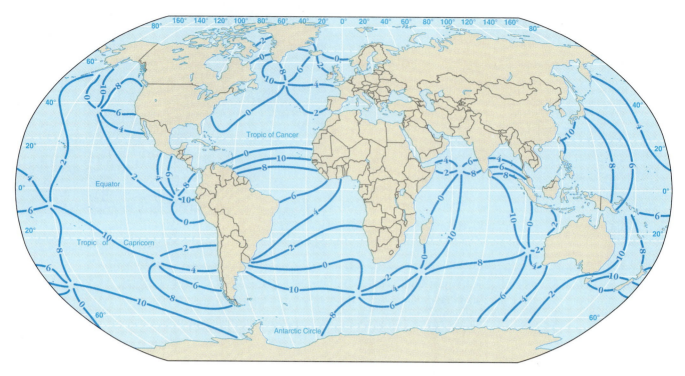

Figure 9–13

Cotidal Map of the World. Contour lines indicate times of the main lunar daily high tide in lunar hours after the moon has crossed the Greenwich Meridian (0°). Tidal ranges generally increase with increasing distance along cotidal lines away from the amphidromic points. Where cotidal lines terminate at both ends in amphidromic points, maximum tidal range will be near the midpoints of the lines. (Base map courtesy of National Ocean Survey. After von Arx, 1962; original by H. Poincaré 1910, *Leçons de Mécanique Céleste,* a Gauther-Crofts, Vol. 3.)

exactly the same at that location. Semidiurnal tides are common along the Atlantic Coast of the United States. The tidal period is 12 hours and 25 minutes.

The **mixed tide** may have characteristics of both diurnal and semidiurnal tides. Successive high tides and/or low tides will have significantly different heights. Mixed tides commonly have a tidal period of 12 hours and 25 minutes, which is a semidiurnal characteristic, but they may also possess diurnal periods. This is the tide that is most common throughout the world and the type that is found along the U.S. Pacific Coast.

Inequalities are greatest when the moon is at its maximum declination, and such tides are called *tropical tides* because the moon is over one of the tropic regions. When the moon is over the equator, the inequality is minimal; and tides with this characteristic are called *equatorial tides.*

Tides in Narrow Bays

When tide waves enter coastal waters, they are subject to reflection. In some cases, the standing waves set up by reflections may have periods near that of the forced tide wave. Under such conditions, constructive interference can produce significant increases in the tidal range.

Nova Scotia's Bay of Fundy is such a place. With a length of 258 kilometers (160 miles), it has a wide opening into the Atlantic Ocean. The Bay of Fundy splits into two narrow basins at its northern end, Chignecto Bay and Minas Basin (Figure 9–15). The period of free oscillation in the Bay of Fundy is very nearly that of the tidal period. The resulting constructive interference, along with the narrowing of the bay toward the north end and the shoaling in that direction, produces maximum tidal ranges in the extreme northern end of Minas Basin.

The maximum tidal range when the moon is closest is about 17 meters (56 feet) at the northern end of the bay to about 2 meters (6.6 feet) at its opening to the sea. The tidal range progressively increases from the mouth of the bay northward.

Coastal Tide Currents

The current that accompanies the slowly turning tide crest in a Northern Hemisphere basin will turn in a

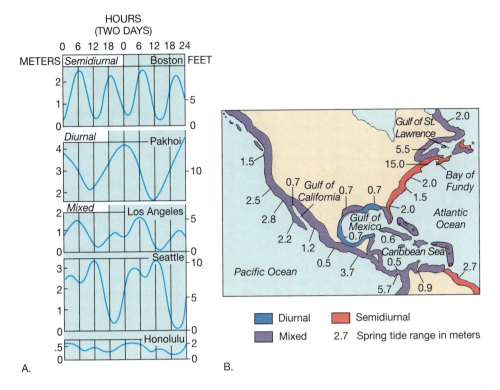

Figure 9–14

Types of Tides. *A:* Three types of tides. In a *semidiurnal* (twice daily) tide, there are two highs and lows during each lunar day, and the heights of each successive high and low are about the same. (Note that the diagrams show *two* days, not one.) In the diurnal (daily) type of tide, there is only one high and one low each lunar day. In the mixed type of tide, both diurnal and semidiurnal effects are detectable, and the tide is characterized by a large difference in high water heights, low water heights, or both during one lunar day. Even though a tide at a place can be identified as one type, it still may pass through stages of one of both of the other types. *B:* The types of tides observed along North and South American coasts. The numbers give the spring tide range in meters and are therefore near the maximum tidal range that can be expected. Storm waves, lower barometric pressure, ocean currents, and coincidence with extremes of distance from the sun could increase the range. (After C. Hauge, *Tides, currents, and waves.* California Geology, July 1972.)

counterclockwise direction, producing a *rotary current* in the open portion of the basin. Because of increased effects of friction in shoaling, nearshore waters, the rotary current is changed to an alternating or *reversing current* that moves in and out rather than along the coast, as would a rotary current. The rising and falling of the tides along the coast are called *flood tide* and *ebb tide,* respectively.

Velocities of the rotating currents in the open ocean are usually well below 1 kilometer/hour (0.6 miles/hour). However, these reversing currents are of the greatest concern to coastal navigators, for in this setting they are known to reach velocities of 44 kilometers/hour (28 miles/hour) in restricted channels between islands of coastal British Columbia.

Even in deeper waters of the ocean, tidal currents can be significant. In July of 1986, shortly after the discovery of the remains of the Titanic, strong tidal currents were encountered. At a depth of 3795 meters (12,448 feet) on the continental slope south of Newfoundland's Grand Banks, strong tidal currents forced researchers to abandon the use of the camera-equipped robot, *Jason Jr.* This experience demonstrated that significant tidal currents occur at all depths in the ocean.

Tides in Rivers

The Amazon River probably possesses the longest estuary (river mouth) that is affected by oceanic tides. Tides can be measured as far as 800 kilometers (500 miles) from the river's mouth, although the effects are quite small at this distance. Tide waves that move up river mouths lose their energy due to the decreasing depth of water and the flow of the river water against the tide. As a tide-crest wave moves up the river, it becomes more and more asymmetrical, developing a steep front (Fig-

A.

B.

Figure 9–15

Tidal Range in the Bay of Fundy. The largest tidal range known in the world occurs at the northern end of the Minas Basin in Nova Scotia's Bay of Fundy (see map). Because of its dimensions, this bay has a natural free-standing-wave period about equal to that of the forced tide wave. This, combined with the fact that the bay narrows and becomes shallower toward its head, causes a maximum tidal range at the northern end of 17 meters (56 feet). High tide (*A*) and low tide (*B*) in Minas Basin. (Photos courtesy of Nova Scotia Department of Tourism.)

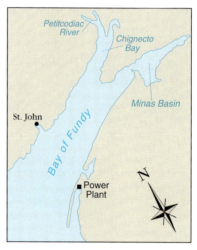

C.

ure 9–16). This front produces a rapidly rising tide that falls slowly.

An extreme development of this type produces a **tidal bore** in which a very steep wave front surges up the river. In the Amazon it is called *pororoca* and appears as a waterfall up to 5 meters (16.4 feet) in height moving upstream at speeds up to 22 kilometers/hour (13.7 miles/hour)! Other rivers that experience tidal bores are the Chientang in China, where bores may reach 8 meters (26 feet); the Petitcodiac in New Brunswick, Canada; the Seine in France; and the Trent in England.

Tides as a Source of Power

The history of human efforts to harness tidal energy dates at least to the Middle Ages. Pursuit of this renewable energy source waned with the availability of cheap fossil fuels. But today, as we realize that cheap fossil fuels will run out, there is increased interest in generating electricity with tidal energy.

Basic Considerations of Tidal Power

The most obvious benefit of generating electrical power with tidal energy is reduced operating costs compared with conventional thermal power plants that require fossil fuels or radioactive isotopes. Even though the initial cost of building a tidal power-generating plant may be higher, there would be no ongoing fuel bill.

A negative consideration involves the periodicity of the tides. Power could be generated only through a portion of a 24-hour day. People operate on a solar period, but tides operate on a lunar period; thus, the energy available through generating power from the tides would coincide with need only part of the time. To get around this, power would have to be distributed to the point of need somewhere on Earth at the moment; this is an expensive transmission problem. Or power could be stored, but this presents a large and expensive technical problem.

Electrical turbines (generators) must run at constant speed yet use the flow of the tidal current in two directions (flood tide and ebb tide). This requires spe-

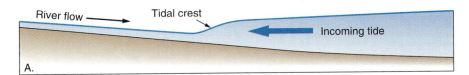

Figure 9–16

River Bores. *A:* As the tidal crest moves upriver, it develops a steep forward slope due to resistance to its advance by the river, which is flowing in the opposite direction to the ocean. Such crests (called bores) may reach heights of 5 meters (16.4 feet) and move at speeds up to 22 kilometers/hour (13.7 miles/hour). *B:* A tidal bore moves up a river flowing into Chignecto Bay on the coast of New Brunswick, Canada. (Courtesy New Brunswick Department of Tourism.)

cial design that would allow both advancing and receding water to spin the turbine blades.

La Rance Tidal Power Plant

A successful tidal power plant is operating in the estuary of La Rance River off the English Channel in France. The estuary, shown in Figure 9–17, has a surface area of approximately 23 kilometers squared (8.9 miles squared), and the maximum tidal range at La Rance is 13.4 meters (44 feet). Usable tidal energy is proportional to the area of the basin and to the amplitude of the tide.

A barrier was built across the estuary a little over 3 kilometers (1.9 miles) upstream, where it is 760 meters (2500 feet) wide, to protect it from storm waves. The deepest water ranges from just over 12 meters (39 feet) at low tide to more than 25 meters (82 feet) at high tide. To allow water to flow through the barrier when the generating units are shut down, sluices (artificial channels) were built into the barrier.

Twenty-four electrical generating units operate beneath the power plant. Each unit can generate 10 million watts (10 megawatts) of electricity, enough to serve over 1500 homes for a year.

The plant generates electricity only when sufficient water height exists between the pool and the ocean—about one-half of the tidal period. Annual power production of about 540 million kilowatt-hours without pumping can be increased to 670 million kilowatt-hours by using the turbine-generators as pumps at the proper times.

One tide cycle= 12 h 25 min

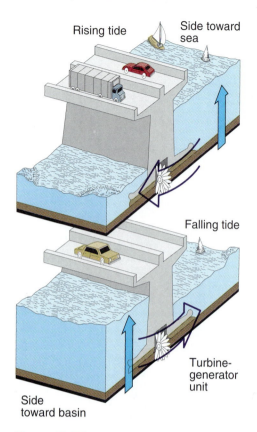

Figure 9–17

La Rance Tidal Power Plant at St. Malo, France. Block diagram shows the barrier between the open ocean to the right and the La Rance estuary to the left. The relative water levels during rising and falling oceanic tides are also shown. (Photo courtesy of Phototeque/Electricite de France.)

Other Tidal Power Plants

Within the Bay of Fundy, the province of Nova Scotia has constructed a tidal power plant that has generated up to 40 million kilowatt-hours per year since its completion in 1984. It is built on the Annapolis River, where maximum tidal range is 8.7 meters (26 feet).

Some engineers think that a tidal power plant could be made to generate electricity constantly if located on the Passamaquoddy Bay near the U.S.-Canadian border at the south end of the Bay of Fundy. Others are less optimistic. Potentially, the usable tidal energy seems great compared with La Rance, because the flow volume is about 117 times greater.

Whether or not tidal power stations ever are constructed on a large scale, this potential source of energy will receive increased attention as the cost of generating electricity by conventional means increases. Figure 9–18 shows locations of some sites that have potential for tidal power generation.

Any use of coastal waters for the tidal generation of electricity will have environmental costs resulting from the modification of current flow. It will interfere with many traditional uses of coastal waters, such as transportation and fishing.

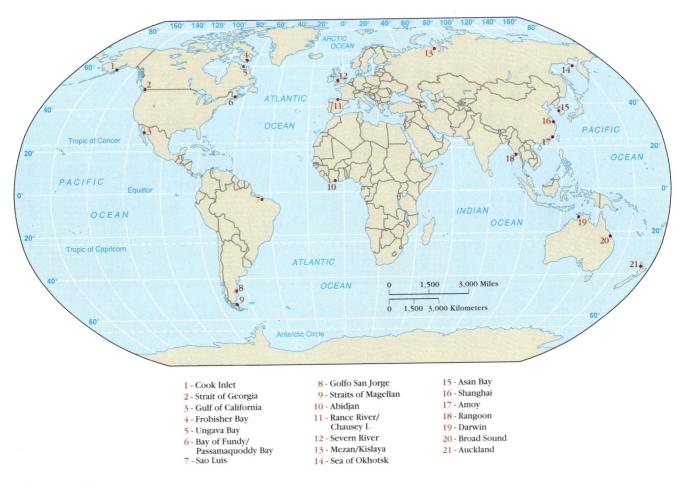

1 - Cook Inlet
2 - Strait of Georgia
3 - Gulf of California
4 - Frobisher Bay
5 - Ungava Bay
6 - Bay of Fundy/
    Passamaquoddy Bay
7 - Sao Luis

8 - Golfo San Jorge
9 - Straits of Magellan
10 - Abidjan
11 - Rance River/
     Chausey I.
12 - Severn River
13 - Mezan/Kislaya
14 - Sea of Okhotsk

15 - Asan Bay
16 - Shanghai
17 - Amoy
18 - Rangoon
19 - Darwin
20 - Broad Sound
21 - Auckland

Figure 9–18

Sites with Major Potential for Tidal Power Generation. Twenty-one locations world-wide where tidal ranges are great enough to have potential for generating electricity. Where a large area exists for storing water behind a dam, a tidal range of 3 meters (10 feet) would suffice. Where smaller storage areas exist, greater tidal ranges would be required.

S U M M A R Y

The tides of Earth are caused by gravitational attraction of the sun and moon. The moon has about twice the tide-generating effect of the sun because it so much closer. Small horizontal forces tend to push water into two bulges on opposite sides of Earth, one directly facing the tide-generating body and the other directly opposite. Since the tidal bulges due to the moon's gravity are dominant, the tides observed on Earth have periods dominated by lunar motions. They are modified by the changing position of the solar bulges.

If Earth were a uniform sphere covered with an ocean of uniform depth, tides would be easily predicted. Such tides would have a period of 12 hours and 25 minutes, or half a lunar day. Tides with maximum tidal range (spring tides) would occur each new moon and full moon,

and tides with minimum range (neap tides) would occur with the first and third quarter phases of the moon.

Since the moon may have a declination up to 28.5° north or south of the equator, and the sun is directly over the equator only two times per year, tidal bulges usually are located to create two high tides of unequal height per lunar day. The same could be said for the low tides. Tidal ranges are greater when Earth is nearest the sun and moon.

Because Earth has an irregular surface, with continents dividing the world ocean into irregularly shaped basins, the tides actually observed on Earth are explained by the dynamic theory of tides. Open-ocean tides rotate counterclockwise around an amphidromic point, a point of zero tidal range, in the Northern Hemisphere.

The basic types of tides observed on Earth are a diurnal tide (period of 1 lunar day,) a semidiurnal tide (period of half a lunar day, like that predicted for the equilibrium tide), and a mixed tide, with characteristics of both. Mixed tides are usually dominated by semidiurnal periods and display significant inequality. The inequalities are greatest when the moon is over the tropics and least when it is over the equator.

The effects of constructive interference and the shoaling and narrowing of coastal bays can be seen in the extreme tidal range at the northern end of Nova Scotia's Bay of Fundy.

Tidal currents follow a rotary pattern in open-ocean basins but are converted to reversing currents along continental margins. The maximum velocity of reversing currents occurs during ebb and flood currents when the water is halfway between high and low standing waters. Tidal bores are tide waves that force their way up rivers. They are common in such rivers as the Amazon, Chientang, Seine, and Trent.

Because tides can be used to generate power without need for fossil or nuclear fuel, the possibility of constructing such generating plants has always attracted engineers. One such plant is operating satisfactorily in the estuary of the La Rance River in France.

KEY TERMS

Autumnal equinox (p. 208)
Centripetal force (p. 203)
Cotidal lines (p. 211)
Declination (p. 206)
Diurnal tide (p. 211)
Dynamic theory (p. 204)
Ecliptic (p. 206)
Full moon (p. 206)

Law of gravitation (p. 202)
Lunar day (p. 205)
Mixed tide (p. 212)
Neap tide (p. 206)
New moon (p. 206)
Newton, Sir Isaac (p. 202)
Quarter moon (p. 206)

Semidiurnal tide (p. 211)
Spring equinox (p. 206)
Spring tide (p. 206)
Summer solstice (p. 208)
Tidal bore (p. 214)
Tidal range (p. 206)
Winter solstice (p. 208)

QUESTIONS AND EXERCISES

1. Explain why the sun's influence on Earth's tides is only 46 percent that of the moon's, even though the sun exerts a gravitational force on Earth 177 times greater than that of the moon.
2. Discuss why the length of the lunar day is 24 hours and 50 minutes of solar time.
3. Explain why the maximum tidal range (spring tide) occurs during new and full moon phases and the minimum tidal range (neap tide) at first quarter and third quarter moons.
4. Discuss the length of cycle and degree of declination of the moon and sun relative to Earth's equator.
5. Describe the effects of the declination of the moon and sun on the tides.
6. Diagram the Earth-moon system's orbit about the sun. Label the positions on the orbit at which the moon and sun are closest to and farthest from Earth, stating the terms used to identify them. Discuss the effects of the moon's and Earth's positions on Earth's tides.
7. Describe the period and inequality of the following: diurnal tide, semidiurnal tide, and mixed tide.
8. Discuss factors that help produce the world's greatest tidal range in the Bay of Fundy.
9. Discuss the difference between rotary and reversing tidal currents.
10. Discuss at least one positive and one negative factor related to tidal power generation.

REFERENCES

Clancy, E. P. 1969. *The tides: Pulse of the earth*. Garden City, N.Y.: Doubleday.

Defant, A. 1958. *Ebb and flow: The tides of earth, air, and water*. Ann Arbor: University of Michigan Press.

Gill, A. E. 1982. *Atmosphere and ocean dynamics. International Geophysics Series*, Vol. 30, Orlando, Fla.: Academic Press.

Pond, S., and Pickard, G. L. 1978. *Introductory dynamic oceanography*. Oxford: Pergamon Press.

Sverdrup, H. U.; Johnson, M. W.; and Fleming, R. H. 1942. Renewal 1970. *The oceans: Their physics, chemistry, and biology*. Englewood Cliffs, N.J.: Prentice Hall.

von Arx, W. S. 1962. *An introduction to physical oceanography*. Reading, Mass.: Addison-Wesley.

SUGGESTED READING

Sea Frontiers

Canove, P. 1989. The reclamation of Holland. 35:3, 154–164. A comprehensive history of how the Dutch have reclaimed and protected coastal lands from the threat of rising water.

Holloway, T. 1989. Eling tide mill. 35:2, 114–119. The tide mill at Eling Toll Bridge near Southampton, England, has been in existence since at least A.D. 1086. It is now a working museum, and the article describes its history and how it works.

Sobey, J. C. 1982. What is sea level? 28:3, 136–142. The role of tides and other factors in changing the level of the ocean surface are discussed.

Zerbe, W. B. 1973. Alexander and the bore. 19:4, 203–208. An account of Alexander the Great's encounters with a tidal bore on the Indus River.

Scientific American

Goldreich, P. 1972. Tides and the earth-moon system. 226:4, 42–57. The tide-generating force of the sun and moon on Earth and the effect of transfer of angular momentum from Earth to the moon as a result of tidal friction are discussed. Also considered are theories of lunar origin.

Greenberg, D. A. 1987. Modeling tidal power. 257:3, 128–131. The effects of using the large tidal ranges experienced in the Bay of Fundy to generate electricity are modeled on a computer.

Lynch, D. K. 1982. Tidal bores. 247:4, 146–157. Tidal bores can be spectacular walls of water rushing up rivers when the tide rises.

C H A P T E R   1 0
Coastal Geology

Throughout our history, we have been attracted to seacoasts for their moderate climate, seafood, recreational opportunities, and commercial benefits. In the United States, our migration toward the Atlantic, Pacific, and Gulf coasts has continued; 80 percent of the U.S. population now lives within easy access of these regions. The coastal regions are enduring increasing stress that threatens to damage this national resource so badly that future generations may not have the opportunity to enjoy it.

Before discussing our beautiful coastal regions and their processes, I want to present to you a dismaying situation that is increasing the magnitude of this problem. I hope to stir your interest in improving the situation.

You may not have heard of the National Flood Insurance Program (NFIP). Let me introduce it to you, for it is one of the saddest examples of our federal government's inability to meet its responsibility to protect the health and safety of U.S. citizens.

In 1965 Hurricane Betsy killed 75 people and did $1.4 million in property damage. The federal government's response was to develop the National Flood Insurance Program in 1968. Its original goal was to encourage people to build safely inland of flood-threatened areas in return for federally subsidized flood insurance (in case some very unusual circumstance resulted in flood damage to property).

The goal was to reduce deaths and property damage from flooding. However, by the time the lobbyists for the banking, construction, and real estate industries had their fears alleviated by members of Congress, the program turned into a vast subsidy that supported risky overdevelopment of the U.S. coastal region.

During the decade prior to passage of the plan, hurricanes killed 186 people and did $2.2 billion in damage. But following the passage of the flood insurance plan, coastal development boomed. Consequently, during the 10 years following passage of the plan, there were 411 deaths and $4.7 billion in property damage from coastal storms. This was a 121 percent increase in deaths and a 114 percent increase in damage. The reason, of course, was that the NFIP had encouraged overdevelopment, placing more people and buildings in harm's way.

And the problem continues to worsen. Studies conducted by the General Accounting Office in 1983 into the soundness of the program showed that many insurance premiums should be increased by 800 percent. However, only new properties were subject to these higher rates. Rates have increased further in recent years, but the government still pays out billions with each flooding disaster in the form of grants and loan subsidies—to cover damage that would not have happened if our government had a sensible policy on seacoast development.

In addition to these benefits, which help owners replace structures built in locations that are clearly in danger of storm damage, the federal government subsidizes development of these areas by paying most of the cost of infrastructure development. (Infrastructure refers to water systems, sewage treatment facilities, highways, and bridges.) Further, the federal government assumes from 50 to 70 percent of the cost of erosion-control programs, and erosion is what coastal geology is all about.

It is clear that U.S. taxpayers are subsidizing imprudent development decisions that benefit a small number of individuals.

In 1982 Congress passed the Coastal Barrier Rezoning Act. It excludes construction in undeveloped coastal areas from receiving federal assistance. The act also encourages states to develop "setback" regulations that require construction to be set back at some distance from the shore, realistically incorporating the threat of storm damage into all coastal building programs.

The insurance remains a real bargain. The owner of a $200,000 coastal home can purchase for about $1000 insurance that would cost $18,000 from a private insurance company.

The National Academy of Sciences is warning of possible accelerated sea level rise in coming years. If this happens, the problem with existing homes and businesses could become immense in as few as 10 years.

What can be done to further reverse the direction of such unsound public policy? What general guidelines should be followed in developing a national program for conservative use of our coastal areas, to preserve them for everyone for generations to come?

Of course, politics is the major hurdle. But understanding the coastal environment is an important first step. In the following pages, you will learn about the major features of the seacoast and shore and the processes that modify them. You will also see examples of how people interfere with these processes, creating hazards to themselves and to the environment.

The Coastal Region

Traveling by boat in the ocean, as you approach land, you encounter the shore. The **shore** is a zone that lies between the lowest tide level (low tide) and the highest elevation on land that is affected by storm waves. From this point landward is the **coast,** which extends inland as far as ocean-related features can be found (Figure 10–1).

The width of the shore varies between a few meters and hundreds of meters. The width of the coast may vary from less than a kilometer (0.6 miles) to many tens of kilometers. As the waves beat against the shore, they cause erosion. This erosion produces sediment that is transported along the shore and deposited in areas where wave energy is low.

All shores experience some degree of both *erosion* and *deposition.* Shores often described as primarily erosional have well-developed cliffs and generally are in areas where tectonic uplift of the coast is occurring. Much of the U.S. Pacific Coast is classified as primarily erosional.

In contrast, along the U.S. southeastern Atlantic Coast and Gulf Coast, sand deposits and offshore barrier islands are common. These coasts are experiencing a gradual subsidence of the shore. Although we classify such coasts as primarily depositional, erosion can be a major problem when human developments interfere with natural coastal processes. This can cause easily eroded deposits to be washed away.

Features of Erosional-Type Shores

Refer to Figure 10–1 to see the features that are described here. The landward limit of a shore that is dom-

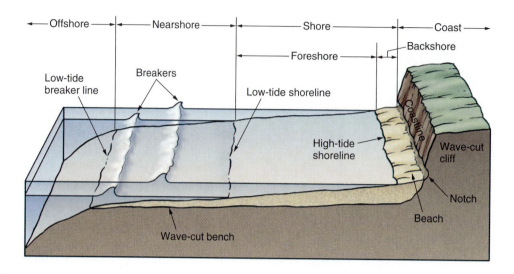

Figure 10–1

Landforms and Terminology of Coastal Regions. A *coastline* is the most landward evidence of direct erosion by ocean waves. The coastline separates the *shore* from the *coast*. The shore extends from the coastline to the *low-tide shoreline* (water's edge). It is divided into the *backshore*, above the *high-tide shoreline*, which is covered with water only during storms, and the *foreshore* (also called the *intertidal* or *littoral zone*). Never exposed to the atmosphere, but affected by waves that touch bottom, is the *nearshore*, which extends seaward to the low-tide *breaker line*. The greatest amount of sediment transport (as beach deposit) occurs within the shore and nearshore zones. Beyond the nearshore lies the *offshore* region, where depths are such that waves rarely affect the bottom.

inated by erosion is commonly marked by a cliff. The **coastline,** which marks the boundary between the shore and the coast, is a line along the cliff that connects points at which the highest effective wave action takes place.

The shore is divided into the **foreshore,** that portion exposed at low tide and submerged at high tide, and the **backshore,** which extends from the normal high tide to the coastline. The **shoreline** migrates back and forth with the tide and is the water's edge. The **nearshore** zone is that region between the low-tide shoreline and breakers. Beyond the low-tide breakers is the **offshore** zone.

A **beach** is a deposit of the shore area. It consists of wave-worked sediment that moves along the **wave-cut bench** (a flat, wave-eroded surface). A beach may continue from the coastline across the nearshore region to the line of breakers.

Wave Erosion. Due to refraction (the bending of waves; see Figure 8–13), wave energy is concentrated on any **headlands** that jut out from the continent, while the amount of energy reaching the shore in bays is reduced. As the waves concentrate their energy on the headlands, erosion occurs and the shoreline retreats.

The greatest concentration of wave energy, on a day-to-day basis, is in the foreshore region (Figure 10–1).

However, during rare periods when storm waves batter the shore, more erosion may occur across the entire shore in one day than may be achieved by average wave conditions over a period of years.

The cliff shown in Figure 10–1 is referred to as a **wave-cut cliff,** produced by wave action relentlessly pounding away at its base. The cliff develops as the upper portions collapse after being undermined by wave action. The undermining may be evident in the form of a notch at the base of the cliff that may be characterized by **sea caves.** Such caves are most commonly cut into hard sedimentary rock.

Further wave action eroding the softer portions of the rock outcrops may develop the caves into openings running through the headlands, called **sea arches** (Figure 10–2). Continued erosion and crumbling of these arches will produce **sea stacks** (Figure 10–2). Such remnants rise from the relatively smooth wave-cut bench cut into the bedrock by wave erosion (Figure 10–2).

The *rate of wave erosion* is determined by these variables:

1. *Degree of exposure* of the coastal region to the open ocean is very important. Coasts that are fully exposed receive higher-energy wave action and are likely to have rugged cliffs in areas of high topographic relief.

Figure 10–2

Sea Arch and Sea Stack along the Coast of Iceland. If one ever doubted the erosive power of ocean waves, here is ample evidence. (Photo by Bruce F. Molnia, Terra-photographics/BPS.)

2. *Tidal range* is another important variable. Given the same amount of wave energy, a region with a small tidal range will erode much more rapidly than one with a large tidal range. This may sound contradictory, but the smaller tidal range intensifies wave energy over a much narrower shore. (Although high-velocity tidal currents may develop in areas where a large tidal range exists, such currents are of limited importance in eroding the coastline.)

3. *Composition of coastal bedrock* is very significant. Crystalline igneous rocks such as granite, metamorphic rocks, and hard sedimentary rocks are relatively resistant to erosion and produce a rugged shoreline topography. Weak sedimentary rocks such as sandstone and shale are more easily eroded, producing a gentler topography associated with more extensive beach deposits.

Regardless of the erosion rate, all coastal regions follow the same developmental path. As long as there is no change in the elevation of the landmass relative to the ocean surface, the cliffs will continue to erode and retreat until the beaches widen sufficiently to prevent waves from reaching them. The eroded material is carried from the high-energy areas and deposited in the low-energy areas.

Features of Depositional-Type Shores

The coastal erosion just discussed produces large amounts of sediment. Adding to it is sediment eroded inland by running water and delivered to the ocean by rivers. All of this sediment must be distributed along the continental margin. The agent that does so is the *longshore current.*

As waves strike the shore at an angle, they set up a longshore movement of water, called the **longshore current**. This current moves parallel to the shore between the shoreline and the breaker line, carrying with it the sediments that make up the beach. At the landward margin of the surf zone the **swash,** a thin sheet of water, moves sediment up onto the exposed beach at an angle. But gravity pulls the backwash with its sediment load straight down the beach face. As a result, swash-transported pebbles and sand grains move in a zigzag pattern along the shore in the same direction as the longshore current within the surf zone (see the zigzag "longshore drift" line in Figure 10–3). The net direction of annual longshore current is southward along both the Atlantic and Pacific shores.

The velocity of the longshore current increases with beach slope, with the angle of breakers to the beach, and with wave height. The current's velocity decreases with wave period.

Longshore drift refers to the movement of sediment by the process just described. The amount of longshore drift in any coastal region is determined by an equilibrium between erosional and depositional forces. Any interference with the movement of sediment along the shore will destroy this equilibrium, causing a new erosional and depositional pattern to form.

Rip Currents. As noted, longshore-current water flows up onto the shore and then runs back into the ocean. This backwash of water generally finds its way into the open ocean as a thin "sheet flow" across the ocean bottom. However, some of this water flows back in local **rip currents.** These currents occur either perpendicular to or at an angle to the coast. They happen where topographic lows or other conditions allow their formation.

Rip currents may be less than 25 meters (80 feet) wide and can attain velocities of 7 to 8 kilometers/hour (4 to 5 miles/hour). They do not travel far from shore before they break up. If a light-to-moderate swell is breaking, numerous rip currents may develop, moderate in size and velocity. A heavy swell will usually produce fewer, more concentrated rips (Figure 10–3).

The rip currents that occur during heavy swell are a significant hazard to coastal swimmers. However, swimmers who realize they are caught in a rip current and do not panic can escape by swimming parallel to the shore for a short distance (simply swimming out of

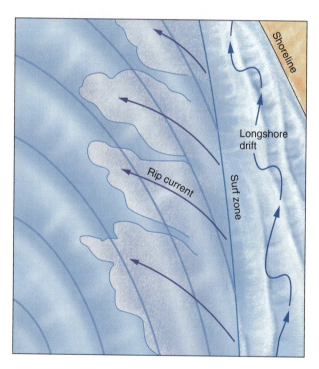

Figure 10–3

Longshore Currents and Rip Currents. In this example, as waves (curving blue lines) approach shore from a southerly direction, they produce a *longshore current,* and longshore drift, that flow northward. Water in the current follows a zigzag path as waves push it up the beach slope from the direction of approach. The water runs back down the slope under the influence of gravity. The longshore drift of sediment particles carried by the longshore current follows a similar zigzag path. Water of the longshore current finds its way offshore by passing through topographic lows as strong seaward flows called *rip currents.* These are visible in the photograph as jets of water that appear light in color because of the turbidity from sediment they have resuspended from the ocean floor. The red arrows in the diagram show the path of the four rip currents. (Photo courtesy of Scripps Institute of Oceanography, University of California/San Diego.)

the current). Another option is to ride out the current until it becomes less intense, because rip currents extend only a short distance from shore.

Beach Composition. The material that composes a beach deposit depends on the source of sediment. The sediment may be locally available, or it may transported by longshore drift. Where sediment is provided by coastal mountains, beaches will be composed of mineral particles from the rocks of those mountains. This sediment may relatively coarse in texture.

If the sediment is provided primarily by rivers that drain lowland areas, sediment that reaches the coastal regions normally will be finer in texture. Often, mud flats develop along the shore because only tiny clay-sized and silt-sized particles are emptied into the ocean.

In low-relief, low-latitude areas such as southern Florida, where there are no mountains or other sources of rock-forming minerals nearby, most beach material is derived from the remains of the organisms that live in the coastal waters. Beaches in these areas are composed predominantly of shell fragments and the remains of microscopic animals.

Many beaches on volcanic islands in the open ocean are composed of dark fragments of the basaltic lava that makes up the islands, or of coarse debris from coral reefs that develop around islands in low latitudes.

Beach Slope. The slope of beaches is closely related to the size of the particles of which they are composed. Waves washing onto the beach carry sediment, thereby increasing the slope of the beach. If the backwash returns as much sediment as the waves carried in, the beach has reached equilibrium and will not steepen.

A beach composed of fine-grained sand that is relatively angular will have a gently sloping, firm surface. Because these small grains interlock closely, little of the swash sinks down among the grains. Most of it runs back down the slope to the ocean, possessing enough energy to maintain equilibrium on a gentle slope. The backshores of such beaches are usually nearly horizontal.

Beaches composed of coarse sands or pebbles usually contain particles that are more rounded and more loosely packed. The swash quickly percolates into such deposits as it moves up the beach slope. Deposition of particles by the swash will continue until the beach slope becomes steep enough that the backwash running to the ocean has sufficient energy to maintain equilibrium. Such beaches are usually much less firm than beaches composed of finer material, and the backshores will slope significantly toward the coastline (Table 10–1).

Special Geomorphic Features. Numerous depositional features exist that are partially or wholly separated from the shore. These are deposited by the

Table 10–1
The Relation of Particle Size to Beach Slope

	Wentworth Particle Size (mm)	Mean Slope of Beach
Cobble	256	24°
Pebble	64	17°
Granule	4	11°
Very coarse sand	2	9°
Coarse sand	1	7°
Medium sand	0.5	5°
Fine sand	0.125	3°
Very fine sand	0.063	1°

Source: After Table 9, "Average beach face slopes compared to sediment diameters" from *Submarine geology*, 3rd ed. by Francis P. Shepard, p. 127. Copyright © 1948, 1963, 1973 by Francis P. Shepard. Reprinted by permission of Harper & Row, Publishers, Inc.

longshore drift and through other processes that are not well understood. Some of these are shown in Figure 10–4 and are briefly described here.

A **spit** is a linear ridge of sediment attached to land at one end. The other end points in the direction of longshore drift and ends in the open water. Spits are simply extensions of beaches into the deeper water near the mouth of a bay. The open-water end of the spit normally curves into the bay as a result of current action.

If tidal currents, or currents from river runoff, are too weak to keep the mouth of the bay open, the spit may eventually extend across the bay and tie to the mainland, thus completely cutting off the bay from the open ocean. The spit then has become a **bay barrier.**

A **tombolo** is a sand ridge that connects an island with another island or to the mainland. Tombolos usually are aligned at a large angle to the main shore. This is because they are usually sand deposits laid down in the wave-energy shadow of the island. Thus, their alignment is at right angles to the average direction of wave approach.

Barrier Islands. Long offshore deposits of sand lying parallel to the coast are called **barrier islands** (Figure 10–4). They are well named, for they do constitute a barrier to storm waves that otherwise would severely assault the shore. Their origin is complex, and several explanations have been proposed for their existence. However, it appears that many barrier islands developed during the worldwide rise in sea level that began with the melting of the most recent major glaciers some 18,000 years ago.

Barrier islands are nearly continuous along the Atlantic Coast of the United States. They extend around Florida and along the Gulf of Mexico coast, where they exist well south of the Mexican border. Barrier islands may exceed 100 kilometers (60 miles) in length and

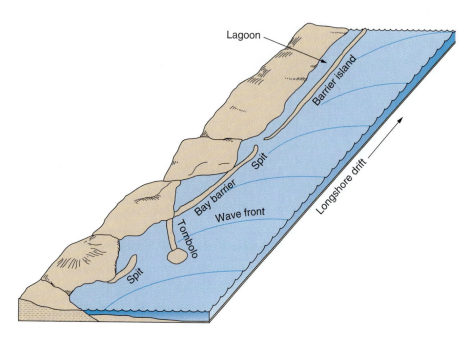

Figure 10–4

Coastal Depositional Features. A *tombolo* is a deposit that connects an island to the mainland or another island. A *spit* is a deposit that extends from land into open water. A deposit that extends across a bay and closes the bay from open water is a *bay barrier*. A *barrier island* is a long deposit separated from the mainland by a *lagoon*.

have widths of several kilometers. Examples of such features are Fire Island off the New York coast, North Carolina's Outer Banks, and Padre Island off the coast of Texas.

A typical barrier island has the following physiographic features, shown in Figure 10–5A. From the ocean landward, they are (1) ocean beach, (2) dunes, (3) barrier flat, (4) high salt marsh, (5) low salt marsh, and (6) lagoon between the barrier island and the mainland. We will look briefly at each portion.

The **ocean beach** is typical of the beach environment discussed earlier. During the summer, as gentle waves carry sand to the beach, it widens and becomes steeper. Higher-energy winter waves carry sand offshore and produce a narrow, gently sloping beach.

Winds blow sand inland during dry periods to produce **dunes,** which are stabilized by dune grasses. These plants can withstand salt spray and burial by sand. Dunes are the lagoon's primary protection against excessive flooding during storm-driven high tides. Numerous passes exist through the dunes, particularly along the southeastern Atlantic Coast, where dunes are less well developed than to the north.

Behind the dunes, the **barrier flat** forms as the result of deposition of sand driven through the passes during storms. These flats are quickly colonized by grasses. During storms, these low barriers are washed over by seawater. If for some reason the frequency of overwash by storms decreases, the plants will undergo natural biological succession, with the grasses successively being replaced by thickets, woodlands, and even forests.

Salt marshes typically lie inland of the barrier flat. They are divided into the *low marsh,* extending from about mean sea level to the high neap-tide line,

and the *high marsh,* extending to the highest spring-tide line. Biologically, the low marsh is by far the most productive part of the salt marsh.

New marshland is formed as overwash carries sediment into the lagoon, filing portions so they become intermittently exposed by the tides. Marshes may be poorly developed on parts of the island that are far from floodtide inlets. Their development is greatly restricted on barrier islands where people perform artificial dune enhancement and fill inlets, activities that prevent overwashing and flooding.

Because sea level is gradually rising along the eastern North American coast, barrier islands are migrating landward. This is clearly visible to those who build structures on these islands. Other evidence for such migration can be seen in peat deposits, remnants of old marshes that lie beneath the barrier islands. Since the only barrier island environment in which peat forms is the salt marsh, the island must have moved inland over previous marsh development.

A possible cycle of salt marsh formation and destruction by landward migration of a barrier island is presented in Figure 10–5B.

Deltas. Some rivers carry more sediment than can be distributed by the longshore current. Such rivers dump sediment at their mouths, developing a **delta** deposit. One of the largest such features on Earth is that produced by the Mississippi River where it empties into the Gulf of Mexico (Figures 10–6A and 10–6B). Deltas are fertile, flat areas that are subject to periodic flooding.

Delta formation begins when a river has filled its mouth with sediment. Once the delta forms, it grows through the distribution of sediment by forming **dis-**

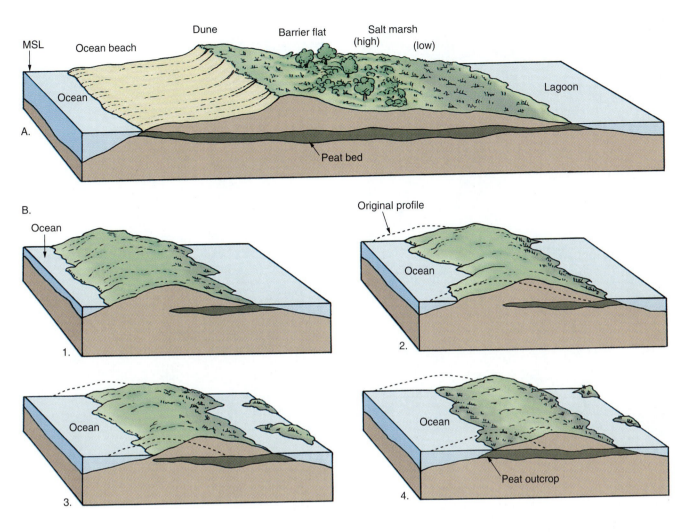

Figure 10–5

Barrier Islands. *A:* Cross section through a barrier island. The major physiographic zones of a barrier island are the *ocean beach, dunes, barrier flat, high salt marsh,* and *low salt marsh. MSL* is mean sea level. The *peat bed* represents ancient marsh environments that have been covered by the island as it migrates toward the mainland with rising sea level. *B:* (1) *Barrier island* before overwash, with salt marsh development on mainland side toward the lagoon to the right; (2) overwash erodes the seaward margin of the barrier island, cuts inlets, and carries sand into the lagoon, covering much of the existing marsh; (3) new marsh forms in the lagoon that has retreated toward the mainland; (4) with continued rise in sea level and migration of the barrier island toward the mainland, peat formed from previous marsh deposits may be exposed by erosion of the foreshore of the ocean beach.

tributaries, branching channels that radiate out over the delta (Figures 10–6A and 10–6B). Distributaries lengthen as they deposit sediment and produce finger-like extensions to the delta. When the fingers get too long, they become choked with sediment. At this point, a flood may easily shift the distributary course and provide sediment to low-lying areas between the fingers. Where depositional processes dominate over coastal erosion and transportation processes, a "bird foot" Mississippi-type delta results.

Where erosion and transportation processes dominate, instead of erosion, a delta shoreline will be smoothed to a gentle curve, like that of the Nile River Delta in Egypt (Figure 10–6C). The Nile Delta is presently eroding, owing to the intervention of people: Sediment is being entrapped behind the Aswan High Dam, completed in 1964. Prior to building of the dam, generous volumes of sediment were carried by the Nile into the Mediterranean Sea.

Erosion is only one negative effect of this dam. The problem of reduced sediment availability may become very significant along the eastern end of the delta, where subsidence has occurred at a rate of 0.5 centimeters (0.2 inches) per year for 7500 years. At the present

A.

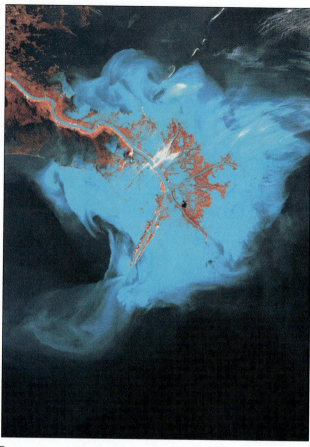

B.

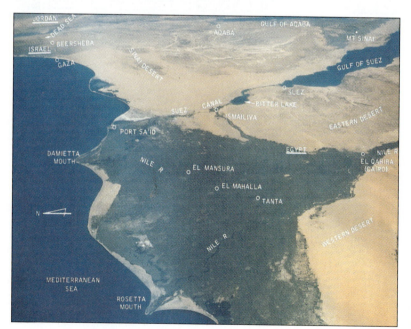

C.

Figure 10–6

Deltas. *A and B:* Digitate (fingerlike) structure of the Mississippi River Delta results from the low-energy environment. These images show the growth of the delta at the mouth of the Mississippi River's main channel during the 16 years between January 16, 1973 (*A*) and March 3, 1989 (*B*). *C:* The relatively smooth, curved shoreline of the Nile River Delta dominates this view. The Mediterranean Sea is to the left, and the Gulf of Suez is in the upper right. (*A:* Photo by GEOPIC®, Earth Satellite Corporation; *B:* photo courtesy of NASA.)

rate of inundation, the sea may move inland 30 meters (100 feet) during the next century.

On a positive note, however, this new, human-induced erosion may have made possible the discovery of the ruins of the ancient city of Alexandria beneath the Mediterranean waters at the edge of the delta.

Emerging and Submerging Shorelines

Along some coasts are flat platforms, called **marine terraces,** backed by cliffs. *Stranded beach deposits* and other evidence of marine processes may exist many meters above the present shoreline. These features characterize **emerging shorelines,** shorelines that are emerging above sea level. They have reached their present positions relative to the existing shoreline by an uplift of the continent, a lowering of sea level, or a combination of the two.

In other areas, we find underwater **drowned beaches** and *submerged dune topography.* These features, along with the drowned river mouths along the present shoreline, indicate **submerging shorelines,** shorelines that are sinking below sea level. This submergence must be due to subsidence of the continent, a rise in sea level, or a combination of the two.

Attempts to determine the causes of sea level change (the relative level of the ocean compared with the land) have not met with great success. While a shoreline is in the act of emerging or submerging, it has definite characteristics. But once either process ceases, erosion reduces either type to the same end result. Thus, whether the shoreline has become submerged because of a rising sea level or a subsiding continent in a particular region cannot be determined by examining the coastal features in that area (Figure 10–7).

Tectonic and Isostatic Movements of Earth's Crust

Changes in sea level relative to a continent may occur because of movement of the land, which is *tectonic movement.* Such movement could mean large-scale uplift or large-scale subsidence of major portions of continents or ocean basins. Or such movement could mean a more localized deformation of the continental crust involving folding, faulting, and tilting.

Earth's crust also responds isostatically, by sinking under the accumulation of heavy loads of ice, sediment, or lava, or rising when such a load is removed (see Chapter 2).

There is evidence that, during the last 2.5 to 3 million years, at least four major accumulations of glacial ice developed in high latitudes. Although Antarctica is still covered by a very large, thick glacial accumulation, much of the ice cover that once existed over northern Asia, Europe, and North America has disappeared. The most recent period of melting began about 18,000 years ago. Accumulations of ice that were up to 3 kilometers (2 miles) thick have disappeared from northern Canada and Scandinavia.

While these areas were beneath the thick ice sheet, the crust beneath them sank. Today, they are still recovering, slowly rising after the melting of the ice. Some geologists believe that by the time this isostatic rebound is finished, the floor of Hudson Bay, which is now about 150 meters (500 feet) deep, will be above sea level. There is evidence in the Gulf of Bothnia, between Sweden and Finland, of 275 meters (900 feet) of isostatic rebound during the last 18,000 years (Figure 10–8).

Generally, tectonic and isostatic changes in the level of the shoreline are local or regional, confined to a segment of the shoreline of a given continent.

Eustatic Changes in Sea Level

A worldwide change in sea level can be caused by an increase/decrease of seawater volume or an increase/decrease in ocean basin capacity. Such a change in sea level is called **eustatic.** This term refers to a highly idealized phenomenon in which all of the continents remain static while only the sea rises or falls.

For example, small eustatic changes in sea level could be created by formation or destruction of large inland lakes. But more important is the effect of glaciers. As they freeze during a glacial stage, large glaciers tie up a vast volume of water, causing a eustatic lowering of sea level. During interglacial stages, such as we are in at present, the glaciers melt, releasing great volumes of water that drain to the sea, causing a eustatic rise of sea level.

Changes in seafloor spreading rates also can change sea level. Fast spreading produces larger rises, such as the East Pacific Rise, that displace more water than slow-spreading ridges such as the Mid-Atlantic Ridge. Thus, fast spreading produces a rise in sea level. Significant changes of sea level resulting from changes in spreading rate typically occur over hundreds of thousands or millions of years.

During the Pleistocene Epoch, when glacial growth occurred, the amount of water in the ocean basin fluctuated considerably. We know this from the present elevations of marine features. Because the climate was colder during glacial times, we might account for some of the lowering of the shoreline by the contraction of the ocean volume as its temperature decreased. It has been calculated that, for every 1°C (1.8°F) decrease in the mean temperature of the ocean water, sea level would drop 2 meters (6.6 feet). Temperature indications derived from study of fossils in Pleistocene ocean sediments suggest that ocean surface temperature may have been as much as 5°C (9°F) lower than at present. Therefore, contraction of the ocean water may have lowered sea level by about 15 meters (40 feet).

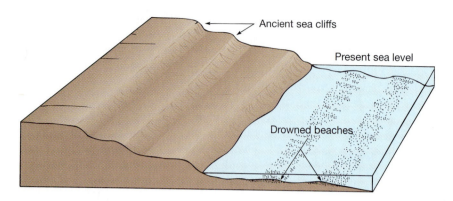

B.

Figure 10–7

Evidence of Changing Shoreline Levels. *A:* Marine terraces result from exposure of ancient sea cliffs and wave-cut benches above present sea level. They mark ancient shorelines, as do the drowned beaches that lie below sea level. *B:* Wave erosion coast, Santa Cruz Island, California. *C:* Depositional coast—barrier coast, spit, and bay barrier along the coast of Martha's Vineyard, Massachusetts. (*A:* Photo by James R. McCullagh; *B:* photo by USDA-ASCS.)

C.

Although it is difficult to state definitely the range of shoreline fluctuation during the Pleistocene, there is cause to believe that it was at least 120 meters (400 feet) below the present shoreline (Figure 10–9). It is also estimated that, if all the remaining glacial ice on Earth were to melt, sea level would rise another 60 meters (200 feet). This would give a minimum possible range of sea level during the Pleistocene of 180 meters (600 feet), most of which must be explained through the capture and release of Earth's water by glaciers.

During the last 18,000 years, ocean volume has increased due to expansion of seawater resulting from warmer temperatures plus the melting of polar sea ice and glacial ice. The combination of tectonic and eustatic changes in sea level is very complex and difficult to sort out, so it is hard to classify coastal regions as purely

Figure 10–8

Extent of Ice Coverage during Most Recent Glacial Age. Arrows show the general direction of ice flow. Coastlines are shown as they may have been at that time, when sea level may have been 120 meters (390 feet) lower than at present. The present Hudson Bay and the Gulf of Bothnia are areas that were depressed beneath the weight of the glacial ice. With the ice now melted, these areas still are rebounding slowly and eventually may rise above sea level—if another ice advance does not occur before rebound is complete.

emergent or submergent. Most coastal areas show evidence of having experienced both submergence and emergence in the recent past. It is believed, however, that sea level has not risen significantly as a result of melting glacial ice during the last 3000 years.

There is at present much discussion about combined gleanings from historical research. There is clear evidence that atmospheric carbon dioxide has increased

by 15 percent since 1958, when monitoring began. There has also been an apparent global warming of about 0.5°C (0.9°F) and a eustatic rise in sea level of 10 centimeters (4 inches) since 1880 (Figure 10–10).

The obvious question is this: Are all of these observations related, and are they the result of an increased greenhouse effect? The answer is unknown. But the great lesson for humankind from these observations is that we cannot dominate nature; we must live within it. To do otherwise is futile. Some of the consequences of a rising sea level for U.S. coastal communities are discussed in the following sections.

Plate Tectonics and Coasts

The most dramatic examples of sea level changes during the past 3000 years have been due to tectonic processes. To understand the underlying tectonic cause of emergence or submergence, we briefly return to the concept of global plate tectonics.

Atlantic-Type (Passive) Coast

Considering only the coasts of North America, we find the Atlantic Coast to be subsiding and the Pacific Coast to be emerging. If we begin with the breakup of Pangaea and the formation of the Atlantic Ocean, this emer-

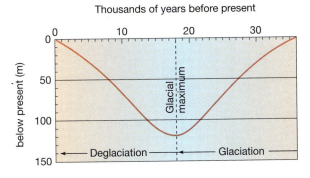

Thousands of years before present

Figure 10–9

Sea Level Change during the Most Recent Advance and Retreat of Pleistocene Glaciers. Sea level dropped worldwide by about 120 meters (400 feet) as the last glacial advance removed water from the oceans and transferred it to continental ice sheets. About 18,000 years ago, sea level began to rise as the glaciers melted and water was returned to the oceans.

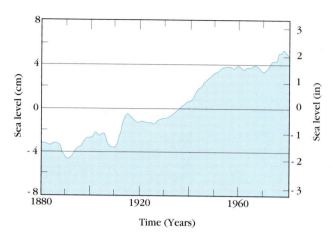

Figure 10–10

Sea Level Change from 1880 to 1980. Tide-gauge data, averaged over 5-year periods, show that global mean sea level has increased about 10 centimeters (4 inches) over the past 100 years. However, more recent data that remove the effect of isostatic recovery indicate that global sea level has risen 1.75 millimeters (0.07 inches) per year during the last 50 years. This rate is 1.5 times the rate shown in this figure. (After V. Gornits, S. Lebedeff, and J. Hausen, 1982. *Science* 215:1611–1614.)

gence-submergence pattern can be readily understood. Recall that the lithosphere thickens, the ocean deepens, and heat flow decreases as lithospheric plates cool with age or with increasing distance from the spreading centers (Chapter 3). As a result of this process, the ocean floor of this passive margin, the Atlantic-type margin, is thought to have subsided about 3 kilometers (2 miles) over the past 150 million years.

Further, erosion of the continent has produced a tremendous volume of sediment. It has accumulated to a maximum thickness of 15 kilometers (9.3 miles) along the Atlantic Coast. Such a thick deposit has been made possible by the plastic asthenosphere, which has allowed the lithosphere to flex downward under its growing sediment burden. This is isostatic sinking, as described pre-

viously. It is this thick *sedimentary wedge* underlying the continental shelf, slope, and rise that has been exploited for large petroleum reserves in the Gulf of Mexico.

These two processes—seafloor spreading and isostatic sinking of the crust—account for the subsidence of the Atlantic and Gulf coasts. Although the rate of subsidence due to cooling decreases with time and the rate of subsidence due to sediment loading depends on sediment supply, it is not reversed unless the tectonic or isostatic conditions change (Figure 10–11).

Pacific-Type (Active) Coast

In contrast to passive, subsiding Atlantic-type coasts, the Pacific-type coast, or active coast, is the scene of intense tectonic activity. Along such margins, volcanoes erupt and earthquakes rumble. The thick, broad sediment wedge characteristic of the Atlantic Coast is poorly developed along a Pacific-type coast. The reason is this: A Pacific-type coast is not the product of the creation of a new ocean basin, but is the scene of lithospheric plate subduction, the compressive forces of which push crust upward, causing an emergent coast.

You can see the characteristic alignment of mountain ranges parallel to the continental margin. In all cases where continental margins are associated with plate convergence, the compressive forces produce uplift. Along the California coast, the process has been further complicated by the development of the San Andreas Fault. This fault has made possible the local subsidence of the depositional basins in the Los Angeles area. But on the broader scale, emergence along Pacific-type margins is the characteristic condition (Figure 10–12).

U.S. Coastal Conditions

Whether the dominant process along a coast is erosion or deposition depends on the combined effect of the variables we have been discussing—degree of exposure to ocean waves, tidal range, composition of

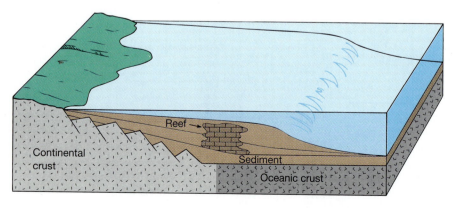

Figure 10–11

Atlantic-Type Margin. This schematic shows why the U.S. Atlantic Coast is slowly subsiding. When a continent is moving away from a spreading center such as the Mid-Atlantic Ridge, its trailing edge subsides due to cooling, plus the weight of accumulating sediment. A low level of tectonic deformation, earthquakes, and volcanism accompanies such conditions, making the coast far more quiet and stable than a Pacific-type margin.

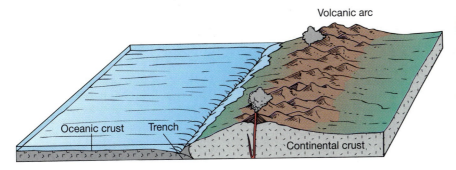

Figure 10–12

Pacific-Type Margin. This schematic shows why the U.S. Pacific Coast is slowly rising, or emerging. Uplift is typical of continental margins where plate collisions occur. Earthquakes, volcanic action, and mountain chains paralleling the coast are common characteristics of such margins.

coastal bedrock, tectonic subsidence or emergence, isostatic subsidence or emergence, and eustatic sea level change.

In 1971 the U.S. Army Corps of Engineers reported on erosion of the U.S. coastline, which is 135,870 kilometers (84,240 miles) long, including Alaska but not Hawaii. They found over 24 percent of it to be "seriously eroding." Subsequent studies supported by the U.S. Geological Survey along the shores of the contiguous coastal states and Alaska produced the rates of shoreline change presented in Figure 10–13.

The Atlantic Coast

The U.S. Atlantic coast is quite a study of coastal processes in action (refer to Figure 10–13):

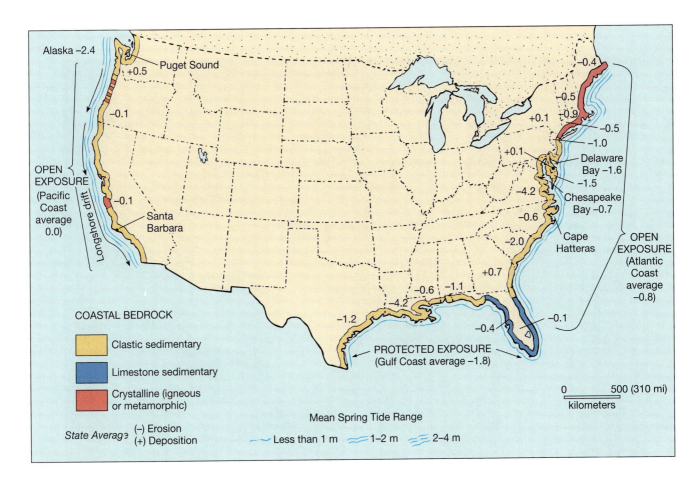

Figure 10–13

Coastal Erosion and Deposition, 1979–1983. For the Pacific, Atlantic, and Gulf coasts, this map shows coastal bedrock type (red, yellow, blue), the mean spring-tide range (light blue lines), degree of exposure, and rates of erosion/deposition. Net erosion (-) or net deposition (+) is shown for specific locations along with average condition for each coast.

- Most of the Atlantic coastline is exposed to storm waves from the open ocean. However, barrier island development from Massachusetts southward protects the mainland from large storm waves.
- Tidal ranges generally increase from less than 1 meters (3.3 feet) along the Florida coast to more than 2 meters (6.5 feet) in Maine.
- Bedrock for most of Florida is a resistant limestone. But from Florida northward through New Jersey, most bedrock is poorly consolidated sedimentary rocks formed in the recent geologic past. These rocks have poor resistance to erosion. As they erode, they become sand sources for barrier islands and other depositional features common along the coast.
- From New York northward, continental glaciers affected the coastal region directly. Many coastal features, including Long Island and Cape Cod, are glacial deposits, moraines left behind when the glaciers melted.

North of Cape Hatteras in North Carolina, the coast is subject to very high energy conditions during fall and winter when storms called "nor'easters" (northeasters) blow in from the North Atlantic. The great energy of these storms is manifested in waves up to 6 meters (20 feet) high, with a 1 meter (3.3 feet) rise in sea level that follows the low pressure as it moves northward. Such high-energy conditions significantly change coastlines that are predominantly depositional and may be expected to cause considerable erosion.

Most of the Atlantic coast appears to be experiencing a sea level rise of about 0.3 meters (1 foot) per century. This condition may be reversed in northern Maine because of isostatic rebound due to the melting of the Pleistocene ice sheet.

The Atlantic Coast has an average annual rate of erosion of −0.8 meters (−2.6 feet). In everyday terms, this means that the sea is migrating landward each year by about the length of your legs. The majority of East Coast states are slowly losing real estate. Virginia leads the way with a loss of 4.2 meters (13.7 feet) per year, which is largely confined to barrier islands.

Erosion rates for Chesapeake Bay are about average for the Atlantic coast, but Delaware Bay shows av-

Figure 10–14

Barge Channels through Louisiana Marshland. The construction of barge channels such as these have accelerated the loss of salt marsh in southern Louisiana. (Photo by J. Carl Ganter/Contact Press Images.)

erage erosion rates about twice (–1.6 meters per year) the Atlantic coast average. Of the observations made along the Atlantic coast, 79 percent showed some degree of erosion. Delaware, Georgia, and New York display depositional coasts despite serious erosion problems in these states as well.

The Gulf Coast

The erosion/deposition story of the Gulf Coast is simpler. The Louisiana-Texas coastline is dominated by the Mississippi River Delta, which is being deposited in a microtidal and generally low-energy environment. The tidal range is normally less than 1 meter (3.3 feet). Except for the hurricane season (June to November), wave energy is generally low. Tectonic subsidence is general throughout the Gulf Coast, and the average rate of sea level rise is similar to that of the southeast Atlantic coast, about 0.3 meters (1 foot) per century.

The average rate of erosion is –1.8 meters (–6 feet) per year in the Gulf Coast. The Mississippi River Delta experiences the greatest rate, so the state of Louisiana is losing an average of 4.2 meters (13.7 feet) per year. Made worse by dredging of barge channels through marshlands (Figure 10–14), Louisiana has lost more than 1 million acres of delta since 1900. The Pelican State is now losing marshland at a rate exceeding 130 square kilometers (50 square miles) per year.

Although all Gulf states show a net loss of land, and the Gulf Coast has a greater erosion rate than the Atlantic Coast, only 63 percent of the shore is receding because of erosion. The high average rate of erosion reflects the heavy erosion loss in the Mississippi River Delta area.

The Pacific Coast

In general, the Pacific Coast is experiencing less erosion than the Atlantic and Gulf Coasts. Along the Pacific coast, relatively young, weak marine deposits dominate the bedrock, with local outcrops of more resistant granite, metamorphic rock, and volcanic rock. Tectonically, the coast is rising, as evidenced by wave-cut terraces (Figure 10–15). Sea level still shows at least small rates of rise, except for a segment along the coast of Oregon and the Alaskan coast. The tidal range is mostly between 1 and 2 meters.

The Pacific Coast is open to large storm waves. High-energy waves may strike the coast in winter, with 1 meter (3.3 feet) waves being normal. Frequently, the wave height will increase to 2 meters (6.6 feet), and a few times per year 6 meters (20 feet) waves hammer the shore. These high-energy waves erode sand from many beaches. The exposed beaches, which are composed primarily of pebbles and boulders during the winter months, regain their sand as low-energy summer waves beat more gently against the shore.

Many Pacific Coast rivers have been dammed for flood control and hydroelectric power generation. This has reduced the amount of sediment supplied by rivers to the sea for longshore transport. As a result, some areas now experience a severe erosion threat.

With an average erosion rate of only 0.005 meters (0.016 feet) per year and only 30 percent of the coast showing erosion loss, the Pacific Coast appears to be under much less of an erosion threat than the Atlantic and Gulf coasts. Yet some areas along the coast experience high rates of erosion; Alaska is losing an average of 2.4 meters (7.9 feet) per year.

Only Washington shows a net deposition. The long, protected Washington shoreline within Puget Sound

Figure 10–15

Marine (Wave-Cut) Terraces on San Clemente Island South of Los Angeles, California. Once at sea level, the highest terraces in the background are now about 400 meters (1320 feet) above it. (Photo by John S. Shelton.)

helps skew the Pacific Coast figures (Figure 10–13). Although the average erosion rate for California is only 0.1 meters (0.33 feet) per year, over 80 percent of the coast is experiencing erosion, with local rates of up to 0.6 meters (2 feet) per year.

Erosion Rates and Landform Types

The variability we see in erosion rates along coasts is closely correlated to coastal landform types (Table 10–2):

- Deltas and mud flats composed of fine-grained sediment erode most rapidly, with mean erosion rates of 2 meters (6.6 feet) per year. This is not surprising, and it accounts for the fact that the Gulf Coast is eroding at a greater rate than other coasts.
- Coarser deposits of sandy beaches and barrier islands erode more slowly (0.8 meters, or 2.6 feet, per year).
- The rock shore of Maine, which is experiencing isostatic rebound, shows great resistance to erosion, with a deposition rate of 1 meter (3.3 feet) per year.
- On the Pacific Coast, the rocks are not so durable and erode at an average rate of 0.5 meters (1.6 feet) per year.

Table 10–2
Rates of Deposition (+) and Erosion (–) for Coastal Landform Types

Region	Change (m/yr)
Mud flats	-2.0
Florida	-0.1
Louisiana–Texas	-2.1
Gulf of Mexico	-1.9
Sand Beaches	-0.8
Maine–Massachusetts	-0.7
Massachusetts–New Jersey	-1.3
Atlantic Coast	-1.0
Gulf Coast	-0.4
Pacific Coast	-0.3
Barrier islands	-0.8
Louisiana–Texas	-0.8
Florida–Louisiana	-0.5
Gulf of Mexico	-0.6
Maine–New York	+0.3
New York–North Carolina	-1.5
North Carolina–Florida	-0.4
Atlantic Coast	-0.8
Rock shorelines	+0.8
Atlantic Coast	+1.0
Pacific Coast	-0.5

Effects of Artificial Barriers

People continually modify coastal sediment erosion/deposition in attempts to improve or preserve their property. One device for doing so is a **jetty,** a structure built at an angle to the shore. Jetties constructed to protect harbor entrances from wave action are the most common barriers to the longshore transport of sediment (Figure 10–16).

Many examples of beach destruction resulting from the jetties can be cited. Along the southern California coast, excessive beach erosion related to the development of local harbor facilities is especially common. California's longshore drift is predominantly southward, so construction of a jetty traps sediment on its northern side, which caused increased erosion on its southern side.

The harbor at Santa Barbara, California, provides an interesting example. In Figure 10–17 you can see the jetty constructed as a **breakwater** to protect the harbor. This jetty greatly disturbed the equilibrium established in this coastal region. The barrier created by the breakwater on the western side of the harbor caused an accumulation of sand that was migrating eastward along the coast. The beach to the west of the harbor continued to grow until finally the sand moved around the breakwater and began to fill in the harbor (Figure 10–17).

While abnormal deposition occurred to the west, erosion proceeded at an alarming rate east of the harbor. The wave energy available east of the harbor was no greater than before, but the sand that had formerly moved down the coast was no longer available, because it was entrapped behind the breakwater.

To compensate for this deficiency of sand downcurrent from the harbor, the harbor is dredged to keep it from filling in. Dredged sand is pumped down the coast so it can reenter the longshore drift and replenish the eroded beach. The dredging operation has stabilized the situation, but at a considerable expense. It seems obvious that any time people interfere with natural processes in the coastal region, they must provide the energy needed to replace what they have misdirected through modification of the shore environment.

Along many coasts, small jettylike structures called **groins** have been constructed at intervals along the beach (Figure 10–16). They help beach development by causing sand to accumulate on their upcurrent sides. Being shorter than jetties built to protect harbors, groins eventually allow sand to migrate around their ends. An equilibrium may be reached that allows sufficient sand transport along the coast before excessive erosion occurs downcurrent from the last groin. Although some serious erosional problems have developed, groins usually cause much less beach erosion than a single large jetty.

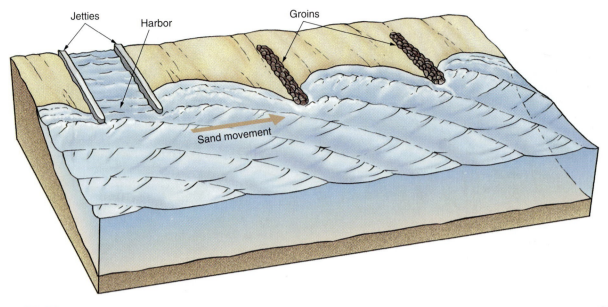

Figure 10–16

Jetties and Groins. Jetties and groins trap sand that would otherwise be moved along the shore by wave action. (From Tarbuck, E. J., and Lutgens, F. K., *Earth science,* 5th ed. Columbus: Merrill, 1988. Reprinted by permission.)

One of the most destructive of structures built to "stabilize" eroding shores is the **seawall** (Figure 10–18). Instead of extending out into the longshore current, as do groins and jetties, seawalls are built parallel to the shore. The idea is to protect developments landward of them from the action of ocean waves.

However, once waves begin breaking against a seawall, turbulence generated by the abrupt release of wave energy quickly erodes the sediment on its seaward side, causing it to collapse into the surf (Figure 10–18C). Where they have been used to protect property on barrier islands, the seaward slope of the island beach has steepened and the rate of erosion increased.

Although most state laws state that beaches belong to the public, government action allows owners of shore property to destroy beaches in an effort to provide short-term protection to their property. To add to the public insult, taxpayers subsidize property owners in many shore

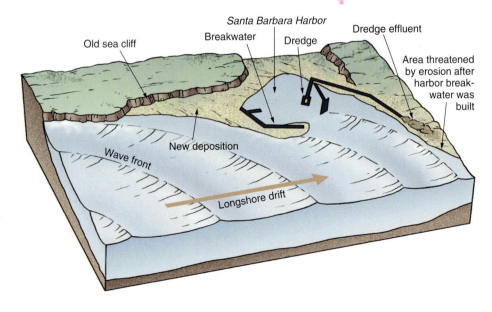

Figure 10–17

Santa Barbara Harbor. Construction of a breakwater to the west of Santa Barbara Harbor interfered with the eastward-moving longshore drift, creating a broad beach. Sand being deposited against the breakwater was no longer available to replace sand being removed by wave erosion to the east. As the beach extended around the breakwater into the harbor, dredging operations were initiated to keep the harbor open and to put the sand back into the longshore drift. This helped reduce coastal erosion east of the harbor. (U.S. Coast and Geodetic Survey Chart 5161.)

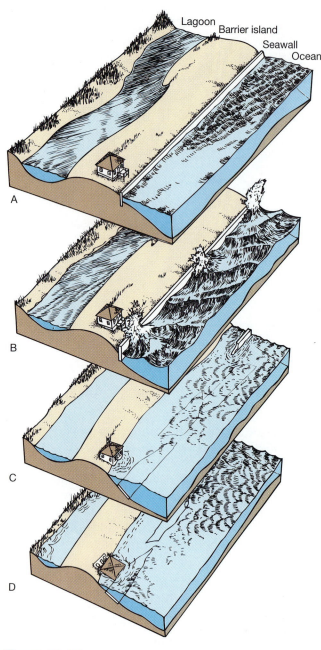

Lagoon
Barrier island
Seawall
Ocean

A

B

C

D

Figure 10–18

Seawalls. When a seawall is built along a beach to protect beachfront property (*A*), the first large storm will remove the beach from the seaward side of the wall and steepen its seaward slope (*B*). Eventually, the wall is undermined and falls into the sea (*C*). The property is lost as the oversteepened beach slope advances landward in its effort to reestablish a natural slope angle (*D*).

developments by helping to rebuild their storm-damaged properties at great public expense through the programs discussed at the beginning of this chapter.

There is an interesting alternative to the solid structures used to protect local shore areas from ocean waves. It is the **tethered-float breakwater,** developed

at the Scripps Institution of Oceanography (Figure 10–19). Tested successfully in San Diego Harbor, these inverted pendulums protect against coastal wave erosion by absorbing energy from passing waves, yet still letting the sediment pass through unhindered. If they prove durable, they would be a welcome substitute for traditional breakwaters.

Plastics: An Artificial Beach Material

During the past 30 years, the use of plastics has increased at a tremendous rate. Disposing of this "throw-away" component of Western culture has already strained the capacity of our land-based solid-waste disposal systems. Plastic waste now is an increasingly abundant component of oceanic *flotsam* (refuse).

Small pellets ranging in size from a BB to a pea are used to produce essentially all plastic products. They are transported in bulk aboard commercial vessels and are found throughout the oceans. The pellets probably find their way into the oceans as a result of spillage at loading terminals. In coastal waters, plastic products used in fishing and thrown overboard by recreational and commercial vessels are common. They also find their way into the open ocean waters from careless dumping by commercial vessels.

The best documentation of negative effects of plastics in the ocean on marine organisms is found in the strangulation of seals and birds caught in plastic netting and packing straps (Figure 10–20). Marine turtles are known to mistake plastic bags for jellyfish or other transparent plankton on which they typically feed.

It is believed that plastics eventually are removed from seawater. This happens as ocean currents wash against the shorelines of islands and continents, "filtering" plastic waste. Beaches throughout the world are probably increasing their plastic pellet content as a result of this filtering process. Some Bermuda beaches have up to 10,000 pellets per square meter. The beaches of Menemsha Harbor on Martha's Vineyard in Massachusetts yielded 16,000 plastic spherules per square meter.

These data are from a survey conducted between 1984 and 1987. Researchers also discovered more than 10,000 plastic pieces and 1500 pellets per square kilometer in the northern Sargasso Sea between 28°N and 40°N latitude. The 1987 survey found that since 1972 the concentration of plastic pellets had doubled.

The U.S. Congress is considering a bill to ban disposal of plastics in the 200-meter-wide Exclusive Economic Zone. Internationally, the Convention for the Prevention of Pollution from Ships prevents disposal of all plastics into the oceans. The acceptance of both of these regulations would aid greatly in reducing the increasing concentration of plastics in the world's oceans.

A.

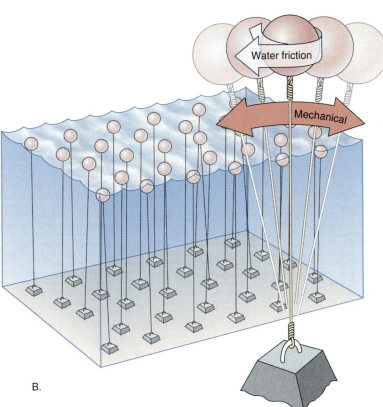

B.

Figure 10–19

Tethered-Float Breakwaters. *A:* Deployment of tethered-float breakwater in San Diego Harbor. *B:* Artist's sketch of a tethered breakwater system incorporating 1.5 meters (4.8 feet) spheres of steel placed 1.5 meters apart. The spheres are anchored to the ocean floor and held just beneath the ocean surface. This system has the advantage over traditional breakwater structures of removing energy from waves without interfering with transport of sediment along the shore. (*A:* Photo courtesy of Scripps Institute of Oceanography, University of California, San Diego.)

Figure 10–20

Elephant Seal with Plastic Packing Strap Tight around Its Neck. This animal was found on a beach of San Clemente Island off the southern California coast. It was saved from strangulation by the plastic packing strap when scientists from the National Marine Fisheries Service removed the strap. (Photo by Wayne Perryman, NMFS.)

SUMMARY

Since passing the National Flood Insurance Program (NFIP) in 1968, the federal government has encouraged development of shores that are subject to storm damage. The result has been increased death, destruction, and cost. The direction of this and complementary programs must be reversed to improve the situation. One helpful piece of legislation was the Coastal Barrier Rezoning Act, which excludes construction in undeveloped coastal areas from receiving federal assistance.

The region of contact between the oceans and the continents is marked by the shore, lying between the lowest low tides and the highest elevation on the continents affected by storm waves. The coast extends inland from the shore as far as marine-related features can be found. The shore is divided into the foreshore, extending from low tide to high tide, and the backshore, extending beyond the high-tide line to the coastline. Seaward of the low-tide are the nearshore zone, extending to the breaker line, and the offshore zone beyond.

Wave erosion of the shore produces a wave-cut cliff that constantly retreats, leaving behind such features as wave-cut benches, sea caves, sea arches, and sea stacks. Wave erosion increases with greater exposure of the shore to the open ocean, decreasing tidal range, and decreasing strength of bedrock.

As waves break at an angle to the shore, a longshore current is set up that produces a longshore drift of sediment along the shore. Deposition of sediment thus transported may produce beaches, spits, tombolos, and barrier islands. Viewed from ocean side to lagoon side, barrier islands commonly have an ocean beach, dunes, barrier flat, and salt marsh. Deltas form at the mouths of rivers that carry more sediment to the ocean than can be distributed by the longshore current.

A sea level drop may be indicated by ancient wave-cut cliffs and stranded beaches well above the present shoreline. A sea level rise may be indicated by old submerged beaches, wave-cut cliffs, or drowned river valleys. Such sea level changes may result from tectonic processes causing local movement of the landmass or from eustatic processes changing the amount of water in the oceans or the capacity of ocean basins. Melting of continental ice caps during the past 18,000 years has caused a eustatic rise in sea level of about 120 meters (400 feet).

Along the Atlantic Coast, sea level is rising about 0.3 meters (1 feet) every century along most of the coast, and the average annual erosion rate is −0.8 meters (−2.6 feet). Along the Gulf Coast, sea level is rising at 0.3 meters (1 foot) per century, and the average rate of erosion is −1.8 meters (−6 feet) per year. The Mississippi River Delta is eroding at a rate of 4.2 meters (13.7 feet) per year, resulting in a large annual loss of wetlands. Along the Pacific Coast, the average erosion rate is only −0.005 meters (−0.016 feet) per year. Based on coastal landform types, mud flats erode more rapidly than sand beaches, barrier islands, or rocky shores.

Structures built along the shore, such as jetties to protect harbors and groins used to widen beaches, trap sediment on their upcurrent side, but erosion may then become a problem downcurrent. A less-damaging replacement for fixed breakwater structures might be the tethered-float breakwater.

The amount of plastic accumulating in the oceans has increased dramatically. Certain forms of plastic are known to be lethal to marine mammals and turtles, and national and international legislation is being considered to ban the disposal of plastic in the oceans.

KEY TERMS

Backshore (p. 222)
Barrier flat (p. 226)
Barrier islands (p. 225)
Bay barrier (p. 225)
Beach (p. 222)
Breakwater (p. 236)
Coast (p. 221)
Coastline (p. 222)
Delta (p. 226)
Distributary (p. 226)
Drowned beach (p. 229)
Dunes (p. 226)
Emerging shorelines (p. 229)

Eustatic (p. 229)
Foreshore (p. 222)
Groins (p. 236)
Headlands (p. 222)
Longshore current (p. 223)
Longshore drift (p. 224)
Marine terrace (p. 229)
Nearshore (p. 222)
Ocean beach (p. 226)
Offshore (p. 222)
Rip currents (p. 224)
Salt marshes (p. 226)
Sea arches (p. 222)

Sea caves (p. 222)
Sea stacks (p. 222)
Seawall (p. 237)
Shore (p. 221)
Shoreline (p. 222)
Spit (p. 225)
Submerging Shorelines (p. 229)
Swash (p. 224)
Tethered-float breakwater (p. 238)
Tombolo (p. 225)
Wave-cut bench (p. 222)
Wave-cut cliff (p. 222)

QUESTIONS AND EXERCISES

1. Why has the National Flood Insurance Program had a negative impact on the nation's shores?
2. To help reinforce your knowledge of the shore, construct and label your own diagram similar to Figure 10–1.
3. Discuss the formation of such erosional features as sea cliffs, sea caves, sea arches, and stacks.
4. List and discuss three factors in the rate of wave erosion.
5. What variables affect the velocity of the longshore current?
6. What is the longshore drift, and how is it related to the longshore current?
7. Discuss the composition of beaches and the relation of beach slope to particle size.
8. Define these depositional features: spit, tombolo, bay barrier, and barrier island. Discuss how some barrier islands develop peat deposits running through them from ocean shore to marsh.
9. Discuss why some rivers have deltas and others do not. Also include the factors that determine whether a

"bird foot" or a smoothly curved Nile-type delta will form.
10. Compare the causes and effects of tectonic versus eustatic changes in sea level.
11. List the two basic processes by which coasts advance seaward, and list their counterparts that lead to coastal retreat.
12. Describe the tectonic and depositional processes causing subsidence along Atlantic-type margins.
13. How does the Pacific-type margin differ from the Atlantic-type margin?
14. Discuss the Atlantic Coast, Gulf Coast, and Pacific Coast by describing the conditions and features of emergence-submergence and erosion-deposition that are characteristic of each.
15. Describe the effect on erosion and deposition caused by putting a structure such as a breakwater or jetty across the longshore current and drift.

REFERENCES

Bird, E. C. F. 1985. *Coastline changes: A global review.* Chichester, U.K.: John Wiley & Sons.

Burk, K. 1979. The edges of the ocean: An introduction. *Oceanus* 22–3:2–9.

Coates, R., ed. 1973. *Coastal geomorphology. Publications in Geomorphology.* Binghamton, N.Y.: State University of New York.

Kuhn, G. G., and Shepard, F. P. 1984. *Sea cliffs, beaches, and coastal valleys of San Diego County: Some amazing histories and some horrifying implications.* Berkeley: University of California Press.

Leatherman, S. P. 1983. Barrier dynamics and landward migration with Holocene sea-level rise. *Nature* 301:5899, 415–417.

May, S. K.; Kimball, H.; Grandy, N.; and Dolan, R. 1982. The Coastal Erosion Information System. *Shore Beach* 50, 19–26.

Peltier, W. R., and Tushingham, A. M. 1989. Global sea level rise and the greenhouse effect: Might they be connected? *Science* 244:4906, 806–810.

Roemmich, D. 1992. Ocean warming and sea level rise along the southwest U.S. coast. *Science* 257:5068, 373–375.

Sahagian D. L.; Schwartz, F. W.; and Jacobs, D. K. 1994. Direct anthropogenic contributions to sea levels rise in the twentieth century. *Nature* 367:6458, 54–56.

Shepard, F. P. 1973. *Submarine geology,* 3rd ed. New York: Harper & Row.

———. 1977. *Geological oceanography.* New York: Crane, Russak & Company.

Stanley, D. J., and Warne, A. G. 1993. Nile delta: Recent geological evolution and human impact. *Science* 260:5108, 628–634.

SUGGESTED READING

Sea Frontiers

Carr, A. P. 1974. The ever-changing sea level. 20:2, 77–83. A discussion of the causes of sea level change is very well presented.

Emiliani, C. 1976. The great flood. 22:5, 256–270. An interesting discussion of the possible relationship of the rise in sea level resulting from the melting of glaciers 11,000 to 8000 years ago and biblical and other ancient accounts of a great flood.

Feazel, C. 1987. The rise and fall of Neptune's kingdom. 33:2, 4–11. A discussion of factors that change the level of the sea, including atmospheric conditions, currents, climate, ocean topography, and seafloor spreading.

Fulton, K. 1981. Coastal retreat. 27:2, 82–88. The problems of coastal erosion along the southern California coast are considered.

Grasso, A. 1974. Capitola Beach. 20:3, 146–151. The destruction of the beach of Capitola, California, shortly after the construction of a harbor by the Army Corps of Engineers at Santa Cruz to the north.

Mahoney, H. R. 1979. Imperiled sea frontier: Barrier beaches of the east coast. 25:6, 329–337. The natural alteration of barrier beaches is considered.

Pilkey, O. H. 1990. Barrier islands. 36:6, 30–39. The origin, evolution, and types of barrier islands are discussed.

Schumberth, C. J. 1971. Long Island's ocean beaches. 17:6, 350–362. This is a very informative article on the nature of barrier islands. The specific problems observed on the Long Island barriers serve as examples.

Wanless, H. R. 1989. The inundation of our coastlines: Past, present and future with a focus on South Florida. 35:5, 264–267. This is an informative overview of the evidence that is useful in evaluating the past and future movements of the shorelines.

Wanless, H. R., and Tedesco, L. P. 1988. Sand biographies. 34:4, 224–232. Discusses how we can tell where beach sand came from and how it traveled to its present location.

Westgate, J. W. 1983. Beachfront roulette. 29:2, 104–109. The problems related to the development of barrier islands are discussed.

Scientific American

Bascom, W. 1960. Beaches. 203:2, 80–97. A comprehensive consideration of the relationship of beach processes, both large and small scale, to release of energy by waves.

Broecker, W. S., and Denton, G. H. 1990. What drives glacial cycles? 262:1, 48–107. The most up-to-date information is included in this consideration of the causes of glacial cycles.

Dolan, R., and Lins, H. 1987. Beaches and barrier islands. 257:1, 68–77. A discussion dealing with the ultimate futility of trying to develop and protect structures on beaches and barrier islands.

Fairbridge, R. W. 1960. The changing level of the sea. 202:5, 70–79. A discussion of what is known of the causes of the changing level of the sea, which seems to be related mostly to the formation and melting of glaciers and changes in the ocean floor.

C H A P T E R   1 1
The Coastal Ocean

The term *fishery* refers to fish caught from the ocean by commercial fishermen. Of the world fishery, 95 percent is obtained within about 320 kilometers (200 miles) of shore. Indeed, it is these coastal waters that support about that same percentage of the total mass of life in the oceans. These waters also are the locus of most shipping routes, oil and gas production, and recreational activities, and the final destination of much of the residue of those living on the adjacent land.

The coastal ocean is a very busy place, filled with life, commerce, and waste.

The mass of the world fishery decreased by 5 percent from 1989 to 1993. This decline resulted primarily from overfishing by a vastly increased fishing effort in waters that are becoming more polluted, according to a 1994 report released by the Worldwatch Institute. This decreased catch, along with increasing human population, has sadly diminished the per capita catch from 19.4 kilograms (42.8

pounds) in 1988 to 18.0 kilograms (39.7 pounds) in 1992, according to the U.N. Food and Agriculture Organization.

Clearly, the human activity most damaging to the ecology of the oceans is commercial fishing. Beyond decreasing the fish yield, overfishing may strongly impact the overall ecology of the oceans.

Along with overfishing is another major worry: marine pollution. Marine pollution in coastal waters results from accidental spilling of petroleum and the accumulation of sewage, certain chemicals (such as DDT and PCBs), and the element mercury. Such pollution is both acute and chronic.

Pollution in coastal waters is both natural and anthropogenic (human caused). To better predict the effects of pollution on coastal waters, we must learn much more about the physical processes that give these waters their high biological production and an amazing resiliency to the onslaught of contamination.

Coastal Waters

The primary difference between coastal waters and the open ocean is *depth*. Because of the shallowness of coastal waters, river runoff and tidal currents have a far more significant effect on them than on the open ocean.

Salinity

Along the coast, river runoff reduces the salinity of the surface layer where mixing is not significant. Runoff also reduces salinity throughout the water column where mixing does occur. Where precipitation on land is mostly rain, river runoff peaks in the same season that precipitation does. However, if runoff is largely fed by melting snow and ice, runoff always peaks in summer. In general, salinity is lower in coastal regions than in the open ocean due to the runoff of fresh water from the continents (Figures 11–1A and 11–1C).

Counteracting the salinity-reducing effect of runoff in some coastal regions is the presence of prevailing offshore winds. They usually lose most of their moisture over the continent. These winds evaporate considerable water as they move across the surface of the coastal waters. The increase in evaporation rate in these areas also tends to increase surface salinity (Figure 11–1B).

Temperature

In coastal regions where the water is relatively shallow, very great ranges in temperature may occur over a year. Sea ice forms in many high-latitude coastal areas where temperatures are determined by the water's freezing point, generally warmer than −2°C (28.4°F) (Figure 11–1D). Maximum surface temperature in low-latitude coastal waters may approach 45°C (113°F), where the water is restricted in its circulation with the open ocean and thus is protected from strong mixing (Figure 11–1E). The seasonal change in temperature can be most easily detected in coastal regions of the midlatitudes, where surface temperatures are coolest in winter and warmest in late summer.

Figure 11–1F shows how strong thermoclines may develop where mixing does not occur. Very high temperature surface water may form a relatively thin layer. Mixing reduces the surface temperature by distributing the heat through a greater vertical column of water, thus pushing the thermocline deeper and making it less pronounced.

Tidal currents can considerably influence vertical mixing of shallow coastal water. Prevailing winds also can significantly affect surface temperatures, if they blow from the continent. These air masses, previously mentioned regarding salinity, usually have relatively high temperature during the summer. They increase the ocean surface temperature and seawater evaporation. During winter, they are much cooler than the ocean surface and will absorb heat from the ocean surface, cooling surface water near shore.

Coastal Geostrophic Currents

Recall from Chapter 7 the discussion of the geostrophic effect associated with current gyres. Geostrophic currents also develop in coastal waters, and the causes are basically the same. There are two specific agents: wind and runoff.

Wind blowing parallel to the coast piles up water along the shore. Gravity then eventually pulls this piled-up water back downslope toward the open ocean. As it runs downslope away from the shore, the Coriolis effect causes it to veer. In the Northern Hemisphere, the current veers northward on the western coast and southward on the eastern coast of continents. In the Southern Hemisphere, the directions are reversed.

The other condition that produces geostrophic flow along continental margins is the high-volume runoff of fresh water that gradually mixes with oceanic water. This produces a surface slope of water away from the shore (Figure 11–2). The seaward slope has salinity and density gradients because both properties increase seaward. These variable currents, which depend on the wind and the amount of runoff for their strength, are bounded on the ocean side by the more steady boundary currents resulting from the open-ocean gyres.

These local geostrophic currents frequently flow in a direction opposite the boundary current. Such is the case with the *Davidson Current* that develops along the coast of Washington and Oregon during the winter. Heavy precipitation occurs in the Pacific Northwest dur-

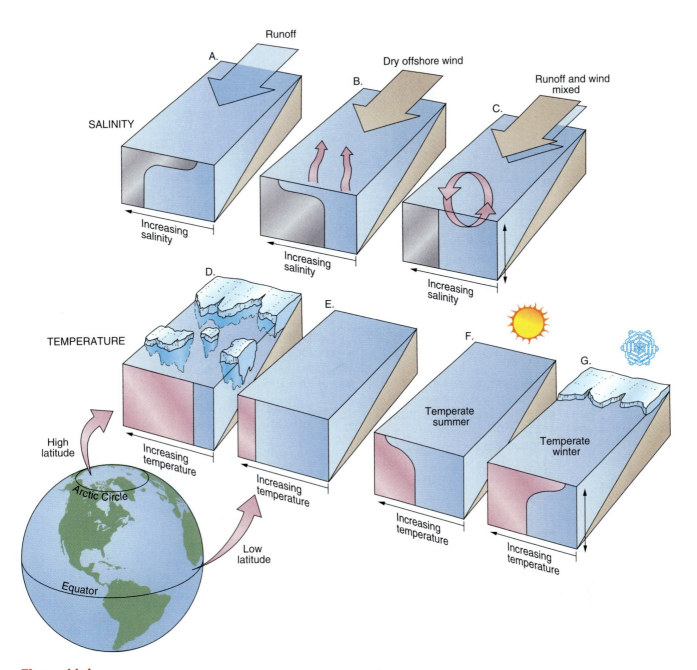

Figure 11–1

Temperature and Salinity in the Coastal Ocean. *A:* Freshwater runoff from streams and rivers is not mixed into the water column. This forms a low-salinity surface layer and well-developed halocline. *B:* Warm, dry offshore winds cause a high rate of evaporation that may offset the effects of runoff. This produces a halocline with a reversed gradient from *A.* *C:* Runoff is mixed with deeper water, producing an isohaline ("equally salty") water column of generally lower salinity than the open ocean. *D:* In high latitudes where sea ice is forming or thawing throughout the year, the temperature of coastal water remains uniformly near freezing. *E:* Water in shallow, low-latitude coastal regions, protected from free circulation with the open ocean, may become uniformly warm. *F:* In midlatitudes, coastal water is significantly warmed during summer. A strong seasonal thermocline may develop. *G:* Winter may produce a low-temperature surface layer. The thermocline shown may develop. But cooling may cause surface water to sink by increasing its density, causing a well-mixed isothermal water column. Mixing from strong winds may drive the thermoclines in *F* and *G* deeper and even mix the entire water column, producing an isothermal condition.

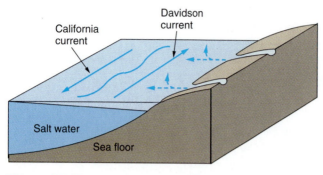

Figure 11-2

Coastal Geostrophic Current. During the winter rainy season, fresh runoff water produces a seaward slope away from low-salinity surface water near the shore. This causes a surface flow of low-salinity water from the shore toward the open ocean. As this flow is acted upon by the Coriolis effect, it veers right. The example shown here is the north-flowing Davidson Current along the coast of Washington and Oregon. It flows between the shore and the southbound California Current.

ing winter, and southwesterly winds are strongest then. Their combined effect produces a relatively strong northward-flowing geostrophic current. It flows between the shore and the southward-flowing California Current, which is part of the open-ocean circulation.

Estuaries

The most common estuary is a river mouth, where it empties into the sea. Many bays, inlets, gulfs, and sounds

may be considered estuaries, too. Thus, we define an **estuary** as a partially enclosed coastal body of water in which salty ocean water is significantly diluted by fresh water from land runoff.

The mouths of large rivers form the most economically significant estuaries, because many are seaports and centers of ocean commerce—Baltimore, New York, San Francisco, Buenos Aires, London, Tokyo, and so on. Many estuaries support important commercial fisheries as well. Environmental changes caused by heavy commercial use of estuaries is a major concern as their future development is planned.

Origin of Estuaries

Essentially all estuaries in existence today owe their origin to sea level rise over the past 18,000 years. Sea level has risen approximately 120 meters (400 feet) due to extensive melting of major continental glaciers. These glaciers covered portions of North America, Europe, and Asia during the Pleistocene Epoch, more commonly referred to as the Ice Age. Four major classes of estuaries can be identified on the basis of their origin (Figure 11-3):

1. A **coastal plain estuary** formed as the rising sea level caused the oceans to invade existing river valleys. These estuaries are appropriately called *drowned river valleys*. The Chesapeake Bay in Maryland and Virginia is an example.
2. A **fjord** (pronounced fee-yord) is a glaciated valley estuary. Unlike a water-carved valley, which has a V shape, glacially carved fjords are U-shaped valleys with steep walls. They usually have a glacial deposit

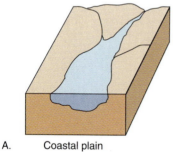

A.   Coastal plain

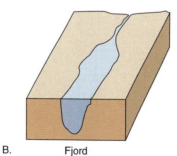

B.   Fjord

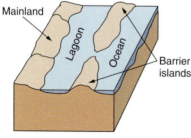

C.   Bar-built

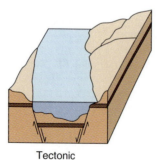

D.   Tectonic

Figure 11-3

Classifying Estuaries by Origin. Four classes of estuaries (with examples): *coastal plain estuary* (Chesapeake Bay in Maryland and Virginia), *fjord* (Strait of Juan de Fuca in Washington State), *bar-built estuary* (Laguna Madre in Texas), and *tectonic estuary* (San Francisco Bay in California).

that forms a sill near the ocean entrance. Fjords are common along the coasts of Norway (the word *fjord* is Norwegian), Canada, Alaska, and New Zealand (Figure 11–4).

3. A **bar-built estuary** is shallow and is separated from the open ocean by sandbars that are deposited parallel to the coast by wave action. *Lagoons* that separate barrier islands from the mainland are bar-built estuaries. Examples abound along the U.S. East Coast, including Laguna Madre along the Texas coast and Chincoteague Bay in Maryland.

4. A **tectonic estuary** is produced by faulting or folding of rocks. This forms a restricted down-dropped area into which rivers flow. San Francisco Bay is in part a tectonic estuary.

Water Mixing in Estuaries

Generally, freshwater runoff that flows into an estuary moves as an upper layer of low-density water across the estuary toward the open ocean. Beneath this upper layer, an opposite inflow of denser, salty seawater occurs. Mixing takes place at the contact between these water masses.

Estuaries are classified into four patterns based on mixing of freshwater and seawater (shown in Figure 11–5):

1. **Vertically mixed estuary**—a shallow, low-volume estuary where the net flow always proceeds from the river head of the estuary toward the estuary's mouth. Salinity at any point in the estuary is uniform from surface to bottom because river water mixes evenly with ocean water at all depths. Salinity simply increases from the river head to the mouth of the estuary, as shown in the figure. Note that the Coriolis effect skews the salinity.

2. **Slightly stratified estuary**—a somewhat deeper estuary in which salinity increases from the head to the mouth at any depth, as in vertical mixing. However, two basic water layers can be identified: less saline, less dense upper water provided by the river, and denser, deeper marine water. They are separated by a zone of mixing. It is in this type of estuary that we begin to see the typical **estuarine circulation** pattern develop, because there is a net surface flow of low-salinity water toward the ocean and an opposite net subsurface flow of marine water toward the head of the estuary.

3. **Highly stratified estuary**—a deep estuary in which upper-layer salinity increases from the head to the mouth, reaching a value close to that of open-ocean water. The deep-water layer has a rather uniform marine salinity at any depth throughout the length of the estuary. Net flow of the two layers is similar to that

Figure 11–4

Fjord. A fjord is a glacially formed U-shaped valley estuary.

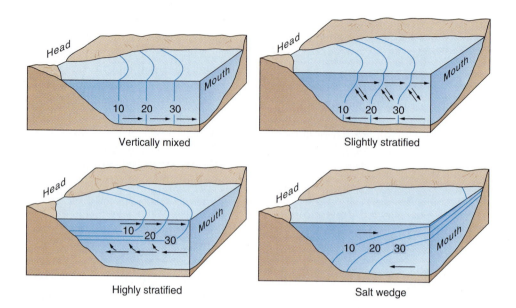

Figure 11–5

Classifying Estuaries by Mixing. The basic flow pattern in an estuary is a surface flow of less dense freshwater toward the ocean and an opposite flow in the subsurface of salty seawater into the estuary. The dimensions of each flow, and the degree of mixing between the two, depend on specific conditions in each estuary. Note that surface freshwater flow toward the mouth of the estuary extends seaward much farther along the righthand shore as one faces seaward. This is due to the Coriolis effect. The opposite side of the estuary experiences a greater marine (saltwater) influence. In most estuaries, the marine water inflow occurs in the subsurface.

in a slightly stratified estuary, but mixing at the interface of the upper water and the lower water is such that net movement is from the deep-water mass into the upper water. Less saline surface water simply moves from the head toward the mouth of the estuary, growing more saline as water from the deep mass joins it. Relatively strong haloclines develop in such estuaries at the contact between the upper and lower water masses. It is not unusual for these haloclines to reach magnitudes approaching 20 percent near the head of the estuary at maximum river flow.

4. **Salt wedge estuary**—an estuary in which a saline wedge of water intrudes from the ocean beneath the river water; this is typical of the mouths of deep, high-volume rivers. No horizontal salinity gradient exists at the surface in these deep estuaries. Water is essentially fresh throughout the length of, and even beyond, the estuary. There is, however, a horizontal salinity gradient at depth and a very pronounced vertical salinity gradient manifested as a strong halocline at any station throughout the length of the estuary. This halocline is shallower and more highly developed near the mouth of the estuary.

The mixing patterns described often cannot be applied to an estuary as a whole. Mixing within an estuary may change with distance, season, or tidal conditions.

Estuaries and Human Activities

Estuaries are vital natural ecosystems that have evolved over millennia. Their health is essential to the world fishery and to coastal environments worldwide. However, they are exploited heavily by economic activities that do not depend on the health of the estuary. Estuaries support shipping, logging, manufacturing, waste disposal, and others that can potentially damage the estuarine environment.

Obviously, estuaries are most threatened where human population is high and growing. But even where population pressure is still modest, estuaries can be severely damaged. A good example is the *Columbia River estuary,* which forms most of the border between Washington and Oregon.

Columbia River Estuary. The Columbia River estuary is flushed by tides that drive a salt wedge 42 kilometers (26 miles) upstream and raise its water level by over 3.5 meters (12 feet) (Figure 11–6). When the tide falls, up to 28,000 meters^3/s (1,000,000 feet^3/s) of fresh water push out hundreds of kilometers into the Pacific Ocean to aid returning salmon to find their "home" waters. Once they find their streams, they swim up them to spawn (see Chapter 14).

In the late 19th century, farmers and dairymen moved onto the rich floodplains and eventually diked

Figure 11-6

Columbia River Estuary. This high-volume salt wedge estuary has suffered more from interference with its flow by hydroelectric dams and diked floodplains than from chemical pollution. Like the Mississippi River, the estuary's large flow volume aids it in reducing the effects of chemical pollution.

them to prevent the inevitable natural flooding. Now these lands are removed from the natural ecosystem and, ironically, they no longer are suitable for agriculture. The periodic flooding was needed to maintain soil richness.

The river has been the principal conduit for the logging industry, which has dominated the region's economy through most of its modern history. The river has survived some terrible insults inflicted by the logging industry, but the hydroelectric dams have dealt the system a more serious and permanent blow. Many of these dams did not include *salmon ladders,* which help fish "climb" in short vertical steps around the dams to reach the headwaters of their home streams. This closed off many previous spawning grounds. As is the case with most interactions between nature and humans, these structures were almost essential to the development of the region, but a high price was paid.

At present, in an effort to diversify the local economy, shipping facilities are being developed, with the accompanying dredges and increased pollution potential. If such problems have developed at such sparsely populated areas as the Columbia River estuary, what must be the conditions at more highly stressed estuaries?

Chesapeake Bay Estuary. Chesapeake Bay is a classic example of a coastal plain estuary. It was produced by the drowning of the Susquehanna River and its tributaries (Figure 11–7). Most of the fresh water entering the bay enters along the western margin via rivers that drain the slopes of the Appalachian Mountains.

Figure 11–7 shows the salinity of the bay. Note the salinity pattern, which not only increases oceanward but also reflects the Coriolis effect. Coriolis effect acts on marine water entering the bay from the south and the southward flow of river water.

With maximum river flow in the spring, a strong halocline (pycnocline) develops, preventing mixing of the fresh surface water and saltier deep water. Beneath the pycnocline, which can be as shallow as 5 meters (16 feet), waters may become *anoxic* (without oxygen) from

May through August as dead organic matter decays in the deep water. Major kills of commercially important blue crab, oysters, and other bottom-dwelling organisms occur during these events.

For unknown reasons, the degree of stratification and kills of bottom-dwelling animals have increased over the last 40 years. However, the bay has experienced intense pressure from human activities. It is suggested that increased nutrients have been added to the bay as a result of human activities. The nutrients are from sewage and farm fertilizers. Increased production of microscopic plants resulting from this could increase the deposition of organic particulates on the bottom.

That much is yet to be learned about estuarine circulation is illustrated by the following observations, made over an extended period in Chesapeake Bay.

- Researchers have recorded intervals of *upstream surface flow,* accompanied by *downstream deep flow,* just the opposite of the expected pattern.
- Periods of total downstream flow have been observed.
- Even more complex patterns known to develop involve a surface and bottom flow in one direction separated by a mid-depth flow in the opposite direction, or landward flows along the shores and seaward flows in the central portions of the estuaries.

Estuaries and the Continental Shelf Plankton Community

Plankton are microscopic plant cells and animals that abound in nutrient-rich waters. Being tiny floaters, plankton are carried about by ocean currents. *Plankton communities* were studied in four estuarine environments and out onto their adjacent continental shelf areas, as shown in Figure 11–8A. The purpose was to determine the influence of these estuaries on the plankton communities. All studies revealed similar conditions.

Estuarine phytoplankton are microscopic cells that, like plants, make their own food through photosynthesis. These phytoplankton are the basis of the food chain in an estuary. They are carried seaward by the estuarine flow. As they travel, their rate of photosynthetic production increased sixfold because of improved water clarity, as observed in these studies. Beyond 5 kilometers (3 miles) out to sea, production gradually dropped across the shallow shelf as nutrient-poor Gulf Stream waters were approached (Figure 11–8B).

Seasonal changes in photosynthetic production were similar in the estuaries, in nearshore shallow shelf waters (less than 20 meters depth), and in deeper offshore shelf waters. Photosynthetic productivity in all regions peaked in late summer (green curve in Figure 11–8C). Similarly was the abundance of zooplankton (microscopic animal populations, such as fish eggs and larvae) (purple, blue, and red curves in Figure 11–8C).

These populations have different reproduction rates, and would not be expected to peak together. So their similar changes in photosynthetic production and the abundance of populations of animal plankton may be explained by the fact that all were produced in the estuaries and flushed into the shelf waters by estuarine outflow. If this is true, these estuaries have a powerful influence on the plankton populations and overall ecology of the continental shelf.

We must understand the dynamics of estuarine circulation if we are to meaningfully describe water quality, suspended particulate matter and bottom sediment transport, and biological activity, particularly that of phytoplankton and animals, fish eggs, and larvae.

Figure 11–7

Chesapeake Bay. *A:* Map shows average surface salinity. The dark blue area is a region of anoxic (oxygenless) waters, from Baltimore to the mouth of the Potomac River. *B:* Comparison of dissolved oxygen levels for the summers of 1950 and 1980. Deeper midbay waters are essentially without oxygen in 1980. (After Officer et al., 1984.)

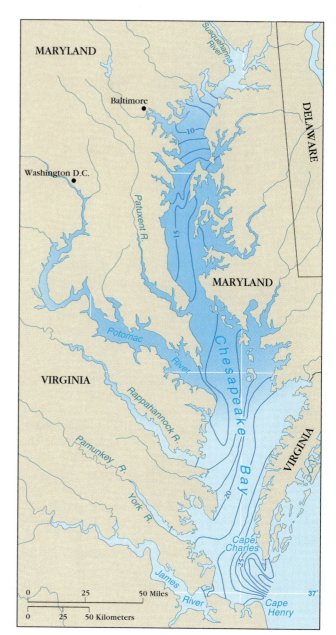

A.

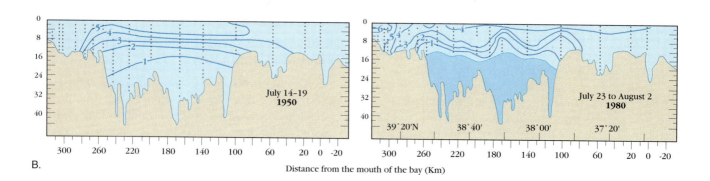

B.

Distance from the mouth of the bay (Km)

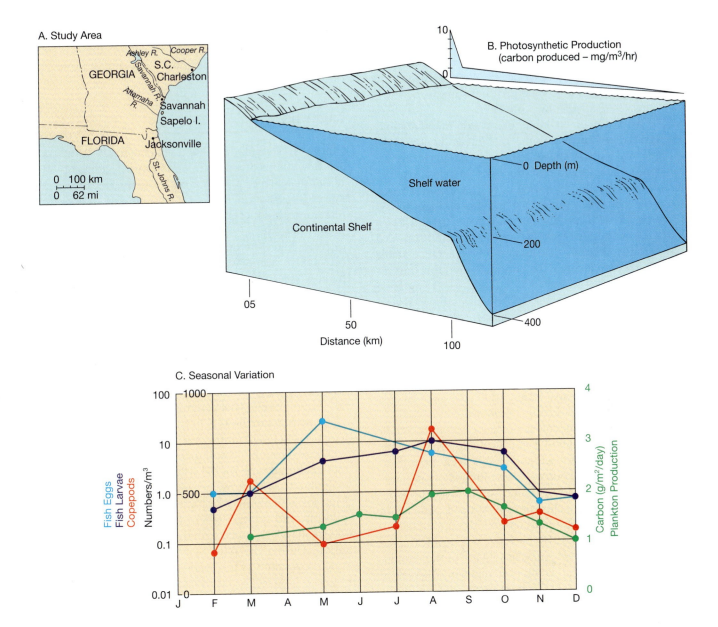

Figure 11–8

Estuaries and Shelf Waters. Along the southeastern U.S. coast, a study of the relationship between estuaries and the coastal ocean had the following results: *A:* Along this coastal area, relatively small-volume rivers flow onto a gently sloping continental shelf at least 100 kilometers (60 miles) wide. At the four locations shown, transects from estuaries across the continental shelf revealed photosynthetic production (measured as the amount of carbon in phytoplankton) and plankton abundance. *B:* The graph (upper right) simulates the average photosynthetic production pattern along the transects. The graph aligns with the continental shelf view below it. Zero on the distance scale is the nearshore shelf waters adjacent to the estuary mouth. *C:* Seasonal changes in four biological indicators: photosynthetic production, copepods (an important population of small zooplankton that eat photosynthetic cells), fish eggs, and fish larvae. The similar pattern for these quite varied life forms indicates they may be produced in the estuary and flushed together into the shelf waters by estuarine outflow. (After Turner, R. E. et al., 1979.)

Wetlands

Wetlands border estuaries and other shore areas that are protected from the open ocean. Wetlands are strips of land delicately in tune with natural shore processes. They are very biologically productive.

Two types of wetlands exist: **salt marshes** and **mangrove swamps.** Both are intermittently submerged by ocean water and are characterized by oxygen-poor mud and peat deposits. Marshes, characteristically inhabited by a variety of grasses, are known to occur from the equator to latitudes as high as 65° (Figures 11–9A and 11–9B). Mangrove trees are restricted to latitudes below 30° both north and south of the equator (Figures 11–9A and 11–9C). Once mangroves colonize an area, they normally outgrow and replace marsh grasses.

Research is showing that wetlands have a very high economic value when left alone. Salt marshes are believed to serve as nurseries for over half the species of commercially important fishes in the southeastern United States. Other fishes such as flounder and bluefish use marshes for feeding and overwintering. Fisheries of oysters, scallops, clams, and fishes such as eels and smelt are located directly in the marshes.

A very important characteristic of wetlands is their ability to "scrub" polluted water. Wetlands remove inorganic nitrogen compounds and metals from groundwater polluted by land sources. (The nitrogen compounds are from fertilizers.) Most such removal is probably achieved through attachment to clay-sized particles in the wetlands. Some nitrogen compounds trapped in sediment are decomposed by bacteria, which release the nitrogen to the atmosphere as gas. Many of the remaining nitrogen compounds support plant production in this environment, one of the most productive on Earth.

As plants die in marshes, their organic nitrogen compounds are either incorporated into the sediment to become peat or broken up to become food for bacteria, fungi, or fish.

Serious Loss of Valuable Wetlands

You may be surprised to learn that more than half of the nation's wetlands have been lost. Of the original 215 million acres of wetlands that once existed in the conterminous United States (excluding Alaska and Hawaii), only about 90 million acres remain. Both marsh and mangrove wetlands have been filled in and developed for housing, industry, and agriculture. All of this is a consequence of our strong desire to live near the oceans.

To help prevent the loss of remaining wetlands, the U.S. Environmental Protection Agency established an Office of Wetlands Protection (OWP) in 1986. At that time, wetlands were being lost to development at a rate of 300,000 acres per year. The OWP is actively enforcing regulations against wetlands pollution and is identifying the most valuable wetlands that can be protected or restored.

Lagoons

Landward of the barrier island salt marshes lie protected, shallow bodies of water called **lagoons** (Figure 11–3C). Recall that a lagoon is a bar-built estuary. Because of restricted circulation between lagoons and the ocean, three distinct zones usually can be identified within a lagoon. Typically, a *freshwater zone* exists near the mouths of rivers that flow into the lagoon. A *transitional zone* of brackish water (water with a salinity between that of fresh water and ocean water) occurs. A *saltwater zone* is present close to the entrance, where the maximum tidal effects within the lagoon can be observed.

Farther away from the saltwater zone near the mouth of the lagoon, tidal effects diminish and are usually undetectable in the freshwater region, which is well protected from tidal effects. Figure 11–10 shows the characteristics of a typical lagoon.

Lagoon salinity is determined by another factor beyond nearness to the sea. In latitudes that have seasonal variations in temperature and precipitation, ocean water will flow through the entrance during a warm, dry summer to compensate for the volume of water lost through evaporation. This flow increases salinity in the lagoon. Lagoons actually may become *hypersaline* (excessively salty) in arid regions where the inflow of seawater is too small to keep pace with the lagoon's surface evaporation. During the rainy season, the lagoon becomes much less saline as freshwater runoff increases.

Laguna Madre

Laguna Madre is one of the best-known lagoons. It is along the Texas coast between Corpus Christi and the mouth of the Rio Grande River. This long, narrow body of water is protected from the open ocean by Padre Island, an offshore island 160 kilometers (100 miles) long. The lagoon probably formed about 6000 years ago as sea level was approaching its present height.

Much of the lagoon is less than 1 meters (3.3 feet) deep. The tidal range of the Gulf of Mexico in this area is about 0.5 meters (1.6 feet). The inlets at each end of the barrier island are quite small (Figure 11–11). There is, therefore, very little tidal interchange of water between the lagoon and the open sea.

Laguna Madre is a *hypersaline lagoon*. The shallowness of its waters makes possible a very great seasonal range of temperature and salinity in this semiarid region. Water temperatures are high in the summer and may fall below 5°C (41°F) in winter. Salinities range

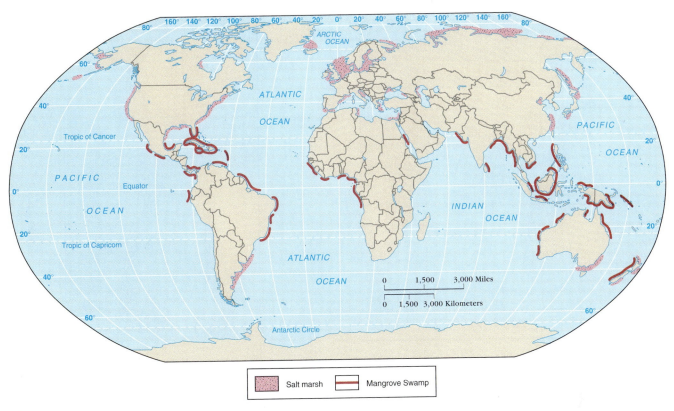

A.

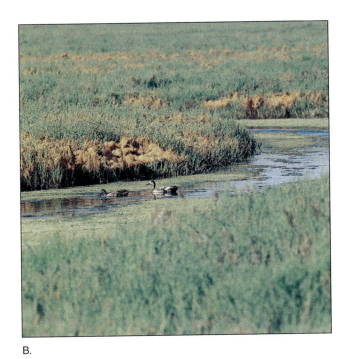

B.

C.

Figure 11–9

World Distribution of Salt Marshes and Mangrove Swamps. *A:* Note dominance of mangrove swamps in low latitudes and salt marshes in high latitudes. *B:* Salt marsh on San Francisco Bay at Shoreline Park, California. *C:* Mangrove trees on Lizard Island, Great Barrier Reef, Australia. (*A:* Base map courtesy of National Ocean Survey; *B:* photo © Eda Rogers; *C:* photo © R. N. Mariscal/Bruce Coleman, Inc.)

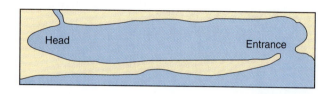

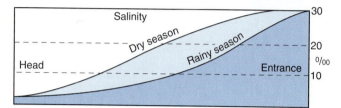

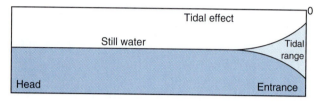

Figure 11–10

Lagoons. Typical shape, salinity, and tide conditions of a lagoon.

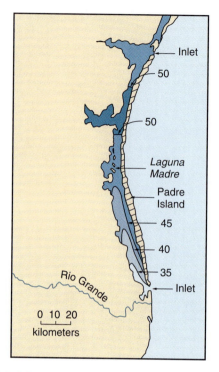

Figure 11–11

Laguna Madre, Texas. Typical summer surface salinity distribution (in parts per thousand, ‰).

from 2‰ during dry periods to over 100‰ when infrequent local storms provide large volumes of fresh water. High evaporation generally keeps salinity well above 50‰.

Because even the salt-tolerant marsh grasses cannot withstand such high salinities, the marsh is replaced by an open sand beach on Padre Island. At the inlets, ocean inflow occurs as a surface wedge *over* the denser water of the lagoon. In turn, the outflow from the lagoon occurs as a *subsurface* flow. This is just the opposite of the circulation described for estuaries.

Mediterranean Sea

The Mediterranean Sea is a most unusual body of water. It actually is a number of small seas connected by narrow necks of water into one larger sea. It is the remains of an ancient sea (the Tethys Sea) that once separated the old continents of Laurasia and Gondwanaland hundreds of millions of years ago.

The Mediterranean has a very irregular coastline, which divides it into subseas such as the Aegean Sea and Adriatic Sea (Figure 11–12A). Each of these seas has a separate circulation pattern.

The Mediterranean is bounded by Europe and Asia Minor on the north and east and Africa to the south. It is surrounded by land except for three very narrow connections to other water bodies: to the Atlantic Ocean through the *Strait of Gibraltar* (about 14 kilometers, or 9 miles, wide), to the Black Sea through

the *Bosphorus* (roughly 1.6 kilometers, or 1 mile, wide), and the Suez Canal, a waterway of 160 kilometers (100 miles) that connects to the Red Sea.

The Mediterranean has two major basins. They are separated by an underwater *sill* (ridge) that extends from Sicily to the coast of Tunisia at a depth of 400 meters (1300 feet). Strong currents run between Sicily and the Italian mainland through the Strait of Messina (Figure 11–12A).

Mediterranean Circulation

Atlantic Ocean water enters the Mediterranean through the Strait of Gibraltar as a surface flow. It enters to replace water that rapidly evaporates in the very arid eastern end of the sea. The water level in the eastern Mediterranean is generally 15 centimeters (6 inches) lower than at the Strait of Gibraltar. The surface flow follows the northern coast of Africa throughout the length of the Mediterranean and spreads northward across the sea (Figure 11–12A).

The remaining Atlantic Ocean water continues eastward to Cyprus. Here, during winter, it sinks to form what is called the *Mediterranean Intermediate Water*, which has a temperature of 15°C (59°F) and a salinity of 39.1‰. This water flows westward along the North African coast at a depth of 200 to 600 meters (650 to 2000 feet) and passes into the North Atlantic as a subsurface flow through the Strait of Gibraltar.

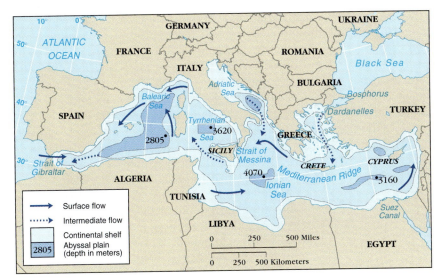

A.

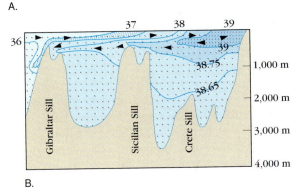

B.

Figure 11–12

Mediterranean Sea—Subseas, Depths, Circulation, and Salinity. *A:* Map shows the subseas, depths, sills (underwater ridges), surface flow, and intermediate flow. *B:* Vertical distribution of salinity (‰). Most of the mass of Mediterranean water has salinities between 38 and 39‰. Maximum salinities (exceeding 39‰) occur in surface waters at the eastern end. After sinking and moving as Intermediate Water toward the Atlantic, salinity is reduced to about 37.3‰ as it flows over the sill at Gibraltar. (*A:* Adapted from *Encyclopedia of oceanography,* edited by Rhodes Fairbridge, © 1966. Reprinted by permission of Dowden, Hutchinson, & Ross, Inc., Stroudsburg, Pa.)

By the time Mediterranean Intermediate Water passes through Gibraltar, its temperature has dropped to 13°C (55°F) and its salinity to 37.3‰. It is still much denser than water at this depth in the Atlantic Ocean, so it moves downward along the continental slope until it reaches a depth of approximately 1000 meters (3300 feet), where it encounters Atlantic Ocean water of the same density. At this level, it spreads in all directions into the Atlantic water and becomes a detectable water mass over a broad area because of its high salinity.

Circulation between the Mediterranean Sea and the Atlantic Ocean is typical of closed, restricted basins where evaporation exceeds precipitation. Such restricted basins always lose water rapidly to surface evaporation, and this water must be replaced by surface inflow from the open ocean. Evaporation of inflowing water from the open ocean increases its salinity to very high values. This denser water eventually sinks and returns to the open ocean as a subsurface flow (Figure 11–12B).

You will recognize this circulation pattern as being opposite the pattern characteristic of estuaries, where surface freshwater flow goes into the open ocean and saline subsurface flow enters the estuary. The Mediterranean Sea is the classic area for such circulation, which is called **Mediterranean circulation.** Its opposite, *estuarine circulation,* develops between a marginal water body and the ocean when freshwater input exceeds the water loss to evaporation.

Pollution Stresses the Coastal Waters

As the use of coastal areas has increased for residences, recreation, and commerce, pollution of coastal waters has increased as well. The following sections present some well-documented cases of chemical pollution of coastal waters by petroleum, sewage, halogenated hydrocarbons, and mercury.

Petroleum

Major oil spills into the ocean are a fact of our modern oil-powered economy. The spills result from tanker accidents (collisions, running aground) and the blowout of undersea oil wells being drilled or producing in the coastal ocean. It is useful to know how much oil enters the oceans and from what sources, but our primary concern is the effect of this pollution on marine organisms and on the marine environment in general.

Oil is a hydrocarbon, meaning that it is composed of the elements hydrogen and carbon. Hydrocarbons are organic substances, and thus are broken down by microorganisms (biologically degraded, or "biodegraded"). For this reason, some consider them to be among the least damaging pollutants of the ocean, yet hydrocarbons are complex. But most crude oils also include chemicals containing oxygen, nitrogen, sulfur, and trace metals. When this complex chemical mixture is combined with seawater—another complex chemical mixture, which also contains organisms—the result can indeed be negative to the organisms.

Florida *Spill in West Falmouth Harbor.* "How long does it take for a shore to recover from an oil spill?" Because of the complexity of petroleum, it is difficult to answer this question. The oldest well-studied oil spill in the United States occurred in 1969 near West Falmouth Harbor in Buzzards Bay, Massachusetts. When the barge *Florida* came ashore and ruptured, currents carried its load of 700 metric tons of number-2 fuel oil northward into Wild Harbor, where the most severe damage occurred (Figure 11–13).

The initial kill was nearly complete for intertidal and subtidal animals in the most severely oiled area. A severe reduction in species diversity was accompanied by rapid increases in the population of polychaete worms (close relatives of segmented earthworms found on land), because they are resistant to oil. The species

Figure 11–13

Florida **Oil Spill at West Falmouth Harbor, Massachusetts.** When the barge *Florida* came ashore and ruptured, currents carried its load of number-2 fuel oil northward into Wild Harbor, where the most severe damage occurred.

Capitella capiata accounted for up to 99.9 percent of the individuals taken in samples of the most severely oiled locations during the first year. Species diversity did not appreciably increase until well into the third year after the spill.

During a period from 3 to 5 years after the spill, marsh grasses and animals reentered the area. No visible damage was to be seen after 10 years. After 20 years, there was virtually no oil in the subtidal sediments. The intertidal marsh sediments were more than 99 percent free of *Florida* oil.

At the most heavily oiled site in Wild Harbor, oil was still present at a depth of 15 centimeters (6 inches). In this region, there was enough oil to kill animals that burrow into the sediments. However, considering the state of the environment at Wild Harbor in 1969, it is clear that even in quiet protected marsh environments, recovery from an oil spill can occur much faster than some researchers thought possible.

Argo Merchant *Off Nantucket.* When oil spills do not come ashore, their effects are less obvious. The sinking of the *Argo Merchant* after it ran aground on Fishing Rip Shoals off Nantucket in 1976 provides the best-studied effects of such a spill. A full load of 19,000 metric tons (7.7 million gallons) of number-6 fuel oil was dumped into the ocean 40 kilometers (25 miles) southeast of Nantucket Island. Fortunately, the winds were such that no oil came ashore. The surface slick moved eastward out to sea and was gone by mid-January.

Although the oil was not visible for long, it did significant biological damage that was investigated by scientists from Woods Hole Oceanographic Institution, the National Marine Fisheries Service, and local institutions (Figure 11–14).

The primary observable effect of the *Argo Merchant* spill on marine organisms was in fish eggs (pollock and cod) in plankton samples taken shortly after the spill. Of the 49 pollock eggs recovered, 94 percent were oil fouled, and 60 percent of the 60 cod eggs were in that state. Dead or dying were 20 percent of the cod eggs and 46 percent of the pollock eggs (Figure 11–15). Little is known about the natural mortality rate of such fish eggs. But for comparison, of a sample of cod eggs spawned in a laboratory, only 4 percent were dead or dying at the same development stage.

Because most of the oil floated in a surface slick, contamination in subsurface water samples did not exceed more than 250 parts per billion. Other than the described damage to fish eggs and significant damage to other plankton, little direct evidence of major biological damage was recovered. Numerous oiled birds did wash ashore at Nantucket and Martha's Vineyard.

Because each pollock female spawns about 225,000 eggs each season, and cod about 1 million, and they do so over a range from New Jersey to Greenland,

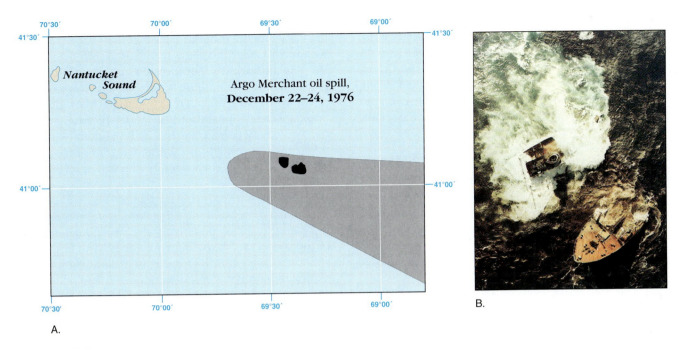

Figure 11–14

***Argo Merchant* Oil Spill off Nantucket, Massachusetts.** *A:* The ship's grounding site (black) and oiled area of the ocean (gray). *B: Argo Merchant,* after breaking up and spilling much of its cargo into the Atlantic Ocean southeast of Nantucket, Massachusetts. (*B:* Photo courtesy of U.S. EPA/EMSL.)

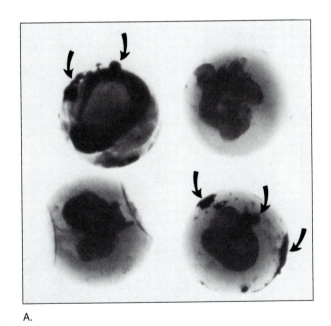

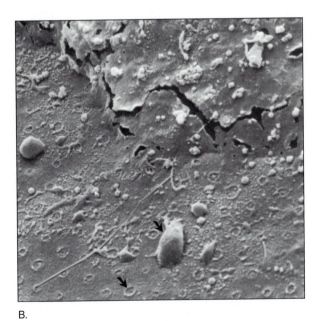

Figure 11–15

Pollock Eggs Affected by the *Argo Merchant* Spill. *A:* These eggs, in the tail-bud and tail-free embryo stages, were taken from the edge of the oil slick. The outer membranes of the eggs at the upper left and lower right are contaminated with a tarlike oil; arrows point to some of the oil masses. The uncontaminated egg at the upper right has a malformed embryo; that at the lower left is collapsed and also has an abnormal embryo. The actual size of pollock eggs is about 1 millimeter (0.04 inches); of cod eggs, about 1.5 millimeter (0.06 inches). *B:* A portion of the surface of an oil-contaminated egg. The upper arrow points to one of many oil droplets, and the lower arrow to one of the membrane pores. (*B:* Magnified about 5000X with scanning electron microscope.) (Courtesy of A. Crosby Longwell.)

it is unlikely that a single oil spill would significantly affect those fisheries. However, it is easy to see why the fishing communities of the northwest Atlantic are so happy about the poor results of oil exploration on Georges Bank. It may be that a fishery already severely stressed by overexploitation could not survive even the lowest level of pollution brought about by the establishment of oil production on the bank.

Exxon Valdez *in Prince William Sound, Alaska.* The first major spill resulting from development of petroleum reserves on the North Slope of Alaska occurred in 1989, when the California-bound *Exxon Valdez* went aground on rocks 40 kilometers (25 miles) out of Valdez, Alaska (Figure 11–16). The 32,000 metric tons of oil spilled makes this the largest spill in the history of U.S. waters.

This spill in the pristine waters of Prince William Sound spread to the Gulf of Alaska, and 1775 kilometers (1000 miles) of shoreline were damaged. The U.S. Fish and Wildlife Service reported that at least 994 sea otters and 34,434 birds were killed by the spill. The actual kill could be 10 times that, by some official estimates. The total death toll will never be known. It was

predicted that affected waters would have a long, slow recovery, but the fisheries closed in 1989 bounced back with record takes in 1990.

Encouraged by good results on small test plots, Exxon spent $10 million to spread fertilizers rich in phosphorus and nitrogen on Alaskan shorelines to boost the development of indigenous oil-eating bacteria. This resulted in a cleanup rate more than twice that under natural conditions. The method is the most successful one yet developed for coping with the effects of oil spills.

Oil spills are a problem that will be with us for many years, and they may become more common as petroleum reserves underlying the continental shelves of the world are increasingly exploited. The largest oil spill on record occurred June 3, 1979, as a result of this exploitation. A Petroleus Mexicanos (PIMEX) oil-drilling station named *Ixtoc #1* in the Bay of Campeche off the Yucatàn peninsula, Mexico, blew out and caught fire. Before it was capped nearly 10 months later, it spewed 468,000 metric tons of oil into the Gulf of Mexico (Figure 11–17).

Oil can be broken down by microorganisms such as bacteria and fungi. Certain of these organisms are ef-

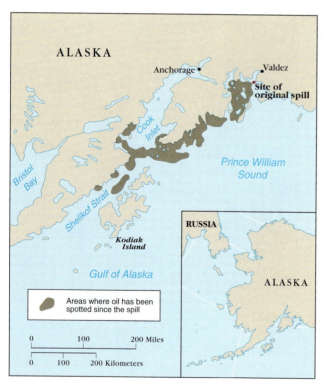

A.

B.

Figure 11–16

Exxon Valdez Oil Spill, Alaska. :A: Location and extent of contamination resulting from the spill of heavy crude oil from the supertanker *Exxon Valdez*. B: Oil from the spill damaged shorelines and killed sea otters and birds. (Photo courtesy of U.S. Coast Guard.)

A.

B.

Figure 11–17

Ixtoc #1 Blowout, Mexico. *A:* Location of the Gulf of Campeche blowout and oil slick that threatened the Texas coast. *B:* Firefighting equipment spraying water around the flaming oil directly above the wellhead. Oil can be seen spreading across the ocean surface. (Photo courtesy of NOAA.)

fective in breaking down only a particular variety of hydrocarbon, and none is effective against all forms. In 1980 Dr. A. M. Chakrabarty, now a microbiologist at the University of Illinois, worked at the General Electric Research and Development Center in Schenectady, New York. There he produced a microorganism capable of breaking down nearly two-thirds of the hydrocarbons in most crude oil spills.

A similar strain was released into the Gulf of Mexico to test its effectiveness in cleaning up the crude oil spilled after a 1990 explosion disabled the tanker *Mega Borg*. Preliminary results indicate that the bacteria are effective. No negative effects were reported on the ecology of the Gulf waters.

Sewage Sludge

Each year over 500,000 metric tons of sewage sludge have been dumped through sewage outfalls into the coastal waters of southern California. This sludge contains a toxic brew of human waste, oil, zinc, copper, lead, silver, mercury, PCBs, and pesticides. A more massive 8 million metric tons have been dumped into the New York Bight each year. (The Bight is the coastal ocean between Long Island and the New Jersey shore.)

Although the Clean Water Act of 1972 prohibited dumping of sewage into the ocean after 1981, the high cost of treating and disposing of sludge on land resulted in extension waivers being granted to these areas. Then an unrelated event forced a change in the law. Non-biodegradable debris, probably carried into the ocean

through storm drains by heavy rains, was washed up on Atlantic coast beaches during the summer of 1988. Ironically, this event, completely unrelated to sewage disposal, hurt the tourist business, bringing about passage of new legislation to terminate sewage disposal in the ocean.

It now appears that political pressure to clean up coastal waters is sufficient to ensure that land treatment and disposal will be the primary disposal procedure of the future. For the present, Los Angeles has stopped pumping sewage into the sea. If it is allowed to resume this practice, the sewage sludge pumped through the lines will have to be scrubbed of toxic chemicals and pathogens.

New York's Sewage Sludge Disposal at Sea. Off the East Coast, the more than 8 million metric tons of sewage sludge dumped by barge each year during the past few years was spread over sites totaling 150 kilometers^2 (58 miles^2). These dumpings occurred at the New York Bight Sludge Site and the Philadelphia Sludge Site, shown in Figure 11–18.

The New York Bight water depth is about 29 meters (95 feet), and Philadelphia's site is 40 meters (130 feet) deep. In such shallow water, the water column displays relatively uniform characteristics from top to bottom. Even the smallest sludge particles reach the bottom without undergoing much horizontal transport, and the ecology of the dump site can be destroyed.

At the very least, such concentration of organic and inorganic nutrients seriously disrupts biogeochemical cycling. Greatly reduced species diversity results,

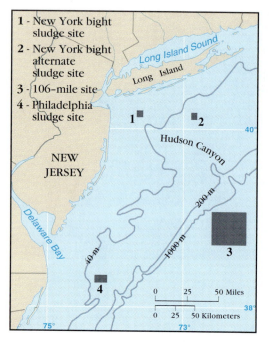

Figure 11–18

Atlantic Sewage Sludge Disposal Sites. More than 8 million metric tons of sewage sludge were dumped by barge annually during the past few years, spread over sites totaling 150 kilometers^2 (58 miles^2). These dumpings occurred at the New York Bight Sludge Site (1) and the Philadelphia Sludge Site (4).

and in some locations the environment becomes devoid of oxygen.

The shallow-water sites have been abandoned, and sewage is now being transported to a deep-water site 171 kilometers (106 miles) out to sea (Figure 11–19). At the deep-water site, beyond the continental shelf break, there is usually a well-developed density gradient that separates low-density, warmer surface water from high-density, colder deep water. Internal waves moving along this density gradient can retard the sinking rate of particles and may allow a horizontal transport rate 100 times greater than the sinking rate.

Local fishermen have been concerned about the deep-water dumping and were reporting adverse effects on their fisheries soon after it began. As it stands now, this East Coast dumping program for sewage is also doomed, and municipalities will have to find the money for land-based processing.

Boston Harbor Sewage Project. Some 48 different communities have deposited their sewage into Boston Harbor, making it one of the most polluted water bodies in the country. A new system is to stop this dumping and carry sewage through a tunnel 15.3 kilometers (9.5 miles) long into Massachusetts Bay (Figure 11–19A). By the year 2000, all of the effluent will be treated with bacteria-killing chlorine. To pay the $4 bil-

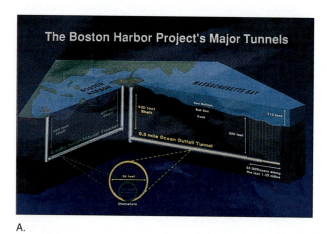

A.

Figure 11–19

Boston Harbor Sewage Project. *A:* Main components of the sewage treatment and disposal system. Sewage will be treated at the Deer Island facility. Effluent will be pumped into the ocean through diffuser lines rising from the main tunnel 76 meters (250 feet) beneath the ocean floor. *B:* Computer-generated bathymetric map of coastal ocean in Boston—Cape Cod area. There is concern that the discharge of sewage effluent in the outfall area will degrade the environment of Stellwagen Bank and Cape Cod Bay to the south. (Vertical exaggeration = 100.) (*A:* Courtesy of Massachusetts Water Resources Authority; *B:* courtesy of U.S. Geological Survey, Woods Hole, Massachusetts.)

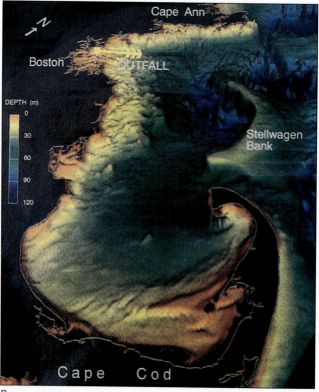

B.

lion system cost, the average sewage bill for a Boston-area household will be about $1200.

Some fear that the project will degrade the environment in Cape Cod Bay and Stellwagen Bank, an important whale habitat. The area was recently designated a national marine sanctuary.

DDT and PCBs

The pesticide DDT (dichlorodiphenyltrichlorethane) and industrial chemicals called PCBs (polychlorinated biphenyls) are now found throughout the marine environment. They are persistent, biologically active chemicals that have been put into the oceans entirely as a result of human activities.

DDT became a widely used pesticide during the 1950s. It was solely responsible for the rapid improvement in crop production throughout the developing countries. However, its extreme effectiveness and persistence as a toxin in the environment eventually wrought a host of environmental problems.

PCBs are industrial chemicals found in a variety of products from paint to plastics. They have been indicated as causes of spontaneous abortions in sea lions and the death of shrimp in Escambia Bay, Florida.

DDT and Eggshells. A near-total ban on U.S. production of the pesticide DDT was emplaced in 1971. By that time, 2 billion kilograms (4.4 billion pounds) had been manufactured, most by the United States. Since 1972 the use of DDT in the Northern Hemisphere has virtually ceased.

The danger of excessive use of DDT and similar pesticides first became manifest in the marine environment, noted in its effects on marine bird populations. During the 1960s there was a serious decline in the brown pelican population of Anacapa Island off the coast of California (Figure 11–20). High concentrations of DDT in the fish eaten by the birds had caused them to produce excessively thin eggshells.

The osprey is a common bird of prey in coastal waters, similar to a large hawk. A decline in the osprey population of Long Island Sound began in the late 1950s and continued throughout the 1960s. This also was caused by thinning eggshells brought on by DDT contamination.

Studies showed a 1 percent increase in eggshell thickness for brown pelicans and ospreys from 1970 to 1976, as the concentration of DDT residue decreased. Because there was an increased egg hatching rate associated with these changes, it was concluded that DDT was the cause of the decline in the bird populations during the 1960s.

DDT and PCBs Linger in the Environment. The main routes by which DDT and PCBs enter the ocean are through the atmosphere and river runoff.

Figure 11–20

Brown Pelican. Pelican populations in the Channel Islands off southern California are increasing, following a serious decline in the 1960s caused by DDT toxicity.

They are concentrated initially in the thin surface slick of organic chemicals at the ocean surface. Then they gradually sink to the bottom, attached to sinking particles. A study off the coast of Scotland indicated that open-ocean concentrations of DDT and PCBs are 10 and 12 times less, respectively, than in coastal waters. Long-term studies have shown that DDT residue in mollusks along the U.S. coasts peaked in 1968.

The pervasiveness of DDT and PCBs in the marine environment can best be demonstrated by the fact that Antarctic marine organisms contain measurable quantities of them. Obviously, there has been no agriculture or industry on that continent to explain this presence. These substances have been transported from distant sources by winds and ocean currents.

Mercury and Minamata Disease

The metal mercury is a liquid at everyday temperatures, and has many industrial uses. Unfortunately, when it enters the ecosystem, mercury forms an organic compound that is toxic to living things.

The stage was set for the first tragic occurrence of mercury poisoning with the establishment of a chemical factory on Minamata Bay, Japan, in 1938. A product of this plant was acetaldehyde, which requires mercury in its manufacture. The first ecological changes in Minamata Bay were reported in 1950, human effects were noted in 1953, and the mercury poisoning known as **Minamata disease** became epidemic in 1956, when the plant was only 18 years old. Minamata disease involves a breakdown of the nervous system. This was the first major human disaster resulting from ocean pollution.

It was not until 1968, however, that the Japanese government declared mercury the cause of the disease. The plant was immediately shut down, but by 1969 over 100 people were known to suffer from the disease. Almost half of these victims died. A second occurrence of mercury poisoning resulted from pollution by another acetaldehyde factory, closed in 1965, in Niigata, Japan. Between 1965 and 1970, 47 fishing families contracted Minamata disease.

During the 1960s and 1970s mercury contamination in seafood received considerable focus. Studies done on the amount of seafood consumed by various human populations led to establishment of safe levels of mercury in fish to be marketed. To establish these levels, three variables were considered:

1. Fish consumption rate of each group of people under consideration
2. Mercury concentration in the fish being consumed by that population
3. Minimum ingestion rate of mercury that induces disease symptoms

Assuming that we have dependable data, and applying a safety factor of 10, a maximum allowable mercury concentration can be established. We can have a high degree of confidence that, if everyone respects this maximum level, the health of the human population will be safeguarded.

Figure 11–21 shows how we use such data to derive a safe level of mercury in fish to be consumed by different groups of people. For the general populations of Japan, Sweden, and the United States, the average individual fish consumption is as follows:

Japan, 84 g/day

Sweden, 56 g/day

United States, 17 g/day

By comparison, members of the Minamata fishing community averaged 286 and 410 g/day during winter and summer seasons, respectively.

The minimum level of mercury consumption that causes poisoning symptoms, as determined by Swedish scientists using Japanese data, is 0.3 mg/day over a 200-day period. The three populations shown in Figure 11–21 would begin to show symptoms if they ate fish with the following mercury concentrations:

Japan, 4 ppm

Sweden, 6 ppm

United States, 20 ppm

If a safety factor of 10 is applied, the maximum concentration of mercury in fish that could be safely consumed by these populations would be as follows:

Japan, 0.4 ppm

Sweden, 0.6 ppm

United States, 2.0 ppm

Although the U.S. Food and Drug Administration (FDA) initially established an extremely cautious limit of 0.5 ppm, the present limit of 1 ppm adequately protects the health of U.S. citizens.

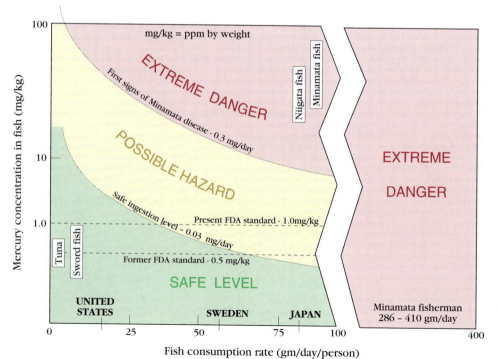

Figure 11–21

Mercury Concentrations in Fish versus Consumption Rates for Various Populations. Curve levels of mercury ingestion and their risk. The first signs of Minamata disease can be expected at 0.3 mg/day, allowing a safety factor of 10 above the 0.03-mg/day safe level. For the United States, it appears that tuna and swordfish are safe to consume, although some swordfish always will be banned because their mercury concentrations are above the present 1.0 parts per million limit imposed by the U.S. Food and Drug Administration.

Essentially all tuna fish falls below this concentration, and most swordfish is acceptable. The present limit amounts to the use of a safety factor of 20 instead of 10. Thus, unless you eat an unusually large amount of tuna or swordfish, you should have little concern about consuming these fishes.

SUMMARY

Characteristics of the coastal ocean, such as temperature and salinity, vary over a greater range than the open ocean. This is because the coastal ocean is shallow and experiences river runoff, tidal currents, and seasonal changes in solar radiation. Coastal geostrophic currents are produced as a result of runoff of fresh water and coastal winds.

Estuaries are semienclosed bodies of water where fresh water runoff from the land mixes with ocean water. Estuaries are classified by their origin as coastal plain, fjord, bar built, or tectonic. Mixing patterns of fresh and saline water in estuaries include vertically mixed shallow estuaries, slightly stratified, highly stratified, and salt wedge, which is characteristic of large-volume river mouths. Typical estuarine circulation consists of a surface flow of low-salinity water toward the mouth and a subsurface flow of marine water toward the head.

Chesapeake Bay is a classic example of a coastal plain estuary with most of its stream inflow entering along its western margin. This fact—along with the Coriolis effect acting on the estuarine circulation—makes the eastern side of the bay saltier.

Wetlands are of two types: salt marsh and mangrove swamp. These biologically productive regions are alternately covered and exposed by the tide. Marshes scrub the water of land-derived pollutants before they reach the ocean. Despite their ecological importance, wetlands are fast disappearing from our coasts as they are destroyed by human activities.

Long offshore deposits called barrier islands protect marshes and lagoons. Some lagoons have restricted circulation with the ocean and may exhibit a great range of salinity and temperature conditions as a result of seasonal change.

The Mediterranean Sea has a classical circulation characteristic of restricted bodies of water in areas where evaporation greatly exceeds precipitation. It is the reverse of estuarine circulation and is called Mediterranean circulation.

Oil pollution reduces species diversity and persists longer on muddy bottoms than on sandy or rocky bottoms. Oil spills that do not come ashore, such as that of the *Argo Merchant,* do considerably less environmental damage than those that do. The greatest damage resulting from the *Argo Merchant* spill was probably to plankton, especially fish eggs.

Although 1972 legislation required an end to dumping of sewage in the coastal ocean by 1981, exceptions continued to be made. Increased public concern resulted in new legislation to prohibit sewage dumping in the ocean.

DDT pollution produced a decline in the Long Island osprey population in the 1950s and the brown pelican population of the California coast in the 1960s. Virtual cessation of DDT use in the Northern Hemisphere in 1972 allowed the recovery of both populations. The DDT thinned the eggshells and reduced the number of successful hatchings.

The first major human disaster resulting from ocean pollution was Minamata disease in Japan. Mercury contamination levels have been set by the FDA, and they appear to have been effective in preventing further poisonings.

KEY TERMS

Bar-built estuary (p. 247)
Coastal plain estuary (p. 246)
DDT (p. 261)
Estuarine circulation (p. 247)
Estuary (p. 246)
Fjord (p. 246)

Highly stratified estuary (p. 247)
Lagoon (p. 252)
Mangrove swamp (p. 252)
Mediterranean circulation (p. 255)
Minamata disease (p. 261)
PCBs (p. 261)

Salt marsh (p. 252)
Salt wedge estuary (p. 248)
Slightly stratified estuary (p. 247)
Tectonic estuary (p. 247)
Vertically mixed estuary (p. 247)
Wetlands (p. 252)

QUESTIONS AND EXERCISES

1. For coastal oceans where deep mixing does not occur, discuss the effect that offshore winds and freshwater runoff will have on salinity distribution. How will the winter and summer seasons affect the temperature distribution in the water column?

2. How does coastal runoff of low-salinity water produce a longshore geostrophic current?

3. Describe the difference between vertically mixed and salt wedge estuaries in terms of salinity distribution,

depth, and volume of river flow. Which displays the more classical estuarine circulation pattern?

4. Discuss the factors that may explain why the surface salinity of Chesapeake Bay is greater on its east side, and periods of summer anoxia in the deep-water mass are increasing in severity with time.

5. Name the two types of wetland environments and the latitude ranges where each will likely develop. How do wetlands contribute to the biology of the oceans and the cleansing of polluted river water?

6. What factors lead to a wide seasonal range of salinity in Laguna Madre?

7. Describe the circulation between the Atlantic Ocean and the Mediterranean Sea, and explain how and why it differs from estuarine circulation.

8. Describe the effect of oil spills on species diversity and recovery in the benthos of Wild Harbor.

9. Discuss whether oil spills that wash ashore or those that do not, such as that of the *Argo Merchant,* are more destructive to marine life.

10. How would dumping sewage in deeper water off the East Coast help reduce the degradation of the ocean bottom?

11. Discuss the animal populations that clearly suffered from DDT pollution and the way in which this negative effect was manifested.

12. Refer to Figure 11–21 and discuss the desirability of setting a mercury contamination level for fish at the 1.0-milligram/kilogram level for the citizens of the United States, Sweden, and Japan.

REFERENCES

Borgese, E. M., and Ginsburg, N., eds. *1988 Ocean Yearbook,* Vol. 7. Chicago: University of Chicago Press.

Cherfas, C. 1990. The fringe of the ocean: Under siege from land. *Science* 248:4952, 163–165.

Clark, W. C. 1989. Managing planet earth. *Scientific American* 261:3, 47–54.

Manheim, T., and Butman, B. 1994. A crisis in waste management, economic vitality, and a coastal marine environment: Boston Harbor and Massachusetts Bay. *GSA Today* 4:7, 197–199.

Officer, C. B. 1976. Physical oceanography of estuaries. *Oceanus* 19:5, 3–9.

Officer, C. B.,; Briggs, R. B.; Taft, J. L.; Tyler, M. A.; and Bovnton, W. R. 1984. Chesapeake Bay anoxia: Origin, development, and significance. *Science* 223:4631, 22–27.

Pickard, G. L. 1964. *Descriptive physical oceanography: An introduction.* New York: Macmillan.

Stommel, H. M., ed. 1950. *Proceedings of the colloquium on "The flushing of estuaries."* Woods Hole: Woods Hole Oceanographic Institution.

Valiela, I., and Vince, S. 1976. Green borders of the sea. *Oceanus* 19:5, 10–17.

SUGGESTED READING

Sea Frontiers

Baird, T. M. 1983. Life in the high marsh. 29:6, 335–341. The ecology of the salt marsh is discussed.

Baker, R. D. 1972. Dangerous shore currents. 18:3, 138–143. A discussion of the hazards to swimmers of various types of currents found near the shore.

Cardoza, Y., and Hirsch, B. 1991. Tidal creeks and scuba gear. 37:4, 32–37. The life forms inhabiting Florida tidal creeks are discussed.

deCastro, G. 1974. The Baltic: To be or not to be? 20:5, 269–273. A review of the pollution threat to life in the Baltic Sea and what has been done to solve the problem.

Edwards, L. 1982. Oyster reefs: Valuable to more than oysters. 28:1, 23–25. The value of oyster reefs to various estuarine life forms is detailed.

Heidorn, K. C. 1975. Land and sea breezes. 21:6, 340–343. A discussion of the causes of land and sea breezes.

Sefton, N. 1981. Middle world of the mangrove. 27:5, 267–273. The life forms associated with Cayman Island mangrove swamps are described.

Smith, F. 1982. When the Mediterranean went dry. 28:2, 66–73. The role of global plate tectonics in causing the Mediterranean Sea to dry up some 5 million years ago.

Sousa Schaefer, F. 1993. To harvest the Chesapeake: Oysters, blue crabs, and softshell clams take a dive. 39:1, 36–49. The effects of excessive nutrients entering the bay from sewage and acid rain have depleted the biological resources of Chesapeake Bay.

White, I. C. 1985. An organization to help combat oil spills. 31:1, 15–21. The efforts of the International Tanker Owners Pollution Federation Limited to increase the efficiency of cleanup operations and reduce the negative effects of oil spills have resulted in more intelligent responses to oil spills.

Scientific American

Halloway, M. 1994. Nurturing nature. 270:4, 98–108. Focusing on the Florida Everglades, the potential to restore damaged wetlands is investigated.

Hsu, K. H. 1972. When the Mediterranean dried up. 227:6, 26–45. Evidence is presented that shows that the Mediterranean Sea was a dry basin 6 million years ago.

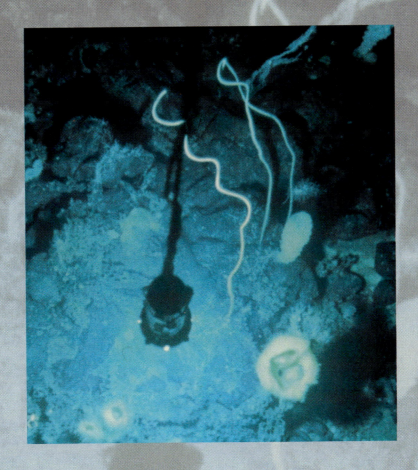

C H A P T E R   1 2
The Marine Habitat

In Chapter 2 we discussed evidence indicating that life probably originated in the oceans. Primary among this evidence is that the oceans are Earth's biggest water environment. Considering the universal need organisms have for water, you can see that life could not exist on the continents were it not for the hydrologic cycle that continually circulates water worldwide. Water evaporates from the ocean and precipitates onto land, from whence gravity pulls it back to the sea.

Obviously, water availability is not a problem for marine organisms. But maneuvering around in this seemingly open and fluid environment can be difficult. The individual success of species depends on their ability to cope with many physical variables that form barriers to their movement.

The diverse species on our planet, and their nutrient sources, are unevenly distributed:

- Over half of the known terrestrial species live in tropical rain forests that cover only 7 percent of the continental surface.
- Coral reefs possess the greatest known diversity in the marine environment, yet they exist in waters with the most limited supply of nutrients.
- Salt marshes, with their high levels of nutrients, support an amazingly productive community of organisms, but one of low diversity.

What are the reasons for this varied distribution? We continually seek the answer, and it surely lies in better understanding of relations between organisms and their physical environment. This chapter focuses on that relation.

Marine Organisms and Their Environment

The ocean environment is far more stable than the terrestrial environment. This is particularly true regarding temperature. As a result of this stability, ocean-dwelling organisms generally have not developed highly specialized regulatory systems to combat sudden changes that might occur within their environment. They are, therefore, affected to various degrees by quite small changes in salinity, temperature, turbidity, and other environmental variables.

Water constitutes over 80 percent of the mass of **protoplasm,** the substance of living matter. Over 65 percent of your weight, and 95 percent of a jellyfish's weight, is simply water (Table 12–1). Water carries dissolved within it the gases and minerals needed by organisms. Water itself is a raw material used in photosynthesizing food, as is done by plants and marine phytoplankton.

Land plants and animals have developed complex "plumbing systems" and devices to distribute water throughout their bodies and to prevent its loss. The threat of atmospheric desiccation (drying out) does not exist for the inhabitants of the open ocean, as they are abundantly supplied with water.

Table 12–1
Water Content of Organisms

Organism	Water Content
Human	65
Herring	67
Lobster	79
Jellyfish	95

Physical Support

You may have not thought about it, but a basic need of plants and animals is for simple physical support. Land plants don't just sit on the ground; they have vast root systems that securely anchor the plant to Earth. In the animal kingdom, a number of support systems are used. Each requires some combination of appendages—legs, arms, fingers, toes—that must support the entire weight of the land animal.

In the ocean, this is not the case. The water of the oceans, as it so lavishly bathes the organisms with their needed gases and nutrients, serves as their physical support as well. Organisms that live in the open ocean depend primarily upon *buoyancy* and *frictional resistance to sinking* to maintain their desired position. Photosynthetic phytoplankton are an example.

This is not to say that they maintain their positions without difficulty. Some have special adaptations developed to increase their efficiency at positioning themselves. We will discuss these adaptations in this and succeeding chapters.

Effects of Salinity

As noted, marine animals are significantly affected by relatively small changes in their environment. Sensitivity to salinity varies. For example, some oysters live in estuarine environments at the mouths of rivers, and they are capable of withstanding a considerable range of salinity. After all, the daily tides force salty ocean water into the river mouth and draw it out again, changing the salinity considerably. Further, during floods salinity is extremely low. Consequently, oysters and most other organisms that inhabit coastal regions have evolved a tolerance for a wide range of salinity conditions. They are called **euryhaline,** from the Greek root word *eury-,* meaning wide or broad; they are "widely salty."

By contrast, other marine organisms, particularly those that inhabit the open ocean, are seldom exposed to much salinity variation. Consequently, they are adapted to steady salinity and can withstand only very small changes. These organisms are called **stenohaline,** from the Greek root word *steno-,* meaning narrow; they are "narrowly salty."

Some phytoplankton and animals cause important alterations in the amount of dissolved material in ocean water. They do so by extracting these minerals to construct hard parts of their bodies, which serve as protective coverings. The compounds primarily used by phytoplankton and animals for this purpose are silica (SiO_2) and calcium carbonate ($CaCO_3$).

Silica is used by a very important phytoplankton population in the ocean—diatoms—and by microscopic animals called radiolaria. *Calcium carbonate* is used by the foraminifera, most mollusks, corals, and some algae that secrete a calcium carbonate skeletal structure.

How do nutrients get into marines phytoplankton and animals? Soluble substances, such as nutrients, diffuse through water, meaning that as they dissolve they spread until they are uniformly distributed. The outer membrane of a living cell is permeable to many molecules. Cells may take in nutrients they need from the surrounding water by **diffusion**—the nutrients simply diffuse through the cell wall. Nutrient compounds are plentiful in seawater. They pass through the cell wall into its interior, where nutrients are less concentrated (Figure 12–1).

After a cell uses the energy stored in nutrients, it must dispose of waste. Waste passes out of a cell by the same method: diffusion through the cell membrane. As the concentration of waste materials becomes greater within the cell than in the fluid medium surrounding it, these materials pass from this area within the cell and into the surrounding fluid. The waste products are then carried away by circulating fluid that services cells in higher animals, or by the surrounding water medium that bathes the simple one-celled organisms of the oceans.

Osmosis

When water solutions of unequal salinity are separated by a water-permeable membrane, such as the membrane around a living cell, water molecules diffuse through the membrane. They move from the *less concentrated solution* into the *more concentrated solution*. This process is called **osmosis,** from a Greek word meaning to push (Figure 12–2). The membrane allows molecules of some substances to pass through while screening out others. **Osmotic pressure** is pressure that must be applied to the more concentrated solution to prevent passage of water molecules into it from a supply of pure water.

Osmosis is important to organisms that live in water, either marine or fresh. Their "skin," in whatever form, is a semipermeable membrane that separates internal body fluids from seawater. If the salinity of an organism's body fluid equals that of the ocean, they are said to be **isotonic** and to have equal osmotic pressure. No net transfer of water will occur through the membrane in either direction.

If the seawater has a lower salinity than the body fluid within an organism's cells, seawater will pass through the cell walls into the cells (always toward the more concentrated solution). This organism is **hypertonic** (saltier) relative to the seawater.

Should the opposite be true, with salinity in an organism's cells less than that of surrounding seawater, water from the cells will pass through the cell membranes out into the seawater (toward the more concentrated solution). This organism is described as **hypotonic** (less salty) relative to the external medium (Figure 12–2).

To simplify, consider osmosis simply as a diffusion process. If you consider the relative concentration of water molecules on either side of the semipermeable

A.   DIFFUSION

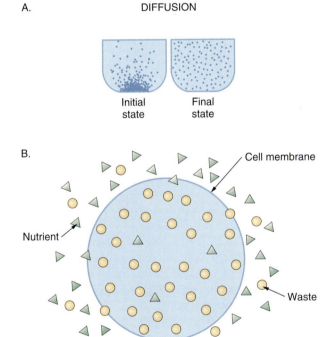

Initial state   Final state

B.

Figure 12–1

Diffusion. *A:* If a water-soluble substance is piled on the bottom of a container of water, it eventually becomes evenly distributed throughout the water. This is due to random molecular motion. Diffusion means that molecules of a substance move from areas of high concentration of the substance to areas of low concentration until the distribution is uniform. *B:* Nutrients (triangles) are in high concentration outside the cell in this example. They diffuse into the cell through the cell membrane. Waste (circles) is in high concentration inside the cell and diffuses out of the cell through the membrane.

OSMOSIS

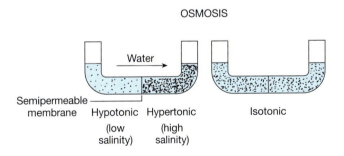

Semipermeable membrane   Hypotonic (low salinity)   Hypertonic (high salinity)   Isotonic

Figure 12–2

Osmosis. If we separate two water solutions of different salinities with a semipermeable membrane, we set the stage for osmosis. The semipermeable membrane must allow water molecules to pass through it, but not the ions of dissolved substances. Water molecules will diffuse through the membrane from the less concentrated (hypotonic) solution (left) into the more concentrated (hypertonic) solution (right). If the salinity of the two solutions is the same (isotonic), there is no net movement of water through the membrane.

membrane, you will see that the net transfer of water molecules is from the side where the greatest concentration of water molecules exists, through the membrane, to the side that contains a lesser concentration of water molecules.

In summary, you can observe three things happening simultaneously across the cell membrane:

1. Water molecules move through the semipermeable membrane toward the lesser water concentration.
2. Nutrient molecules or ions move from where they are more concentrated into the cell, where they will be used to maintain the cell.
3. Waste molecules move from the cell into the surrounding seawater.

Keep in mind that, during this process, molecules or ions of all the substances in the system are passing through the membrane in both directions. Nevertheless, there will be a net transport of molecules of a given substance from the side on which they are most highly concentrated to the side where the concentration is less, until equilibrium is attained.

In the case of the marine invertebrates such as worms, mussels, and octopi, the body fluids and the seawater in which they are immersed are in a nearly isotonic state. Because no significant difference exists, these creatures have not had to evolve special mechanisms to maintain their body fluids at a proper concentration. Thus, they have an advantage over their freshwater (Table 12–2) relatives, whose body fluids are hypertonic in relation to the low-salinity freshwater.

Marine fish have body fluids that are only slightly more than one-third as saline as ocean water. This low level may be the result of their having evolved in low-salinity coastal waters. They are, therefore, hypotonic (less salty) in relation to the surrounding seawater.

This difference creates a problem in that saltwater fish, without some means of regulation, would dehydrate by osmosis, with loss of water from their body fluids into the surrounding ocean. But this loss is counteracted because marine fish drink ocean water and excrete the salts through "chloride cells" in their gills. They also help maintain their body water by discharging a very small amount of very highly concentrated urine.

To summarize, marine fish drink large quantities of water, extract the salts of that water, and dispose of the salts by excreting them. They conserve body water by excreting a highly concentrated urine solution (Figure 12–3).

Freshwater fish are hypertonic (internally more saline) in relation to the very dilute water in which they live. The osmotic pressure of the body fluids of such fish may be 20 to 30 times greater than that of the freshwater that surrounds them. A freshwater fish's problem is to avoid osmosing into its body cells excessive quantities of water that would eventually rupture the cell walls.

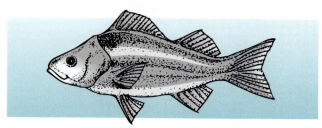

FRESHWATER FISH
(Hypertonic)

Do not drink
Cells absorb salt
Large volume of dilute urine

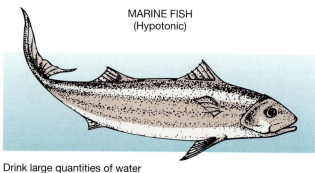

MARINE FISH
(Hypotonic)

Drink large quantities of water
Secrete salt through special cells

Figure 12–3

Freshwater Fish versus Marine Fish. Each is adapted to the salinity of its environment.

To prevent this, the freshwater fish does not drink water. Its cells have the capacity to absorb salt. Water that may be gained as a result of body fluids being greatly hypertonic is disposed of with large volumes of very dilute urine.

To summarize, freshwater fish do not drink, have cells that can absorb salt to maintain their osmotic pressure, and dispose of large quantities of water through copious flow of dilute urine (Figure 12–3).

This also explains why humans cannot drink saltwater when at sea. The salinity of our own body fluids is about one-fourth that of seawater. To drink seawater would cause a drastic loss of water through the membrane surrounding our digestive tract by osmosis, causing eventual dehydration.

Availability of Nutrients

The distribution of life throughout the ocean's breadth and depth depends on the availability of phytoplankton nutrients. Where the conditions are right for the supply of large quantities of nutrients, marine populations reach their greatest concentration. Where are these areas in which we find the greatest mass of marine organisms, or *biomass?* To answer this question, we must consider the sources of nutrients.

Previously, we discussed the role of water in eroding the continents, carrying the eroded material to the oceans, and depositing it as sediment at the margins of the continents. During this process, the water dissolves nitrate and phosphate compounds that serve as the basic nutrient supply for aquatic plants and phytoplankton. (Nitrates and phosphates are the basic ingredients in garden and farm fertilizer; the same chemicals also nourish aquatic plants and phytoplankton.)

Through photosynthesis, aquatic plants and phytoplankton combine these nutrients with carbon dioxide and water to produce the carbohydrates, proteins, and fats that store the energy. The rest of the biological community depends upon this storehouse of energy in plants.

Because the continents are the major sources of these nutrients, you might expect the greatest concentrations of marine life to be at the continental margins, and this is the case. As we head for the open sea from the continental margins, we travel through waters containing a progressively lower concentration of marine life. This is basically due to the vastness of the world's oceans and the great distance between the open ocean and the coastal regions where nutrients are concentrated.

Availability of Solar Radiation

To use these nutrients, phytoplankton must meet another requirement. Photosynthesis cannot proceed unless it is powered by light energy, provided by solar radiation. Sunlight penetrates the atmosphere quite readily, so plants on the continents would seem to have an advantage over those in the ocean. There is almost never a shortage of solar radiation to drive the photosynthetic reactions for these land-based plants.

By contrast, ocean water is a significant barrier to the penetration of solar radiation. In the clearest water, solar energy may be detected to depths of about 1 kilometer (0.6 miles). However, the amount of solar energy reaching these depths is extremely small and cannot power photosynthesis.

Photosynthesis in the ocean is restricted to a very thin surface layer, approximately the upper 100 meters (330 feet). This is the depth to which sufficient light energy can penetrate to carry on photosynthesis. Near the coast, this photosynthetic zone is much thinner because the water contains more suspended material that restricts light penetration.

Considering these two factors necessary for photosynthesis—the supply of nutrients and the presence of solar radiation—we find that in the open ocean away from the continental margins the column of water having solar energy available extends deeper, but that this column contains only a small concentration of nutrients. Conversely, in coastal regions where the water is more

turbid, light penetrates to much shallower depths, but the nutrient supply is quite rich. Because the coastal zone is by far the most productive, you can see the overriding importance of nutrient availability to life in the oceans.

Margins of the Continents

We said that the stability of the ocean environment makes it ideal for the continuation of life processes, yet we now are indicating that the richest concentration of marine organisms is in the very margins of the oceans, where conditions are least stable. The coastal region has many physical conditions that appear at first to be deterrents to the establishment of life:

- Water depths are shallow, allowing seasonal variations in temperature and salinity that are much greater than would be found in the open ocean. You might expect this to stress organisms.
- The thickness of water column varies in the nearshore region as a result of tide movements that periodically cover and uncover a thin strip along the margins of the continents. You might expect this to stress organisms.
- The surf condition represents a sudden release of energy that has been carried for great distances across the open ocean. You might expect this to stress organisms.

Such a great concentration of biomass in an environment that contains so many stress factors highlights the importance of the *evolutionary process* in developing new species by *natural selection.* Over the eras of geologic time, billions of years, new life forms have developed to fit every imaginable biological niche. Many of these life forms have adapted to live under adverse conditions within which life can exist so long as nutrients are available.

Along continental margins, some areas have more abundant life than others. What characteristics of these areas create such an uneven distribution of life? Again, we need consider only those basic requirements for the production of food for the answer. If you were to measure the properties of the water along continental coasts and compare these measurements from place to place, you would find that *areas having greatest biomass are those where water temperatures are lower.*

Due to the low temperature, water keeps in solution greater amounts of the essential gases, oxygen and carbon dioxide, than warmer waters do. Of particular importance is the increased availability of carbon dioxide. This again is an example of the greater availability of the basic requirements of phytoplankton affecting the distribution of life in the oceans.

Upwelling Brings Nutrients to the Surface.

In certain areas of the coastal margins we find an additional factor that enhances conditions for life—upwelling. Upwelling, as the name implies, is a flow of subsurface water toward the surface. This phenomenon occurs along the western margins of continents where surface currents are moving toward the equator (Figure 12–4A).

Along the western margins of continents in both Northern and Southern hemispheres, the Ekman transport moves surface water away from the coast. As this happens, water rises to replace it, from depths of 200 to 1000 meters (660 to 3300 feet) (Figure 12–4B).

This water comes up from depths below where photosynthesis occurs. It is rich in phytoplankton nutrients because there are no phytoplankton in deeper waters to use them. This constant replenishing of nutrients at the surface enhances the conditions for life in areas upwelling. Upwelling water is usually of low temperature, which produces the additional benefit of having a high capacity for dissolved gases (Figure 12–4C).

The combined availability of solar radiation and nutrients results in maximum biomass concentrations in shallow coastal waters. Biomass decreases with greater distance from shore and with greater depth.

Water Color and Life in the Oceans

Usually you can determine areas of high and low organic production in the ocean from the water's color. Coastal waters are almost always greenish in color because they contain more large particulate matter. This disperses solar radiation in such a way that the wavelengths most scattered are those for greenish or yellowish light. This condition is also partly the result of yellow-green microscopic marine plants in these coastal waters.

In the open ocean, where particulate matter is relatively scarce and marine life exists in low concentration, the water appears blue due to the size of the water molecules and their scattering effect on the solar radiation, which is similar to the process that produces the apparent blueness of the skies.

Green color in water usually indicates the presence of a lush biological population, such as might correspond to the tropical forests on the continents. The deep indigo blue of the open oceans, particularly between the tropics, usually indicates an area that lacks abundant life and could be considered a biological desert.

Importance of Organism Size

Marine phytoplankton are quite simple compared with the specialized plant forms that exist on land. How can we explain why phytoplankton remain as single-celled organisms and have not become multicellular?

The major requirements of phytoplankton are that they stay in upper ocean water where solar radiation is available, that nutrients are available, and that plants efficiently take in nutrients from surrounding waters and expel waste materials. These requirements lead us to consider phytoplankton "design," specifically their size and shape.

Phytoplankton in the upper layers of the ocean have no roots and no means of swimming, yet they maintain their general position in the water. How they do so is closely related to the ratio between their surface area and body mass.

Greater surface area per body mass means a greater resistance to sinking. This is because more surface is in frictional contact with the surrounding water. An examination of Figure 12–5 will help you see how this ratio of surface area to mass increases as size shrinks. You can see that it benefits one-celled organisms, which make up the bulk of photosynthetic marine life, to be as small as possible. They are, in fact, microscopic.

The efficiency with which photosynthetic cells take in nutrients from surrounding water and expel waste through their cell membranes also is related to the ratio of surface area to mass. Both the intake of nutrients and waste disposal depend to a large extent on diffusion, so there is a point at which increased size would reduce the ratio of surface area to body mass to where the cell could not function properly and would die. This is why we find cells in all plants and animals to be microscopic, regardless of the overall size of the organism, from bacteria to whales. Each cell must take in nutrients and dispose of wastes by diffusion.

Diatoms are an important group of phytoplankton. You can see representative diatoms in Figures 12–13C through 12–13L. Examining diatoms carefully, we see that they commonly have long, needlelike extensions. These increase a diatom's ratio of surface area to mass. Other members of the microscopic marine community display a similar strategy.

Small organisms use another contrivance to stay in the upper layers of the ocean: Some produce a drop of oil, which lowers their overall density and increases buoyancy.

Despite these adaptations to improve floating ability, the organisms still are denser than water, and tend to sink, if ever so slowly. But this is not a serious handicap, for the tiny cells are carried readily with water movements. Near the surface, wind causes considerable mixing and turbulence. Turbulence dominates the movement of these small organisms, keeping them positioned to bask in the solar radiation needed to photosynthesize, producing the energy needed by other members of the marine community.

Viscosity (Thickness) of Seawater

You know that liquids flow with various degrees of ease. For example, syrup flows less readily than water. This property is **viscosity,** defined as internal resistance to flow. It is a characteristic of all fluids—water, syrup,

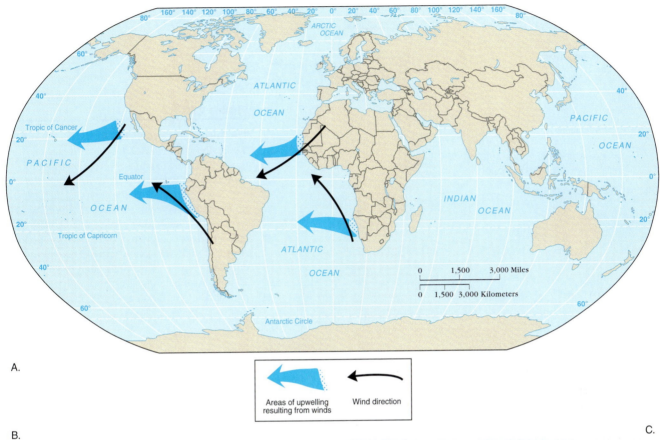

A.

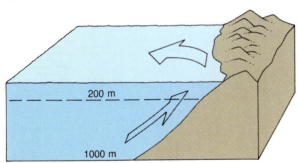

B.

C.

⬅ Areas of upwelling resulting from winds	← Wind direction

Figure 12–4

Upwelling. *A:* Coastal winds (black arrows) usually drive currents southward along the western margins of continents in the Northern Hemisphere. In the Southern Hemisphere, such currents usually are driven northward. The Ekman transport moves water away from the western margins of the continents. The net current directions are shown with blue arrows. *B:* As the surface water moves away from the continent, water rises to replace it from depths of 200 to 1000 meters (660 to 3300 feet). Because this water comes from below the zone of photosynthesis, it is cold (unheated by sunlight) and rich in nutrients. *C:* In this false-color Coastal Zone Color Scanner (CZCS) image from the *Nimbus-7* satellite, wind-driven upwelling along the coast of northwestern Africa results in high phytoplankton biomass and productivity. (Highest productivity = red and orange, decreasing through yellow, green, and blue.) (*A:* Base map courtesy of National Ocean Survey; *C:* image courtesy of NASA/Goddard Space Flight Center.)

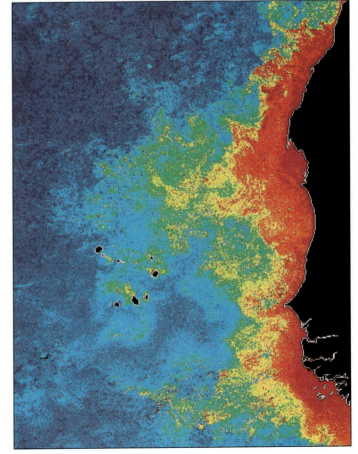

CUBE *A*

Linear dimension - 1 cm
on each side

Area - 6 cm² (6 sides)
Volume - 1 cm³

Ration of surface area to volume:

$$\frac{6 \text{ cm}^2}{1 \text{ cm}^3} = 6 : 1$$

Cube *A* has 6 units of surface area
per unit of volume.

CUBE *B*

Linear dimension - 10 cm
on each side

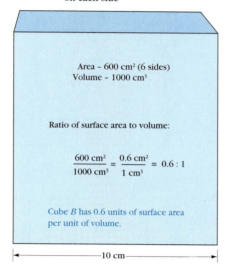

Area - 600 cm² (6 sides)
Volume - 1000 cm³

Ratio of surface area to volume:

$$\frac{600 \text{ cm}^2}{1000 \text{ cm}^3} = \frac{0.6 \text{ cm}^2}{1 \text{ cm}^3} = 0.6 : 1$$

Cube *B* has 0.6 units of surface area
per unit of volume.

10 cm

Figure 12–5

The Importance of Size. Obviously, cube *A* is far smaller than cube *B*. But cube *A* has 10 times the surface area *per unit of interior volume* as cube *B*. In practical terms, if cubes *A* and *B* were plankton, cube *A* would have 10 times the resistance to sinking *per unit of mass* as cube *B*. Therefore, it could stay afloat by exerting far less energy than cube *B*. Also, if cube *A* were a planktonic alga, it could take in nutrients and dispose of waste through its cell wall 10 times as efficiently as an alga with the dimensions of cube *B*.

---

motor oil, air, and molten lava. Loosely speaking, viscosity is the "thickness" of a fluid; the higher the viscosity, the thicker it is. Viscosity is particularly affected by temperature (compare the viscosity of refrigerated syrup with heated syrup).

The viscosity of ocean water is affected by two variables—temperature and salinity. Greater salinity increases seawater viscosity. But temperature has an even greater effect, with lower temperatures "thickening" seawater.

As a result of this temperature relationship, phytoplankton and animals that float in colder waters have less need for extensions to aid them in floating, because the water is more viscous. In fact, members of the same species of floating crustaceans are very ornate, with featherlike appendages, where they occupy warmer waters, whereas these appendages are missing in colder, more viscous environments. Thus, high-viscosity water benefits floating members of the marine biocommunity, helping them maintain their position near the surface.

Viscosity and Streamlining. Not all sea organisms are one-celled floaters. With increasing organism size and swimming ability, viscosity ceases to enhance survival and becomes an obstacle instead. This is particularly true of the large organisms that swim freely in the open ocean. They must pursue prey or flee predators, which means they must displace water to move forward. The faster they swim, the greater is the stress on them. Not only must water be displaced ahead of the swimmer, but water also must move in behind it to occupy the space that the animal has vacated. These are important considerations in **streamlining.**

The familiar shape of free-swimming fish, and of mammals such as whales and dolphins, exemplifies the

streamlining adaptations of organism to move with minimum effort through water. A common shape to meet this need is a flattened body, which presents a small cross section at the front end and a gradually tapering back end. This form is characteristic of fish. It reduces stress from movement through water and allows minimal energy to be expended in overcoming viscosity (Figure 12–6).

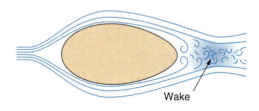

Wake

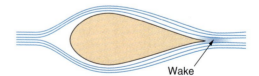

Wake

Figure 12–6

Streamlining. Due to the viscosity (thickness, or resistance to flow) of water, any body moving rapidly through it must produce as little stress as possible as it displaces the water through which it moves. After the water has moved past the body, it must flow in behind the body with as little eddy action as possible.

Temperature

Of all the conditions we can measure, none is more important to the life in the ocean than temperature. The marine environment is much more stable than that on land, and a comparison of temperature ranges proves it. Ocean temperatures remain in a far narrower range, and change far more slowly, than land temperatures.

- The minimum surface temperature observed in the open sea is seldom much below −2°C (28.4°F). The maximum seldom exceeds 32°C (89.6°F). In contrast, record continental temperatures are −88°C (−127°F) and 58°C (136°F).
- Daily ocean surface variations rarely exceed 0.2C° (0.4F°) or 0.3C° (0.5F°), although in shallower coastal waters they may vary daily by 2° or 3C° (3.6 or 5.4F°). In contrast, continental temperatures vary widely from day to day and from season to season.
- Annual ocean water temperature variations are also small. They range from 2C° (3.6F°) at the equator to 8C° (14.4F°) at 35° and 45° latitude and decrease again in the higher latitudes. Annual variations of temperature in shallower coastal areas may be as high as 15°C (27°F). In contrast, continental temperatures can vary dramatically over a year.

The temperature variations in surface waters are reduced in magnitude with depth. In the deep ocean, daily or seasonal temperature variation becomes insignificant. Throughout the deeper parts of the ocean, the temperature remains uniformly low. At the bottom of the ocean basin, where depth exceeds 1.5 kilometers (0.9 miles), temperatures hover around 3°C (37.4°F) regardless of latitude.

Why Temperature Is Important. Recall that reducing temperature will increase density, increase viscosity, and increase the capacity of water to hold gases in solution. All these changes significantly affect organisms inhabiting the ocean.

A direct consequence of the increased capacity of colder water to contain dissolved gases is seen in the vast phytoplankton communities that develop in high latitudes during summer, when solar energy becomes available for photosynthesis. A major factor in this phenomenon is the abundance of dissolved gases, specifically carbon dioxide, which phytoplankton need for photosynthesis, and oxygen, needed by animals that feed upon the phytoplankton.

Tropical populations have more species, increasing the biomass. But in colder waters, floating organisms are physically larger, which also increases the biomass. However, the total biomass of floating organisms in colder, high-latitude planktonic environments greatly exceeds that of the warmer tropics.

The fact that organisms in the tropics are smaller than those observed in the higher latitudes may be related to the lower viscosity and density of seawater in lower-latitude waters. Being smaller, tropical species can expose more surface area per body mass. They are also characterized by ornate plumage to increase surface area (Figure 12–7). These plumose adaptations are strikingly absent in the larger cold-water species.

Warmer temperatures increase the rate of biological activity, which more than doubles with an increase of 10C° (18F°). Tropical organisms apparently grow faster, have a shorter life expectancy, and reproduce earlier and more frequently than their counterparts in the colder waters. Some animal species can live only in cooler waters, whereas others can live only in warmer waters. Many of these can withstand only very small temperature change and are called **stenothermal** ("narrowly thermal"). Other varieties apparently are little affected by temperature and can withstand changes over a large range. These are classified as **eurythermal** ("widely thermal").

Stenothermal organisms are found predominantly in the open ocean, and at depths where large ranges of temperature do not occur. Eurythermal organisms are

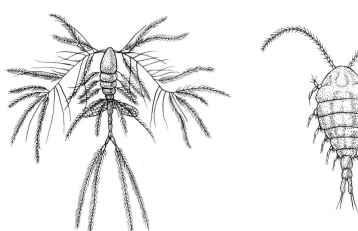

A.

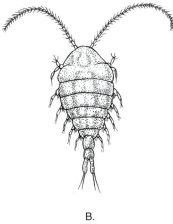

B.

Figure 12–7

Water Temperature and Appendages. *A:* Copepod (*Oithona*) displays the ornate plumage characteristic of warm-water varieties. *B:* Copepod (*Calanus*) displays the less ornate appendages found on temperate and cold-water forms. (After Sverdrup, Johnson, and Fleming.)

more characteristic of the shallow coastal waters, where the largest ranges of temperature are found, and surface waters of the open ocean.

Divisions of the Marine Environment

We can readily divide the marine environment into two basic units. The ocean water itself is the **pelagic environment** (pelagic means "of the sea"). Here floaters and swimmers play out their lives in a complex food web. The ocean bottom constitutes the **benthic environment** (benthic means "bottom"). Plants and animals that do not float or swim (or at least not very well) spend their lives here.

Pelagic (Open Sea) Environment

There are a great many terms for the different divisions of the pelagic environment. You may not remember all of them, but it is important to understand that each labels a zone in the ocean that has distinctive biological characteristics. All of these divisions are shown in Figure 12–8A.

The pelagic environment is divided into two provinces. The **neritic province** extends from the shore seaward, including all water overlying an ocean bottom that is less than 200 meters (660 feet) in depth. The word *neritic* means "shallow."

Seaward, where depth increases beyond 200 meters (660 feet), is the **oceanic province** (Figure 12–8). It includes water with a very great range in depth from the surface to the bottom of the deepest ocean trenches.

The oceanic province is further subdivided into four zones:

- The **epipelagic zone** ("top of ocean") from the surface to a depth of 200 meters (660 feet).
- The **mesopelagic zone** ("middle depth of ocean") from 200 to 1000 meters (660 to 3280 feet).
- The **bathypelagic zone** ("deep ocean") from 1000 to 4000 meters (3,300 to 13,000 feet).
- The **abyssopelagic zone** ("without bottom") includes all the deepest parts of the ocean below 4000 meters (13,000 feet) depth.

An important factor in determining the distribution of life in the oceanic province and its zones is availability of light:

- The **euphotic zone** ("good light") extends from the surface to a depth where enough light still exists to support photosynthesis. This rarely is deeper than 100 meters (330 feet).
- Below this zone, there are small but measurable quantities of light within the **disphotic zone** ("apart from light"), to a depth of about 1000 meters (3300 feet).

- Below about 1000 meters (3300 feet), no light exists in the **aphotic zone** ("without light").

Epipelagic Zone. The upper epipelagic zone is the only oceanic biozone in which there is sufficient light to support photosynthesis. The boundary between it and the mesopelagic zone, at 200 meters (660 feet) depth, is also the approximate depth at which the level of dissolved oxygen begins to decrease significantly (red curve in Figure 12–8B).

This decrease in oxygen results because no plants occur below about 150 meters (500 feet), and dead organic tissue descending from the biologically productive upper waters is undergoing decomposition by bacterial oxidation. Nutrient content also increases abruptly below 200 meters (600 feet) (green curve in Figure 12–8B). This depth also serves as the approximate bottom of the mixed layer, seasonal thermocline, and surface water mass.

Mesopelagic Zone. Within the mesopelagic zone, a dissolved oxygen minimum occurs at a depth of about 700 to 1000 meters (2300 to 3280 feet) (Figure 12–8B). The intermediate-water masses that move horizontally in this depth range often possess the highest levels of plant nutrients in the ocean.

Within the mesopelagic zone, although sunlight from the surface is very, very dim, we still see evidence that animals sense this light. The mesopelagic zone is inhabited by fish that have unusually large and sensitive eyes, capable of detecting light levels 100 times lower than humans can sense.

Another important inhabitant of this zone is the bioluminescent group, which literally glows in the dark. This group includes especially shrimp, squid, and fish. Approximately 80 percent of these organisms carry light-producing *photophores*. These are glandular cells containing luminous bacteria surrounded by dark pigments. Some contain lenses to amplify the light radiation.

This "cold" light is produced by a chemical process involving the compound *luciferin*. Molecules of luciferin are excited and emit photons of light in the presence of oxygen. Only a 1 percent loss of energy is required to produce this illumination. This system is similar to that used by fireflies and glow worms.

Deep Scattering Layer (DSL). Also in the mesopelagic layer is the phenomenon called the **deep scattering layer (DSL).**

When the U.S. Navy was testing sonar equipment early in World War II, it observed a sound-reflecting surface that changed depth daily. This indicated a sound-reflecting mass at a depth of 100 to 200 meters (330 to 660 feet) during the night, which sank as deep at 900 meters (3000 feet) during the day.

They determined that this echo, from the deep scattering layer, was produced by masses of migrating

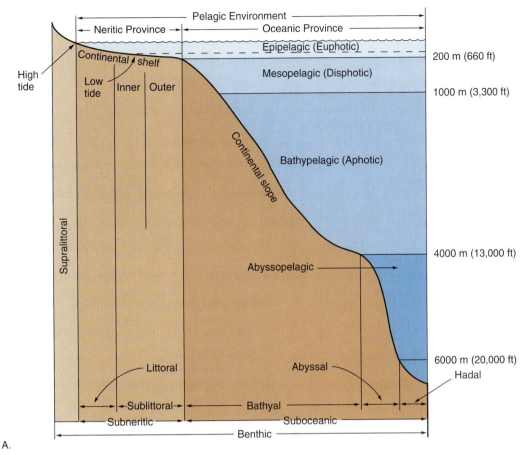

A.

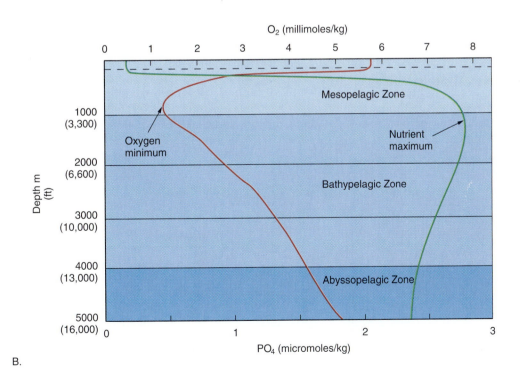

B.

Figure 12–8

Oceanic Biozones. *A:* Biozones of the pelagic and benthic environments. *B:* Distribution of oxygen (O_2) and plant nutrient phosphate (PO_4) in the water column at mid-to-low latitudes. In surface water, oxygen is abundant due to continual mixing with the atmosphere and continual plant photosynthesis, and nutrient content is low due to continual uptake by plants. At the very top of the chart, below the *euphotic zone,* oxygen content abruptly decreases and nutrient content abruptly increases. At or near the base of the mesopelagic zone, an oxygen minimum and nutrient maximum are recorded. Nutrient levels remain high to the bottom. Oxygen increases with depth as deep-water and bottom-water masses carry it into the deep ocean.

marine life. These organisms moved closer to the surface at night and then to a greater depth during the day. The DSL appeared to respond to the changing intensity of light (Figure 12–9).

Research with plankton nets and submersibles has found the DSL to contain layers of various organisms, including small fish. A very small concentration of fish is sufficient to reflect sonar signals. Fish are predators, so organisms on which they prey are probably responsible for the daily rise and fall of the DSL.

Bathypelagic Zone and Abyssopelagic Zone. The aphotic (lightless) bathypelagic and abyssopelagic zones represent over 75 percent of the living space in the oceanic province. In this region of total darkness, many totally blind fish exist. Bizarre-looking small predaceous species make up the total fish population.

Many species of shrimp that normally feed on detritus become predators at these depths, where the food supply is greatly reduced from that available in shallower waters. Animals that live in the these deep zones feed mostly upon one another. They have evolved impressive warning devices and unusual apparatuses to make them more efficient predators. They are characterized by small expandable bodies, extremely large mouths relative to body size, and very efficient sets of teeth (Figure 12–10).

Oxygen content increases with depth, as the deep- and bottom-water masses carry oxygen from the cold surface waters where they formed to the deep ocean. The abyssopelagic zone is the realm of the bottom-water masses that commonly move in the opposite direction of the overlying deep-water masses of the bathypelagic zone.

Benthic (Sea Bottom) Environment

The benthic, or seafloor, environment is subdivided into two larger units. These correspond to the neritic/oceanic provinces of the pelagic environment overhead:

- The **subneritic province** extends from the spring high tide shoreline to a depth of 200 meters (660 feet) approximately encompassing the continental shelf.
- The **suboceanic province** includes all the benthic environment deeper than 200 meters (660 feet) depth.

We will briefly look at the benthic environment, starting from the shore and moving out to sea.

The word *littoral* comes from a Latin word meaning shore, and there are several areas or zones that incorporate the term—supralittoral, sublittoral, and others.

The transitional region from land to seafloor that is above the high tide line is called **supralittoral** ("high shore"). Commonly called the *spray zone,* it is covered with water only during periods of extremely high tides and when tsunamis or large storm waves break on the shore.

The intertidal zone (zone between high and low tides) is the **littoral zone** (shore zone). From low tide shoreline out to 200 meters (660 feet) water depth, the subneritic province is called the **sublittoral zone** ("lower than the shore"), or shallow subtidal zone and is essentially the continental shelf.

The **inner sublittoral zone** includes the sublittoral to a depth of approximately 50 meters (160 feet). This seaward limit varies considerably because it is determined by the depth at which we find no plants growing attached to the ocean bottom. This is controlled primarily by the amount of solar radiation that penetrates the surface water.

The **outer sublittoral zone** includes that portion of the sublittoral zone from the inner sublittoral zone out to a depth of 200 meters (660 feet) or the shelf break, which is the seaward edge of the continental shelf.

With increased depth in the suboceanic system, we find the **bathyal zone** extending from a depth of 200 to 4000 meters (660 to 13,000 feet) and corresponding generally to the continental slope. From a depth of 4000 to 6000 meters (13,000 to 20,000 feet) stretches the

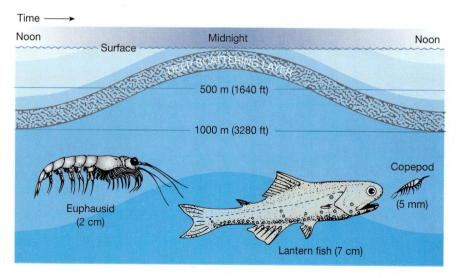

Time ⟶

Noon　　　　　　Midnight　　　　　　Noon

Surface

DEEP SCATTERING LAYER

500 m (1640 ft)

1000 m (3280 ft)

Euphausid (2 cm)

Lantern fish (7 cm)

Copepod (5 mm)

Figure 12–9

Deep Scattering Layer. The deep scattering layer (DSL) scatters and reflects sonar signals well above the bottom. It varies from 100 to 200 meters (328 to 660 feet) deep in daylight to as deep as 900 meters (2950 feet) at night. It may be caused by large numbers of euphausids and lantern fish (myctophid). They are predators that feed on smaller planktonic organisms, like the copepod shown, which daily migrate vertically in the water column.

Figure 12–10

Deep-Sea Anglerfish. The anglerfish, *Lasiognathus saccostoma,* attracts two deep-water shrimp to its bioluminescent "lure." The fish is about 10 centimeters (4 inches) long. (Photo by Peter Arnold, Inc.)

abyssal zone, including over 80 percent of the benthic environment. The **hadal zone** is a restricted environment, including all depths below 6000 meters (20,000 feet), found only in deep trenches along the margins of continents.

When we reach the end of the sublittoral region, we see no more attached plants. All photosynthesis seaward of the inner sublittoral zone is carried on by microscopic algae plankton floating above.

The ocean floor representing the abyssal zone is covered by soft oceanic sediment, primarily abyssal clay. The tracks and burrows of animals that live in this sediment are frequently recorded in bottom photographs (Figure 12–11). The abyssal zone represents over 60 percent of the surface area of the benthic environment, or almost 43 percent of Earth's surface.

The hadal zone below 6000 meters (19,700 feet) is primarily ocean trenches. Isolation in these deep, linear depressions allows the development of animal communities unique to these trenches.

Distribution of Life in the Oceans

It is difficult to describe the degree to which the sea is inhabited because of its immense volume and the paucity of our knowledge. We do know, however, that some populations fluctuate greatly each season. This fact increases the difficulty of describing the extent to which the marine environment is populated. However, we may compare the marine and terrestrial environments by comparing the number of marine species with land species.

Figure 12–12 shows the distribution of Earth's known animal species. Well over 1,410,000 are known, but only about 17 percent of these live in the ocean. Many biologists believe there may be from 3 to 10 million additional unnamed and undescribed animal species living on Earth. A large number may inhabit the oceans. Yet we expect fewer species of marine animals to exist than terrestrial animals. The following theory explains why.

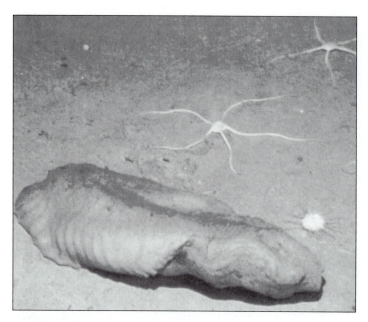

Figure 12–11

Benthic Organisms. Deep-sea cucumber, brittle stars, and sea urchin. (Photograph by Fred Grassle.)

Why Are There Fewer Marine Species?

If the ocean is such a prime habitat for life, and if life originated in this environment, why do we now see such a small percentage of the world's animals species living here? This lesser number may well result because the marine environment is more stable than the terrestrial environment.

The relatively uniform conditions of the open ocean do not produce pressures for organisms to adapt. Also, below the surface layers of the ocean, temperatures are not only stable but also relatively low. Chemical reactions are retarded by this lower temperature. This in turn may reduce the tendency for variation to occur.

Considering the great variety of organisms on the continents, we can assume that this development was the product of a less stable environment, one presenting many opportunities for natural selection to produce new species to inhabit varied new niches. At least 75 percent of all land animals are insect species that have evolved the capability of inhabiting very restricted environmental niches. If we ignore the insects, the sea does possess over 45 percent of the remaining animal species living in the marine and terrestrial environments.

Based on our incomplete knowledge of marine species diversity, the distribution of the 235,000 known marine species shows that only about 4700 (about 2 percent) inhabit the pelagic environment. The other 98 percent are benthic, inhabiting the ocean floor. These numbers are certainly minimums, as recent discoveries indicate that many more benthic species may exist than previously thought (Figure 12–12).

Before I conclude this book with three chapters on organisms of the oceans, I must provide a more comprehensive definition of terms. These describe organisms based on the portion of the ocean they inhabit and the means by which they move: *plankton* (floaters), *nekton* (swimmers), and *benthos* (bottom dwellers).

Plankton (Floaters)

Plankton include all organisms—algae, animals, and bacteria—that drift with ocean currents. An individual organism is called a *plankter*. The fact that plankters drift does not mean that all are unable to swim. Many plankters have this capacity but either move only weakly or are restricted to vertical movement and cannot determine their horizontal position within the ocean.

Among plankton, the algae (microscopic photosynthetic cells) are called **phytoplankton.** The animals are **zooplankton.** Representative members of each group are shown in Figure 12–13.

Plankton also include bacteria. It has recently been discovered that these free-living *bacterioplankton* are much more abundant in the plankton community than previously thought. Having an average dimension of only 0.5 micrometers (µm) (0.00002 inches), they were missed in earlier studies because of their small size.

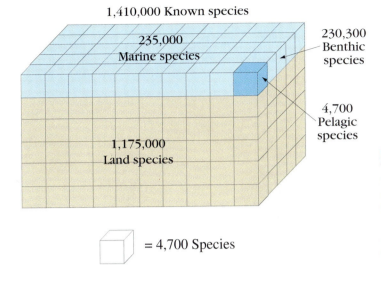

1,410,000 Known species

235,000 Marine species

230,300 Benthic species

4,700 Pelagic species

1,175,000 Land species

☐ = 4,700 Species

Figure 12–12

Distribution of Animal Species on Earth. The entire large cube represents the 1,410,000 animal species known to be living on Earth. The blue-shaded portion indicates the proportion (16.7 percent) that lives in the oceans. Of the marine species, 98 percent live on or in the ocean floor, and 2 percent are plankton or nekton.

You probably have never seen plankton, and you may not even have heard of them before. But the fact is, *most of the biomass of Earth is found adrift in the oceans as plankton.* The volume of Earth's space inhabited by animals that either drift or swim (nekton) greatly exceeds that occupied by all animals that live on land or on the ocean's bottom.

Plankton range greatly in size. They include large floating animals and algae, such as jellyfish and Sargassum (a seaweed). These are called *macroplankton,* measuring 2 to 20 centimeters (0.8 to 8 inches). At the other extreme are bacterioplankton that are too small to be filtered from the water even by a fine silk net. They must be removed by other types of microfilters. These very tiny floaters are called *picoplankton,* measuring 0.2 to 2 μm (0.000008 to 0.00008 inches).

We typically classify plankton as phytoplankton, zooplankton, or bacterioplankton. An additional scheme for classifying plankton is based on the portion of their life cycle spent within the plankton community. Organisms that spend their entire life as plankton are *holoplankton.*

Many organisms we normally consider to be nekton or benthos, because they spend their adult life in one of these living modes, actually spend a portion of their life cycle as plankton. Many nekton and very nearly all the benthos make the plankton community their home during their larval stages. These organisms, which as adults either sink to the bottom to become benthos or begin to swim freely as nekton, are called *meroplankton.*

Nekton (Swimmers)

Nekton include all those animals capable of moving independently of the ocean currents, by swimming or other means of propulsion. They are not only capable of determining their own positions within relatively small areas of the ocean but are also capable, in many cases, of long migrations. Included in the nekton are most adult fish and squid, marine mammals, and marine reptiles (Figure 12–14). When you go swimming, you become nekton, too.

Most freely moving nekton are unable to move at will throughout the breadth of the ocean. Certain nonvisible barriers effectively limit their lateral range. These barriers are gradual changes in temperature, salinity, viscosity, and availability of nutrients. As an example, the death of large numbers of fish frequently has been caused by temporary horizontal shifts of water masses in the ocean. Vertical range is normally determined by water pressure.

Fish appear to be everywhere, but normally are considered to be more abundant near continents and islands and in colder waters. Some fish, such as the salmon, ascend freshwater rivers to spawn. Many eels do just the reverse, growing to maturity in freshwater and then descending the streams to breed in the great depths of the ocean.

Benthos (Bottom Dwellers)

Benthos live on or in the ocean bottom. *Epifauna* live on the surface of the seafloor, either attached to rocks or moving over the bottom. *Infauna* live buried in the sand, shells, or mud. Some benthos not only live on the bottom but also move with relative ease through the water above the ocean floor. They are called *nektobenthos.* Examples of benthos are shown in Figure 12–15.

The littoral (shore) zone and inner sublittoral zone have a great diversity of physical and nutritive conditions. Animal species have developed in great numbers

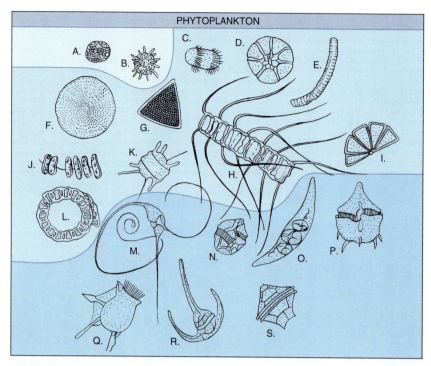

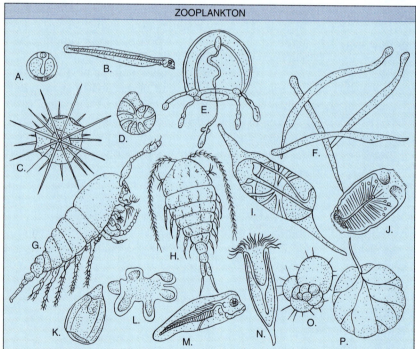

Figure 12–13

Phytoplankton and Zooplankton (Floaters). *Plankton* comes from a Greek word meaning to drift, roam, or wander. (Maximum dimensions in parentheses.) **Phytoplankton:** *A* and *B:* Coccolithophoridae (15 µm, or 0.0006 inches). *C–L:* Diatoms (80 µm, or 0.0032 inches): *C: Corethron; D: Asteromphalus; E: Rhizosolenia; F: Coscinodiscus; G: Biddulphia favus; H: Chaetoceras; I: Licmophora; J: Thalassiorsira; K: Biddulphia mobiliensis; L: Eucampia. M–S:* Dinoflagellates (100 µm, or 0.004 inches): *M: Ceratium recticulatum; N: Goniaulax scrippsae; O: Gymnodinium; P: Goniaulax triacantha; Q: Dynophysis; R: Ceratium bucephalum; S: Peridinium.* **Zooplankton** *A:* Fish egg (1 mm, or 0.04 inches); *B:* Fish larva (5 centimeters, or 2 inches); *C:* Radiolaria (0.5 mm, or 0.02 inches); *D:* Foraminifera (1 mm, or 0.04 inches); *E:* Jellyfish (30 m, or 98 feet); *F:* Arrowworms (3 centimeters, or 1.2 inches); *G* and *H:* Copepods (5 mm, or 0.2 inches); *I:* Salp (10 centimeters, or 4 inches); *J:* Doliolum; *K:* Siphonophore (30 m, or 98 feet); *L:* Worm larva (1 mm, or 0.04 inches); *M:* Fish larva (5 centimeters, or 2 inches); *N:* Tintinnid (1 mm, or 0.04 inches); *O:* Foraminifera (1 mm, or 0.04 inches); *P:* Dinoflagellate (*Noctiluca*) (1 mm, or 0.04 inches).

Figure 12-14

Nekton (Swimmers). Not drawn to scale; typical maximum dimension is in parentheses. *A:* Bluefin tuna (2 m, or 6.6 feet); *B:* Bottlenosed dolphin (4 m, or 13 feet); *C:* Nurse shark (3 m, or 10 feet); *D:* Barracuda (1 m, or 3.3 feet); *E:* Striped bass (0.5 m, or 1.6 feet); *F:* Sardine (15 centimeters, or 6 inches); *G:* Deep-ocean fish (8 centimeters, or 3 inches); *H:* Squid (1 m, or 3.3 feet); *I:* Angler fish (5 centimeters, or 2 inches); *J:* Lantern fish (8 centimeters, or 3 inches); *K:* Gulper (15 centimeters, or 6 inches).

within this nearshore benthic community as a result of the variations existing within the habitat. As you move across the bottom from the shore into the deeper benthic environments, the number of *species* per square meter may remain relatively constant throughout the benthic environment. However, an inverse relation can be observed between the distance from shore and the number of benthic *individuals* and biomass.

The littoral and inner sublittoral zones are the only areas where we find large algae (seaweeds) attached to the bottom, because these are the only benthic zones to which sufficient light can penetrate.

Throughout most of the benthic environment, animals live in perpetual darkness where photosynthetic production cannot occur. They must feed on each other, or on whatever outside nutrients fall from the productive zone near the surface. The deep-sea bottom is an environment of coldness, stillness, and darkness. Under these conditions, life moves slowly. For animals that move around on the deep bottom, the streamlining that is so important to nekton is of little concern.

Organisms that live in the deep sea normally are widely distributed because physical conditions vary little on the deep-ocean floor, even over great distances. A few species appear to be extremely tolerant of pressure changes in that members of the same species may be found in the littoral province and at depths of several kilometers.

Hydrothermal Vent Biocommunities. In 1977 the first biocommunity at a hydrothermal vent was discovered in the Galápagos Rift off South America. This has shown us that high concentrations of deep-ocean benthos are possible. It appears that the primary limiting factor for life on the deep-ocean floor is the availability of food.

At these hydrothermal vents, food is abundant. It is produced not by photosynthesis, for no sunlight is available, but by chemosynthesis in bacteria. The size of individuals and the total biomass in the hydrothermal communities far exceed that previously known for the deep-ocean benthos. These biocommunities will be discussed in Chapter 15.

Figure 12–15

Benthos (Bottom Dwellers)—Some Intertidal and Shallow Subtidal Forms. Not drawn to scale. Typical maximum dimension shown in parentheses. *A:* Sand dollar (8 centimeters, or 3 inches); *B:* Clam (30 centimeters, or 12 inches); *C:* Crab (30 centimeters, or 12 inches); *D:* Abalone (30 centimeters, or 12 inches); *E:* Sea urchins (15 centimeters, or 6 inches); *F:* Sea anemones (30 centimeters, or 12 inches); *G:* Brittle star (20 centimeters, or 8 inches); *H:* Sponge (30 centimeters, or 12 inches); *I:* Acorn barnacles (2.5 centimeters, or 1 inches); *J:* Snail (2 centimeters, or 0.8 inches); *K:* Mussels (25 centimeters, or 10 inches); *L:* Gooseneck barnacles (8 centimeters, or 3 inches); *M:* Sea star (30 centimeters, or 12 inches); *N:* Brain coral (50 centimeters, or 20 inches); *O:* Sea cucumber (30 centimeters, or 12 inches); *P:* Lamp shell (10 centimeters, or 4 inches); *Q:* Sea lily (10 centimeters, or 4 inches); *R:* Sea squirt (10 centimeters, or 4 inches).

S U M M A R Y

The relatively stable marine environment is thought to have given rise to all living things. Those organisms that have established themselves on land have had to develop complex systems for support and for acquiring and retaining water. Pelagic marine organisms depend for support (not sinking) primarily on buoyancy and frictional resistance to sinking.

If the body fluids of an organism and ocean water are separated by a membrane that allows water molecules to pass through, problems related to osmosis may result.

Osmosis is the passing of water molecules from a region in which they are in higher concentration through a semipermeable membrane into a region where they are in lower concentration.

Marine invertebrates and sharks are essentially isotonic, having body fluids with a salinity similar to that of ocean water. Most marine vertebrates are hypotonic, having body fluids with a salinity lower than that of ocean water, and tend to lose water through osmosis.

Freshwater organisms are essentially all hypertonic, having body fluids much more saline than the water in which they live, so they must compensate for a tendency to take water into their cells through osmosis.

For life to flourish in any environment, there must be a sufficient food supply. The basic producers of food are algae, so the requirements of algae must be met first if food is to be plentiful for all. The availability of nutrients and solar radiation makes algal life possible.

Because solar radiation is available only in the surface water of the ocean, algal life is restricted to a thin layer of surface water, usually no more than 100 meters (330 feet) deep. Nutrients (nitrates and phosphates) derived ultimately from continental erosion are much more abundant near continental features.

Biomass concentration decreases away from the continents and with increased depth. The color of the oceans ranges from green in highly productive regions to blue in areas of low productivity.

The algae that must stay in surface water to receive sunlight and the small animals that feed on them do not have effective means of locomotion. They depend, therefore, on their small size and other adaptations to give them a high ratio of surface area per unit of body mass, which results in a greater frictional resistance to sinking. Large animals that swim freely face an altogether different problem and generally have streamlined bodies to reduce frictional resistance to motion.

Surface temperatures of the world ocean range between −2°C (28.4°F) and 32°C (89.4°F). Daily surface temperatures rarely very by more than 0.2C° (0.4F°). Annual surface temperature variation occurs in the 35° to 45° latitude range, where the range may be as great as 8C° (14.4F°).

Compared with life in colder regions, organisms living in warm water tend to be individually smaller, comprise a greater number of species, and constitute a much smaller total biomass. Warm-water organisms also tend to live shorter lives and reproduce earlier and more frequently than their cold-water counterparts.

The marine environment is divided into two basic units—the pelagic (ocean water) and the benthic (ocean bottom) environments. These regions, which are further divided primarily on the basis of depth, are inhabited by organisms that can be classified into three categories on the basis of lifestyle: plankton (free-floating forms with little power of locomotion), nekton (free swimmers), and benthos (bottom-dwellers).

K E Y   T E R M S

Abyssal zone (p. 277)
Abyssopelagic zone (p. 274)
Aphotic zone (p. 274)
Bathyal zone (p. 276)
Bathypelagic zone (p. 274)
Benthic environment (p. 274)
Benthos (p. 279)
Deep scattering layer (p. 274)
Diffusion (p. 267)
Disphotic zone (p. 274)
Epipelagic zone (p. 274)
Euphotic zone (p. 274)
Euryhaline (p. 266)
Eurythermal (p. 273)

Hadal zone (p. 277)
Hypertonic (p. 267)
Hypotonic (p. 267)
Inner sublittoral zone (p. 276)
Isotonic (p. 267)
Littoral zone (p. 276)
Mesopelagic zone (p. 274)
Nekton (p. 279)
Neritic province (p. 274)
Oceanic province (p. 274)
Osmosis (p. 267)
Osmotic pressure (p. 267)
Outer sublittoral zone (p. 276)

Pelagic environment (p. 274)
Phytoplankton (p. 278)
Plankton (p. 278)
Protoplasm (p. 266)
Stenohaline (p. 266)
Stenothermal (p. 273)
Streamlining (p. 272)
Sublittoral zone (p. 276)
Subneritic province (p. 276)
Suboceanic province (p. 276)
Supralittoral (p. 276)
Viscosity (p. 270)
Zooplankton (p. 278)

QUESTIONS AND EXERCISES

1. Discuss the major differences between marine and terrestrial photosynthetic organisms, and explain the need for greater complexity of land plants.
2. What do the prefixes *eury-* and *steno-* mean? Define the terms *euryhaline/stenohaline* and *eurythermal/stenothermal*. Where in the marine environment will organisms displaying a well-developed degree of each characteristic be found?
3. What is the problem requiring osmotic regulation that is faced by hypotonic fish in the ocean? How have these animals adapted to meet this problem?
4. An important variable in determining the distribution of life in the oceans is the availability of nutrients. What are the relations among the continents, nutrients, and the concentration of life in the oceans?
5. Another important determinant of plant productivity is the availability of solar radiation. Why is biological productivity relatively low in the tropical open ocean where the penetration of sunlight is greatest?
6. Discuss the characteristics of the coastal ocean where unusually high concentrations of marine life are found.
7. What factors create the color difference between coastal waters and the less productive open-ocean water?
8. Compare the ability to resist sinking of an organism whose average linear dimension is 1 centimeters (0.4 inches) with that of an organism whose average linear dimension of 5 centimeters (2 inches). Discuss some adaptations other than size used by organisms to increase their resistance to sinking.
9. Changes in water temperature significantly affect the density, viscosity of water, and ability of water to hold gases in solution. Discuss how decreased water temperature changes these variables and may affect marine life.
10. Describe how higher water temperatures in the tropics may account for the greater number of species in these regions compared with low-temperature, high-latitude areas.
11. Construct a table listing the subdivisions of the benthic and pelagic environments and the physical factors used in assigning their boundaries.
12. Describe the vertical distribution of oxygen and nutrients in the oceanic province, and discuss the factors that are responsible for this distribution.
13. Discuss the probable cause and composition of the deep scattering layer.
14. List the relative number of species of animals found in the terrestrial, pelagic, and benthic environments, and discuss the factors that may account for this distribution.
15. Describe the lifestyles of plankton, nekton, and benthos. Why is it proper to consider that plankton account for a relatively larger percentage of the biomass of the oceans than the benthos and nekton?
16. List the subdivisions of plankton and benthos and the criteria used for assigning individual species to each.

REFERENCES

Borgese, E. M.; Ginsburg, N.; and Morgan, J. R., eds. 1991. *Ocean yearbook 9.* Chicago: The University of Chicago Press.

Coker, R. E. 1962. *This great and wide sea: An introduction to oceanography and marine biology.* New York: Harper and Row.

Genin, A.; Dayton, P. K.; Lonsdale, P. F.; and Spiess, F. N. 1986. Corals on seamount peaks provide evidence of current acceleration over deepsea topography. *Nature* 322:6074, 59–61.

Grassle, J. F., and Maciolek, N. J. 1992. Deep-sea species richness: Regional and local diversity estimates from quantitative bottom samples. *American Naturalist* 139:2, 313–341.

Hedpeth, J., and Hinton, S. 1961. *Common seashore life of southern California.* Healdsburg, Calif.: Naturegraph.

Isaacs, J. D. 1969. The nature of oceanic life. *Scientific American* 221:65–79.

May, R. M. 1988. How many species are there on Earth? *Science* 241:4872, 1441–1448.

Sieburth, J. M. N. 1979. *Sea microbes.* New York: Oxford University Press.

Sumich, J. L. 1976. *An introduction to the biology of marine life.* Dubuque, Iowa: Wm. C. Brown.

Sverdrup, H.; Johnson, M.; and Fleming R. 1942. Renewal 1970. *The oceans.* Englewood Cliffs, N.J.: Prentice Hall.

Thorson, G. 1971. *Life in the sea.* New York: McGraw-Hill.

Wilson, E. O. 1992. *The diversity of life.* Cambridge, Mass.: Belknap Press of Harvard University Press.

Wishner, K.; Levin, L.; Gowing, M.; and Mullineaux, L. 1990. Involvement of the oxygen minimum in benthic zonation on a deep seamount. *Nature* 346:6279, 57–59.

SUGGESTED READING

Sea Frontiers

Burton, R. 1977. Antarctica: Rich around the edges. 23:5, 287–295. The high level of biological productivity around the continent of Antarctica is the topic.

Grubner, M. 1970. Patterns of marine life. 16:4, 194–205. Many varieties of life in the ocean are discussed in terms of how their form and size fit them for life in a particular environmental niche.

Hammer, R. M. 1974. Pelagic adaptations. 16:1 2–12. A comprehensive discussion of the adaptations of pelagic organisms to reduce the energy required to maintain their position in the open ocean.

Patterson, S. 1975. To be seen or not to be seen. 21:1, 14–20. A discussion of the possible role of color in the protection and behavior of tropical fishes.

Perrine, D. 1987. The strange case of the freshwater marine fishes. 33:2, 114–119. Explains how marine crevalle jacks are able to inhabit the fresh waters of Crystal River, Florida.

Schellenger, K. 1974. Marine life of the Galápagos. 20:6, 322–332. A discussion of the unique life forms of the Galápagos Islands, 950 kilometers (590 miles) from South America.

Thresher, R. 1975. A place to live. 21:5, 258–267. An interesting discussion of how bottom-dwelling animals compete for space on the ocean floor.

Williams, L. B., and Williams, E. H., Jr. 1988. Coral reef bleaching: Current crisis, future warning. 34:2, 80–87. Corals and related reef animals underwent "bleaching" along the Central American coast in 1983 and in the Caribbean Sea in 1987.

Wu, N. 1990. Fangtooth, viperfish, and black swallower. 36:5, 32–39. Strange adaptations help fish survive in the food-scarce and dark waters below 1000-m depths.

Scientific American

Denton, E. 1960. The buoyancy of marine animals. 203:1, 118–129. The means by which some marine animals reduce the energy expenditure required to live in the ocean water far above the ocean floor are discussed.

Eastman, J. T., and DeVries, A. L. 1986. Antarctic fishes. 255:5, 106–114. Explains how one group of fish survived when the Antarctic turned cold.

Horn, M. H., and Gibson, R. N. 1988. Intertidal fishes. 258:1, 64–71. Intertidal fishes have undergone remarkable adaptation to survive this physically harsh environment.

Isaacs, J. D. 1969. The nature of oceanic life. 221:3, 146–165. A well-developed survey of the conditions for life in the ocean as they relate to the variety and distribution of marine life forms.

Isaacs, J. D., and Schwartzlose, R. A. 1975. Active animals of the deep-sea floor. 233:4, 84–91. A surprisingly large population of large fishes on the deep-sea floor is suggested by automatic cameras dropped to the ocean bottom.

Palmer, J. D. 1975. Biological clock and the tidal zone. 232:2, 70–79. This article investigates the mechanism of biological clocks set to rhythm of the tides, which are found in organisms from diatoms to crabs.

Partridge, B. L. 1982. The structure and function of fish schools. 246:6, 114–123. Schooling benefits and the means by which fish maintain contact with the school are considered.

Vogel, S. 1978. Organisms that capture currents. 239:2, 128–139. The manner in which sponges use ocean currents is an important part of this discussion.

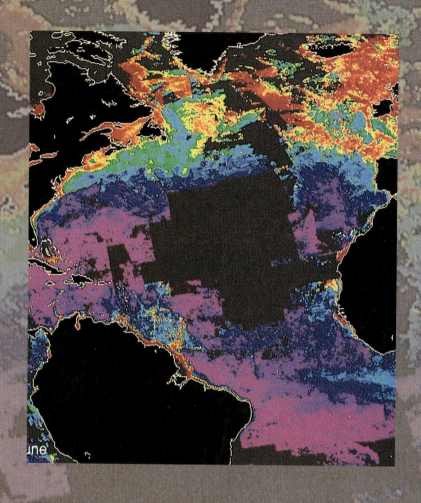

C H A P T E R   1 3
Biological Productivity
and Energy Transfer

Producers are plants and algae that photosynthesize their own food from carbon dioxide, water, and sunlight. Their ability to capture solar energy and bind it into their food sugars is the basis for all nutrition in the ocean (except around hydrothermal vents). The ocean's producers are the foundation of the ocean food web. The major primary producers of the oceans are marine algae. They capture most of the solar energy used to support the marine biological community.

When you think of marine algae, you probably picture large macroscopic seaweeds you may have see growing near shore in many coastal areas. However, these large plants play only a minor part in the production of energy for the ocean population as a whole. Instead, marine organisms depend primarily on the small planktonic varieties of marine algae that inhabit the near-surface sunlit water of the world's oceans. These algae are not so obvious. All are microscopic, but they are scattered throughout the breadth of the ocean surface layer. They represent the largest biomass community in the marine environment—*phytoplankton*.

In addition to the food production of algae living near the ocean surface, another food-production method operates in the deep ocean. In the total darkness of the deep sea, where no measurable sunlight penetrates, certain bacteria use energy released by oxidizing hydrogen sulfide or methane to synthesize food. This food-production method supports a vast array of unusually large deep-sea benthos.

The chemical energy stored by both of these producers of organic matter—surface phytoplankton and deep-sea bacteria—is passed to the various populations of animals that inhabit the oceans through a series of feeding relationships called food chains and food webs.

Classification of Living Things

All living things belong to one of the five kingdoms shown in Figure 13–1—Monera, Protista, Mycota, Metaphyta, and Metazoa.

Kingdom **Monera** are the simplest of all organisms. These organisms are single-celled but lack a discrete nucleus. Their nuclear material is spread throughout the cell. Included in this kingdom are the *cyanobacteria* (blue-green algae) and heterotrophic bacteria. Recent discoveries have shown bacteria to be much more important to marine ecology than previously believed. They are found throughout the breadth and depth of the oceans.

Kingdom **Protista** represent a higher stage of evolutionary development. It includes single-celled organisms that have a nucleus. Protista include single-celled algae that produce food for most marine animals and single-celled animals called **protozoa.**

Kingdom **Mycota,** or the fungi, appear to be poorly represented in the oceans. Less than 1 percent of the known 50,000 species are sea-dwellers. Although fungi exist throughout the marine environment, they are much more common in the intertidal zone. Here, they live in a relationship with cyanobacteria or green algae. This relationship creates what we call *lichen,* in which fungus provides a protective covering that retains water during periods of exposure, while algae provide food for the fungus through photosynthesis. Other

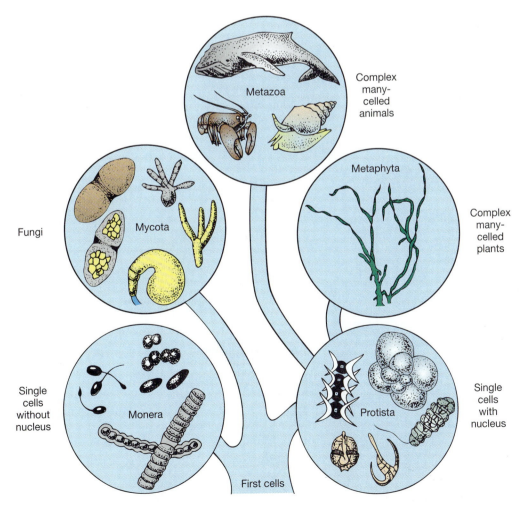

Figure 13–1

The Five Kingdoms of Organisms.

fungi function primarily as decomposers in the marine ecosystem: They remineralize organic matter.

Kingdom **Metaphyta** comprises the multicelled plants. They are mostly restricted to the shallow coastal margins of the ocean and are not a major component of the marine productive community. However, they are very important as producers within restricted communities, such as mangrove swamps and salt marshes.

Kingdom **Metazoa** comprises the multicelled animals. Metazoans range in complexity from the simple sponges to complex vertebrates (animals with backbones).

The five kingdoms are divided into increasingly specific groupings. The system of classification used today was introduced by Carl von Linné (Linnaeus) in 1758. It includes the following major categories:

Kingdom
Phylum
Class
Order
Family
Genus ⎫
⎬ Species name
Species ⎭

Every organism has a scientific name that includes its genus and species—for example, *Delphinus delphis*. Most organisms also have a common name, which is generally used by people who don't need the precision of the "scientific" name. *Delphinus delphis* is better known to most of us as the common dolphin.

Before examining productivity and energy transfer among organisms in the ocean, let us first take a quick tour of the algae and plants of the sea.

Macroscopic (Large) Algae and Plants

We will start with a group of plants and algae with which you are most likely to be familiar. These are the attached forms of macroscopic algae and metaphyta (multicelled plants) found in shallow waters along the ocean margins. In classifying algae, we use color criteria, based partly on pigment they contain (Figure 13–2).

Brown Algae (Phaeophyta)

The brown algae include the largest members of the marine algal community that are attached to the bottom. Their color ranges from very light brown to black.

Brown algae occur primarily in temperate and cold water areas. Their sizes range widely. Smallest is the small black encrusting patch of *Ralfsia* of the upper and middle intertidal zones, where the algae may become crisp and dry in the sun without dying. Largest is the bull kelp (*Pelagophycus*), which may grow in water deeper than 30 meters (100 feet) (Figures 13–2A and 13–2C).

Green Algae (Chlorophyta)

In fresh water, you commonly see these green algae, but they are not well represented in the ocean. Most species are intertidal, or grow in shallow bay waters. They contain the pigment chlorophyll, which makes most of them grass green in color. They grow only to moderate size, seldom exceeding 30 centimeters (12 inches) in the largest dimension. Forms range from finely branched filaments to flat thin sheets.

Various species of sea lettuce (genus *Ulva*), a thin membranous sheet only two cell layers thick, are widely scattered throughout cold-water areas. Sponge weed (genus *Codium*), a two-branched form more common in warm waters, can exceed 6 meters (20 feet) in length (Figure 13–2A and 13–2D).

Red Algae (Rhodophyta)

Red algae are the most abundant and widespread of marine macroscopic algae. Over 4000 species occur from the very highest intertidal levels to the outer edge of the inner sublittoral zone. Many are attached to the bottom. They are very rare in fresh water. Red algae range from just visible to the unaided eye to lengths up to 3 meters (10 feet). While found in both warm and cold water areas, the warm water varieties are relatively small.

The color of red algae varies considerably depending on their depth in the intertidal or inner sublittoral zones. In the upper, well-lighted areas it may be green to black or purplish, changing through a brown to a pinkish red in deeper-water zones, where less light is available.

The bulk of marine photosynthetic productivity is believed to occur above water depths where the amount of light is reduced to 1 percent of that available at the surface. This depth is about 100 meters, or 330 feet. However, a red alga has been observed growing at a depth of 268 meters (880 feet) on a seamount near San Salvadore, Bahamas. Available light at this sighting was thought to be only 0.05 percent of the light available at the ocean's surface.

Seed-Bearing Plants (Spermatophyta)

As noted, the metaphyta generally are confined to coastal areas. The only metaphyta observed in the marine environment belong to the highest group of plants, the seed-bearing Spermatophyta. Two seed-bearing plants found in the marine environment are eelgrass (*Zostera*) and surf grass (*Phyllospadix*). Eelgrass, a grasslike plant with true roots, exists primarily in quiet waters of bays and estuaries from the low tide zone to a depth of some 6 meters (20 feet). Surf grass prefers the high-energy environment of an exposed rocky coast and can be found from the intertidal zones down to a depth of 15 meters (50 feet).

Both of these plants are important sources of the detrital food for the marine animals that inhabit their environment. Found in salt marshes are grasses (most of the genus *Spartina*), whereas mangrove swamps contain primarily mangroves (genera *Rhizophora* and *Avicennia*).

Microscopic (Small) Algae

Now we will introduce all the members of the important phytoplankton that produce over 99 percent of the food supply for marine animals. As the *plankton* part of their name indicates, they are primarily floating forms, although some live on the bottom in the nearshore environment.

Golden Algae (Chrysophyta)

Containing predominantly the yellow pigment **carotin,** these microscopic plants store food in the form of a carbohydrate and oils. There are two types of golden algae: diatoms and coccolithophores.

Diatoms. The *diatoms* are a class of algae contained in a shell. The shell is composed of *opaline silica,* which means that it is like the semiprecious mineral opal, containing considerable water locked into its silicate structure ($SiO_2 \cdot nH_2O$). These silica housings are important geologically because they accumulate on the ocean bottom and produce a siliceous sediment called *diatomite.* Some deposits of diatomite, now elevated and on land, are mined and used in filtering devices. They are probably the most important marine algae in terms of production.

The shell of the diatom resembles a microscopic pillbox. The top and bottom are called valves. The protoplasm of the single cell is contained within this shell, and it exchanges nutrients and waste with the surrounding water through slits or pores in the valves.

Coccolithophores. **Coccolithophores** are covered with small calcareous plates called **coccoliths,** made of calcium carbonate ($CaCO_3$). The name of the group means "bearers of coccoliths" (Figure 13–2F). The individual plates are about the size of a bacterium, and the entire organism is too small to be captured in plankton nets (cone-shaped nylon nets towed by research vessels). Coccolithophores contribute significantly to calcareous deposits in all the temperate and warmer oceans.

Dinoflagellate Algae (Pyrrophyta)

A group second in importance to the diatoms in marine productivity is the **dinoflagellates.** They possess **flagella** (whiplike structures) for locomotion, giving them a slight capacity to move into more favorable areas for photosynthetic productivity. Dinoflagellates are rarely important geologically because many have no long-lasting hard covering. Many of the 1100 species undergo structural changes in response to changes in their environment. Many are luminescent (Figure 13–2E).

Red Tides. Sometimes up to 2 million dinoflagellates may be found in 1 liter (1 quart) of water, causing a **red tide.** Mainly responsible for red tides are two genera of dinoflagellates, named *Ptychodiscus* and *Gonyaulax.* Both produce water-soluble toxins. *Gonyaulax* toxin is not poisonous to shellfish, but it concentrates in their tissue and is poisonous to humans who eat the shellfish. *Ptychodiscus* toxin kills fish and shellfish.

April through September are particularly dangerous months for red tides in the Northern Hemisphere. In most areas, there is a quarantine during these months against harvesting of shellfish that feed on these microscopic organisms and concentrate the poisons that they secrete to levels that are dangerous to humans.

A potentially tragic epidemic of paralytic shellfish poisoning from a *Gonyaulax* red tide occurred in the coastal waters of Massachusetts in the fall of 1972. Fortunately, no deaths occurred, although 30 cases of poisoning were reported. Symptoms of paralytic shellfish poisoning are similar to those of drunkenness, including incoherent speech, uncoordinated movement, dizziness, and nausea. Documented cases throughout the world include 300 deaths and 1750 nonfatal cases. There is no known antidote for the toxin, which attacks the human nervous system, but the critical period usually passes in 24 hours.

A new type of poisoning caused by domoic acid, a toxin produced by a diatom, occurred along the eastern coast of Canada in 1987. One hundred people were poisoned from eating mussels; of these, 3 people died and 10 still suffer from memory loss. Domoic acid poisoning has been called *amnesic shellfish poisoning* because of the memory loss suffered by a number of the victims.

In September 1991 the first known occurrence of marine bird poisoning by domoic acid occurred along the northern California coast. Brown pelicans and Brandt's cormorants died in a sudden mass mortality. The diatom was ingested by unaffected anchovies that were eaten by the birds, the only marine organisms known to have been killed by the toxin.

Red tides are occurring more frequently, more toxic species are showing up, and larger areas are being affected. More marine species, including dolphins and

RHODOPHYTA
Red Algae

detail

Corallina

1 cm
(0.4 in)

Dinoflagellates
PYRROPHYTA

0.1 mm
(0.004 in)

PHAEOPHYTA
Brown Algae

Macrocystis

10 cm
(4 in)

CHLOROPHYTA
Green Algae

Ulva

1 cm
(0.4 in)

CHRYSOPHYTA

Diatoms

0.1 mm
(0.004 in)

A.

B.

C.

humpback whales, are succumbing to the toxins. Human activities may be contributing to this increasing problem. For example, sewage and fertilizer that make their way into coastal waters can harm organisms by causing an overabundance of nutrients.

Primary Productivity

Primary productivity is the amount of carbon fixed by organisms through the synthesis of organic matter using energy derived from solar radiation or chemical

D.

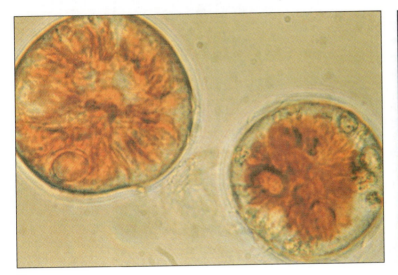

E.

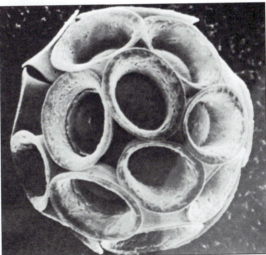

F.

Figure 13-2

Algae. *A:* Several types of algae. *B:* Encrusting red algae, *Lithothamnion,* on sea floor beneath a sea star, *Henricia. C:* Brown algae, *Laminaria,* oarweed at low tide. *D:* Green algae, *Codium fragile.* The prominent fingerlike plant is called *sponge weed. E:* Photomicrograph of living *Gonyaulax polyedra,* magnified 1665 times. *Gonyaulax* is a large genus of phosphorescent marine dinoflagellates. In great abundance, they cause "red tide" along the shoreline. *F:* Coccoliths, disc-shaped calcium carbonate ($CaCO_3$) plates that cover coccolithophore cells. Diameter is 0.06 millimeters. (*A:* Line drawing by Phil David Weatherly; *B:* photo by Shane Anderson; *C:* photo by D. J. Wroebel, Monterrey Bay Aquarium/BPS; *E:* Photo from Biophoto Associates/Science Source/Photo Researchers, Inc.; *F:* Courtesy of Deep Sea Drilling Project, Scripps Institution of Oceanography, University of California, San Diego.)

reactions. The major process through which primary productivity occurs is photosynthesis.

We have new knowledge of the role of *chemosynthesis* in supporting hydrothermal vent communities along oceanic spreading centers. But chemosynthesis is much less significant in worldwide marine primary production than is photosynthesis. We will look at both methods, but we will study photosynthesis in more detail because of its far greater importance and our greater knowledge of it.

Photosynthetic Productivity

Photosynthesis is a chemical reaction in which energy from the sun is stored in organic molecules, in this manner:

Water + Carbon dioxide + Light energy →
Carbohydrate food sugar + Oxygen

$$6H_2O + 6CO_2 + \text{Light energy} \rightarrow$$
$$C_6H_{12}O_6 \text{ (glucose)} + 6O_2$$

The total amount of organic matter produced by photosynthesis per unit of time is the **gross primary production** of the oceans. Algae use some of this organic matter for their own maintenance, through respiration. What remains is **net primary production,** which is manifested as growth and reproduction products. It is the net primary production that supports the rest of the marine population—animals, protozoa, and bacteria.

Gross primary production has two components, new production and regenerated production. **New production** is that part supported by nutrients brought in from outside the local ecosystem by processes such as upwelling. The higher the ratio of new production to gross primary production in an ecosystem, the greater its ability to support animal populations that we depend on for fisheries, such as pelagic fishes and benthic scallops. **Regenerated production** results from nutrients being recycled within the ecosystem.

How is primary productivity measured? In the 1920s the **Gran method** was developed. It is based on the fact that photosynthesis releases oxygen in proportion to the amount of organic carbon that is synthesized. The method places equal quantities of phytoplankton and seawater into a series of bottles, all of which contain the same amount of dissolved oxygen. The bottles are then arranged in pairs, one being transparent and the other fully opaque. These bottle pairs are suspended on a line through the light zone where photosynthesis can occur. Here they are left for a specific period of time. After the bottles are brought to the surface, the oxygen concentration is determined for each bottle.

Respiration will consume oxygen in both the transparent and opaque bottles. But photosynthesis is confined to the transparent bottles, where it releases oxygen to the water. Increased oxygen concentration in the transparent bottles represents the net production of the plants within the transparent bottle. Decreased oxygen content within the opaque bottles simply corresponds to the respiration rate. For any depth, gross production can be estimated by adding the oxygen gain in the clear bottles to the oxygen loss in the opaque bottles.

The depth at which the oxygen production and the oxygen consumption are equal is called the **oxygen compensation depth.** This represents the level of light intensity below which algae do not survive.

Respiration goes on 24 hours a day. But it is during the daylight hours that algae must produce biomass through photosynthesis. Biomass must be produced in excess of that which is consumed by respiration in any 24-hour period if the total biomass of the community is to increase (Figure 13–3). An analogy can be made with a paycheck: Gross photosynthesis (gross pay earned) = oxygen change in clear bottle (take-home pay) + oxygen loss in dark bottle (income tax and other withholding).

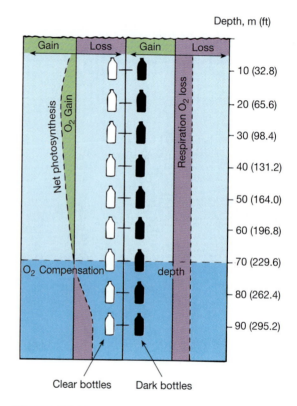

Clear bottles   Dark bottles

Figure 13–3

Oxygen Compensation Depth. This is a comparison of phytoplankton production and consumption of oxygen (O_2), under two conditions—light and no light. Clear bottles are used to let in light needed for photosynthesis. Dark bottles are used to prevent light from entering, to form a control group. The comparison reveals how deeply light can penetrate water to sustain enough photosynthesis to increase phytoplankton mass. This phytoplankton growth can occur only above the *oxygen compensation depth.* This is the depth at which oxygen produced by photosynthesis equals oxygen consumption by respiration. For the clear bottles, the green represents net photosynthesis (gross photosynthesis, less the consumption of oxygen by respiration). In the dark bottles, the loss of oxygen gives a value for the amount of respiration that occurred in the clear bottles (violet). This loss of oxygen, added to the net photosynthesis that occurred in the clear bottles, gives a value for the gross photosynthesis in the clear bottles.

During the 1950s a method involving the use of radioactive carbon (carbon 14) was developed. It has been refined and is currently the most often used method for determining marine primary productivity.

A third method used in studies of marine primary productivity is to measure chlorophyll in living phytoplankton samples from surface waters. This method is less precise than carbon 14, but there is a direct relation between the amount of chlorophyll and primary productivity. The satellite-borne Coastal Zone Color Scanner (CZCS) senses the effect of phytoplankton pigment on the color of ocean water. CZCS data were used to prepare the pigment concentration map in Figure 13–7.

Patterns of Biological Production and Temperature

Primary photosynthetic production in the oceans varies from about 0.1 *gram of carbon per square meter per day* (gC/m^2/d) in the open ocean to more than 10 gC/m^2/d in highly productive coastal areas. This variability is the result of the uneven distribution of nutrients throughout the photosynthetic zone and seasonal changes in the availability of solar energy.

About 90 percent of the biological mass generated in the light zone of the open ocean is decomposed into inorganic nutrients before descending below this zone. Approximately 10 percent of this organic matter sinks into deeper water, where most of it is decomposed. About 1 percent of the light zone production reaches the deep-ocean floor. We call this process the *biological pump,* because it removes ("pumps") carbon dioxide and nutrients from the upper ocean and concentrates them in the deep-sea waters and sediments.

The permanent thermocline prevents the return of these nutrients to the light layer throughout much of the subtropical gyres. The thermocline and resulting pycnocline develop only during the summer season in temperate latitudes and are absent in polar regions.

In the following sections, you will see that the degree to which waters have thermal layers profoundly affects the patterns of biological production observed at different latitudes in the world ocean.

Productivity in Polar Oceans.

As an example of seasonal productivity in a polar sea, consider the Barents Sea, north of the Arctic Circle, off the northern coast of Europe. Diatom productivity peaks here during May and tapers through July (Figure 13–4A). This feeds a zooplankton development, mostly small crustaceans (Figure 13–4D). The zooplankton biomass peaks in June and continues at a relatively high level until winter darkness begins in October. In this region above 70°N latitude, there is continuous darkness for about 3 months of winter and continuous illumination for about 3 months during summer.

In the Antarctic region, particularly at the southern end of the Atlantic Ocean, productivity is somewhat greater. The most likely explanation is the continual upwelling of water that has sunk in the North Atlantic. Moving southward as a deep-water mass, this North Atlantic Deep Water surfaces hundreds of years later, carrying high concentrations of nutrients (Figure 13–4B).

As an illustration of the very great productivity that occurs during the short summer season in polar oceans, consider the growth rate of blue whales. The largest of all whales, the blue whale (see Figure 14–12) migrates through temperate and polar oceans at times of maximum zooplankton productivity. This excellent timing enables the whales to develop and support large calves (following a gestation of 11 months, the calves can exceed 7 meters (23 feet) length at birth).

The mother blue whale suckles the calf for 6 months with a teat that actually pumps the youngster full of rich milk. By the time the calf is weaned, it is over 16 meters (50 feet) long. In 2 years, it will attain 23 meters (75 feet). In 3+ years, a 60-ton blue whale has developed. This phenomenal growth rate gives some indication of the enormous biomass of copepods and krill that these large mammals feed upon.

Polar waters have little density or temperature difference (Figure 13–4C). Thus, in most polar areas, surface waters freely mix with deeper, nutrient-rich waters. However, some density segregation of water masses occurs when summer ice melts. This lays down a thin, low-salinity layer that does not readily mix with the deeper waters. Such stratification is crucial to summer production, because it helps prevent phytoplankton from being carried into deeper, darker waters. The result is that they are concentrated instead in the sunlit surface waters. Without this density barrier, the summer bloom could not develop.

It is becoming increasingly clear that high levels of biological productivity occur only under these conditions: when periods of deep mixing, which create high nutrient levels in the sunlit surface waters, are followed by periods of density stratification.

Nutrient concentrations (phosphates and nitrates) usually are adequate in surface waters. Thus, photosynthetic productivity in the high latitudes is more commonly limited by solar energy availability than by nutrients. The productive season in these waters is relatively short but outstandingly productive.

Productivity in Tropical Oceans.

In direct contrast with high productivity during the summer season in the polar seas, low productivity is the rule in tropical regions of the open ocean. This may sound contradictory, for warm, sunny tropical waters sound very productive—but there is a key limiting factor. It is true that light penetrates much deeper into the open tropical ocean than into the temperate and polar waters,

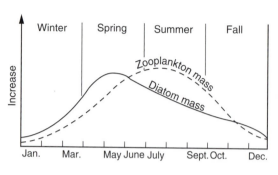

A. Barents Sea productivity

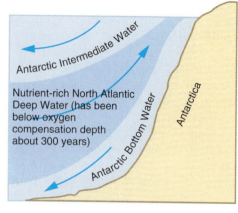

B. Antarctic upwelling

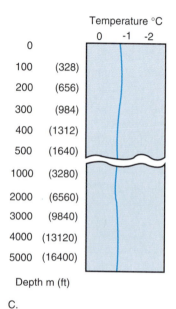

C.

D.

Figure 13-4

Productivity in Polar Oceans. *A:* Diatom mass increases rapidly in the spring when the sun rises high enough in the sky to cause deep penetration of sunlight. As soon as this diatom food supply develops, the zooplankton population begins feeding on it, and the zooplankton biomass peaks early in the summer. *B:* The continuous upwelling of North Atlantic Deep Water keeps Antarctic waters rich in nutrients. When the summer sun provides sufficient radiation, there is an explosion of biological productivity. *C:* Polar water shows nearly uniform temperature. *D:* Copepods, *Calanus,* each about 8 millimeters (0.3 inches) in length. (*D:* Photo courtesy of Scripps Institution of Oceanography, University of California, San Diego.)

and this produces a very deep oxygen compensation depth. However, in the tropical ocean a permanent thermocline produces a stratification of water masses. This prevents mixing between the surface waters and the nutrient-rich deeper waters (Figure 13–5).

At about 20° latitude north and south, phosphate and nitrate concentrations are commonly less than $\frac{1}{100}$ of the concentrations of these nutrients in temperate oceans during winter. In fact, nutrient-rich waters within the tropics lie for the most part below 150 meters (500 feet), and the highest nutrient concentration occurs

between 500 and 1000 meters (1640 and 3300 feet) depth.

Primary productivity in tropical oceans generally is at a steady, low rate. However, when we compare the total annual productivity of tropical oceans with that of the more productive temperate oceans, we find that the tropical productivity is generally at least half of that found in the temperate region on an annual basis.

Within tropical regions, three environments have unusually high productivity—regions of equatorial upwelling, coastal upwelling, and coral reefs:

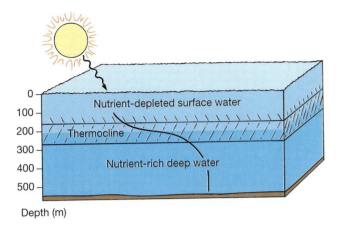

```
        0
      100   Nutrient-depleted surface water
      200   Thermocline
      300   Nutrient-rich deep water
      400
      500
      Depth (m)
```

Figure 13–5

Productivity in Tropical Oceans. In normal tropical regions, deep penetration of sunlight produces a deep oxygen compensation depth with a good supply of solar radiation available for photosynthesis. A permanent thermocline serves as an effective barrier to mixing of surface and deep water. As plants consume nutrients in the surface layer, productivity is retarded because the thermocline prevents replenishment of nutrients from deeper water.

- *Equatorial upwelling*—Where trade winds drive westerly equatorial currents on either side of the equator, surface water diverges as a result of the Ekman spiral. The surface water that moves off toward higher latitudes is replaced by nutrient-rich water that surfaces from depths of up to 200 meters (660 feet). This condition of equatorial upwelling is probably best developed in the eastern Pacific Ocean.
- *Coastal upwelling*—Where the prevailing winds blow toward the equator and along western continental margins, surface waters are driven away from the coast. They are replaced by nutrient-rich waters from depths of 200 to 900 meters (660 to 2950 feet). This nutrient-rich upwelling promotes high primary productivity in these areas, which supports large fisheries. In the Pacific, such conditions exist along the southern coast of California and the southwestern coast of Peru; in the Atlantic, they exist along the northwestern coast of Morocco and the southwestern coast of Africa.
- *Coral reefs*—The relatively high productivity of coral reef environment is not related to the upwelling process. It is discussed in Chapter 15.

Productivity in Temperate Oceans. We have discussed general productivity in the polar regions, where it is limited by available sunlight, and in the tropical low-latitude areas, where the limiting factor is nutrients. Now let us consider the temperate regions, where an alteration of these factors controls productivity in a more complex pattern. We will look at production season by season.

Winter. Productivity in temperate oceans is very low during winter, despite high nutrient concentrations in the surface layers. Ironically, nutrient concentration is highest during winter. The limiting factor is solar energy. In Figure 13–6 (winter), you can see that the sun is at its lowest elevation above the horizon during this season. A high percentage of solar energy is reflected, leaving a smaller percentage to be absorbed into surface waters. The oxygen compensation depth for basic producers such as diatoms is so shallow that it does not allow growth of the diatom population.

Spring. In Figure 13–6 (spring), as the sun rises higher in the sky, the oxygen compensation depth deepens as solar radiation is transmitted deeper into the surface water. Eventually, sufficient water volume exists above the oxygen compensation depth to permit exponential diatom growth. This places a tremendous demand on the nutrient supply in the light zone. In most Northern Hemisphere areas, decreases in the diatom population occur by May due to insufficient nutrients.

Summer. In Figure 13–6 (summer), as the sun rises higher, surface waters in temperate parts of the ocean are warmed. The water becomes separated from deeper water masses by a seasonal thermocline that may develop around 15 meters (50 feet) depth. As a result, little or no exchange of water occurs across this discontinuity. Nutrients that are depleted from surface waters cannot be replaced by those from deep waters. Throughout summer, the phytoplankton population remains at a relatively low level, but it increases again in some temperate areas during autumn.

Fall. The autumn increase is much less spectacular than that of the spring. In Figure 13–6 (fall), solar radiation diminishes as the sun drops lower in the sky. This lowers surface temperature, and the summer thermocline breaks down. A return of nutrients to the surface layer occurs as increased wind strength mixes it with the deeper water mass in which the nutrients have been trapped throughout the summer months.

This bloom is very short-lived. The phytoplankton population declines rapidly. The limiting factor in this case is the opposite of that which reduced the population of the spring phytoplankton. In the case of the spring bloom, solar radiation was readily available, and the decrease in nutrient supply was the limiting factor.

How Accurate Are Productivity Measurements? Coastal waters with high nutrient levels are highly productive, but most of the open ocean has a low nutrient level and productivity. Figure 13–7 shows this pattern based on the concentration of phytoplankton pigment.

A reminder of how sparse is our knowledge of the vast world ocean comes from new studies of photosyn-

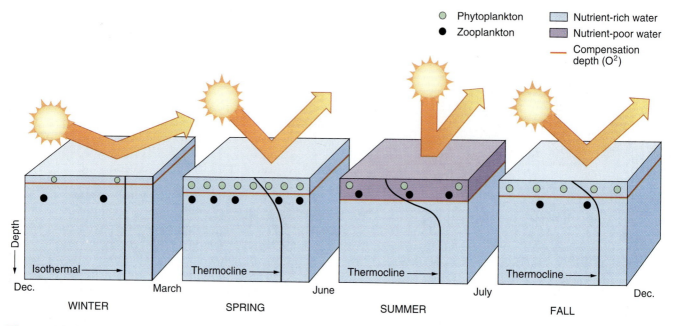

Figure 13–6

Productivity in Temperate Oceans. In winter, the sun is low in the sky. Much solar energy is reflected, and little is absorbed into the ocean. The water column is *isothermal. Nutrients* are present throughout the water column. Phytoplankton and zooplankton populations are at low levels due to lack of solar energy for photosynthesis (shallow oxygen compensation depth). In spring, the sun is higher in the sky. More solar energy is absorbed by the ocean. A *thermocline* begins to develop. A "spring bloom" of plants occurs as a result of the availability of both solar energy and *nutrients*. The oxygen compensation depth increases. In summer, the sun is high in the sky. Surface water warms and strong thermal stratification prevents mixing of surface and deep water. The *thermocline* reaches maximum development. When *nutrients* in the surface water are used up, the supply cannot be replenished from deep water. Even though the oxygen compensation depth is at its maximum, phytoplankton become scarce in late summer. The increase in animal population that followed "spring bloom" is followed by a decrease. In fall, the sun is lower in the sky. Surface water cools. Fall winds aid in mixing surface and deep water. The *thermocline* begins to disappear. A small "fall bloom" of phytoplankton, initiated by fresh nutrients from mixing, is terminated because too much solar energy is lost by reflection as the sun is lower in the sky.

thetic productivity in less productive waters of the North Atlantic and North Pacific oceans. These new data indicate that these areas may be from two to seven (or even more) times as productive as previous measurements indicated.

These are very large differences. How could so much productivity have been missed in earlier studies? Mats of diatoms have been found floating in the North Pacific gyre and the Sargasso Sea. Averaging 7 centimeters (2.8 inches) in length, these mats possess symbiotic bacteria and cyanobacteria that store nitrogen in a form usable as a nutrient by diatoms (Figure 13–8). Also, bundles of nitrogen-fixing cyanobacteria have been verified in the Sargasso Sea and in the tropical Indian and Pacific Oceans. Such aggregations are readily missed by traditional ocean survey methods.

Pacific and Atlantic studies indicate that a large percentage (60 percent in one study) of photosynthesis in poorly productive areas is achieved by tiny *picoplankton* (0.2 to 2.0 micrometers, or 0.000008 to 0.00008 inches) that slip through many filters unnoticed. Continued investigation may reveal that the overall productivity of "less productive" ocean waters has been greatly underestimated.

Chemosynthetic Productivity

Another source of potentially significant biological productivity in the oceans does not occur in the surface light zone. It is occurring in the rift valleys of the oceanic spreading centers at depths of over 2500 meters (8200 feet), where there is no light for photosynthesis.

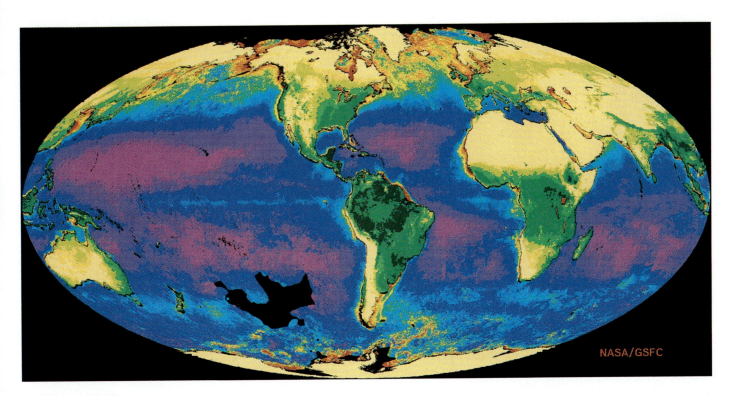

NASA/GSFC

Figure 13–7

Photosynthetic Production in the World Ocean. Data from the Coastal Zone Color Scanner (CZCS) aboard the satellite *Nimbus 7* between November 1978 and June 1981 were used to produce this false-color image. The CZCS senses changes in seawater color caused by changing concentrations of photosynthetic pigment. An increasing photosynthetic pigment concentration correlated with increasing photosynthetic productivity. It shows low values (magenta: below 0.1 milligrams per cubic meter (mg/m^3)) in the oligotrophic open ocean and high values (red: exceeding 10.0 mg/m^3) along eutrophic continental margins. Productivity of 1 mg/m^3 is represented by the boundary between yellow and green. Productivity values may be off as much as 50 percent of their magnitude. Black indicates insufficient data. (Image is courtesy of Jane A. Elrod and Gene Feldman. NASA/Goddard Space Flight Center.)

In some places, ocean water seeps down fractures into the ocean crust deep enough to become heated by underlying magma chambers. This creates hydrothermal springs that support remarkable benthic communities.

As the heated water rises to the ocean floor, it dissolves minerals from the crustal rocks that it passes through. These minerals then become deposited on the seafloor. Thus, potentially significant mineral deposits are associated with biotic communities along hydrothermal vents.

Tube worms are a significant component of these communities (Figure 13–9). These 1-meter (3-foot) worms, clams, and mussels grow to unusual size. They live in association with bacteria that produce their own food, not from sunlight, but by deriving energy from hydrogen sulfide (H_2S) gas dissolved in the hydrothermal spring water. By oxidizing this gas, the bacteria release chemical energy to carry on chemosynthesis.

Because these bacteria depend on the release of chemical energy, their food production is called *chemosynthesis*. The true significance of bacterial productivity on the deep-ocean floor cannot be fully understood until much more research is conducted. Such research has the potential to increase our estimates of the ocean's biological productivity.

Biochemists recently discovered that bacteria can obtain chemical energy for the synthesis of their food through oxidation of various compounds containing the metals iron, manganese, copper, nickel, and cobalt. These microorganisms may be important in the formation of ore-quality deposits of oxides of these metals on the ocean floor in the form of manganese nodules.

Energy Transfer

We have looked at nutrient availability. Now we will turn our attention to the cycling of important classes of nutrients and the flow of energy.

A.

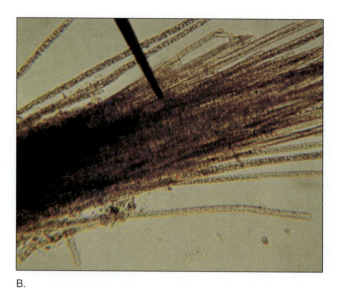

B.

Figure 13–8

Nitrogen Fixation by Aggregates of Diatoms and Cyanobacteria in Oligotrophic Ocean Waters. *A:* Typical mat of the diatom *Rhizosolenia* about 5 centimeters (2 inches) long. It is composed of intertwining chains of *R. castracanei* (wider cells) and *R. imbricata* (narrower cells). Within these cells, symbiotic bacteria fix nitrogen for uptake by the plant cells. *B:* An aggregation of the nitrogen-fixing cyanobacteria *Oscillatoria* spp. with an oxygen probe (dark object at upper left) inserted into it. The probe has a maximum diameter of about 5 micrometers. (*A:* Photo by James M. King; *B:* photo courtesy of Hans W. Paerl, University of North Carolina.)

Marine Ecosystems

The term **biotic community** refers to the assemblage of organisms that live together within some definable area. An **ecosystem** includes the biotic community plus the environment with which it interacts in the exchange of energy and chemical substances.

Within an ecosystem there are generally three basic categories of organisms—**producers, consumers,** and **decomposers.** Algae and some bacteria are the autotrophic producers, with the capacity to nourish themselves through chemosynthesis or photosynthesis. The consumers and the decomposers are heterotrophic organisms that depend on the organic compounds produced by the autotrophs for their food supply.

Animals may be divided into three categories: **herbivores,** which feed directly on plants (herbs); **carnivores,** which feed only on other animals (*carni-* = meat); and **omnivores,** which feed on both (*omni-* = all). As the role of bacteria in the marine ecosystem becomes better understood, a fourth category of animals, the **bacteriovores,** which feed on bacteria, may be identified as an important component of the marine ecosystem.

The *decomposers,* such as bacteria, break down the organic compounds of dead plants and animals and animals' excretions for their own energy requirements.

They characteristically release simple inorganic salts that are used by the algae as nutrients.

Symbiosis is a relationship in which two or more organisms are closely associated in a way that benefits at least one of the participants, and sometimes both. Such relationships are classified as commensalism, mutualism, or parasitism:

- In **commensalism,** a smaller or less dominant participant benefits without harming its host, which affords subsistence or protection to the other. An example is the *remora,* a fish that attaches to a shark or other fish to obtain food and transportation without harming its host (Figure 13–10A).
- In **mutualism,** both participants benefit. Such a relationship exists between the large reef fish and the "cleaner fish" that eat their parasites (Figure 13–10B).
- In **parasitism,** one participant, the parasite, benefits at the expense of the host. Many fish are hosts to isopods, which attach to the fish and derive their nutrition from the body fluids of the fish, thereby robbing the host of its energy supply (Figure 13–10C).

Energy Flow

Before considering the biogeochemical cycles that transfer organic and inorganic matter, we need to con-

A.

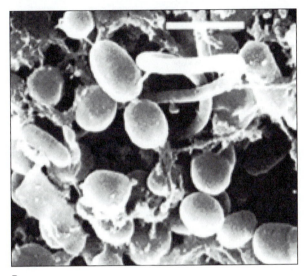

B.

Figure 13–9

Chemosynthetic Life of the Galápagos Rift. *A:* Tube worms up to 1 meter long (3.3 feet) found at Galápagos Rift and other deep-sea hydrothermal vents. These worms possess sulfur-oxidizing bacteria, which chemosynthetically produce their own food by combining inorganic nutrients dissolved in the deep-ocean water. Similar communities have been discovered near cold-water seeps and hydrocarbon seeps. They are discussed in Chapter 15. *B:* Enlarged 20,000 times, these are the sulfur-oxidizing bacteria that live symbiotically within the tissue of tube worms, clams, and mussels found at hydrothermal vents. They chemosynthetically produce food to support these organisms. (White bar at top is 1 micrometer.) (Courtesy of Woods Hole Oceanographic Institution; *A:* by Fred Grassle.)

sider the flow of energy in general. It is important to understand that energy flow is not a cycle but a unidirectional flow, from solar input, continually dissipating until gone. Most energy in ecosystems begins as solar energy and enters a biotic community through algae. From the algae, the energy is diminished continually, culminating in *entropy,* or a state where energy is so dissipated that it no longer can do work.

Figure 13–11 depicts the flow of energy through an algae-supported biotic community, from radiant energy to a state of entropy. As you can see, energy enters the system as solar energy, which is absorbed by the algae. Photosynthesis converts this energy into a chemical energy, which is used for the algae's respiration. It also is passed on to animal consumers for their growth and other life functions. Energy is expended by the animals as mechanical and heat energy, which are progressively less recoverable forms of energy. Finally, the residual energy becomes biologically useless as entropy increases.

Composition of Living Things

Having discussed the noncyclic, unidirectional flow of energy flow through the biotic community, let us now consider the flow of nutrients, which is cyclic. These are the *biogeochemical cycles,* so called because they involve *biological, geological* (Earth processes), and *chemical* elements. In these cycles, matter does not dissipate but is cycled from one chemical form to another by the various members of the community.

Biological mass is made up of chemical compounds, which are composed of Earth's elements. Because living organisms are made up of carbohydrates, fats, and proteins, let us first concern ourselves with the approximately 20 elements that make up these substances:

1. **Major elements in organic materials,** each of which make up 1 percent or more (dry weight) of organic material are, in order of abundance, carbon, oxygen, hydrogen, nitrogen, and phosphorus (primary constituents).
2. **Secondary constituents** occur in concentrations from 500 to 10,000 parts per million (dry weight). They are the elements sulfur, chlorine, potassium, sodium, calcium, magnesium, iron, and copper.
3. **Tertiary constituents** occur in concentrations of less than 500 parts per million (dry weight). They are boron, manganese, zinc, silicon, cobalt, iodine, and fluorine.

The *primary constituents* make up the bulk of organisms—the visible tissue, bone, and body fluids.

The *secondary constituents* usually are sufficiently concentrated so as not to limit productivity. However, recent observations show that some ocean production regions may be limited in iron. In Antarctic and equatorial surface waters near the Galápagos Islands, photosynthetic production is low even though the concentration of all nutrients is high, except iron. Production is high only in regions of shallow water downcurrent from islands or landmasses where iron from rocks and sediments was dissolved into the water to significant levels.

A.

B.

C.

Figure 13–10

Symbiotic Relationships. *A: Commensalism*—a remora swims to attach itself below a Caribbean jewfish in hopes of sharing the food of its host. The remora uses a sucking organ on the top of its head to attach itself to larger fish. *B: Mutualism*—juvenile bluehead wrasses clean parasites from a stoplight parrot fish, *Sparisoms viride,* on a reef at Bonaire Island in the Netherlands Antilles. *C: Parasitism*—a parasitic isopod on the head of a blackbar soldierfish. The parasite usually does not damage the fish's health much. If the fish dies, so does the parasite. (Photos *A* and *C* © Marty Snyderman; *B* © Fred Bavendam/Peter Arnold, Inc.)

The *tertiary constituents* occur in very low concentrations in organisms. They also occur in very low concentrations in the marine environment. It seems probable that the tertiary constituents might create by their absence conditions that could limit productivity.

For example, low levels of zinc have been demonstrated in laboratory tests to inhibit the uptake of carbon by relatively large species of diatoms. Further investigation may reveal that low zinc concentrations can indeed limit photosynthetic production in the oceans, much as iron does.

Returning to the major constituents, we find that carbon dioxide and water are so widely available that enough carbon, oxygen, and hydrogen is at hand to ensure that these elements would never limit productivity. However, nitrates and phosphates we need to consider in more detail because they do, under many conditions, limit marine productivity.

Biogeochemical Cycling

In biogeochemical cycles, elements follow a definite pattern in which an inorganic form is taken in by a organisms that synthesize it into organic molecules (food sugars). This food is passed through a food web that usually ends in bacterial decomposition of the organically produced compounds, back into inorganic forms that may again be used in plant production.

Three cycles are especially important: carbon, nitrogen, phosphorus. We will now look at each.

Carbon, Nitrogen, and Phosphorus Cycles

The element carbon is the basic component of all organic compounds (including carbohydrates, protein, and fats). There is no scarcity of carbon for photosynthetic productivity: only about 1 percent of the total carbon in the sea is involved in photosynthetic productivity. (The rest is in seawater as carbonate ions or is bound into calcium carbonate shells.)

Comparatively, nitrogen compounds involved in plant productivity may be 10 times the total nitrogen compound concentration that can be measured as a yearly average. This level implies that the soluble nitrogen compounds must be cycled completely up to 10 times per year. Available phosphates may be turned over up to 4 times per year.

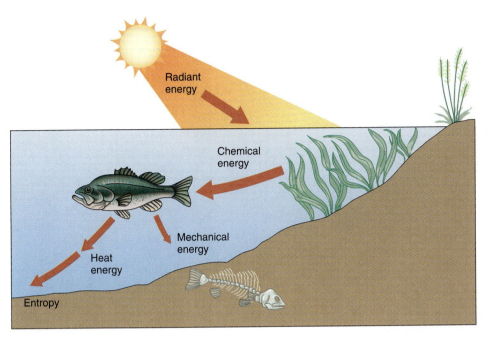

Figure 13–11

Energy Flow through a Photo-synthetic Ecosystem. Energy enters the ecosystem as radiant solar energy. It is converted to chemical energy through photosynthesis. Metabolism in the fish then releases the chemical energy for conversion to mechanical energy. Energy also is lost from the biotic community (the algae and fish) as heat, which increases the entropy of the ecosystem.

Comparing the ratio of carbon to nitrogen to phosphorus in dry weights of diatoms, we find that the proportions are 41:7:1. This ratio is also observed in the zooplankton that feed on the diatoms, and in ocean water samples taken worldwide. Thus, phytoplankton take up nutrients in the ratio in which they are available in the ocean water and pass them on to zooplankton in the same ratio. When these plankton and animals die, carbon, nitrogen, and phosphorus are restored to the water in this ratio.

The Carbon Cycle. Figure 13-12A shows the carbon cycle. This cycle involves the uptake of carbon dioxide by algae and plants for their use in the photosynthetic process. Carbon dioxide is returned to the ocean water primarily through respiration of algae, animals, and bacteria and secondarily by breakdown of dissolved organic materials.

The oceans are believed to remove about 25 to 50 billion metric tons of carbon from the atmosphere each year. We refer to the "biological pump" that is believed to photosynthesize this atmospheric CO_2 into organic matter and then transport it to ocean sediments, where it may be held for millions of years. But the magnitude of this "biological pump" is poorly understood. With the threat of an increased greenhouse effect resulting in part from increasing CO_2 concentration in the atmosphere, the marine carbon cycle has become an important research focus.

The Nitrogen Cycle. Figure 13-12B shows the nitrogen cycle. Nitrogen is essential in producing amino acids, the building blocks of proteins that are synthesized by algae. These photosynthetic products are consumed by animals and free-living bacteria. They are then passed on to the decomposing bacteria, along with dead algae tissue, dead animal tissue and excrement.

The decomposing bacteria gain energy from breaking down these compounds. This breakdown liberates inorganic compounds, such as nitrates, that are the basic nutrients used by algae.

Different bacteria involved in the nitrogen cycle make it somewhat complicated. Although some bacteria consume dissolved organic matter or convert organic compounds into inorganic substances, others can bind nitrogen into useful nutrients. These are called **nitrogen-fixing bacteria.** Another special group is **denitrifying bacteria,** whose metabolism depends upon the breakdown of nitrates and the liberation of molecular nitrogen.

Studies of nutrient cycles clearly indicate that nitrogen availability is a limiting factor in productivity during summer. One reason is that the process of converting organic substances to useful nutrients through bacterial action may require up to 3 months. The conversion begins in the lower portion of the photosynthetic zone as particulate matter sinks toward the ocean bottom.

By the time this conversion is completed, the nitrogen compounds usually are below the light zone and thus are unavailable for photosynthesis. They cannot readily be returned to the light zone during summer due to strong thermostratification throughout much of the ocean surface. For nitrates to again become available to phytoplankton, the thermostratification must disappear, allowing upwelling and mixing during winter.

The Phosphorus Cycle. Figure 13-12C shows the phosphorus cycle. The phosphorus cycle is simpler than the nitrogen cycle. It is simpler primarily because the bacterial action involved in breaking down organic

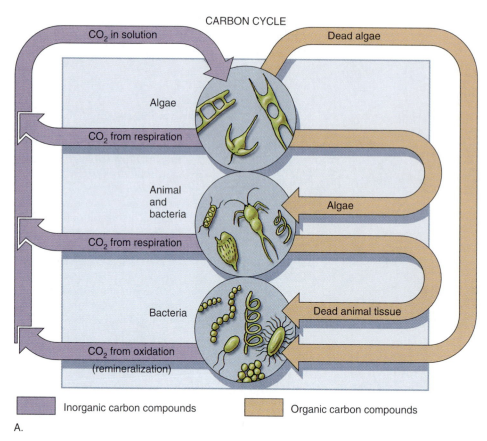

CARBON CYCLE

CO$_2$ in solution

Dead algae

Algae

CO$_2$ from respiration

Animal and bacteria

Algae

CO$_2$ from respiration

Bacteria

Dead animal tissue

CO$_2$ from oxidation (remineralization)

Inorganic carbon compounds

Organic carbon compounds

A.

Figure 13–12

Biogeochemical Cycling.
Overview: The chemical components of organic matter enter the biological system through photosynthesis (or through bacteria chemosynthesis at hydrothermal vents). These chemical components are passed on to animal populations through feeding. When algae and animals die, their organic remains are converted to inorganic form by bacterial or other decomposition processes. In this form, they are again available for uptake by plants and algae. *A: Carbon cycle.* There is a large supply of inorganic carbon (purple) in the oceans. Only about 1 percent of the carbon dissolved in the ocean is involved in photosynthetic productivity (orange). Algae store carbon dioxide through photosynthesis. They pass this carbon on to animals that eat them. Bacteria decompose dead algae and animals to release carbon dioxide. All release carbon dioxide through respiration.

phosphorus compounds is simpler. This difference can be studied by comparing Figures 13–12B and 13–12C.

The rate at which the organic phosphate compounds can be decomposed into useful nutrients is much faster than for nitrogen. Consequently, much of it can be completed above the oxygen compensation depth. Thus, phosphorus is made available for photosynthesis within the photosynthetic zone. Lack of phosphorus is rarely a limiting factor of algae productivity.

The Silicon Cycle. A fourth cycle, for silicon, is also important. Silicon is not a nutrient, but as you learned, the shells of diatoms are composed of silica (SiO$_2$). Availability of silica can be a limiting factor in the productivity of diatoms, but it is rarely a limiting factor of total primary productivity, because not all phytoplankton require silica as a protective covering. Silicon probably never will be a limiting nutrient in total productivity because it is so abundant.

Silicon concentrations range from unmeasurable up to 400 milligrams per cubic meter. Fluctuations in concentration roughly coincide with those observed in nitrogen and phosphorus. However, the fluctuation displays a much greater amplitude than those for nitrogen and phosphorus. This condition is probably due to the fact that silica does not undergo bacterial decay and is taken directly into solution.

A New Role for Bacteria. It had been widely accepted that zooplankton are the primary grazers on phytoplankton. Now, new evidence indicates that free-living bacteria may consume up to 50 percent of phytoplankton production. These bacteria are thought to consume the dissolved organic matter that is lost from the conventional food web by three processes:

1. Phytoplankton *exudate.* As phytoplankton age they lose some of their material directly into the ocean.
2. Phytoplankton *munchates.* As phytoplankton are eaten by zooplankton, some material is "spilled" into the ocean.
3. Zooplankton *excretions.* The liquid excretions of zooplankton become dissolved into ocean water.

Free-living bacteria absorb this dissolved organic matter and reenter the conventional food web, primarily when they are eaten by microscopic flagellates. Bacteria may have more varied roles in the biogeochemical cycling of matter in the oceans than had previously been known.

Trophic Levels and Biomass Pyramids

As producers make food (organic matter) available to the consuming animals of the ocean, it passes from one

Figure 13–12 *continued.*

B: Nitrogen cycle. The total nitrogen fixed into organic molecules (orange) at any given time may be 10 times as great as the yearly average of soluble nitrogen compounds dissolved in ocean water. Therefore, each nitrogen atom must be recycled biogeochemically about 10 times per year. Also, the decomposition of organic nitrogen compounds back into the preferred inorganic form of nitrogen, nitrate (NO_3), requires three steps of bacterial decomposition. Nitrogen is considered the nutrient most likely to limit biological productivity as a result of its depletion.

C: Phosphorus cycle. Each phosphorus atom may need to be recycled up to 4 times per year to maintain biological productivity in the oceans. Yet phosphorus is seldom depleted to where it limits biological productivity. Organic phosphorus is quickly returned as a usable nutrient to algae through breakdown and the single step of bacterial decomposition.

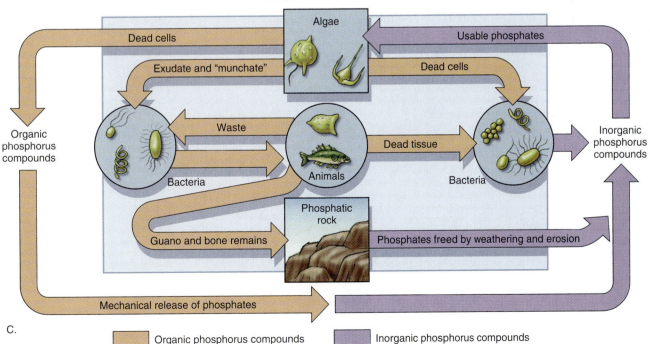

NITROGEN CYCLE

Molecular nitrogen

Animals

Free-living bacteria

Denitrifying bacteria

Dead tissue

Waste

Exudate and "munchate"

Algae

Nitrogen fixing bacteria

Dead cells

Bacteria

NH_3 Ammonium nitrogen

NO_2 Nitrate

Nitrous oxides

Bacteria

NO_2 Nitrate

Inorganic nitrogen compounds

Organic nitrogen compounds

B.

PHOSPHORUS CYCLE

Algae

Dead cells

Usable phosphates

Exudate and "munchate"

Dead cells

Organic phosphorus compounds

Waste

Animals

Dead tissue

Bacteria

Inorganic phosphorus compounds

Bacteria

Phosphatic rock

Guano and bone remains

Phosphates freed by weathering and erosion

Mechanical release of phosphates

C.

Organic phosphorus compounds

Inorganic phosphorus compounds

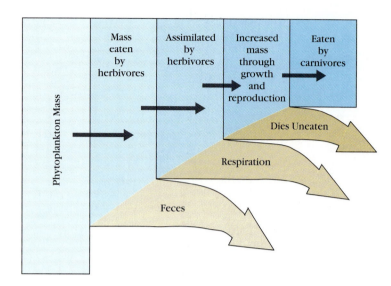

Figure 13–13

Passage of Energy through a Trophic Level. The mass of phytoplankton taken in by an herbivore is processed by the herbivore, with some lost as feces. That which is assimilated is used for running the herbivore's body (respiration, supplying energy for swimming, digestion, and so on). The remaining energy is added to the mass of the herbivore as increased body weight or goes toward producing young. This increased mass is available for a carnivore to eat, but not all herbivores are consumed, and some die uneaten. Thus, a small percentage (about 10 percent) of the food mass consumed by the herbivore is consumed by the carnivore that feeds upon it.

feeding population to the next. Only a small percentage of the energy taken in at any level is passed on to the next because of energy consumption and loss at each level. As a result, the mass of producers in the ocean is many times greater than the mass of the top consumers, such as sharks or killer whales. You may find it difficult to believe that the total mass of large animals such as killer whales is much less than the total mass of tiny diatoms, but the following discussion of trophic levels should make it clear why this must be.

Trophic Levels

Chemical energy that is stored in the mass of the ocean's algae is transferred to the animal community through feeding, in part. Zooplankton eat diatoms and other microscopic marine algae, and in doing so, are *herbivores,* like cows. Larger animals feed on the larger macroscopic algae and "grasses" that grow attached to the ocean bottom near shore.

In turn, the herbivores are eaten by larger animals, the *carnivores.* They in turn are eaten by another population of larger carnivores, and so on. Each of these feeding levels is called a **trophic level.**

Generally, individual members of a feeding population are larger—but not too much larger—than the individuals in the population they eat. But there are outstanding exceptions to this condition. The blue whale, possibly the largest animal known to have existed on Earth, feeds upon krill, which attain maximum lengths of only 6 centimeters (2.4 inches).

Remember that the transfer of energy from one population to another represents a *continuous flow* of energy. This flow is interrupted by small-scale recycling and storage of this energy, which slows the process of converting potential (chemical) energy to kinetic energy, then to heat energy, and finally to entropy. However, despite this cycling of energy, all energy that enters the organic community ultimately is lost to entropy.

Transfer Efficiency

In the transfer of energy between trophic levels, we must be greatly concerned with efficiency. For example, under various laboratory conditions, researchers observe considerable variability in the efficiency of different algae species. As an average, we consider the percentage of light energy absorbed by algae and ultimately synthesized into food made available to herbivores to be only about 2 percent.

The **gross ecological efficiency** at any trophic level is the ratio of energy passed on to the next higher trophic level divided by the energy received from the trophic level below. The ecological efficiency of herbivorous anchovies would be, for example, the energy consumed by carnivorous tuna that feed on the anchovies divided by the energy represented by the phytoplankton that the anchovies consumed.

Figure 13–13 shows that some of the chemical energy taken in as food by herbivores is passed by the animal as feces and the rest is assimilated by the animal. Of the assimilated chemical energy, much is quickly converted through respiration to kinetic energy for maintaining life, and what remains is available for growth and reproduction.

Only a portion of this mass is passed on to the next trophic level through feeding. Figure 13–14 shows the passage of energy between trophic levels through an entire ecosystem, from the solar energy assimilated by phytoplankton to the mass of the ultimate carnivore, which is us. (Think of this diagram the next time you enjoy a fish dinner.)

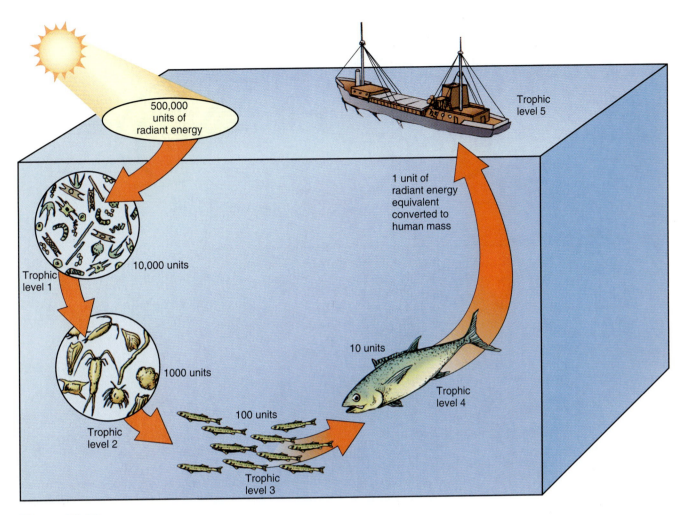

Figure 13–14

Ecosystem Energy Flow—Efficiency. This diagram shows probable effects of energy transfer within an ecosystem. One unit of mass is added to the fifth trophic level (humans) for every 500,000 units of radiant energy input available to the producers (phytoplankton). This value is based on a 2 percent efficiency of transfer by phytoplankton and 10 percent efficiency at all other levels.

The efficiency of energy transfer between trophic levels has been intensively studied. Many variables must be considered. For example, young animals display a higher growth efficiency than older animals. Also, when food is plentiful, animals expend more energy in digestion and assimilation than when food is not readily available.

Most efficiencies range between 6 percent and 15 percent. It is well accepted that ecological efficiencies in natural ecosystems average approximately 10 percent. There is some evidence that, in populations important to our present fisheries, this efficiency may run as high as 20 percent. The true value of this efficiency is of practical importance to us because it determines the size of the fish harvest we can anticipate from the oceans.

Biomass Pyramid

Considering the energy losses that occur in each feeding population, it is obvious that some limit must exist to the number of feeding populations in a food chain. It also is evident that each feeding population necessarily must have less mass than the population it eats. We know that individual members of a feeding population generally are larger than their prey, but it seems that they also should be less numerous.

Food Chains. A **food chain** is a sequence of organisms through which energy is transferred, starting with the primary producer, through the herbivore, and through successive carnivores, up to the "top carnivore" that does not have any predators. In nature, it is

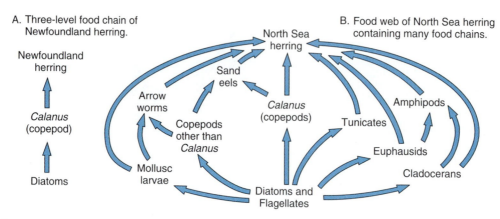

Figure 13-15

A Food Chain and a Food Web. *A:* A food chain is the passage of energy along a single path, such as from diatoms to copepods to Newfoundland herring. *B:* The food sources of the North Sea herring involve multiple paths—a food web—where this fish may be at the third or fourth trophic level. Because the Newfoundland herring is always at the third trophic level, it has a potentially larger biomass available than the North Sea herring. However, the Newfoundland herring also has all of its eggs in one basket: If its food chain becomes disrupted, it would be in serious trouble, with no alternative food sources. In contrast, if one chain of the North Sea herring's food web were disrupted, it would not pose as great a threat because of all the other food chains in the web.

rare to find food chains comprising more than five trophic levels.

Because of the inefficiency of energy transfer between trophic levels, it is beneficial for fishermen to choose a population that feeds as close to the primary producing population as possible. This increases the biomass available for food and the number of individuals available to be taken by the fishery.

An example of an animal population that is an important fishery, and which usually represents the third trophic level in a food chain, is herring that live off the coast of Newfoundland. Although some herring populations are involved in longer food chains, the Newfoundland herring feed primarily on a population of small crustaceans (copepods) that in turn feed upon diatoms (Figure 13–15A).

Food Webs. However, it is uncommon to see feeding relationships as simple as that of the Newfoundland herring. More commonly, top carnivores in a food chain feed on a number of animals, each of which has its own simple or complex feeding relationships. This constitutes a **food web** (Figure 13–15B).

The overall importance of food webs is not well understood, but one consequence for animals that feed through a web rather than a linear chain is their greater likelihood of survival, because they have alternative foods to eat should others fail. Those animals involved in food webs, such as the North Sea herring in Figure 13–15B, are less likely to starve should one of their food sources diminish in quantity or even become extinct.

The Newfoundland herring eat only copepods, so extinction of copepods would have a catastrophic effect on the herring population.

The Newfoundland herring population does, however, have an advantage over its relatives who feed through the broader-based food web. The Newfoundland herring is more likely to have a larger biomass to eat, because the herring are only two steps removed from the producers, whereas the North Sea herring is at the fourth level in some of the food chains within its web.

The ultimate effect of energy transfer between trophic levels can be seen in Figure 13–16. It depicts the *progressive decrease in numbers of individuals and total biomass* at successive trophic levels resulting from decreased amounts of available energy.

A study of food webs in 55 marine ecosystems indicated an average of 4.45 trophic levels:

- Only two trophic levels were encountered at rocky shores in New England and Washington, and at a Georgia salt marsh.
- Seven links were found in the longest food chains in Antarctic seas.
- Two tropical plankton communities in the Pacific Ocean were observed to have seven and ten trophic levels.
- Tropical epipelagic seas and Pacific Ocean upwelling zones had eight trophic levels.

The study found that the amount of primary productivity and the variability of the environment have little influence on the length of food chains within an ecosys-

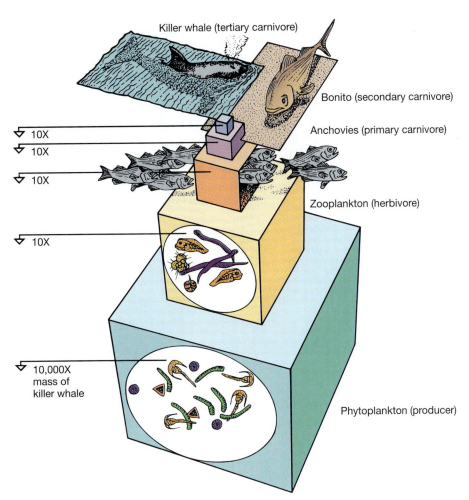

Killer whale (tertiary carnivore)

Bonito (secondary carnivore)

Anchovies (primary carnivore)

10X
10X
10X

Zooplankton (herbivore)

10X

10,000X
mass of
killer whale

Phytoplankton (producer)

Figure 13–16

Biomass Pyramid. The higher on
the food chain an organism lives, the
physically larger it is as an individual.
However, the total biomass represented by a population high on the
food chain is less than that for a population at a lower level.

tem. The only variable that showed a clear relation to
food chain length was whether or not the environment
was two- or three-dimensional. Three-dimensional (pelagic) marine ecosystems averaged 5.6 trophic levels,
compared with an average of 3.5 for two-dimensional
(benthic) ecosystems.

S U M M A R Y

Organisms are divided into five kingdoms: Monera, single-
celled organisms without a nucleus; Protista, single-celled
organisms with a nucleus; Mycota, fungi; Metaphyta,
many-celled plants; and Matazoa, many-celled animals.
Classification of organisms involves dividing the kingdoms
into the increasingly specific groupings: phylum, class, order, family, genus, and species.

Protistan algae include the macroscopic algae kelp
and *Sargassum* (Phaetophyta), green algae (Chlorophyta),
and red algae (Rhodophyta). The microscopic algae include
diatoms and coccolithophores (Chrysophyta) and dinoflagellates (Pyrrophyta). The more complex Spermatophyta
are represented by a few genera of nearshore plants such
as eelgrass (*Zostera*), surf grass (*Phyllospadix*), marsh grass
(*Spartina*), and mangrove trees.

Photosynthetic productivity of the oceans is limited
by the availability of solar radiation and of nutrients. The
depth to which sufficient light penetrates to allow photosynthesis to produce only that amount of oxygen required
for their respiration is the oxygen compensation depth. Algae cannot live successfully below this depth, which may
occur at less than 20 meters (65 feet) in turbid coastal waters or at a probable maximum of 150 meters (500 feet) in
the open ocean.

Nutrients are most abundant in coastal areas due to
runoff and upwelling. In high-latitude areas thermoclines
are generally absent, so upwelling can readily occur, and
productivity is commonly limited more by the availability
of solar radiation than by lack of nutrients. In low-latitude
regions, where a strong thermocline may exist year-round,
productivity is limited except in areas of upwelling.

The lack of nutrients is generally the limiting factor.
In temperate regions, where distinct seasonal patterns are
developed, productivity peaks in the spring and fall and is

limited by lack of solar radiation in the winter and lack of nutrients in the summer.

In addition to the primary photosynthetic productivity, organic biomass is produced through bacterial chemosynthesis. Chemosynthesis, observed on oceanic spreading centers in association with hydrothermal springs, is based on the release of chemical energy by the oxidation of hydrogen sulfide.

Marine ecosystems are composed of populations of organisms called producers, consumers, and decomposers. Some of these organisms live closely together in symbiotic relationships where one organism benefits while the other is unaffected (commensalism), both benefit (mutualism), or one benefits at the expense of the other (parasitism).

Radiant energy captured by algae is converted to chemical energy and passed through the biotic community. It is expended as mechanical and heat energy and ultimately reaches a state of entropy, where it is biologically useless. There is, however, no loss of mass. The mass used as nutrients by algae is converted to biomass. Upon the death of organisms, the mass is decomposed to an inorganic form ready again for use as nutrients for algae.

Of the nutrients required by algae, compounds of nitrogen are most likely to be depleted and restrict productivity. Because the total decomposition of organic nitrogen compounds to inorganic nutrients requires three stages of bacterial decomposition, these compounds may have sunk beneath the photosynthetic zone before decomposition was complete and therefore are unavailable for photosynthesis.

As energy is transferred from algae to herbivore and the various carnivore feeding levels, only about 10 percent of the mass taken in at one feeding level is passed on to the next. The ultimate effect of this decreased amount of energy that is passed between trophic levels higher in the food chain is a decrease in the number of individuals and total biomass of populations higher in the food chain.

KEY TERMS

Bacteriovore (p. 298)
Biotic community (p. 298)
Carnivore (p. 298)
Carotin (p. 289)
Coccoliths (p. 289)
Commensalism (p. 298)
Consumer (p. 298)
Decomposer (p. 298)
Denitrifying bacteria (p. 301)
Dinoflagellates (p. 289)
Ecosystem (p. 298)
Flagella (p. 289)
Food chain (p. 305)

Food web (p. 306)
Gran method (p. 292)
Gross ecological efficiency (p. 304)
Gross primary production (p. 292)
Herbivore (p. 298)
Metaphyta (p. 288)
Metazoa (p. 288)
Monera (p. 287)
Mutualism (p. 298)
Mycota (p. 287)
Net primary production (p. 292)
New production (p. 292)

Nitrogen-fixing bacteria (p. 301)
Omnivore (p. 298)
Oxygen compensation depth (p. 292)
Parasitism (p. 298)
Primary productivity (p. 290)
Producer (p. 298)
Protista (p. 287)
Protozoa (p. 287)
Red tide (p. 289)
Regenerated production (p. 292)
Symbiosis (p. 298)
Trophic level (p. 304)

QUESTIONS AND EXERCISES

1. List the five kingdoms of organisms and the fundamental criteria used in assigning organisms to them.
2. Compare the macroscopic algae in terms of color, maximum depth in which they grow, common species, and size.
3. The golden algae contains two classes of important phytoplankton. Compare their composition and the structure of their hard parts and explain their geologic significance.
4. Discuss and compare the contributions of the Pyrrophyta genera *Ptychodiscus* and *Gonyaulax* to red tide development.
5. Define oxygen compensation depth, and explain the use of the dark and transparent bottle technique for its determination. Discuss how the quantity of oxygen produced by photosynthesis in each clear bottle (gross photosynthesis) is determined.
6. Compare the biological productivity of polar, temperate, and tropical regions of the oceans. Consider seasonal variables, thermal stratification of the water column, and the availability of nutrients and solar radiation.
7. Discuss chemosynthesis as a method of primary productivity. How does it differ from photosynthesis?
8. Describe the components of the marine ecosystem.
9. Describe the flow of energy through the biotic community; include the forms to which solar radiation is converted. How does this flow differ from the manner in which mass is moved through the ecosystem?
10. What are the proportions by weight of carbon, nitrogen, and phosphorus in ocean water, phytoplankton, and zooplankton? Suggest how these amounts may support or refute the idea that life originated in the oceans.
11. Explain why nitrogen is much more likely than phosphorus to be a limiting factor in marine productivity.
12. How is the energy taken in by a feeding population lost so that only a small percentage is made available to the next feeding level? What is the average efficiency of energy transfer between trophic levels?

13. If a killer whale is a third-level carnivore, how much phytoplankton mass is required to add each gram of new mass to the whale? (Assume 10 percent efficiency of energy transfer between trophic levels.) Include a diagram.

14. Describe the probable advantage to the ultimate carnivore of the food web over a single food chain as a feeding strategy.

REFERENCES

Ducklow, H. W. 1983. Production and fate of bacteria in the oceans. *Bioscience* 33:8, 494–501.

Carpenter, E. J., and Romans, K. 1991. Major role of the cyanobacterium, *Trichodesmium,* in nutrient cycling in the North Atlantic Ocean. *Science* 254:5036, 1356–1358.

George D., and George, J. 1979. *Marine life: An illustrated encyclopedia of invertebrates in the sea.* New York: Wiley-Interscience.

Grassle, J. F., et al. 1979. Galápagos '79: Initial findings of a deep-sea biological quest. *Oceanus* 22:2, 2–10.

Jenkins, W. J. 1982. Oxygen utilization rates in North Atlantic subtropical gyre and primary production in oligotrophic systems. *Nature* 300, 246–248.

Little, M. M.; Little, D. S.; Blair, S. M.; and Norris, J. N. 1985. Deepest known plant life discovered on an uncharted seamount. *Science* 227:4683, 57–59.

Martinez, L.; Silver, M.; King, J.; and Alldredge, A. 1983. Nitrogen fixation by floating diatom mats: A source of new nitrogen to oligotrophic ocean waters. *Science* 221:4066, 152–154.

More, F. M. M.; Reinfelder, J. R.; Roberts, S. B.; Chamberlain, C. P.; Lee, J. G.; and Yee, D. 1994. Zinc and carbon co-limitation of marine phytoplankton. *Nature* 369:6483, 740–742.

Paerl, H. W., and Bebout, B. M. 1988. Direct measurement of O_2-depleted microzones in marine *Oscillatoria*: Relation to N_2 fixation. *Science* 241:4864, 442–445.

Parsons, T. R.; Takahashi, M.; and Hargrave, B. 1974. *Biological oceanographic processes,* 3rd ed. New York: Pergamon Press.

Pimm, S. L.; Lawton, J. H.; and Cohen, J. E. 1991. Food web patterns and their consequences. *Nature* 350:6320, 669–674.

Platt, T., and Sathyendranath, S. 1988. Oceanic primary production: Estimation by remote sensing at local and regional scales. *Science* 241:4873, 1613–1619.

Platt, T.; Subba Rao, D. V.; and Irwin, B. 1983. Photosynthesis of picoplankton in the oligotrophic ocean. *Nature* 310:5902, 702–704.

Russell-Hunter, W. D. 1970. *Aquatic productivity.* New York: Macmillan.

Sathyendranath, S.; Platt, T.; Horne, E. P. W.; Harrison, W. G.; Ulloa, O.; Outerbridge, R.; and Hoepffner, N. 1991. Estimation of new production in the ocean by compound remote sensing. *Nature* 353:6340, 129–133.

Sherr, B. F.; Sherr, E. B.; and Hopkinson, C. S. 1988. Trophic interactions within pelagic microbial communities: Indications of feedback regulation of carbon flow. *Hydrobiologia* 159:1, 19–26.

Shulenberger, E., and Reid, J. L. 1981. The Pacific shallow oxygen maximum, deep chlorophyll maximum, and primary productivity reconsidered. *Deep Sea Research* 28A:9, 901–919.

Sullivan, C. W.; Arrigo, K. R.; McClain, C. R.; Comiso, J. C.; and Firestone, J. 1993. Distributions of phytoplankton blooms in the Southern Ocean. *Science* 262:5141, 1832–1836.

SUGGESTED READING

Sea Frontiers

Arehart, J. L. 1972. Diatoms and silicon. 18:2, 89–94. A very readable description of the important role of silicon and other elements in the ecology of diatoms, including microphotographs showing the varied forms of diatoms.

Coleman, B. A.; Doetsch, R. N.; and Sjblad, R. D. 1986. Red tide: A recurrent marine phenomenon. 32:3, 184–191. The problem of periodic red tides along the Florida, New England, and California coasts is discussed.

Cox, V. 1994. Its no snow job. 40:2. The search for marine snow conducted by biologist Alice Alldredge is the focus of this story. This fall of biological debris to the ocean floor plays an important role in helping remove carbon dioxide from the atmosphere and depositing in deep sea sediments.

Idyll, C. P. 1971. The harvest of plankton. 17:5, 258–267. An interesting discussion of the potential of zooplankton as a major fishery.

Jensen, A. C. 1973. Warning—red tide. 19:3, 164–175. An informative discussion of what is known of the cause, nature, and effect of red tides.

Johnson, S. 1981. Crustacean symbiosis. 27:6, 351–360. A description of various symbiotic relationships entered into by tropical shrimps and crabs.

McFadden, G. 1987. Not-so-naked ancestors. 33:1, 46–51. The nature of the coverings of marine phytoplankton cells is revealed by the electron microscope.

Oremland, R. S. 1976. Microorganisms and marine ecology. 22:5, 305–310. The role of such microorganisms as phytoplankton and bacteria in cycling matter in the oceans is discussed.

Philips, E. 1982. Biological sources of energy from the sea. 28:1, 36–46. The potential for converting marine biomass to energy sources useful to society is discussed.

Scientific American

Anderson, D. M. 1994. Red tides. Dense blooms of algae are becoming more frequent in coastal waters. They are also involving species of algae and animals not previously known to be associated with red tide occurrences.

Benson, A. A. 1975. Role of wax in oceanic food chains. 232:3, 76–89. A report on the findings from observations made of the content of wax in the bodies' of many marine animals from copepods to small deep-water fishes and their implications.

Childress, J. J.; Feldback, H.; and Somero, G. N. 1987. Symbiosis in the deep sea. 256:5, 114–121. Deep-sea hydrothermal vent animals have a symbiotic relationship with sulfur-oxidizing bacteria that allows them to live in the darkness of the deep ocean.

Govindjee, and Coleman, W. J. 1990. How plants make oxygen. 262:2, 50–67. The process of oxygen production by plants is explained.

Levine, R. P. 1969. The mechanism of photosynthesis. 221:6, 58–71. Reveals what is known of the process by which energy is captured by plants and converted to useful forms of chemical energy while freeing oxygen to the atmosphere.

Pettit, J.; Drucker, S.; and Knox, B. 1981. Submarine pollination. 244:3, 134–144. Discusses the pollination of sea grasses by wave action.

C H A P T E R   1 4
Animals of the Pelagic Environment

In the preceding chapter, you learned about the phytoplankton that represent well over half of the marine biomass. The next-largest population is the zooplankton, or floating animals. The larger swimming animals, nekton, represent a relatively small percentage of marine biomass. The benthos, who live in the bottom, will be covered in the next chapter.

Phytoplankton depend primarily on their small size to provide a high degree of frictional resistance to sinking below the sunlit surface waters. The large animals possess bodies that are usually more dense than ocean water, have less surface area per unit of body mass, and therefore tend to sink more rapidly into the ocean. To remain in surface waters where the food supply is greatest, they must increase their buoyancy or swim actively.

Various animals apply one or both of these strategies to remaining in the upper pelagic environment. The ones we will discuss range from ctenophores (1 centimeter, or 0.4 inches) to the blue whales (30 meters, or 100 feet). Most ani-mals obtain their oxygen from seawater and are "cold blooded," meaning that their body temperature equals that of the water in which they live, but some fish, and all mammals, are "warm blooded." Of course, the mam-mals must obtain their oxygen by breathing air. This re-sults in a wonderful variety of adaptations and lifestyles that make it possible for these animals to live success-fully in the oceans.

In addition to physical adaptations that make it possible for individual animals to function, other adap-tations involve all members of the species, such as mi-grating and schooling behavior. These may be related to successful feeding and breeding. We will focus on these adaptations in the following pages.

Staying Above the Ocean Floor

Some animals depend on increased buoyancy to main-tain themselves in near-surface waters. Their buoyancy results either from gas containers they possess, which significantly reduce their average density, or from soft bodies void of dense hard parts. Larger animals with bodies denser than seawater must exert more energy to propel themselves through the water as swimmers. In this section, we will elaborate on these strategies.

Gas Containers

Air has approximately 0.001 the density of water at sea level, so even just a small amount inside an organism increases its buoyancy. Some cephalopods have rigid gas containers in their bodies. The genus *Nautilus* has an external shell, whereas the cuttlefish *Sepia* and deep-water squid *Spirula* have an internal chambered struc-ture (Figure 14–1). These animals become *neutrally buoyant,* meaning that the amount of air in their bodies regulates their density so they can maintain a vertical position at a depth of their choosing.

Because the air pressure in their air chambers is always 1 atmosphere, these creatures are limited in the depth to which they may venture. The *Nautilus* must stay above a depth of approximately 500 meters (1640 feet) to prevent collapse of its chambered shell as the external pressure approaches 50 atmospheres. The *Nautilus* is rarely observed below a depth of about 250 meters (800 feet).

Neutral buoyancy is achieved by some slow-moving bony fish through filling a gas bladder, or **swim bladder,** with gases (Figure 14–2). The swim bladder is normally not present in very active swimmers, such as the tuna, or in fish that live on the bottom, because nei-ther need it. Some fish have a **pneumatic duct** that connects the swim bladder to the esophagus (Figure 14–2). These fish can add or remove air through this

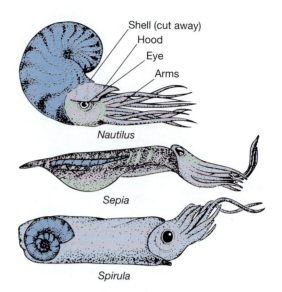

Figure 14-1

Gas Containers in Cephalopods. The *Nautilus* has an ex-ternal chambered shell, and *Sepia* and *Spirula* have rigid in-ternal chambered structures that can be filled with gas to provide buoyancy.

duct. In other fish, the gases of the swim bladder must be added or removed more slowly by an interchange with the blood.

Because change in depth will cause the gas in the swim bladder to expand or contract, the fish removes or adds gas to the bladder to maintain a constant volume. Those fish without the pneumatic duct are limited in the rate at which they can make these adjustments and, therefore, cannot withstand rapid changes in depth.

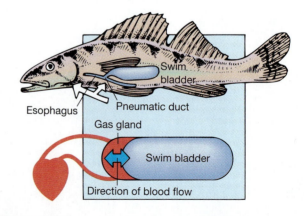

Figure 14-2

Swim Bladder in Some Bony Fishes. Example of a swim bladder connected to the esophagus by the pneumatic duct, allowing air to be added or removed rapidly. In fish with no pneumatic duct, all gas must be added or removed through the blood. This requires more time and is achieved by a net-work of capillaries associated with the gas gland.

The composition of gases in the swim bladders of shallow-water fishes is similar to that of the atmosphere. With increasing depth (fish with swim bladders have been captured from a depth of 7000 meters, or 23,000 feet) the concentration of oxygen in the swim bladder gas increases from the 20 percent common near the surface to more than 90 percent. At the 700-atmosphere pressure that exists at a depth of 7000 meters, gas is compressed to a density of 0.7 grams per cubic centimeter. This is about the density of fat, so many deep-water fish have "air bladders" that are filled with fat instead of compressed gas.

Floating Forms

Floating at the surface are some relatively large plankton, such as the familiar jellyfish. They all have soft, gelatinous bodies with little if any hard tissue.

Coelenterates.   **Coelenterates** are characterized by soft bodies that are more than 95 percent water.

There are two basic types, the siphonophores (Portuguese man-of-war) and the scyphozoans (jellyfish).

Siphonophore coelenterates are represented in all oceans by the Portuguese man-of-war (genus *Physalia*) and "by-the-wind sailor" (genus *Velella*). Their gas floats, called pneumatophores, serve as floats and sails that allow the wind to push these creatures across the ocean surface (Figure 14–3). A colony of tiny individuals is suspended beneath the float.

Portuguese man-of-war tentacles may be many meters long and possess **nematocysts** long enough to penetrate the skin of humans; they have been known to inflict a painful and occasionally dangerous neurotoxin poisoning. The colonies grow from the initial polyp through sexual budding.

Jellyfish (**scyphozoan** coelenterates) are individuals that have a bell-shaped body with a fringe of tentacles and a mouth at the end of a clapperlike extension hanging beneath the bell-shaped float. Ranging in size from nearly microscopic to 2 meters (6.6 feet) in diameter, with long tentacles (60 meters, or 200 feet), most jelly-

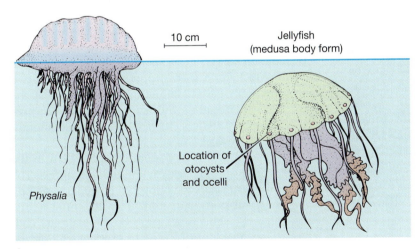

10 cm

Jellyfish (medusa body form)

Location of otocysts and ocelli

Physalia

A.

B.

Figure 14–3

Planktonic Cnidarians. *A: Physalia* and jellyfish. *B:* Medusa. (*B:* Photo by Larry Ford.)

fish have bells with a diameter of less than 0.5 meters (1.6 feet).

Jellyfish move by muscular contraction (see Figure 14–3A). Water enters the cavity under the bell and is forced out by contractions of muscles that circle the bell, jetting the animals ahead in short spurts. To allow the animal to swim generally in an upward direction, sensory organs are spaced around the outer edge of the bell. These may be light sensitive or gravity sensitive. This orientation ability is important because the jellyfish feed by swimming to the surface and sinking slowly through the life-rich surface waters.

Tunicates. Animals called **tunicates** are generally transparent and barrel shaped. They have openings on each end for current to flow in or out (Figure 14–4A). They move by a feeble form of jet propulsion, created by contraction of bands of muscles that force water into the incurrent opening and out of the excurrent opening.

Reproduction of solitary tunicates is complicated, involving alternate sexual and asexual reproduction. Salps (genus *Salpa*) includes solitary forms reaching a length of 20 centimeters (8 inches) and smaller aggregate forms that produce new individuals by budding. Individual members of an aggregate chain may be 7 centimeters (2.8 inches) long, and the chain of newly budded members may reach great lengths (see Figure 14–4B).

The genus *Pyrosoma* is luminescent and colonial. Individual members have their incurrent openings facing the outside surface of a tube-shaped colony that may be a few meters long. One end of the tube is closed, and the excurrent openings of the thousands of individuals all empty into the tube. Muscular contraction forces water out the open end to provide propulsion.

Ctenophores. **Ctenophores** are "comb-bearing" animals closely related to the coelenterates. The body form of most ctenophores is basically spherical, with eight rows of cilia spaced evenly around the sphere. It is from these structures that the name *comb-bearing* is derived; when magnified, they look like miniature combs. Often called "comb jellies," this group of animals is entirely pelagic and confined to the marine environment.

If a ctenophore possesses tentacles, there will be two that contain adhesive organs instead of stinging cells to capture prey. Sea gooseberries (genera *Pleurobranchia*) (Figure 14–4C) and the pink, elongated *Beroe* (Figure 14–4D) range from gooseberry size to over 15 centimeters (6 inches) in the latter.

Chaetognaths. The **arrowworms** are transparent and difficult to see, although they may grow to more than 2.5 centimeters (1 inch) in length (Figure 14–5). The name *chaetognath* means bristle jawed and

refers to the hairlike attachments around their mouths used to grasp prey while they devour it. They are voracious feeders, eating primarily the small zooplankton. In turn, they are eaten by fishes and larger planktonic animals such as jellyfish. These exclusively marine, hermaphroditic animals are usually more abundant in the surface waters some distance from shore.

Swimming Forms

This section examines larger pelagic animals, the *nekton* (swimmers), which have substantial powers of locomotion. They include rapid-swimming invertebrate squids, fish, and marine mammals.

Swimming squid include the common squid (genus *Loligo*), flying squid (*Ommastrephes*), and giant squid (*Architeuthis*). Active predators of small fish, the smaller squid varieties have long, slender bodies with paired fins (Figure 14–6). Unlike the less-active *Sepia* and *Spirula* shown in Figure 14–1, they have no hollow chambers in their bodies and therefore require more energy to remain in the upper water of the oceans without sinking.

These invertebrates can swim about as fast as any fish their size, and do so by trapping water in a cavity between their soft body and penlike shell and forcing it out through a siphon. To capture prey, they use two long arms with pads containing suction cups at the ends. Eight shorter arms with suckers convey the prey to the mouth, where it is crushed by a beaklike mouthpiece (Figure 14–6).

Locomotion in fish is more complex than in the squid because of body motion and the role of fins. The basic movement of a swimming fish is the passage of a wave of lateral body curvature from the front to the back of the fish. This is achieved by the alternate contraction and relaxation of muscle segments along the sides of the body. These muscle segments are called **myomeres.** The backward pressure of the fish's body and fins produced by the movement of this wave provides the forward thrust (Figure 14–7).

Most active swimming fish have two sets of paired fins used in maneuvers such as turning, braking, and balancing. These are the *pelvic fins* and *pectoral (chest) fins* shown in Figure 14–7. When not in use, these fins can be folded against the body. Vertical fins, both dorsal (back) and anal, serve primarily as stabilizers.

The fin most important in propelling the high-speed fish is the tail fin, or *caudal fin*. Caudal fins flare vertically to increase the surface area available to develop thrust against the water. (This is equivalent to humans donning "frog flippers" on their feet to swim more efficiently.) The increased surface area also increases frictional drag. The efficiency of the design of a caudal fin depends on its shape.

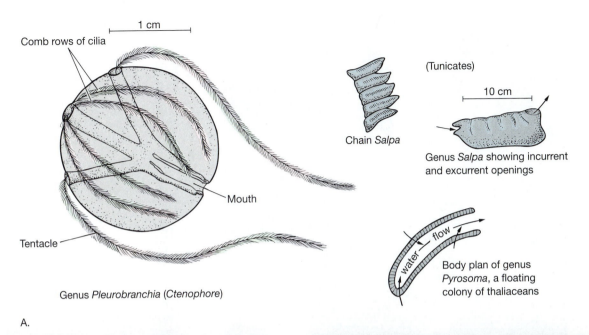

Comb rows of cilia

1 cm

Mouth

Tentacle

Genus *Pleurobranchia* (Ctenophore)

(Tunicates)

Chain *Salpa*

10 cm

Genus *Salpa* showing incurrent and excurrent openings

water → flow

Body plan of genus *Pyrosoma*, a floating colony of thaliaceans

A.

B.

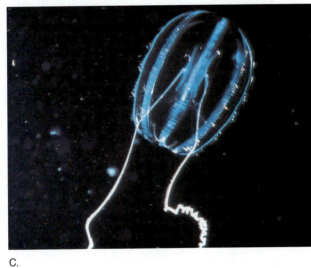

C.

D.

Figure 14–4

Pelagic Tunicates and Ctenophores. *A:* Body structure. *B:* Chain of salps. *C:* Ctenophore, *Pleurobranchia*. *D:* Ctenophore, *Beroe*. (Photos *B, C,* and *D* by James M. King, Graphic Impressions.)

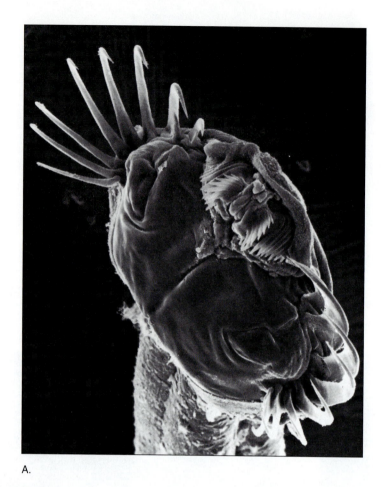

A.

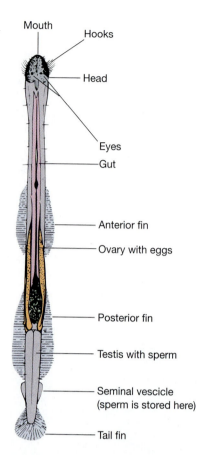

Mouth

Hooks

Head

Eyes

Gut

Anterior fin

Ovary with eggs

Posterior fin

Testis with sperm

Seminal vescicle
(sperm is stored here)

Tail fin

B. General Anatomy of an Arrowworm

Figure 14–5

Arrowworm. *A:* Head of chaetognath *Sagitta tenuis* from Gulf of Mexico, magnified 161 times. *B:* Adult chaetognaths reach lengths ranging from 1 to 5 centimeters (0.4 to 2 inches). (*A:* Photo courtesy of Howard J. Spero.)

There are five basic shapes of caudal fins, illustrated in Figure 14–8 and keyed by letter to the following description:

A. The rounded fin is flexible and useful in accelerating and maneuvering at slow speeds.

B and C. The somewhat flexible truncate tail (B) and forked tail (C) are found on faster fish and still may be used for maneuvering.

D. The lunate caudal fin is found on the fast-cruising fishes such as tuna, marlin, and swordfish; it is very rigid and useless in maneuverability, but very efficient in propelling.

E. The heterocercal fin is asymmetrical (*hetero-* = uneven, *cercal* = tail), with most of its mass and surface area in the upper lobe.

The heterocercal fin produces a significant *lift* to sharks as it is moved from side to side. This lift is important because sharks have no swim bladder and tend to sink when they stop moving. To aid this lifting, the

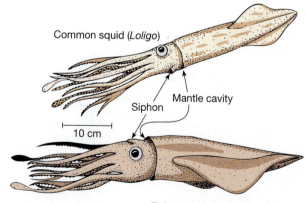

Common squid (*Loligo*)

Mantle cavity

Siphon

10 cm

Flying squid (*Ommastrephes*)

Figure 14–6

Squid. Squid move by trapping water in their mantle cavity between the soft body and penlike shell. They then jettison the water through the siphon for rapid propulsion.

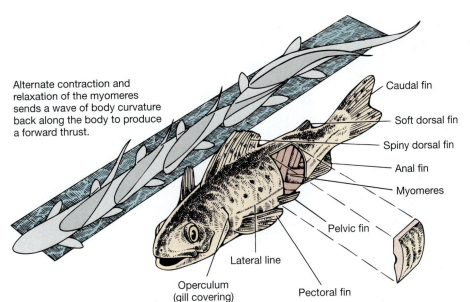

Alternate contraction and relaxation of the myomeres sends a wave of body curvature back along the body to produce a forward thrust.

Caudal fin
Soft dorsal fin
Spiny dorsal fin
Anal fin
Myomeres
Pelvic fin
Lateral line
Operculum (gill covering)
Pectoral fin

Figure 14–7

Fish—Swimming Motions and Fins.

pectoral (chest) fins are large and flat. Positioned on the shark's body like airplane wings, they in fact function like aircraft wings, or a hydrofoil boat, to lift the front of the shark's body to balance the rear lift supplied by the caudal fin. The shark gains tremendous lift but loses a lot in maneuverability as a result of this adaptation of the pectoral fins. This is why sharks tend to swim in broad circles, like a circling airplane, and are not marine acrobats.

Adaptations for Seeking Prey

Several factors affect each species' adaptation for capturing food. These include mobility (lunging versus cruising), speed, body length, body temperature, and circulatory system.

Lungers versus Cruisers. Some fish spend most of their time waiting patiently for prey and exert themselves only in short bursts as they lunge at the prey. Others cruise relentlessly through the water, seeking prey. There is a marked difference in the musculature of fishes that use these different styles to obtain food.

Lungers sit and wait. Figure 14–9A shows a grouper, representative of the lungers. It has a truncate caudal fin for speed and maneuverability, and almost all its muscle tissue is white.

Cruisers, on the other hand, actively seek game, such as the tuna in Figure 14–9B. Less than half of a cruiser's muscle tissue is white; most is red.

What is the significance of red versus white muscle tissue? *Red muscle fibers* are much smaller in diameter (25 to 50 micrometers, or 0.01 to 0.02 inches). *White muscle fibers* are much larger (135 micrometers, or 0.05 inches) and contain lower concentrations of myoglobin, a red pigment with an affinity for oxygen. These quali-

ties allow the red fibers to obtain a much greater oxygen supply than is possible for white fibers. This supports a metabolic rate six times that of white fibers, which is needed for endurance. The red muscle tissue is abundant in cruisers that swim constantly.

Lungers can get along quite well with little red tissue, because they need not constantly move. White tissue, which fatigues much more rapidly than red tissue, is used by the tuna for short periods of acceleration while on the attack. It is also quite adequate for propelling the grouper and other lungers during their quick passes at prey.

Speed and Body Length. The speed of fish is thought to be closely related to body length. For tuna, which are well adapted for sustained cruising and short bursts of high-speed swimming, cruising speed averages about 3 body lengths/second. They can maintain maximum speed of about 10 body lengths/second only for 1 second. A yellowfin tuna, *Thunnus albacares,* has been clocked at 74.6 kilometers/hour (46 miles/hour), which is a rate of more than 20 body lengths/second, but only for about one-fifth of a second (0.19 seconds).

Theoretically, a 4-meter (13-foot) bluefin tuna, *Thunnus thynnus,* could reach speeds up to about 144 kilometers/hour (90 miles/hour). Many of the toothed whales are known to be capable of high rates of speed. The porpoise *Stenella* has been clocked at 40 kilometers/hour (25 miles/hour), and it is believed that the top speed of killer whales may exceed 55 kilometers/hour (34 miles/hour).

Body Temperature. As is true with many chemical reactions, metabolic processes can be faster at higher temperatures. Fish are mostly cold blooded, having the same body temperatures as their environment. But there are some fast swimmers with body tempera-

A.

B.

C.

D.

E.

Figure 14–8

Caudal Fin Shapes. *A:* Rounded on a queen angel (other examples: culpin, flounder). *B:* Truncate on a gray angelfish (also salmon, bass). *C:* Forked on a goatfish in the Red Sea (also herring, yellowtail). *D:* Lunate on a blue marlin (also bluefish, tuna). *E:* Heterocercal on a silvertip shark. (*A:* Photo © Marty Snyderman; *B:* photo © Wayne and Karen Brown; *C:* photo © Pechter Photo/The Stock Market; *D:* photo © Bob Gomel/The Stock Market; *E:* photo © Valerie Taylor/Peter Arnold, Inc.)

A.

B.

Figure 14–9

Feeding Styles—Lunger and Cruiser. *A:* Lungers, such as this tiger grouper, sit patiently on the bottom and capture prey with quick, short lunges. Their muscle tissue is predominantly white. *B:* Cruisers, such as these yellowfin tuna, swim constantly in search of prey and capture it with short periods of high-speed swimming. Their muscle tissue is predominantly red. (*A:* Photo © Fred Bavendam/Peter Arnold Inc.; *B:* photo courtesy of National Marine Fisheries Service.)

tures above that of the surrounding water. Some are only slightly above the temperature of the water: The mackerel (*Scomber*), yellowtail (*Seriola*), and bonito (*Sarda*) have respective body temperature elevations of 1.3°, 1.4°, and 1.8°C (2.3°, 2.5°, and 3.2°F).

Other fast swimmers that have much higher temperature elevations are members of the mackerel shark genera, *Lamna* and *Isurus,* as well as the tuna, *Thunnus.* Bluefin tuna have been observed to maintain a body temperature of 30° to 32°C (86° to 90°F) regardless of water temperature. Although these tuna are more commonly found in warmer water, where the temperature difference between fish and water is no more than 5C° (9F°), body temperatures of 30°C (86°F) have been measured in bluefin tuna swimming in 7°C (45°F) water.

Why do these fish exert so much energy to maintain their body temperatures at high levels, when other fish do quite well with ambient body temperatures? Their mode of behavior is that of a cruiser, and any adaptation (high temperature and metabolic rate) that can increase the power output of their muscle tissue helps them seek and capture prey.

Circulatory System. Mackerel sharks and tuna are aided in maintaining their high body heats by a modified circulatory system (Figure 14–10). Whereas most fish have a **dorsal aorta,** just beneath the vertebral column, that provides blood to the swimming muscles, mackerel sharks and tuna have additional **cutaneous**

arteries just beneath the skin on either side of the body. As cool blood flows into red muscle tissue, its temperature is increased by heat generated by muscle metabolism (muscle contractions). A fine network of tiny blood vessels within the muscle tissue is designed to minimize heat loss.

The vessels that return the blood to the **cutaneous vein,** parallel to the cutaneous artery along the side of the fish, are all paired with small vessels carrying blood into the muscle tissue. In this way, the warm blood leaving the tissue helps to heat the cooler blood entering from the cutaneous artery.

Marine Mammals

Coastal ocean mammals include herbivores, the **sirenans** (sea cows, including dugongs and manatees), and a variety of carnivorous forms (Figure 14–11). Spending all or most of their lives in coastal waters and coming ashore to breed are the **sea otter** and **pinnipeds** (sea lions, seals, and walruses).

The truly oceanic mammals are the **cetaceans**—whales, porpoises, and dolphins (Figure 14–12).

We know that mammals evolved from reptiles on land some 200 million years ago. But no marine mammals are known to have existed earlier than 60 million years ago. Therefore, it is believed that all marine mammals evolved from ancient land-dwelling mammals.

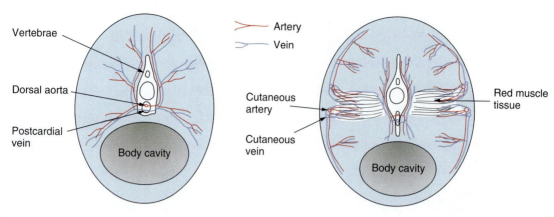

Figure 14–10

Circulatory Modifications in Warm-Blooded Fishes. *A:* Most cold-blooded (poikilothermic) fishes have major blood vessels arranged in the pattern represented on the left. All blood flows to muscle tissue from the dorsal aorta and returns to the postcardial vein beneath the vertebral column. *B:* Warm-blooded (homeothermic) fishes such as the bluefin tuna have cutaneous arteries and veins that help maintain high blood temperature by using heat energy generated by contracting muscle tissue. Its blood vessel pattern is represented on the right.

Whales, Porpoises, and Dolphins (Cetaceans)

Well known as the mammals best adapted to life in the oceans, the 75-plus species of cetaceans are of two basic types. The *toothed whales,* the **odontoceti,** include sperm whales, porpoises, and dolphins. The *baleen whales,* the **mysticeti,** probably evolved from the toothed whales some 30 million years ago, In place of teeth, they have baleen, which are plates of horny material that hang from the upper jaw and operate as a sieve.

The toothed whales are predators that feed mostly on smaller fish and squid, although the killer whale is known to feed on a variety of larger animals, including other whales. The baleen whales feed primarily by filtering crustaceans at depths ranging from the surface down to and including the sediment of shallow ocean basins.

The cetacean body is more or less cigar shaped, nearly hairless, and insulated with a thick layer of blubber. Cetacean forelimbs are modified into flippers that move only at the "shoulder" joint. The hind limbs are vestigial, not attached to the rest of the skeleton, and not externally visible. The skull is highly modified, with one nasal opening (toothed) or two openings (baleen) near the top. Cetaceans propel themselves by vertical movements of a horizontal tail fin called a *fluke.*

Baleen whales include three families:

1. The **gray whale** has short, coarse baleen, no dorsal fin, and only two to five ventral grooves beneath the lower jaw.
2. The **rorqual whales** have a short baleen, many ventral grooves, and are divided into two subfamilies: The *balaenopterids* have long, slender bodies, small sickle-

shaped dorsal fins, and flukes with smooth edges (minke, Baird's, Bryde's, sei, fin, and blue whales); the *megapterids,* or humpback whales, have a more robust body, long flippers, flukes with uneven trailing edges, tiny dorsal fins, and tubercles on the head.
3. The **right whales** have a long, fine baleen, broad triangular flukes, no dorsal fin, and no ventral grooves. The "northern" right whale is the baleen whale most threatened with extinction. The southern right whale (Southern Hemisphere) and the bowhead whale that remains near the Arctic pack-ice edge are the other members of this family.

Modifications to Increase Swimming Speed

Cetaceans' muscles are not vastly more powerful than those of other mammals, so it is believed that their ability to swim at high speed must result from modifications that reduce frictional drag. To illustrate the importance of streamlining in reducing the energy requirements of swimmers, a small dolphin would require muscles five times more powerful than it has to swim at 40 kilometers/hour (25 miles/hour) in turbulent flow.

In addition to a streamlined body, cetaceans are believed to actually modify the flow of water around their bodies to a smooth flow with the aid of a specialized skin structure. The skin is composed of two layers: a soft outer layer that is 80 percent water and has narrow canals filled with spongy material and a stiffer inner layer composed mostly of tough connective tissue. The soft layer tends to reduce the pressure differences at the skin-water interface by compressing under regions of higher pressure and expanding in regions of low pressure.

A.

B.

C.

D.

Figure 14–11

Marine Mammals. (Photo *A:* © Fred Bavendam/Peter Arnold, Inc.; *B:* © Tim Ragen; *C:* © C. Allan Morgan/Peter Arnold, Inc.; *D:* © Jeff Foott/Bruce Coleman, Inc.)

Modifications to Allow Deep Diving

People can free-dive to a maximum recorded depth of about 100 meters (330 feet) and hold their breaths in rare instances for 6 minutes. In contrast, the sperm whale *Physeter* is known to dive deeper than 2200 meters (7200 feet), and the North Atlantic bottle-nosed dolphin *Hyperoodon ampullatus* can stay submerged for up to two hours.

What adaptations do whales have to permit such long, deep dives? Whales can alternate between periods of normal breathing and cessation of breathing. The cessation periods occur while the animal is submerged.

To understand how some cetaceans are able to go for long periods without breathing requires a general knowledge of their lungs and associated structures.

Cetacean Breathing. In Figure 14–13, you can see that inhaled air finds its way to tiny terminal chambers, the alveoli. Alveoli are lined by a thin alveolar membrane that is in contact with a dense bed of capillaries; the exchange of gases between the inhaled air and the blood (oxygen in, carbon dioxide out) occurs across the alveolar membrane. Some cetaceans have an exceptionally large concentration of capillaries surrounding the alveoli (Figure 14–13B), which have mus-

Atlantic Right Whale (*Eubalaena glacialis*)
This cold-water whale was the first recorded target of whaling by Basque seamen during the middle ages. Similar species are found near Japan and in high southern latitudes. Length is to 18 m (60 ft).

SPERM WHALE (*Physeter catodon*)
Found mostly in tropical waters, this deep diver has a huge snout that contains a large amount of oil. Length of male is to 19 m (63 ft). Length of female is to 10.5 m (35 ft).

Figure 14–12

Baleen and Toothed Whales. No sea creatures capture our imaginations like the whales, huge mammals that returned to the sea from land about 60 million years ago. Many great whale populations were pushed to near extinction by whaling in the nineteenth and twentieth centuries, but it appears all will survive this threat. The major threat now may be disruption of their feeding and breeding grounds. Drawn to scale here is a mixture of the two whale subor-

BOTTLE-NOSED DOLPHIN (*Tursiops truncatus*)
Found in the North Atlantic, this is one of the many species of beaked dolphins. Length is to 3 m (10 ft).

Killer Whale (*Grampus Orca*)
Cosmopolitan in distribution, this whale is unique in that it not only feeds on fish but also on seals, birds, and other whales. The male grows to twice the female's size. Length is to 9 m (31 ft).

NARWHAL (*Monodon monoceros*)
The scientific name, which means one tooth, one horn, is accurate for the male of this Arctic whale. Length is to 6 m (20 ft).

PACIFIC GRAY WHALE (*Rhachianectes glaucus*)
This primitive baleen whale with a small head and very reduced dorsal fin stays close to shore. It even enters the surf zone. Length is to 14 m (45 ft).

HUMPBACK WHALE (*Meyaptera nodosa*)
The great communicators of the Cetacea have a cosmopolitan distribution. The genus name is derived from their extraordinary wing-like flippers. Length is to 16 m (52 ft).

BLUE WHALE (*Sibbaldus musculus*)
The largest animal to inhabit the Earth will consume over five tons of krill per day. Length is to 30 m (100 ft).

ders that exist in today's oceans. *The toothed whales (Odontoceti)* (bottle-nosed dolphin, killer whale, narwhal, sperm whale) are active predators and may track their prey with echo location. *The baleen whales (Mysticeti)* (right whale, Pacific gray whale, humpback whale, blue whale) fill their mouths with water and force it out between long baleen slats that hang from the upper jaw. The baleen "sieve" traps small fish, krill, and other plankton. The Pacific gray has short baleen slats and feeds by filtering sediment from the shallow bottom of its North Pacific feeding grounds, straining out a diet of benthic amphipods and other invertebrates.

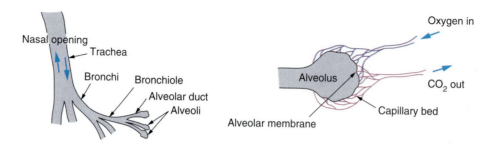

Figure 14-13

Cetacean Modifications to Allow Prolonged Submergence. *A:* Basic lung design. Air enters the lung through the trachea, and oxygen is absorbed into the blood through the walls of the alveoli. *B:* Oxygen exchange in the alveolus. A dense mat of capillaries receives oxygen through the alveolar membrane. Because air is normally held in the lungs of cetaceans for up to 1 minute before exhaling, as much as 90 percent of the oxygen can be extracted. Whales store this oxygen in up to twice as many red blood cells and nine times as much myoglobin in their muscle tissue as terrestrial mammals.

cles that move air against the membrane by repeatedly contracting and expanding.

Cetaceans take from 1 to 3 breaths per minute while resting, compared with about 15 in humans. Because they hold the inhaled breath much longer, and because of the large capillary mass in contact with the alveolar membrane and the circulation of the air by muscular action, many cetaceans can extract almost 90 percent of the oxygen in each breath, compared with only 4 to 20 percent extracted by terrestrial mammals.

To use this large amount of oxygen, which can be taken into the blood, efficiently during long periods under water, cetaceans may apply two strategies: (1) storing the oxygen and (2) reducing oxygen use. The storage of so much oxygen is possible because prolonged divers have a much greater blood volume per unit of body mass than those that dive for only short periods.

Compared with terrestrial animals, some cetaceans have twice as many red blood cells per unit of blood volume and up to nine times as much myoglobin in the muscle tissue. Thus large supplies of oxygen can be chemically stored in the **hemoglobin** of the red blood cells and the **myoglobin** of the muscles.

Additionally, the muscles that start a dive with a significant oxygen supply can continue to function through anaerobic respiration when the oxygen is used up. The muscle tissue is relatively insensitive to high levels of carbon dioxide.

Because the swimming muscles can function without oxygen during a dive, they and other organs, such as the digestive tract and kidneys, may be sealed off from the circulatory system by constriction of key arteries. The circulatory system then services primarily the heart and brain. Because of the decreased circulatory requirements, the heart rate can be reduced by 20 to 50 percent of its normal rate. However, recent research has shown that no such reduction in heart rate occurs dur-

ing dives by the common dolphin (*Delphinus delphis*), the white whale (*Delphinapterus leucas*), or the bottlenosed dolphin (*Tursiops truncatus*).

Nitrogen Narcosis. Another problem with deep and prolonged dives is the absorption of compressed gases into the blood. When humans make dives using compressed air, the prolonged breathing of compressed air, particularly nitrogen, can result in nitrogen narcosis, or decompression sickness (the bends). The effect of nitrogen narcosis is similar to drunkenness, and it can occur when a diver either goes too deep or stays too long at depths greater than 30 meters (100 feet).

If a diver surfaces too rapidly, the lungs cannot remove the excess gases fast enough, and the reduced pressure may cause small bubbles to form in the blood and tissue. The bubbles interfere with blood circulation, and the resulting decompression sickness can cause excruciating pain, severe physical debilitation, or even death.

Cetaceans and other marine mammals do not suffer from these difficulties. Their main defense against absorbing too much nitrogen seems to be a more flexible rib cage. By the time a cetacean has reached a depth of 70 meters (230 feet), the rib cage has collapsed under the 8 atmospheres of pressure. The lungs within the rib cage also collapse, removing all air from the alveoli. Because most absorption of gases by the blood occurs across the alveolar membrane, the blood cannot absorb additional gases, and the problem of nitrogen narcosis is avoided.

It is, however, possible that the collapsible rib cage is not the main defense against the bends. An experiment put enough nitrogen into the tissue of a dolphin to give a human a severe case of the bends. The dolphin suffered no ill effects. This whale, as well as other species, may have simply evolved an insensitivity to this gas that is not present in other mammals.

Use of Sound

It has long been known that cetaceans make a variety of sounds, despite their lack of vocal cords. Speculation about the sounds' purpose range from echo location (clearly true) to a highly developed language (doubtful). In fact, what we know about cetacean use of sound is limited.

All marine mammals have good vision, but conditions often limit its effectiveness. In coastal waters, where suspended sediment and dense plankton blooms make the water turbid, and in the deeper waters, where light is limited or absent, echo location surely would assist pursuit of prey or location of objects.

Using lower-frequency clicks at great distance and higher frequency at closer range, the bottle-nosed dolphin, *Tursiops truncatus,* can detect a school of fish at distances exceeding 100 meters (330 feet). It can pick out an individual fish 13.5 centimeters (5.3 inches) long at a distance of 9 meters (30 feet). It has been estimated that sperm whales can detect their main prey, squid, from a distance of up to 400 meters (1300 feet) by use of their low-frequency scanning clicks.

To locate something with these sounds, the animal's brain involuntarily performs the equivalent of a remarkable mathematical feat: It determines the distance to an object by multiplying the velocity at which sound travels to and returns from the object by the time required for travel, and divides this product by 2 (Figure 14–14).

Baleen whales produce sounds at frequencies generally below 5000 Hertz. Gray whales produce pulses, possibly for echo location, and moans that may maintain contact with other gray whales. Rorqual whales produce moans that last from one to many seconds. These sounds are extremely low in frequency, in the 10- to 20-Hertz range, and are probably used to communicate over distances of up to 50 kilometers (31 miles). Moans of fin whales may be related to reproductive behavior, but much more observation is required to ascertain the message content of all these sounds.

Youthful human ears are sensitive to frequencies from about 16 to 20,000 Hertz. Some toothed whales respond to frequencies as high as 150,000 Hertz. The clicks of the bottle-nosed dolphin are of a frequency range that is partly audible to humans. However, some clicks are at ultrasonic frequencies (above human hearing) and are repeated up to 800 times per second.

How Cetaceans Make Sound. How the pulses of clicks are produced and how the returning sound is received are not fully understood. Two hypotheses are presented in Figure 14–15.

In most mammals, the bony housing of the inner ear structure is fused to the skull. When submerged, sounds transmitted through the water are picked up by the skull and travel to the hearing structure from many directions. This makes it impossible for such mammals to locate accurately the source of the sounds. Obviously such a hearing structure would not work for an animal that depends on echo location to find objects accurately in water.

All cetaceans have evolved structures that insulate the inner ear housing from the rest of the skull. In toothed whales, the inner ear is separated from the rest of the skull and surrounded by an extensive system of air sinuses (cavities). The sinuses are filled with an insulating emulsion of oil, mucus, and air and are surrounded by fibrous connective tissue and venous networks. In many toothed whales, it is believed that sound is picked up by the thin, flaring jawbone and passed to the inner ear via the connecting oil-filled body.

Group Behavior

Depending on the size, feeding behavior, reproductive style, and other requirements for maintaining the species, many animals inhabiting the open ocean have developed patterns of group behavior that allow them to most efficiently exploit their environment. Two of these patterns are **schooling** and **migration.**

Schooling

Obtaining food occupies most of the time of many inhabitants of the open ocean. Some animals are fast and agile, and obtain food through active predation. Other animals move more leisurely as they filter small food particles from the water. Examples of *predators* and *fil-*

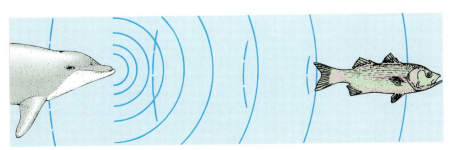

Figure 14–14

Echo Location. Clicking sound signals are generated by cetaceans and bounced off objects in the ocean to determine their size, shape, distance, and movement.

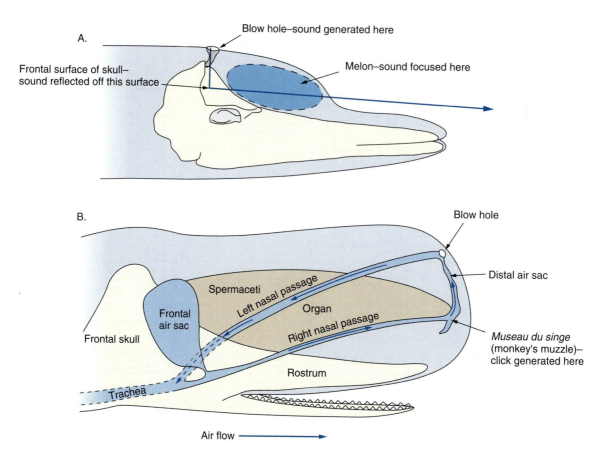

A.
Blow hole—sound generated here

Frontal surface of skull—sound reflected off this surface

Melon—sound focused here

B.
Blow hole

Spermaceti

Frontal air sac

Frontal skull

Left nasal passage

Organ

Right nasal passage

Distal air sac

Museau du singe (monkey's muzzle)—click generated here

Rostrum

Trachea

Air flow

Figure 14-15

Generation of Echo-Location Clicks in Small-Toothed and Sperm Whales. *A:* In small-toothed whales, clicks may be generated within the blowhole mechanism, reflected off the frontal surface of the skull, and then focused by the fatty melon into a forward-directed beam. *B:* Structures that may be related to the generation of clicks by sperm whales. Air may pass from the trachea through the right nasal passage across the *museau du singe,* where clicks are generated. It passes along the distal air sac, past the closed blow hole, and returns to the lungs along the left nasal passage. The initial click is followed by eight progressively weaker clicks that may result from sound energy bouncing back and forth.

ter feeders can be found in populations of pelagic animals, from the tiny zooplankton to the massive whales.

Patches of zooplankton may occur simply because the nutrients that support their food, the phytoplankton, may be highly concentrated in certain coastal waters. We usually do not refer to these patches as schools. *School* is usually reserved for well-defined social organizations of fish, squid, and crustaceans.

The number of individuals in a school can vary from a few larger predaceous fish to hundreds of thousands of small filter feeders. Within the school, individuals of the same size move in the same direction with equal spacing between each. This spacing probably is maintained through visual contact and, in the case of fish, by use of the **lateral line system** that detects vibrations of swimming neighbors (Figure 14–7). The school can turn abruptly or reverse direction as individuals at the head or rear of the school assume leadership positions (Figure 14–16).

Why schooling? The advantage of schooling seems obvious from the reproductive point of view. During spawning, it ensures that there will be males to release sperm to fertilize the eggs shed into the water or deposited on the bottom by females. However, most investigators believe the most important function of schooling in small fish is *protection from predators.*

At first, it may seem illogical that schooling would be protective. Any predator lunging into a school would surely catch something, just as land predators run a herd of grazing animals until one weakens and becomes dinner. So aren't the smaller fish making it easier for the predators by forming a large target? Based more on conjecture than research, the consensus of scientists is no.

Figure 14–16

Schooling. Schools of soldier fish and butterfly fish on a reef, Maldives. (Photo © Thomas Ives/The Stock Market.)

Here is the reasoning: Over 2000 fish species are known to form schools. This fact alone indicates that this behavior has evolved. Schooling is so pervasive today among fish populations because, for fish with no other means of defense, it somehow provides a better chance of survival than swimming alone.

How schooling is protective may grow out of the following considerations:

1. If members of a species form schools, they reduce the percentage of ocean volume in which a cruising predator might find one of their kind.
2. Should a predator encounter a large school, it is less likely to consume the entire unit than if it encounters a small school or an individual.
3. The school may appear as a single large and dangerous opponent to the potential predator and prevent some attacks.
4. Predators may find the continually changing position and direction of movement of fish within the school confusing, making attack particularly difficult for predators, who can attack only one fish at a time.

There may be other, more subtle reasons for schooling. But this gregarious behavior must enhance species survival because it is so widely practiced among pelagic animals.

Migration

Many oceanic animals undertake migrations of various magnitudes. This behavior is observed among sea turtles, fish, and mammals. Some animals, such as sea turtles, return to dry beach deposits to lay their eggs. Other populations, including many of the baleen whales, migrate because the adult physical environment of feeding grounds does not meet the needs of young baleens.

Migratory routes of commercially important baleen whales have been well known since the mid-1800s. The paths of these and other air-breathing mammals are easy to observe, because they must surface periodically for air. However, the migratory paths of many less visible fish have been more difficult to identify, despite their high commercial value. Studies that tag fish have helped us understand their movement patterns. Sampling the distribution patterns of eggs, larvae, young, and adult populations and radio-tracking individuals have also helped to identify migratory routes.

Orientation during Migration. How do migratory species orient themselves in time and space? To put the problem another way, how do they know where they are in relationship to where they want to go? How do they know when to leave so as to arrive at their destination on time?

Answers still are being sought, but researchers believe that all migratory species have an innate sense of time, referred to as a *biological clock*. Evidence of a species' biological clock are the physiological variations that occur independent of changes in the environment. These variations include cyclic respiration rates and body temperatures, and these may tell animals when to begin migration. However, some changes in the external environment can alter the circadian rhythms. Thus, it is possible that changes in food availability, water temperature, and duration of daylight may trigger seasonal migrations.

Orientation in space must be a more complex problem than orientation to time. Animals such as mammals and turtles that migrate at the surface could use their sight for orientation. Gray whales that migrate close to land over a large part of their migratory route might identify landmarks along the shore. When out of sight of land, mammals and turtles could use the relative positions of the sun, moon, and stars to guide them on their way. Fish that migrate beneath the surface are thought to use smell, relative movement within ocean currents, and other small currents related to Earth's magnetic field to orient themselves in ocean space.

Migration Routes. Many important food fish of high northern latitudes deposit pelagic egg masses that are transported by currents. These fish actively migrate upcurrent during spring or summer and spawn where the current will carry the eggs to a nursery ground having an appropriate food supply for newly hatched fry. Migrations of this type are usually over a few tens or hundreds of kilometers.

This behavior is observed in the population of Atlantic cod, *Gadus morhua*. The cod spawn along the southern shore of Iceland (Figure 14–17A, green area). Adults that occupy feeding grounds along the northern

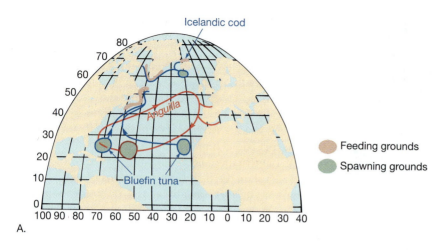

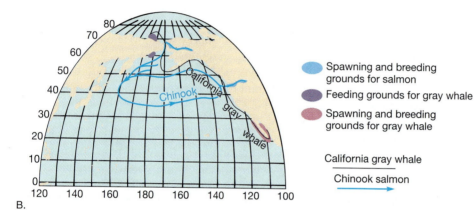

Figure 14–17

Migration Routes. *A:* Migration routes of the Icelandic cod, bluefin tuna, and Atlantic eel (*Anguilla*). *B:* Migration routes of the Chinook salmon and the California gray whale.

and eastern coasts of Iceland and around the southern tip of Greenland migrate to the spawning grounds in the late winter and early spring. This migration involves swimming against the East Greenland current and Irminger current, and includes only mature adults at least 8 years of age. These spawning migrations occur annually until the adults die at 18 to 20 years of age.

The spawning grounds are bathed in the relatively warm water of the Irminger Current, a northward-flowing branch of the Gulf Stream. Each female releases up to 15 million eggs, which float in the surface current. The drifting eggs, carried by the East Greenland and Irminger currents, hatch in about two weeks. The larvae feed at midwater depths while the currents are carrying them to the adult feeding grounds (brown area in figure). They remain there, undertaking only short onshore-offshore migrations, until they mature and make their first spawning migration.

Longer migrations are undertaken by two Atlantic bluefin tuna populations that spawn in tropical waters near the Azores in the eastern Atlantic and the Bahamas in the western Atlantic (green areas in Figure 14–17A). Little is known about where the newly hatched tuna feed, but the adults are known to move northward along the seaward edge of the Gulf Stream in May and June

and feed on the herring and mackerel populations off the coasts of Newfoundland and Nova Scotia (brown area). They may move even farther northward, but their return route to the spawning grounds is unknown.

Catadromous Fish. *Anguilla,* the Atlantic eel, undertakes what appears to be a single round-trip migration during its lifetime. Its behavior is **catadromous:** After being spawned in the ocean, it enters freshwater streams for its adult life and returns near the end of its life to the ocean spawning site.

Anguilla spawning is thought to occur in the Sargasso Sea southeast of Bermuda (Figure 14–17A, green area). The larvae, 5 millimeters (0.2 inches) long, are carried into the Gulf Stream and northward along the North American coast. After one year in the ocean, some metamorphose into young eels, called *elvers,* and move into North American freshwater streams. Others migrate to the European coast before metamorphosing and moving into European freshwater streams.

After spending up to 10 years in the freshwater environment, the mature eels undergo yet another change. They develop a silvery color pattern and enlarged eyes that are typical of fish inhabiting the mesopelagic zone. They swim downriver to the ocean and presumably back to the spawning ground, where they are thought to die.

Anadromous Fish.
An opposite behavior pattern is characteristic of salmon of the north Atlantic and Pacific oceans. This **anadromous** behavior includes spawning in freshwater streams and spending most of their adult lives in the ocean. Pacific species die after spawning; Atlantic salmon return to the ocean.

Six species of Pacific salmon, *Oncorhynchus*, follow similar paths in their migrations (Figure 14–18). Spawning occurs during the late summer and early fall. Eggs are deposited and fertilized in gravel beds far upstream. The chinook salmon spawns from central California to Alaska. The young chinook hatching in the spring may head downstream as a silvery, filter-feeding smolt (a young salmon ready to go to sea) during its first or second year. By the time it reaches the ocean, the young chinook has become a predator, feeding on smaller fish.

After spending about four years in the North Pacific, mature salmon of more than 10 kilograms (22 pounds) in weight and 1 meter (3.3 feet) in length return to their home streams to spawn (Figure 14–18). Although it is not known how salmon accurately return home, most investigators think the salmon's best guides are odors and currents.

California Gray Whale Migration.
Some of the longest migrations known in the open ocean are the seasonal migrations of baleen whales. Many baleen species feed in colder high-latitude waters and breed and calve in warm tropical waters. Feeding occurs during summer when the long hours of sunlight radiate the nutrient-rich waters to produce a vast feast of crustaceans. Only this bountiful food enables the whales to sustain themselves during the long migrations and mating and calving season, when feeding is minimal.

The long migrations that baleen whales now undertake may have resulted from productive feeding grounds that were once near the tropical calving grounds progressively moving poleward because of climate change. If this is the case, the breeding and calving grounds must be very important, for they would surely have been abandoned for locations closer to the present feeding grounds.

An alternative explanation for leaving the colder waters to calve is to avoid killer whales. They are more numerous in colder waters and are a major threat to young whales. Because of the energy demands faced by female baleen whales in producing large offspring (the gestation period is up to 1 year) and providing them with fat-rich milk for several months, it is not uncommon for them to mate only once every two or three years.

The migration route of the *California gray whale* demonstrates how the conditions just discussed are met by whale migration patterns. Gray whales are moderately large, reaching lengths of 15 meters (50 feet) and weighing over 30 metric tons. They feed during summer in the extreme northern Pacific Ocean (purple areas in Figure 14–17B). They are unique among baleen whales in that they do not feed by straining pelagic crustaceans and small fish from the water; instead, they stir up bottom sediment with their snouts and feed on bottom-dwelling amphipods (Figure 14–19).

The western populations winter along the Korean coast, but the population known as the California gray whale migrates from the Chukchi and Bering Seas to winter in lagoons of the Pacific coast of Baja California and the Mexican mainland coast near the southern end of the Gulf of California (Figure 14–17B). The migration usually begins in September when pack ice begins to form over the continental shelf areas that are their feeding grounds. This migration, which is the longest known migration undertaken by any mammal, may involve a round trip distance of 22,000 kilometers (13,700 miles).

First to leave are the pregnant females. They are followed by a procession of mature females that are not pregnant, immature females, mature males, and immature males. After cutting through the Aleutian Islands, they follow the coast throughout their southern journey. Traveling at an average rate of about 200 kilometers/day (125 miles/day), most reach the lagoons of Baja California by the end of January.

In these warm-water lagoons, the pregnant females give birth to 2-ton calves. The calves nurse and put on weight quickly during the next two months. While the calves are nursing, the mature males breed with the ma-

Figure 14–18

Varieties of Salmon. *Top:* chum (*Oncorhynchus keta*); *center:* coho (*O. kisutch*); *bottom:* chinook (*O. tschawyscha*). (Courtesy U.S. Fish and Wildlife Service.)

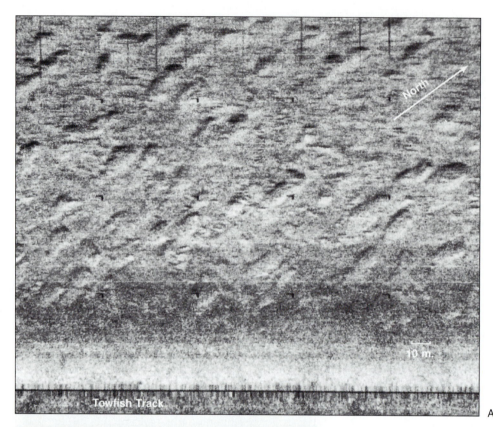

A.

B.

Figure 14-19

California Gray Whale Feeding. *A:* Side-scan sonograph showing pits on the floor of the northern Bering Sea created by California gray whales feeding on bottom-dwelling amphipods and other invertebrates. The smaller pits produced by the initial feeding, 2 to 3 meters (6.5 to 10 feet) long, do not show up well at this scale. All of the readily visible pits have been enlarged by currents that flow in a northerly direction. The pits show this north-south alignment. *B:* Amphipod mat. A mat of mucus-lined burrows is created by ampeliscid amphipods that are the preferred prey of California gray whales. (*A:* Sonograph courtesy of William K. Sacco, Yale University; *B:* photo courtesy of Kirk R. Johnson, Yale University.)

ture females that did not bear calves. Late in March, they return to the feeding grounds, with the procession order reversed. Most of the whales are back in the feeding grounds by the end of June, and they feed on prodigious quantities of amphipods to replenish their depleted store of fat and blubber before the next trip south.

Reproduction

Reproduction of pelagic animals involves bringing the males and females of the species together for this purpose on some periodic basis. These gatherings usually occur during the warmer spring or summer months,

when water temperatures are higher and primary productivity is at its peak.

Oviparous Reproduction. Most of the invertebrates and fish that inhabit the open ocean lay eggs that hatch in the open water or on the bottom into larval forms. The larvae become meroplankters (temporary floating plankton that will become nekton or benthos when mature). These creatures are **oviparous,** which simply means they reproduce by laying eggs as opposed to bearing living young.

Animals that reproduce in this way usually produce enormous numbers of eggs—15 million eggs per

Figure 14–20

Egg Case. Swell shark with yolk sac. (Photo © Alex Kerstitch/Bruce Coleman, Inc.)

season in the migrating Icelandic cod—because most will be consumed by predators before they hatch, or the larvae will be eaten by other zooplankton. However, some oviparous fish may produce only one or a few well-protected eggs per season; the cartilaginous sharks, skates, and rays include many such species (Figure 14–20).

Ovoviviparous Reproduction. The females of other fish species keep their fertilized eggs in their reproductive tract until they hatch. The young come into the world alive, making them *viviparous* (giving birth to living young). However, the process by which the embryos develop is the same as for those that develop from eggs laid in the open water. Thus, this "composite" reproduction scheme is called **ovoviviparous,** reproduction in which eggs are incubated internally.

The sea horse and pipefish exemplify a special twist to internal incubation where the female deposits the eggs in a pouch on the male (Figure 14–21). The male carries the eggs until they hatch, and the young fish enter the ocean directly from this pouch. The additional protection against loss of eggs to predators results in the reduction of the number of eggs that must be produced. For example, the dogfish shark, *Squalus,* usually produces fewer than a dozen at a time.

Viviparous Reproduction. **Viviparous** animals give birth to live young, like mammals. But this method also requires the young to be given more than incubation space in the mother's reproductive tract. Part of the behavior includes feeding the embryo in addition to nourishment in the egg yolk. Some sharks and rays secrete a rich milk in the uterus. With this additional nutrition, the stingray, *Pteroplatea,* produces young that enter the ocean with a mass up to 50 times that of the mass of the initial egg yolk.

Mammals exhibit the highest level of viviparous behavior in that the embryo is encased in the placental sac. The mother's blood flows through the placenta, and the embryo receives nutrition from the mother through an umbilical cord by which it is attached to the placenta. Essentially all nutrition is provided by the mother. This provides strong protection to the developing young but puts a high energy demand on the mother. Thus mammalian births typically involve only one or a few young, which remain dependent on their mothers for protection and food for some time after their birth.

Abduction: A Unique Shield against Predators

You have seen some remarkable animal adaptations in this chapter. We will close with yet another example,

Figure 14–21

Ovoviviparous Behavior in Sea Horses. The female deposits the eggs in the ventral pouch of the male (shown), where they are protected during incubation. (Photo by Larry Ford.)

but an extraordinary one, of how one small marine organism protects itself from predators.

In 1990 pelagic amphipods were observed to capture pteropods (sea butterflies) and carry them around on their backs beneath the Antarctic ice. Some investigation revealed that the 1.25-centimeter (0.5-inch) amphipod, *Hyperiella dilatata,* uses the 0.65-centimeter (0.25-inch) pteropod, *Clione limacina,* as a chemical defense against being eaten by predator fish. The pteropod contains a chemical that tastes awful, and while the amphipod holds it captive, fish will not eat it (Figure 14–22).

The fish that prey on the amphipods depend on sight to find their food. Therefore, it is not surprising that up to 75 percent of the amphipods observed at depths of less than 9 meters (28 feet) were carrying pteropods on their backs. In the darker waters at 50 meters (160 feet), only 6 percent of the amphipods observed were protected by their foul-tasting captives.

This pattern of capturing another organism for the purpose of using its chemical defense against predators does not fit into any of the categories of symbiotic relationships previously observed between marine species. It is believed that the pteropods are eventually released unharmed, although they may experience only a short reprieve until they once again become hostages.

Figure 14–22

Captive Pteropod Protects Pelagic Amphipod from Predators. Scanning electron micrograph of the amphipod, *Hyperiella dilatata,* carrying its captive pteropod, *Clione limacina,* which provides chemical protection against predatory fish. (Photo courtesy of James B. McClintock with permission from *Nature,* Vol. 346, 462–464, 1990. Photo by Phil Oshal.)

S U M M A R Y

Frictional resistance to sinking—a major factor in helping tiny plankton stay near the surface—is not a major factor in keeping the larger nekton from sinking; they depend primarily on buoyancy or swimming. Rigid gas containers found in some cephalopods and the expandable swim bladders of many bony fishes are adaptations that help increase buoyancy. Many invertebrates, such as the jellyfish, tunicates, and arrowworms, have soft, gelatinous bodies that reduce their density. The Portuguese man-of-war also has a gas-filled float that supports this pelagic colony. Many invertebrate forms are weak swimmers and depend primarily on buoyancy to maintain their positions near the surface.

Nekton—squid, fish, and mammals—are strong swimmers. They depend on this expenditure of energy to obtain food in the water column. Squid swim by trapping water in their body cavities and forcing it out through a siphon. Most fish swim by creating a wave of body curvature that passes from the front to the back of the fish and provides a forward thrust.

The caudal (rear) fin is the most important in providing thrust, whereas the paired pelvic and pectoral (chest) fins are used for maneuvering. The vertical dorsal (back) and anal fins serve primarily as stabilizers.

A rounded caudal fin is flexible and can be used for maneuvering at slow speeds. The lunate fin is rigid, is of little use in maneuvering, and is very efficient in producing thrust for fast swimmers such as tuna.

Lungers, such as groupers, are fish that sit motionless and make short, quick passes at passing prey. They have mostly white muscle tissue. Cruisers, such as tuna, are fish that constantly swim in search of prey. They possess mostly red muscle tissue. Red fibers have a great affinity for oxygen and tire less rapidly than white fibers, enabling tuna to maintain a rapid cruising speed.

Although most fish are cold blooded, the tuna, *Thunnus,* is homeothermic: It maintains a body temperature well above water temperature.

The best-adapted mammals for life in the open ocean are the Cetacea (whales, porpoises, dolphins). The toothed whales and baleen whales propel their highly streamlined bodies through the water with vertical movement of their horizontal tails.

Whales can dive deep and stay submerged for unusually long periods because of adaptations that allow them to absorb 90 percent of the oxygen in air they inhale, store large quantities of it, possibly reduce its use, and collapse their lungs below depths of 100 meters (330 feet).

Most whales and many other marine mammals use echo location in finding their way through the ocean and locating prey. The clicking sounds emitted by whales are bounced off objects, and the animal can determine the size, shape, and distance of the objects by the nature of returning signals and the time elapsed.

Schooling of active swimmers such as fish, squid, and crustaceans is not fully understood, but it likely serves a protective function.

Migrations are observed among sea turtles, fish, and mammals and are related to reproductive needs and finding food. It is believed that orientation is maintained during migrations through use of visible landmarks, smell, and Earth's magnetic field. Baleen whales may migrate from their cold-water summer feeding grounds to warm, low-latitude lagoons in winter so that their young can be born into warm water. Fishes such as the Atlantic cod swim upcurrent to deposit their eggs so that when they hatch the cod fry will be in water where suitable food is available.

The North Atlantic eel is catadromous, spawning in the Sargasso Sea and spending its adult life in the freshwater streams of North America and Europe. Atlantic and Pacific salmon are anadromous, spawning in the freshwater streams of North America, Europe, and Asia and spending their adult lives in the open ocean.

Most fish are oviparous, depositing their eggs in the ocean. Some sharks and rays maintain their eggs in a body cavity until they hatch; this ovoviviparous behavior provides a greater protection for the eggs. Where oviparous fish may produce millions of eggs, ovoviviparous fish may produce fewer than a dozen. The stingray, white-tip shark, and mammals are viviparous (giving birth to live young). They not only provide space in their bodies for the eggs to develop but also provide nutrition in addition to the egg yolk. In mammals, the young remain dependent on their mothers for nutrition for some time after birth.

A unique method of defense against predators is practiced by an Antarctic amphipod that captures a foul-tasting pteropod and carries it around on its back.

KEY TERMS

Anadromous (p. 329)
Arrowworm (p. 314)
Catadromous (p. 328)
Cetacea (p. 319)
Coelenterates (p. 313)
Cruiser (p. 317)
Ctenophore (p. 314)
Cutaneous artery (p. 319)
Cutaneous vein (p. 319)
Dorsal aorta (p. 319)
Gray whale (p. 320)
Hemoglobin (p. 324)

Lateral line system (p. 326)
Lunger (p. 317)
Migration (p. 325)
Myoglobin (p. 324)
Myomere (p. 314)
Mysticeti (p. 320)
Nematocyst (p. 313)
Odonticeti (p. 320)
Oviparous (p. 331)
Ovoviviparous (p. 331)
Pinniped (p. 319)

Pneumatic duct (p. 312)
Right whale (p. 320)
Rorqual whale (p. 320)
Schooling (p. 325)
Scyphozoan (p. 313)
Sea otter (p. 319)
Siphonophore (p. 313)
Sirenia (p. 319)
Swim bladder (p. 312)
Tunicate (p. 314)
Viviparous (p. 331)

QUESTIONS AND EXERCISES

1. Discuss how the rigid gas chamber in cephalopods may be more effective in limiting the depth to which they can descend than the flexible swim bladders of bony fish.
2. Describe the body form and lifestyle of the following plankters: Portuguese man-of-war (*Physalia*), jellyfish, tunicates, ctenophorans, and arrowworms.
3. What are the major structural and physiological differences between the fast-swimming cruisers and lungers that patiently lie in wait for their prey?
4. List the modifications that are thought to allow some cetaceans to (1) dive to great depths without suffering the bends and (2) stay submerged for long periods.
5. Describe the process by which the sperm whale may produce echo-location clicks.

6. Discuss how sound may reach the inner ear of toothed whales.
7. Summarize the reasons some investigators believe that schooling increases the safety of fishes from predators.
8. What are the methods believed to be used by migrating animals to maintain their orientation?
9. Why do fishes such as Icelandic cod swim upcurrent to spawn?
10. How are the migrations of the North Atlantic eels and Pacific salmon fundamentally different?
11. Discuss possible reasons for the California gray whales' leaving the cold-water feeding grounds during the winter season.
12. Compare the reproductive behavior of oviparous, ovoviviparous, and viviparous animals.

REFERENCES

Carey, F. G. 1973. Fishes with warm bodies. *Scientific American* 228:2, 36–44.

Denton, E. J., and Shaw, T. I. 1962. The buoyancy of gelatinous marine animals. *Journal of Physiology* 161:14P–15P.

George, D., and George, J. 1979. *Marine life: An illustrated encyclopedia of invertebrates in the sea.* New York: Wiley-Interscience.

Kanwisher, J. W., and Ridgway, S. H. 1983. The physiological ecology of whales and porpoises. *Scientific American* 248:6, 110–121.

Lecomte-Finiger, R. 1992. The early life of the European eel. *Research in Marine Biology* 114, 205–210.

MacGinitie, G. E., and MacGinitie, N. 1968. *Natural history of marine animals.* 2nd ed. New York: Mcgraw-Hill.

McClintock, J. B., and Janssen, J. 1990. Pteropod abduction as a chemical defence in a pelagic Antarctic amphipod. *Nature* 346:6283, 462–464.

Mrosovsky, N.; Hopkins-Murphy, S. R.; and Richardson, J. I. 1984. Sex ratio of sea turtles: Seasonal changes. *Science* 225:4663, 739–740.

Norris, K. S., and Harvey, G. W. 1972. A theory for the function of the spermaceti organ of the sperm whale (*Physeter catodon* L). *Animal Orientation and Navigation,* 397–417. Washington, D.C.: National Aeronautics and Space Administration.

Pike, G. C. 1962. Migration and feeding of the gray whale (*Eschrichtius gibbosus*). *Journal of the Fisheries Research Board of Canada* 19:815–838.

Royce, W.; Smith, L. S.; and Hartt, A. C. 1968. Models of oceanic migrations of Pacific salmon and comments on guidance mechanisms. *Fishery Bulletin* 66:441–462.

Thorson, G. 1971. *Life in the sea.* New York: Mcgraw-Hill.

Würsig, B. 1989. Cetaceans. *Science* 244:4912, 1550–1557.

SUGGESTED READING

Sea Frontiers

Bachand, R. G. 1985. Vision in marine animals. 31:2, 68–74. An overview of the types of eyes possessed by marine animals.

Bleecker, S. E. 1975. Fishes with electric know-how. 21:3, 142–148. A survey of fishes that use electrical fields to navigate, to capture prey, and to defend themselves.

Bushnell, P. G., and Holland, K. N. 1989. Tunas: Athletes in a can. 35:1, 42–48. The physiology of various tuna species, specifically that related to high-speed swimming, is discussed.

Hersh, S. L. 1988. Death of the dolphins: Investigating the east coast die-off. 34:4, 200–207. The 1987–88 East Coast die-off of dolphins is discussed by a marine mammalogist.

Kleen, S. 1989. The diving seal: A medical marvel. 35:6, 370–374. The physiology of seals that allows them to dive deep and stay submerged for long periods of time is covered.

Klimley, A. P. 1976. The white shark: A matter of size. 22:1 2–8. Describes procedure used by Dr. John E. Randall for determining the size of sharks from the perimeter of the upper jaw and height of teeth.

Lineaweaver, T. H., III. 1971. The hotbloods. 17:2, 66–71. Discusses the physiology of the bluefin tuna and sharks that have high body temperatures.

Maranto, G. 1988. The Pacific walrus. 34:3, 152–159. A summary of the natural history of walruses of the Bering Sea and the role of humans as their predators.

McAuliffe, K. 1994. When whales had feet. 40:1, 20–33. Details the work of paleontologists in the Egyptian desert where they found the remains of a 45-million-year-old whale that still possessed feet.

Netboy, A. 1976. The mysterious eels. 22:3, 172–182. An informative discussion describing what is known of the migrations of the catadromous eels and their importance as a fishery.

O'Feldman, R. 1980. The dolphin project. 26:2, 114–118. A description of the response of Atlantic spotted dolphin to musical sounds.

Reeve, M. 1971. The deadly arrowworm. 17:3, 175–183. This important group of plankton is described in terms of body structure and lifestyle.

Volger, G., and Volger, S. 1988. Northern elephant seals. 34:6, 342–347. This article deals primarily with the history of the elephant seal populations off the coast of California and Baja California.

Scientific American

Denton, E. 1960. The buoyancy of marine animals. 303:11, 118–128. How various marine animals use buoyancy to maintain their position in the water column.

Donaldson, L. R., and Joyner, T. 1983. The salmonid fishes as a natural livestock. 249:1, 50–69. The genetic adaptability of the salmonid fishes may help them adapt to "ranching" operations.

Gosline, J. M., and DeMont, M. E. 1985. Jet propelled swimming in squids. 252:1, 96–103. By using jet propulsion resulting from expelling water through their siphons, squids can move as fast as the speediest fishes.

Gray, J. 1957. How fishes swim. 197:2, 48–54. The roles of musculature and fins in the swimming of fishes.

Horn, M. H., and Gibson, R. N. 1988. Intertidal fishes. 258:1, 64–71. A discussion of a wide variety of fishes that inhabit tide pools.

Kooyman, G. L. 1969. The Weddell seal. 221:2, 100–107. The lifestyle and problems related to this mammal's living in water permanently covered with ice.

Leggett, W. C. 1973. Migration of shad. 228:3, 92–100. The routes of the anadromous shad in the rivers and Atlantic waters are described, along with factors that may control the migrations.

O'Shea, T. J. 1994. Manatees. 271:1, 66–73. Manatees evolved from the same ancestors as elephants and aardvarks. The threats to their survival are discussed.

Rudd, J. T. 1956. The blue whale. 195:6, 46–65. A description of the ecology of the largest animal that ever lived.

Sanderson, S. L., and Wassersug, R. 1990. Suspension-feeding vertebrates. 263:3, 96–102. The ecology of filter-feeding fishes and cetaceans is covered.

Shaw, E. 1962. The schooling of fishes. 206:6, 128–136. A consideration of theories seeking to explain the schooling behavior of fishes.

Whitehead, H. 1985. Why whales leap. 252:3, 84–93. Whales seem to communicate with their spectacular lunges above the ocean's surface.

Würsig, B. 1988. The behavior of baleen whales. 258:4, 102–107. A summary of the facts concerning the behavior of baleen whales.

Zapol, W. M. 1987. Diving adaptations of the Weddell seal. 256:6, 100–107. The physiological adaptations that enable the Weddell seal to make deep, long dives are discussed.

CHAPTER 15
Animals of the Benthic Environment

There are 235,000-plus animal species in the ocean. Of these, more than 98 percent live in the two-dimensional world of the ocean floor. Ranging from the rocky, sandy, and muddy environments of the intertidal zone to the muddy deposits of deep-ocean trenches more than 11 kilometers (6.8 miles) deep, the ocean floor provides a varied environment that is home for a diverse benthic community.

Living at or near the interface of the ocean floor and seawater, an organism's suc-

cess is closely tied to its ability to cope with the physical conditions of the water, the ocean floor, and other members of the biological community. The vast majority of known benthic species live on the continental shelf, and approximately 400 of the more than 230,000 benthic species are found in the hadal zone of the deep-ocean trenches.

One of the most prominent variables affecting species diversification may be temperature. We have previously discussed temperature's effect on species diversity with latitude. However, even at the same latitude, a significant difference in the number of benthic species is found on opposite sides of an ocean basin because of the effect of ocean currents on coastal water temperature. For example, along the European coast, where the Gulf Stream warms the water from the northern tip of Norway to the Spanish coast, over three times the benthic species exist than thrive along the similar latitudinal range of the Atlantic coast of North America, where the Labrador Current cools water as far south as Cape Cod.

Figure 15–1 shows that the distribution of benthic biomass is patterned after the distribution of photosynthetic productivity in the surface waters shown in Figure 13–7. This tells us that life on the ocean floor is very much dependent upon the primary photosynthetic productivity of the ocean-surface waters. We will nonetheless discuss some special benthic communities that are an exception to this general condition.

Animals of Rocky Shores

Figure 15–2 shows a typical rocky shore. The *spray zone,* above the spring high tide line, is covered by water only during storms. The **intertidal zone** lies between the high and low tidal extremes.

Along most shores, the intertidal zone can be divided into three subzones (Figure 15–2):

- The *high tide zone,* mostly dry, covered by the highest high tide but not by the lowest high tides.
- The *middle tide zone,* exposed and covered equally—covered by all high tides and exposed during all low tides.
- The *low tide zone,* mostly wet, covered during the highest low tides and exposed during the lowest low tides.

Along rocky shores, these divisions of the intertidal zone are often obvious due to the sharp bound-

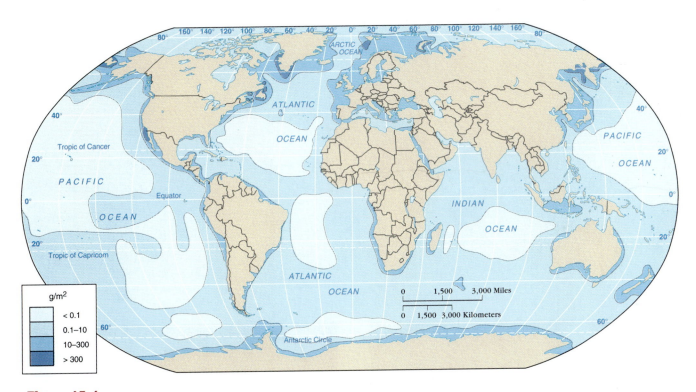

Figure 15–1

Distribution of Benthic Biomass. This map provides a general idea of benthic biomass distribution, but it is based on too few samples to be accurate. The pattern is similar to that of photosynthetic pigment distribution shown in Figure 3–7. This suggests that most of the benthic community depends directly on photosynthetic production in surface waters for nutrition. Not reflected in this distribution are high biomass concentrations around hydrothermal vents and cold seeps. Benthos concentrations beneath subtropical gyre centers are 20,000 times less than those of the continental shelf. Up to 60 percent of organic matter reaching the deep-ocean floor may enter the food chain through bacteria, and deep-sea bacteria may metabolize organic matter at least 10 times slower than continental-shelf bacteria. This could reduce deep-ocean floor biological productivity to 1/200,000 that occurring on the continental shelves. (Units are grams of biomass per square meter.) (After Zenkevitch et al., 1971.)

Figure 15–2

The Rocky Shore. *A:* A typical rocky intertidal zone and some organisms typically found in its subzones (not to scale). *B:* Periwinkles (*Littorina*) nestled in a depression near the upper limit of the high tide zone. This behavior helps reduce exposure to direct sunlight. *C:* Rock louse (*Ligia*). *D:* Rough keyhole limpet (*Diodora aspera*) with encrusting red algae (*Lithothamnion*). *E:* Buckshot barnacles (*Chthamalus*). *F:* Rock weed (*Fucus filiformes*). Sharing the middle tide zone with these rock weeds are *G,* acorn barnacles (*Balanus*), *I:* goose barnacles (*Pollicipes*), and mussels. *H:* Rock weed (*Pelvetia fastigata*). (*D:* Photo © Norbet Wu/Peter Arnold, Inc.; *F:* photo © Breck P. Kent; *H:* photo © Eda Rogers.)

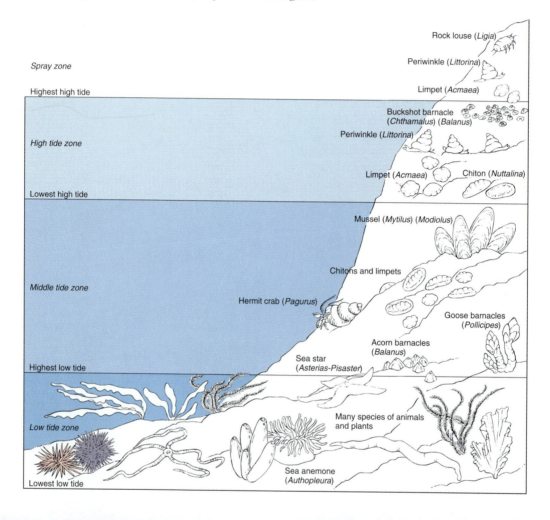

A.

B.

C.

D.

E.

F.

G.

H.

I.

aries between populations of organisms that attach themselves to the surface. Because each centimeter of the littoral rocky shore has a significantly different character from the centimeter above and below it, evolution has been able to produce organisms with the abilities to withstand very specific degrees of exposure to the atmosphere. This results in the most finely defined biozones known in the marine environment.

Rocky intertidal ecosystems have a moderate diversity of species, with the greatest animal diversity being at lower tropical latitudes. Interestingly, the diversity of algae is greater in temperate latitudes.

Spray (Supralittoral) Zone

Throughout the world, the most obvious inhabitant of the rocky supralittoral zone is the periwinkle snail. The **spray zone** can easily be identified if one considers the middle of the periwinkle belt to be the boundary between the intertidal zone and the spray zone. The periwinkle genus *Littorina* includes species able to breathe air, such as land snails (Figure 15–2B).

Hiding among the cobbles and boulders covering the floors of sea caves well above the high tide line are rock lice and sea roaches (isopods of the genus *Ligia*). Neither name is particularly flattering to these little scavengers, which reach lengths of 3 centimeters (1.2 inches) and scurry about at night feeding on organic debris (Figure 15–2C).

Another indicator of the spray zone is a distant relative of the periwinkle snails, a limpet (genus *Acmaea*) with a flattened conical shell that feeds in a manner similar to periwinkles.

High Tide Zone

Some periwinkles of the **high tide zone** can venture into the spray zone, but many of its residents cannot leave the sea for dry land. Thus, unlike the periwinkles, the landward limit for buckshot barnacles is the high tide shoreline. They are confined for two reasons: They filter-feed from seawater, and their larval forms are planktonic.

The most conspicuous algae in the high tide zone are rock weeds, members of the genus *Fucus* in colder latitudes and *Pelvetia* in warmer latitudes (Figure 15–2F). Both have thick cell walls to reduce water loss during periods of low tide.

On a clean, rocky shore, the rock weeds establish themselves before the sessile animal forms. However, once barnacles or mussels move in, the rock weeds seem doomed.

Middle Tide Zone

Rock weeds continue to flourish in the **middle tide zone,** where the variety of life forms is much greater than in the high tide zone. Not only does the variety in-

crease, but the total biomass is much greater. There is, therefore, a greater competition for rock space among forms that attach themselves to the surface. Such forms are called *attached,* or *sessile.*

The barnacle most characteristic of this zone is the goose barnacle (*Pollicipes*) (Figure 15–2I), which attaches itself to the rock surface by a long, muscular neck.

Even more successful than barnacles in the middle tide zone space competition are various mussels (genera *Mytilus* and *Modiolus*). They attach to bare rock, algae, or barnacles. Settling on these surfaces as larval forms, they attach themselves by tough threads. Mussels are fed upon by predators such as sea stars and carnivorous snails.

Two common genera of sea stars are the *Pisaster* and *Asterias*. To get through the calcium carbonate mussel shell to the edible tissue inside, sea stars exert a continuous pull on the shells by using alternate suction tube-like feet, so that some are always pulling while others are resting. The mussel eventually becomes fatigued and can no longer hold its shell halves closed. When the shell opens ever so slightly, the sea star turns its stomach inside out, slips it through the crack in the mussel shell, and digests the mussel without having to take it into its mouth (Figure 15–3).

The dominant feature of the middle tidal zone along most rocky coasts is a mussel bed that thickens toward the bottom until it reaches an abrupt bottom limit. This may be so pronounced that it appears as though an invisible horizontal plane has prevented the mussels from growing below this depth. Protruding from the mussel bed will be numerous goose barnacles, and concentrated in the lower levels of the bed will be the sea stars browsing on the mussels. Less conspicuous forms common to the mussel beds are varieties of algae, worms, clams, and crustaceans.

Where the rock surface flattens out within the middle tidal zone, tide pools trap water as the tide ebbs. These pools support interesting microecosystems containing a wide variety of organisms. The largest member of this community will often be the sedentary relative of the jellyfish, the sea anemone (Figure 15–4).

Shaped like a sack, anemones have a flat foot disc that provides a suction attachment to the rock surface. Directed upward, the open end of the sack is the only opening to the gut cavity, the mouth, surrounded by rows of tentacles. The tentacles are covered with cells that contain a stinging threadlike *nematocyst*. It is automatically released to penetrate any organism that brushes against the tentacles.

The most interesting inhabitant of tide pools is the hermit crab, *Pagurus*. With a well-armored pair of claws and upper body, hermit crabs have a soft, unprotected abdomen. They "adopt" protection for it, usually by moving into an abandoned snail shell. Their abdomen has even evolved a curl to the right to make it fit properly

Figure 15–3

Sea Star Feeding on a Mussel. The sea star pulls apart the two halves of the mussel's shell. The star then turns its own stomach inside out and forces it through the opening between the mussel's shells to digest the mussel's soft tissue in place inside its shell. (Photo © Joy Sparr/Bruce Coleman, Inc.)

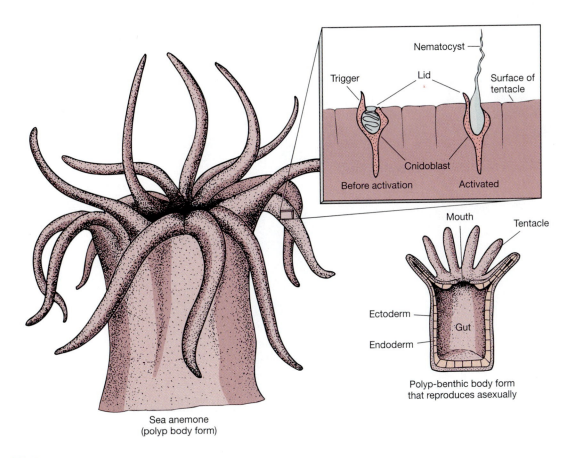

Sea anemone
(polyp body form)

Polyp-benthic body form
that reproduces asexually

Figure 15–4

Sea Anemone. Basic body plan of an anemone and detail of the stinging mechanism.

into snail shells. Once in the snail shell, the crab can close off the opening with its large claws (Figure 15–5A).

In tide pools near the lower limit of the middle tide zone, sea urchins may be found feeding on algae (Figure 15–5B). They have a five-toothed structure centered on the bottom side of a hard spherical covering that supports many spines. The hard protective covering is called a *test* and is composed of fused calcium carbonate plates that are perforated to allow feet and gills to pass through.

Low Tide Zone

Unlike the upper and middle tide zones, the **low tide zone** is dominated by plants rather than animals. A diverse community of animals exists, but they are less ob-

vious because they are hidden by the great variety of seaweeds and surf grass (*Phylospadix*) (Figure 15–6). The encrusting red algae, *Lithothamnion,* also seen in middle zone tide pools, becomes very abundant in the lower tide pools (Figure 15–2D). In temperate latitudes, moderate-sized red and brown algae provide a drooping canopy beneath which much of the animal life is found.

Scampering from crevice to crevice and in and out of tide pools across the full range of the intertidal zone are various species of shore crabs (Figure 15–7). These scavengers help keep the shore clean. Shore crabs spend most of the daylight hours hiding in cracks or beneath overhangs. At night they engage in most of their eating activity, shoveling in algae as rapidly as they can tear it from the rock surface with their large front claws (chelae). These creatures can spend long periods out of the water.

A.

B.

Figure 15–5

Hermit Crab and Sea Urchin. *A:* Hermit crab (*Pagurus*) has taken up residence in a *Maxwellia gemma* shell. *B:* Sea urchins burrowed into the bottom of a tide pool in the lower middle tide zone. (*A:* Photo by James McCullagh.)

Figure 15-6

Algal Colony of the Low Tide Zone Exposed during Extremely Low Tide. The dark-colored sea palms (a brown alga) and green surf grass are common inhabitants of the California low tide zone.

Animals of Sediment-Covered Shores

The sediment-covered shore ranges from steep boulder beaches, where wave energy is high, to the mud flats of quiet, protected embayments. However, the sediment-covered shore we know best is the sand beach typical of areas where wave energy usually is moderate.

The Sediment

The sediment-covered shore includes what are commonly called *beaches, salt marshes,* and *mud flats,* all of which represent lower-energy environments. As the energy level diminishes (meaning the strength of the longshore current), particle size grows smaller and the sediment slope is reduced. Consequently, sediment stability increases.

A.

B.

Figure 15-7

Shore Crabs. *A:* Coral crab, or queen crab (Bonaire Island, Netherlands Antilles). *B:* Shore crab, *Pachygrapsis crassipes,* female with eggs held under her abdomen (tail-like oval structure curled under her body). (*A:* Photo © Fred Bavendam/Peter Arnold, Inc.; *B:* photo © Eda Rogers.)

The water from breaking waves rapidly percolates down through coarse sands to replace the oxygen consumed by animals that live in the sediment. This readily available oxygen also enhances bacterial decomposition of dead tissue.

Life in the Sediment

Life on and in the sediment requires very different adaptations than on the rocky coast. The sandy beach supports fewer species than the rocky shore, and mud flats fewer still; however, the total number of individuals may be as high. In the low tide zone of some beaches and on mud flats, as many as 5000 to 8000 burrowing clams have been counted in only 1 square meter (10.8 square feet).

Burrowing is the most successful adaptation for life in the sediment-covered shore, so life here is less visible. By burrowing a few centimeters beneath the surface, organisms adopt a stable environment where they are not bothered by fluctuations of temperature and salinity and the threat of drying out.

Burrowing in the sediment does not prevent animals from suspension feeding or straining plankton from the clear water above the sediment. With various techniques, the water above the sediment surface can be stripped of its plankton by buried animals. Two of these methods (suspension feeding and deposit feeding) are explained in Figure 15–8.

The Sandy Beach

When you go to the beach, you do not see animals exposed at the surface as you do along a rocky shore.

There is no stable, fixed surface that creatures can attach to, so most of them burrow into the sand, safely hidden from view.

Bivalve Mollusks. A *bivalve* is an animal having two hinged shells, like a clam. Bivalve mollusks are well adapted for life in the sediment. The greatest variety of clams is burrowed into the low tide region of sandy beaches; their numbers decrease where the sands become muddier.

Bivalve mollusks possess a soft body, a portion of which secretes the calcium carbonate valves (shells) that hinge together. A single foot digs into the sediment to pull the creature down into the sand. Siphons protrude vertically for feeding (Figure 15–8A). The procedure used by a bivalve to bury itself is shown in Figure 15–9.

How deeply a bivalve can bury itself depends on the length of its siphons; they must reach above the sediment surface to pull in water from which plankton will be filtered. Oxygen is also extracted in the gill chamber before the water is expelled. Undigestible matter is forced back out the siphon periodically by quick muscular contractions.

Annelid Worms. A variety of annelids, or segmented worms, are also well adapted for life in the sediment. Most common of the sand worms is the lugworm (*Arenicola* species). It lives in a U-shaped burrow, the walls of which are strengthened with mucus (Figure 15–8B). The worm moves forward to feed and extends its proboscis (snout) up into the head shaft of the burrow to loosen sand with quick pulsing movements. A cone-shaped depression forms at the surface over the head

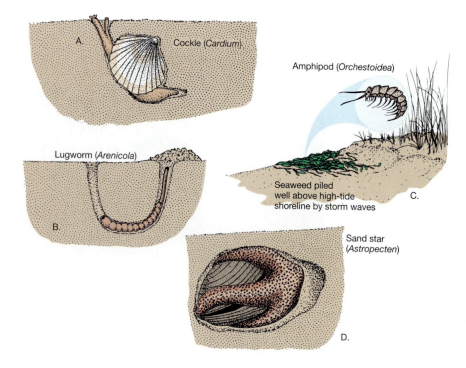

Figure 15–8

Modes of Feeding Along the Sediment-Covered Shore. *A: Suspension feeding.* This method is used by clams that bury themselves in sediment and extend siphons through the surface to pump in overlying water. They feed by filtering plankton and other organic matter that is suspended in the water. *B and C: Deposit feeding.* Some deposit feeders, such as this segmented worm *Arenicola* (*B*) feed by ingesting sediment and extracting organic matter from it. Others, such as the amphipod *Orchestoidea* (*C*), feed on more concentrated deposits of organic matter (detritus) on the sediment surface. *D: Carnivorous feeding.* The sand star, *Astropecten,* cannot climb rocks like its sea star relatives, but it can burrow rapidly into the sand, where it feeds voraciously on crustaceans, mollusks, worms, and other echinoderms.

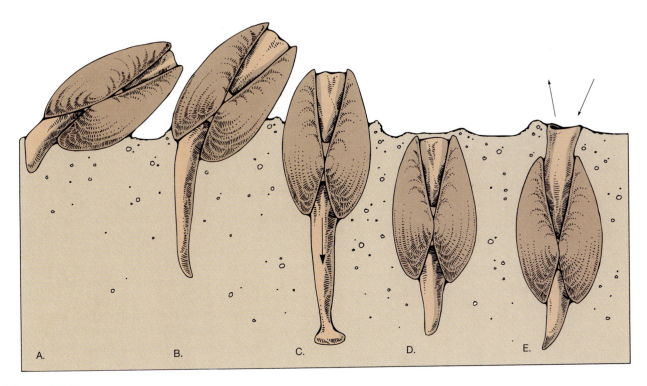

Figure 15–9

How a Clam Burrows. Clams exposed at the sediment surface quickly burrow into the sediment by (*A*) extending their pointed foot into the sediment and (*B*) forcing the foot deeper into the sediment and using this increasing leverage to bring the exposed, shell-clad body toward vertical. When the foot has penetrated deeply enough, a bulbous anchor forms at the tip (*C*), and a quick muscular contraction pulls the entire animal into the sediment (*D*). The siphons are then pushed up above the sediment to pump in water, from which the clam will extract food and oxygen (*E*).

end of the burrow as sand continually slides into the burrow and is ingested by the worm. As the sand passes through its digestive tract, the organic content is digested, and the processed sand is deposited (Figure 15–8B).

Crustaceans. Staying high on the beach and feeding on kelp cast up by storm or high tide waves are numerous crustaceans called *beach hoppers.* They are known to jump distances more than 2 meters (6.6 feet). A common genus is *Orchestoidea,* which usually range from 2 to 3 centimeters (0.8 to 1.2 inches) in length. Laterally flattened, they usually spend the day buried in the sand or hidden in the kelp on which they feed. They become active at night, and so many hop at once that they may form large clouds above the masses of seaweed on which they are feeding (Figure 15–8C).

Figure 15–10 shows a larger crustacean that may be harder to find on sandy beaches: a sand crab (genera *Blepharipoda, Emerita,* and *Lepidopa*). Ranging in length from 2.5 to 8 centimeters (1 to 3 inches), they move up and down the beach near the shoreline. They bury their bodies into the sand and leave their long, curved, V-shaped antennae pointing up the beach slope. These little crabs filter food particles from the water.

Figure 15–10

Sand Crab, *Emerita*. The head of a sand crab emerges from the beach surface. Sand crabs usually are buried just beneath the surface.

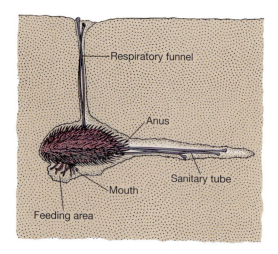

Figure 15–11

Heart Urchin, *Echinocardium*. The heart urchin feeds on the film of organic matter that covers sand grains.

Echinoderms. Echinoderms are represented in the beach deposits by the sand star (*Astropecten*) and heart urchins (*Echinocardium*). Sand stars are well adapted to prey on invertebrates that burrow into the low tide region of sandy beaches. The sand star is well designed for moving through sediment, with five tapered legs with spines and a smooth back (Figure 15–8D).

More flattened and elongated than the sea urchins of the rocky shore, heart urchins live buried in the sand near the low tide line. They gather sand grains into their mouths, where the coating of organic matter is scraped off and ingested (Figure 15–11).

Meiofauna. In Greek, *meio-* means lesser or smaller. Living in the spaces between sediment particles are lesser organisms, only 0.1 to 2 millimeters (0.004 to 0.08 inches) in length (Figure 15–12). They feed primarily on bacteria removed from the surface of sediment particles. The **meiofauna** population, composed primarily of polychaetes, mollusks, arthropods, and nematodes, are found in sediment from the intertidal zone to the deep-ocean trenches.

Intertidal Zonation

A faunal distribution exists across the intertidal range of sediment-covered shore, similar to that observed on rocky shores. The species of animals found in each of the corresponding intertidal zones are appropriately different. However, life forms across both the intertidal rocky and sediment-covered shores have two common characteristics: The maximum number of species and the greatest biomass are found near the low tide shoreline, and both decrease toward the high tide shoreline.

This zonation, shown in Figure 15–13, is best developed on steeply sloping, coarse-sand beaches and is less readily identified on the gentler sloping, fine-sand beaches. On a mud flat, the tiny clay-sized particles form a deposit with essentially no slope, which eliminates the possibility of zonation in this protected, low-energy environment.

The Mud Flat

Two widely distributed plants associated with the mud flat are eelgrass (*Zostera*) and turtle-grass (*Thalassia*). They occupy the low tide zone and adjacent

A.

B.

C.

Figure 15–12

Scanning Electron Micrographs of Meiofauna. *A:* Nematode head (804X). Image width is 0.06 millimeters. The projections and pit on the right side may be sensory structures. *B:* Amphipod (20X). This 3-millimeter-long organism builds a burrow of cemented sand grains. *C:* A 1-millimeter long polychaete (55X) with its proboscis extended. (Courtesy of Howard J. Spero, University of South Carolina.)

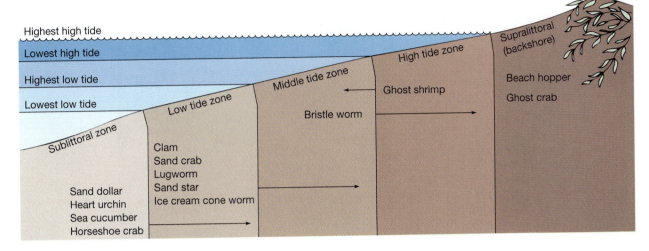

Highest high tide

Lowest high tide

Highest low tide

Lowest low tide

Sublittoral zone

Sand dollar
Heart urchin
Sea cucumber
Horseshoe crab

Low tide zone

Clam
Sand crab
Lugworm
Sand star
Ice cream cone worm

Middle tide zone

Bristle worm

High tide zone

Ghost shrimp

Supralittoral
(backshore)

Beach hopper

Ghost crab

Figure 15–13

Intertidal Zonation on the Sediment-Covered Shore. Zonation is best displayed on
coarse sand beaches with steep slopes. As the sediment becomes finer and the beach slope
decreases, zonation becomes less distinct. It disappears entirely on mud flats.

shallow sublittoral regions bordering the flats. Numerous openings at the surface of mud flats attest
to a large population of bivalve mollusks and other
invertebrates.

Quite visible and interesting inhabitants of the
mud flats are the fiddler crabs (*Uca*), living in burrows
that may exceed 1 meter (3 feet) depth. Relatives of the
shore crabs, they usually measure no more than 2 cen-

timeters (0.8 inches) across the body. Fiddler crabs get
their name because the males have one small claw and
one outsized claw (up to 4 centimeters, or 1.6 inches,
long). This large claw is waved around in such a manner the crab seems to be playing an imaginary fiddle.
The females have two normal-sized claws. The large
claw of the male is used to court females and to fight
competing males (Figure 15–14).

Figure 15–14

**Life of the Mud
Flat.** Eel-grass (*Zostera*)
and fiddler crab (*Uca*) at
Cape Hatteras, North
Carolina. (Photo by
Stephen J. Kraseman ©
Peter Arnold, Inc.)

A.

B.

Figure 15–15

Life of the Sublittoral Zone. *A:* Giant bladder kelp (*Macrocystis*). *B:* Sea hares (*Aplysia californica*) and sea urchins in a kelp forest. (*A:* Photo by Mia Tegner; *B:* by James McCullagh.)

Animals of the Shallow Offshore Ocean Floor

Extending from the spring low tide shoreline to the seaward edge of the continental shelf is an environment that is mainly sediment covered, although bare rock exposures may occur locally near shore. The sediment-covered shelf has a moderate to low species diversity. The diversity of benthos is lowest beneath upwelling regions, because upwelling carries nutrients to the surface, making pelagic production great, which produces an excess of dead organic matter. When this material rains down on the bottom and decomposes, consuming oxygen, the oxygen supply can be locally depleted.

A specialized subtidal community with a higher diversity is the giant kelp bed associated with rocky bottoms.

The Rocky Bottom (Sublittoral)

A rocky bottom within the shallow inner sublittoral region usually will be covered with algae. Along the North American Pacific coast, the giant bladder kelp (*Macrocystis*) attaches to rocks as deep as 30 meters (100 feet) if the water is clear enough to allow sunlight to support plant growth at this depth. The giant bladder kelp and another fast-growing kelp (*Nereocystis*) often form bands of *kelp forest* along the Pacific coast. Smaller tufts of red and brown algae are found on the bottom, living on the kelp fronds (Figure 15–15A).

Two terms from Chapter 12 are important in studying benthic organisms: *Infauna* are organisms that live buried *in* sand or mud (such as clams). *Epifauna* are organisms that live *on* the bottom, either attached to it (such as rock weed) or moving over it (such as crabs).

Epifauna commonly found growing along with the algae tufts on the fronds are smaller life forms that serve as food for many of the animals living within the kelp forest community. Surprisingly, very few animals feed directly on the living kelp plant. Among those that do are the large sea hare (*Aplysia*) and sea urchins (Figure 15–15B).

Lobsters. Large crustaceans that we call *lobsters* are common to rocky bottoms. They are a somewhat varied group, with robust external skeletons. The spiny lobsters are named for their spiny covering and have two very large, spiny antennae with noisemaking devices near their base (Figure 15–16). The genus *Palinurus* is a delicacy, living deeper than 20 meters (65 feet) along the European coast and reaching lengths to 50 centimeters (20 inches). For reasons unknown, the Caribbean species *Panulirus argus* sometimes migrates single-file for several kilometers.

Panulirus interruptus is the spiny lobster of the American West Coast. All spiny lobsters are taken for food, but none are as highly regarded as the so-called true lobsters (genus *Homarus*), which include the American lobster *Homarus americanus*. Although they are scavengers like their spiny relatives, the true lobsters also feed on live animals, including mollusks, crustaceans, and members of their own species (Figure 15–16).

Oysters. Oysters are sessile (anchored) bivalve mollusks found in estuarine environments. They prefer a steady flow of clean water to provide plankton and

A.

B.

Figure 15–16

Spiny and American Lobsters. *A:* Spiny lobster. *B:* American lobster. (*A:* Photo by B. Kiwala; *B:* photo by Harold W. Pratt/Biological Photo Service.)

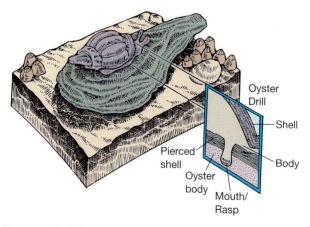

Oyster
Drill

Shell

Pierced
shell

Body

Oyster
body

Mouth/
Rasp

Figure 15–17

An Oyster Drill Feeding on an Oyster. An oyster drill snail drills through the shell of an oyster with a rasp-like mouthpiece to feed on the soft tissue beneath the shell. It seems that, for every type of armor or defense possessed by one species, a weapon is evolved by another species to defeat it.

oxygen. Having great commercial importance throughout the world, they have been closely studied.

Oyster beds are simply the empty shells of many generations cemented to rock bottom or one another, with the living generation on top. Each female produces many millions of eggs each year, which become planktonic larvae when fertilized. After a few weeks as plankton, the larvae settle to attach themselves to the bottom. As a material upon which to anchor, the oyster larvae prefer (in order) live oyster shells, dead oyster shells, and rock.

Oysters are food for a variety of sea stars, fishes, crabs, and boring snails that drill through the shell and rasp away the soft tissue of the oyster (Figure 15–17).

Coral Reefs

The coral reef has the greatest known animal species diversity of any marine biocommunity. Corals are found throughout the ocean, but coral accumulations

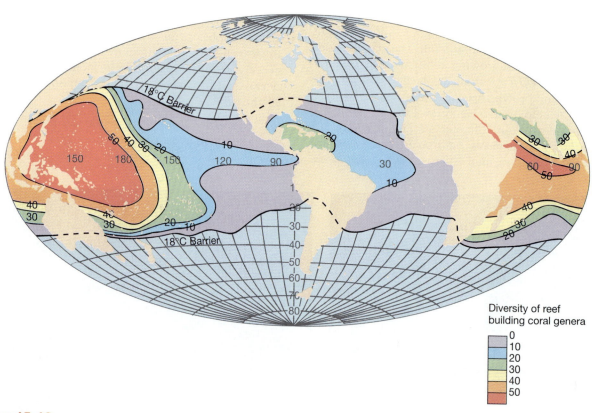

Diversity of reef
building coral genera

0
10
20
30
40
50

Figure 15–18

Coral Reef Distribution and Diversity. Coral reef development is restricted to the low-latitude area between the two 18°C (64°F) temperature lines shown. Minimum water temperatures of 18°C in surface waters of the Northern and Southern hemispheres occur in February and August, respectively. In each ocean basin, the coral reef belt is wider and the diversity of coral genera is greater on the western side. (After Stehli and Wells, 1971.)

that might be classified as *reefs* are restricted to warmer-water regions where the average monthly temperature exceeds 18°C (64°F) throughout the year (Figure 15–18). Such temperature conditions are found primarily between the tropics. However, reefs also grow to latitudes approaching 35° north and south on the western margins of ocean basins, where warm-water masses move into high-latitude areas and raise average temperatures.

Also shown in Figure 15–18 is the greater diversity of reef-building corals on the western side of ocean basins. More than 50 genera of corals thrive in a broad area of western Pacific Ocean and a narrow belt of the western Indian Ocean. Fewer than 30 genera occur in the Atlantic Ocean, with the greatest diversity occurring in the Caribbean Sea. The explanation for this pattern of diversity may be related to past ocean current patterns, but such an explanation is controversial.

Because of changes in wave energy, salinity, water depth, temperature, and other less obvious factors, there is a well-developed vertical and horizontal zonation of the reef slope (Figure 15–19). These zones are readily identified from the assemblages of plant and animal life found in and near them.

Reef growth also requires that the water have a relatively normal salinity and that it be free from particulate matter. Therefore, we see very little coral reef growth near the mouths of rivers, for they lower salinity and carry large quantities of suspended material that would choke the reef colony.

Symbiosis of Coral and Algae. "Coral reefs" are more than just coral. Algae, mollusks, and foraminifers make important contributions to the reef structure. Reef-building corals themselves are **hermatypic.** This means that they have a mutualistic relationship with algae called **zooxanthellae.** These algae live *within the tissue* of the coral polyp. Not only reef-building corals but also other reef animals have a similar symbiotic relationship with algae. Those that derive part of their nutrition from their algae partners are called **mixotrophs.** They include coral, foraminifers, sponges, and mollusks (Figure 15–20).

Because light is essential for algae photosynthesis, reef-building corals are restricted to clear, shallow waters. The algae not only nourish the coral but may contribute to their calcification capability by extracting carbon dioxide from the coral's body fluids.

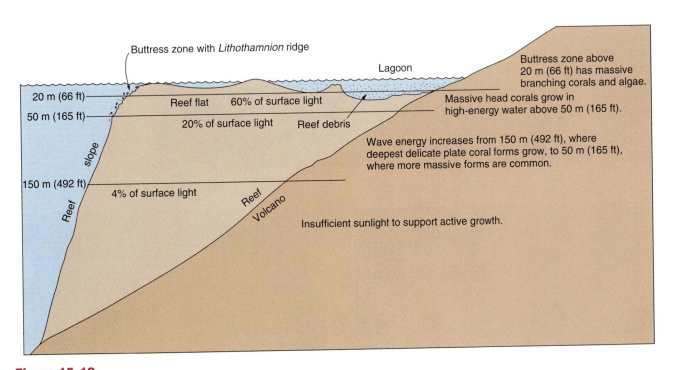

Figure 15–19

Coral Reef Zonation. As depth increases, wave energy decreases and light intensity decreases. Massive branching corals occur above 20 meters (66 feet) where wave energy is great. Corals become more delicate with increasing depth until they die out around 150 meters (500 feet). Below this depth, there is too little solar radiation to support the symbiotic algae that corals must have to survive.

A.

B.

C.

Figure 15–20

Coral Reef Inhabitants that Depend on Algae Symbiotic Partners. *A:* Polyps of a reef coral extend their tentacles to capture tiny planktonic organisms from the surrounding water. However, for most reef-building corals, most nourishment is provided by symbiotic algae, zooxanthellae, that live in the coral tissue. *B:* The blue-gray sponge on the left is *Niphates digitalis,* a totally heterotrophic sponge. On the right is *Angelas* species, which contains some cyanobacterial symbiotic partners. It is a mixotroph, like most reef-building corals. The photo was taken in 20 meters (60 feet) of water at Carrie Bow Cay, Belize. *C:* A giant clam, *Tridacna gigas.* These suspension feeders also depend on symbiotic algae living in the mantle tissue. (*A:* Photo by Christopher Newbert; *B:* C. R. Wilkerson, Australian Institute of Marine Science; *C:* Ken Lucas/Biological Photo Service.)

Corals contribute to the mutual relationship by providing nutrients to the zooxanthellae. This exchange between algae and corals, as well as other mixotrophs, supports high levels of biological productivity within the reef community despite the low nutrient level in the surrounding water.

Coral reefs actually contain up to three times as much algal biomass as animal biomass. The zooxanthellae account for less than 5 percent of the reef's overall algal mass; most of the rest is filamentous green algae. However, zooxanthellae account for up to 75 percent of the biomass of reef-building corals and provide the coral with up to 90 percent of its nutrition.

Because of the sunlight requirement for the algae, the greatest depth to which active coral growth extends is 150 meters (500 feet), below which there is not enough sunlight. Water motion is less at these depths, so relatively delicate plate corals can live on the outer slope of the reef from 150 meters (500 feet) up to about 50 meters (165 feet), where light intensity exceeds 4 percent of the surface intensity (Figure 15–19).

From 50 meters (164 feet) to about 20 meters (66 feet), the strength of water motion from breaking waves increases on the side of the reef facing into the prevailing current flow. Correspondingly, the mass of coral growth and the strength of the coral structure supporting it increase toward the top of this zone, where light intensity exceeds 20 percent of surface value.

The reef flat may be under a few centimeters to a few meters of water at low tide. A variety of beautiful reef fish inhabit this shallow water, as well as sea cucumbers, worms, and a variety of mollusks. In the protected water of the reef lagoon live gorgonian coral, anemones, crustaceans, mollusks, and echinoderms of great variety (Figure 15–21).

An obvious example of commensalism is the behavior of the shrimpfish, which swims head down among the long, slender spines of sea urchins on the reef. The urchin's spines are a significant deterrent to any predator, and it is neither hindered nor aided by the presence of the little fish.

The clown fish receives similar protection by swimming among the tentacles of two species of sea anemones. This relationship is believed to be mutual, because the anemones benefit by the clown fish serving as bait to draw other fish within reach of anemone tentacles (Figure 15–22).

Coral Reefs and Nutrient Levels.

When human populations increase in lands adjacent to coral reefs, the reefs deteriorate. Many aspects of human behavior can damage a reef: fishing, trampling, boat collisions with the reef, sediment increase due to development, and collection of reef inhabitants by visitors. One of the more subtle effects is the inevitable increase in the nutrient levels of the reef waters due to sewage discharge and farm fertilizers.

As nutrient levels increase in reef waters, the dominant benthic community changes:

- At low nutrient levels, hermatypic corals and other reef animals that contain algal symbiotic partners thrive.
- As nutrient levels in the water increase, moderate nutrient levels favor development of fleshy benthic plants, and high nutrient levels favor suspension feeders like clams.
- At high nutrient levels, the phytoplankton mass exceeds the benthic algal mass, so benthic populations that are tied to the phytoplankton food web dominate.

The clarity of water is reduced by increased phytoplankton biomass. The fast-growing members of the phytoplankton-based ecosystem destroy the reef structure in two ways: by overgrowing the slow-growing coral and through **bioerosion,** which is erosion of the reef by organisms. Bioerosion by sea urchins and sponges particularly damages the reef.

The Crown-of-Thorns Phenomenon.

Since 1962 the crown-of-thorns sea star (*Acanthaster planci*) has caused destruction of living coral on many reefs throughout the western Pacific Ocean (Figure 15–23). Some investigators believe this is a modern phenomenon brought about by the activities of humans. However, there is little evidence to point to such a cause.

A 1989 study of the Great Barrier Reef indicated that, during the past 80,000 years, the crown-of-thorns sea star has been even more abundant on the reefs studied than it is today. If this is true, the sea star may be an integral part of the reef ecology in this region rather than a destructive upstart taking advantage of human actions that have modified the reef in some way favorable to its proliferation.

Bleaching of Coral Reef Communities.

Loss of color in coral reef organisms is called **coral bleaching.** The cause of the bleaching is expulsion of the coral's symbiotic partner, the zooxanthellae algae. Recall that essentially all reef-building corals and some other reef mixotrophs are nourished by these algae, which live within their tissue. The loss of this source of nourishment can kill the reef dwellers. If the coral does not regain its zooxanthellae algae, it will die.

Coral bleaching has occurred locally numerous times in the past. However, a mass mortality of at least 70 percent of the corals along the Pacific Central American coast occurred as a result of a bleaching episode associated with the severe El Niño of 1982–1983. This eastern Pacific bleaching is believed to have been caused by warmer water temperatures associated with the El Niño. Whatever the cause, it will take many years

A.

B.

Figure 15–21

Non-reefbuilding Inhabitants. *A:* The puffer is one of the greatly varied, colorful, and strange fishes of coral reef and other shallow-water tropical environments. Feeble swimmers, puffers can't quickly escape a predator. But when a predator comes too close, puffers gulp water to fill a stomach pouch, expanding their loose, scaleless skin to produce a large, spherical shape. Puffers seldom are bothered by predators anyway, because their viscera and skin often contain a deadly neurotoxin. This puffed-up member of the genus *Arothron* and its brethren are virtually free from attack. *B:* Many members of the coral family do not secrete the calcium carbonate of the reef builders. An example is this soft gorgonian coral, shown with feeding polyps extending from its branches. (Photos by Christopher Newbert.)

Figure 15–22

Mutualism. The clown fish, which lives unharmed among the stinging tentacles of the sea anemone, may pay for this protection by attracting predator fish that can be captured by the anemone. This symbiotic relationship is an example of mutualism, which benefits both participants. (Photo by Christopher Newbert.)

for these reefs to recover. Two species of Panamanian coral became extinct during this event.

Figure 15–24 shows the bleaching of a round starlet coral (*Siderastrea siderea*) on Enrique Reef, Puerto Rico. The photograph was taken in November 1987. Normally a rusty brown color, this coral has been bleached on its lower left half.

Whether it is coral bleaching, or increased concentrations of greenhouse gases in the lower atmosphere, or depletion of upper-atmosphere ozone, or polluting the oceans with petroleum, plastics, sewage, chemicals, and sediment, human-induced changes are broad in scope worldwide. All of these may be symptoms of a worldwide pathology—sickness—that will demand major changes to human behavior before it can be cured.

Animals of the Deep-Ocean Floor

We know less of life in the deep ocean than in any of the shallower nearshore environments because of the great difficulty of investigating the deep sea. However, advances in technology are making it possible to observe and sample even the deepest reaches of the ocean. During the next decade, we should learn more

Figure 15–23

Crown-of-Thorns Sea Star. This crown-of-thorns sea star is being attacked by one of its few predators, the Pacific triton. Most of what appears red and thorny is the crown-of-thorns sea star. The sea star's tubelike feet are visible at right (the translucent tubes with yellow suction cups on their ends). The triton is the blue-and-flesh-colored shell at left, with its red-splotched white body wrapped around the sea star. (Photo © Christopher Newbert.)

about life in the deep ocean than we have learned throughout all of history.

The Physical Environment and Species Diversity

The deep-ocean floor includes the bathyal, abyssal, and hadal zones, as described in Chapter 12. Light is present in only the lowest concentrations down to 1000 meters (3300 feet), and absent below this depth. Everywhere the temperature is low, rarely exceeding 3°C (37°F) and

falling as low as –1.8°C (28.8°F) in the high latitudes. Pressure exceeds 200 atmospheres on the oceanic ridges, ranges between 300 and 500 atmospheres on the deep-ocean abyssal plains, and attains over 1000 atmospheres in the deepest trenches. (The pressure is one atmosphere at the ocean surface and increases by one atmosphere for each 10 meters, or 33 feet, of depth.)

Much of the deep-ocean floor is covered by at least a thin layer of sediment. It ranges from the mud-like abyssal clay deposits of the abyssal plains and deep trenches, through the oozes of the oceanic ridges and

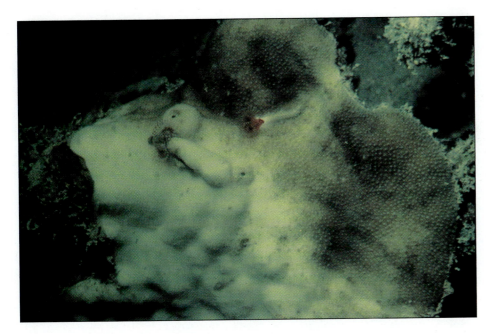

Figure 15–24

Bleached Coral. This round starlet coral (*Siderastrea siderea*) on Enrique Reef, Puerto Rico, has been bleached across the lower left side of the image. (Photo by Dr. Lucy Bunkley Williams.)

rises, to some coarse sediment deposited on the continental rise. Near the crests of the oceanic ridges and rises and down the slopes of seamounts and oceanic islands, sediment is absent and basaltic ocean crust forms the bottom.

A 1989 study of sediment-dwelling animals in the North Atlantic revealed an unexpectedly large diversity of species. An area of 21 square meters (225 square feet) contained 898 species. New to science were 460 of these. After analyzing 200 samples, new species were being revealed at a rate that suggested millions of deep-sea species. However, deep-sea sediment-dwelling benthos have been insufficiently studied to determine accurately their distribution patterns or diversity.

Deep-Sea Hydrothermal Vent Biocommunities

In 1977 the first active hydrothermal vent field was discovered below 2500 meters (8200 feet) in the Galápagos Rift, near the equator in the eastern Pacific Ocean (Figures 15–25 and 15–26). Water temperature immediately around the vents were 8° to 12°C (46° to 54°F), whereas normal bottom-water temperature at these depths is about 2°C (36°F).

These vents were found to support the first known **hydrothermal vent biocommunities,** consisting of unusually large organisms for these depths. The more prominent members are tube worms over 1 meter long, giant clams up to 25 centimeters (10 inches) in length, large mussels, and two varieties of white crabs.

At 21°N latitude on the East Pacific Rise, south of the tip of Baja California (Figure 15–26), tall underwater chimneys were found to belch hot vent water (350°C or 662°F) so rich in metal sulfides that it was black. These chimney vents, first observed in 1979, were composed primarily of sulfides of copper, zinc, and silver and came to be called *black smokers*. A new species of vent fish was first observed here.

The most important members of these hydrothermal vent biocommunities are the **sulfur-oxidizing bacteria.** Through chemosynthesis, they manufacture organic molecules that feed the community and are the base of the food web around the vents. The bacteria use the chemical energy released by sulfur oxidation much as marine algae use the sun's energy to carry on photosynthesis. Although some animals feed on the bacteria and larger prey, many of them depend primarily on a symbiotic relationship with the bacteria. For instance, the tube worms and giant clams depend entirely on sulfur-oxidizing bacteria that live symbiotically within their tissues.

In 1981 the Juan de Fuca Ridge biocommunity was observed (Figures 15–27A and 15–26). Although vent fauna at this site are less robust than at the Galápagos Rift and on the East Pacific Rise, the metallic sulfide deposits from the vents aroused much interest because they are so near the U.S. coast.

The Guaymas Basin, in the Gulf of California (Figure 15–26), was the location of the first observation, in 1982, of hydrothermal vents beneath a thick layer of sediment. Sediment and sulfide samples recovered were saturated with hydrocarbons, which may have entered the food chain through bacterial uptake (Figure 15–27B). The abundance and diversity of life may exceed that of the rocky bottom vents, such as those on

Figure 15–25

Alvin **Approaches a Hydrothermal Vent Community Typical of Those on the East Pacific Rise.** In the lower left corner are a sea anemone and three octacoras, close relatives of the anemone. They may feed primarily on sulfur-oxidizing bacteria suspended in the water near the vents. The two orange grenadier fish (or rattail fish) are common in the deep ocean and are usually the first to arrive at bait placed on the deep-ocean floor. The red-headed tube worms (*Riftia*) and large clams (*Calypotogena*) do not possess guts: They are nourished instead by chemosynthetic bacteria that live in their tissues. White brachyuran crabs swarm over the lava pillows, and a "black smoker" spews hot (350°C, or 662°F), sulfide-rich water from its metallic sulfide chimney.

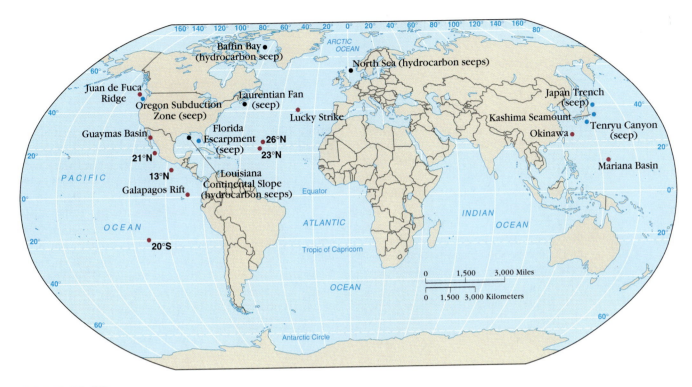

Figure 15–26

Vents and Seeps Known to Support Biocommunities. Hydrothermal vents (red), cold seeps (blue), and hydrocarbon seeps (black).

A.

B.

Figure 15–27

Hydrothermal Vent Biocommunities of the Juan de Fuca Ridge, Guaymas Basin, and Mariana Basin. *A:* Tube worms (*Ridgeia phaeophiale*) found at Juan de Fuca vents. *B:* At the Guaymas Basin in the Gulf of California, these tube worms (*Riftia pachyptila*) are intergrown with a bacteria mat composed of very large bacteria (about 150 micrometers diameter). *C:* This Mariana Back-Arc Basin community includes a new genus and species of sea anemone (*Marianactis bythios*), a new family, genus, and species of gastropod (*Alviniconcha hessleri*), the first known snail to contain chemosynthetic bacteria, and the galatheid crab (*Munidopsis marianica*). (*A:* Photo by Robert W. Embly, courtesy of William Chadwick, Jr., Oregon State University, Hatfield Marine Science Center; *B:* photo by Robert Hessler, courtesy of Scripps Institution of Oceanography, University of California, San Diego; *C:* photo courtesy of Robert Hessler.)

C.

the East Pacific Rise. (There is no sediment on the rise, where the first vents were found.)

A phenomenon similar to that of the Guaymas Basin was reported in 1987 at the Mariana Basin: a small spreading-center system beneath a sediment-filled back-arc basin (Figures 15–27C and 15-26). Subsequent exploration has revealed numerous hydrothermal vent biocommunities in the Pacific Ocean, including one in the Southern Hemisphere on the East Pacific Rise at 20°S (Figure 15–26).

The first active hydrothermal vents with associated biocommunities in the Atlantic Ocean were discovered in 1985 at depths below 3600 meters (11,800 feet) near the axis of the Mid-Atlantic Ridge at 23°N and 26°N. The predominant fauna of these vents consists of shrimp that have no eye lens but can detect levels of light emitted by the black smoker chimneys that is not visible to the human eye (Figure 15–28).

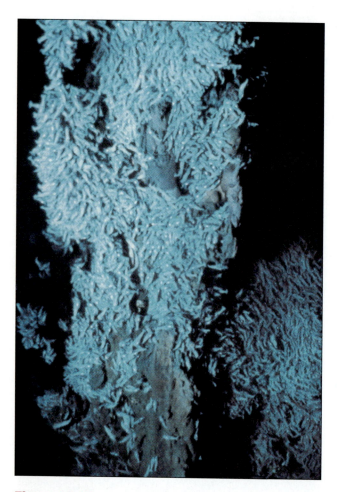

Figure 15–28

Atlantic Ocean Hydrothermal Vent Organisms. Swarm of particulate-feeding shrimp, the predominant animals observed at hydrothermal vents near 23°N and 26°N on the Mid-Atlantic Ridge. This swarm was photographed at 26°N. (Courtesy of Peter A. Rona, NOAA.)

In 1993 a hydrothermal vent community was discovered on a flat-topped volcano rising to 1525 meters (5000 feet)—well above the walls of the Mid-Atlantic Ridge rift valley. Called the Lucky Strike vent field, it is about 1000 meters (3300 feet) shallower than most other sites. It is the only Mid-Atlantic Ridge site to possess the mussels common at many other vent sites and is the only location where a new species of pink sea urchin is found.

An ancient vent area found on the Galápagos spreading center indicates that vent communities may have a relatively short life span. This inactive vent was identified by an accumulation of dead clams. It appears that when the vent becomes inactive and the hydrogen sulfide that serves as the source of energy for the community is no longer available, the community dies.

A low species diversity has been found to date. There are just over 100 known animal species from this very unstable environment. Many species are common to hydrothermal vent communities in general.

Low-Temperature Seep Biocommunities

Three additional submarine spring environments also have been found to chemosynthetically support biocommunities. In 1984 a **hypersaline seep** (46.2 parts per thousand salinity) of ambient temperature was studied at the base of the Florida Escarpment in the Gulf of Mexico (Figure 15–29A). Researchers discovered a biocommunity similar in many respects to the hydrothermal vent communities. The seeping water appears to flow from joints at the base of the limestone escarpment (Figure 15–29B) and move out across the clay deposits of the abyssal plain at a depth of about 3200 meters (10,500 feet).

The hydrogen sulfide-rich waters support a number of white bacterial mats that carry on chemosynthesis in a fashion similar to bacteria at hydrothermal vents. These and other chemosynthetic bacteria may provide most of the support for a diverse community of animals. The community includes sea stars, shrimp, snails, limpets, brittle stars, anemones, tube worms, crabs, clams, mussels, and some fish (Figure 15–29C).

Also observed in 1984 were dense biological communities associated with oil and gas seeps on the Gulf of Mexico continental slope (Figure 15–30). Trawls at depths of between 600 and 700 meters (2000 and 2300 feet) recovered fauna similar to those observed at the hydrothermal vents and the hypersaline seep at the base of the Florida Escarpment. Subsequent investigations identified seeps with associated communities to depths of 2200 meters (7300 feet) on the continental slope.

Carbon-isotope analysis indicates that **hydrocarbon seep biocommunities** are based on a chemosynthetic productivity that derives its energy from hydrogen

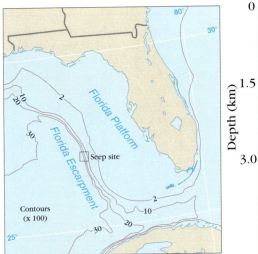

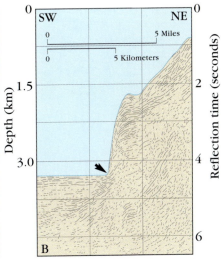

Figure 15–29

Hypersaline Seep Biocommunity at Base of Florida Escarpment. *A:* Location of seep and biocommunity. *B:* Seismic reflection profile of limestone Florida Escarpment and abyssal bedded sediments at its base. Arrow marks location of seep. *C:* Florida Escarpment seep biocommunity of dense mussel beds covers much of the image. White dots are small gastropods on mussel shells. At lower right are tube worms covered with hydrozoans and galatheid crabs. Fractures in the escarpment limestone are visible along the top. (Courtesy of C. K. Paull, Scripps Institution of Oceanography, University of California, San Diego.)

C.

sulfide and/or methane. Bacterial oxidation of methane results in the production of calcium carbonate slabs found here and at other hydrocarbon seeps, shown in Figure 15–26.

Finally, a third environment of **subduction zone seep** communities was observed from *Alvin* during 1984. It is located at a subduction zone of the Juan de Fuca Plate at the base of the continental slope off the coast of Oregon (Figure 15–31). Here the trench is filled with sediments. At the seaward edge of the slope, clastic sediments are folded into a ridge. At the crest of the ridge, pore water escapes from the 2-million-year-old folded sedimentary rocks into a thin overlying layer of soft sediment.

At a depth of 2036 meters (6678 feet), the seeps produce water that is only slightly warmer (about 0.3C°,

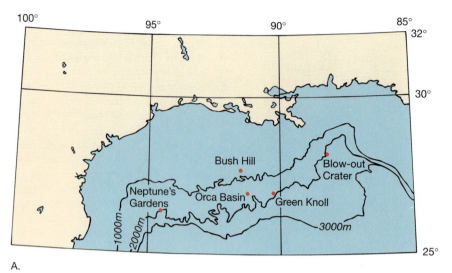

A.

Figure 15–30

Hydrocarbon Seeps on the Continental Slope of the Gulf of Mexico. *A:* Locations of known hydrocarbon seeps with associated biocommunities. *B:* Chemosynthetic mussels and tube worms from the Bush Hill seep. The mussel has demonstrated that it can grow with methane as its sole carbon and energy source. The tube worm is dependent on hydrogen sulfide. *C:* Alaminos Canyon site (Neptune's Garden), discovered in 1990, is a significant oil seep with a new species of mussel that probably depends on methane, and two species of tube worms that depend on hydrogen sulfide. One of the tube worms may be a new species. (*B* and *C:* Courtesy of Charles R. Fisher, Penn State University.)

B.

C.

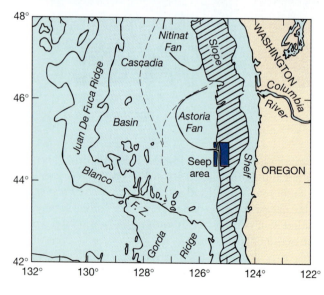

Figure 15–31

Locations of Vent Communities off the Coast of Oregon. These communities are associated with the subduction zone of the Juan de Fuca Plate. Sediment filling the trench is folded into a ridge with vents at its crest. (Courtesy of L. D. Klum, Oregon State University.)

or 0.5F°) than ambient conditions. The vent water contains methane that is probably produced by decomposition of organic material in the sedimentary rocks. The methane is the source of energy for bacteria that oxidize it and chemosynthetically produce food for themselves and the rest of the community, which contains many of the same genera found at other vent and seep sites.

During 1985 similar communities were located in subduction zones of the Japan Trench and Peru-Chile Trench. All the seeps are located on the landward side of the trenches at depths from 1300 to 5640 meters (4265 to 18,500 feet).

S U M M A R Y

Over 98 percent of the more than 235,000 species of marine animals live on the ocean floor. Most of these live in deep-sea sediment, with only 400-odd species found in deep-ocean trenches. The importance of temperature on the development of species diversity is reflected in the fact that there are three times as many species of benthos along the European coast, warmed by the Gulf Stream, than along a similar length of North American coast, cooled by the Labrador Current.

Because of tidal motions, the intertidal zone can be divided into the high tide zone (mostly dry), middle tide zone (equally wet and dry), and the low tide zone (mostly wet). The intertidal zone is bounded by the supralittoral (covered only by storm waves) and the sublittoral, which extends below the low tide shoreline.

The supralittoral zone along rocky coasts is characterized by the presence of the periwinkle snail, the rock louse, and the limpet.

The barnacles most characteristic of the high tide zone are the tiny buckshot barnacles. The most conspicuous algae of the high tide zone are *Fucus* along colder shores and *Pelvetia* in warmer latitudes. These algae become more abundant in the middle tide zone, and the diversity and abundance of the flora and fauna in general increase toward the lower intertidal zone.

Larger species of acorn barnacles are found in the middle tide zone. A common assemblage of rocky middle tide zones includes the goose barnacle, the mussels, and sea stars. Joining the sea stars as predators on the barnacles and mussels are carnivorous snails.

Tide pools within the middle tide zone commonly house sea anemones, fishes, and hermit crabs. At the lower limit of the middle tide zone, sea urchins become numerous.

The low tide zone in temperate latitudes is characterized by a variety of moderately sized red and brown algae providing a drooping canopy for the animal life. Scampering across the entire intertidal zone are the scavenging shore crabs.

The only sediments usually found along rocky shores are a few cobbles and boulders. In more protected segments of the shore, lower levels of wave energy allow deposition of sand and mud. Sand deposits are usually well oxygenated, and mud deposits are anaerobic below a thin surface layer.

Compared with the rocky shore, the diversity of species is reduced on and in beach deposits and is quite restricted in mud deposits. This does not mean that the abundance of life is reduced. Although life is less visible, up to 8000 burrowing clams have been recovered from 1 square meter (10.8 square feet) of mud flat. Suspension feeders (filter feeders) characteristic of the rocky shore are still found in the sediment, but there is a great increase in the relative abundance of deposit feeders that ingest sediment and detritus.

Bivalve (two-shelled) mollusks are well suited for sediment-covered shores. The lugworm is a deposit feeder that ingests sand and extracts whatever organic content is available.

Beach hoppers feed on kelp deposited high on the beach by storm waves. Sand crabs filter food from the water with their long, curved antennae while buried in the sand near the shoreline.

Echinoderms are represented in the sandy beach by the sand star, which feeds on buried invertebrates, and the heart urchin, which scrapes organic matter from sand grains.

As is true for the rocky shore, the diversity of species and abundance of life on the sediment-covered shore increases toward the low tide shoreline. Zonation is best developed on steep, coarse sand beaches and is probably missing on mud flats with little or no slope.

Attached to the rocky sublittoral bottom just beyond the shoreline is a band of algae including large kelp. Growing on the large fronds of the kelp are small varieties of algae and fauna and the sea hare and sea urchin feed on the kelp and other algae. Spiny lobsters are common to rocky bottoms in the Caribbean and along the West Coast, and the American lobster is found from Labrador to Cape Hatteras.

Oyster beds found in estuarine environments consist of individuals that attach themselves to the bottom or to the empty shells of previous generations.

Above a depth of 150 meters (500 feet) in clear, nutrient-poor tropical waters, living coral reef can be found off the shores of islands and continents. Reef-building corals and other mixotrophs are hermatypic, containing algae symbiotic partners (zooxanthellae) in their tissues. Delicate varieties are found at 150 meters, and they become more massive near the surface, where wave energy is higher.

The top 20 meters (66 feet), the buttress zone, is reinforced with calcium carbonate deposited by algae such as *Lithothamnion,* producing an algae ridge. Waves cut surge channels across the ridge, and debris channels extend down the reef front below the surge channels.

Many varieties of commensalism and mutualism are found within the coral reef biological community. There is some evidence that the crown-of-thorns may be a natural phenomenon and the potentially lethal "bleaching" of coral reefs may result from the expulsion of symbiont algae under stress of elevated temperature.

Although little is known of the deep-ocean benthos, it is clear that it is much more varied than previously thought.

An important discovery of the hydrothermal vent communities made in 1977 on the Galápagos Rift has shown that, at least locally, chemosynthesis is an important means of primary productivity at these vents and at cold-water and hydrocarbon seeps.

K E Y   T E R M S

Bioerosion (p. 353)
Coral bleaching (p. 353)
Hermatypic (p. 351)
High tide zone (p. 340)
Hydrocarbon seep
biocommunity (p. 360)

Hydrothermal vent
biocommunity (p. 357)
Hypersaline seep (p. 360)
Intertidal zone (p. 357)
Low tide zone (p. 342)
Meiofauna (p. 346)
Middle tide zone (p. 340)

Mixotroph (p. 351)
Spray zone (p. 340)
Subduction zone seep
biocommunity (p. 361)
Sulfur-oxidizing bacteria (p. 357)
Zooxanthellae (p. 351)

Q U E S T I O N S   A N D   E X E R C I S E S

1. Discuss the general distribution of life in the ocean. Include species diversity (variety of types of organisms) between the pelagic and benthic environments and within the benthic environment.
2. Diagram the intertidal zones of the rocky shore and list characteristic organisms of each zone.
3. Describe the dominant feature of the middle tide zone along rocky coasts, the mussel bed. Include a discussion of other organisms found in association with the mussels.
4. List and describe crabs found within the rocky shore intertidal zone.
5. Discuss how sediment stability and oxygenation of sandy and muddy shores differ.
6. Other than predation, discuss the two types of feeding styles that are characteristic of the rocky, sandy, and muddy shores. One of these feeding styles is rather well represented in all these environments; name it and give an example of an organism that uses it in each environment.
7. How does the diversity of species on sediment-covered shores compare with that of the rocky shore? Can you think of any reasons why this should be so? If you can, discuss them.
8. In which intertidal zone of a steeply sloping, coarse sand beach would you find clams, beach hoppers, ghost shrimp, sand crabs, and heart urchins?
9. What relationships exist among intertidal zonation, sediment particle size, and beach slope?
10. Discuss the dominant species of kelp, their epifauna, and animals that feed on kelp in the Pacific coast kelp forest.
11. Discuss the preferred environment, reproduction, and threats to survival of larval and adult forms of oysters.
12. Describe the environment suited to development of coral reefs.
13. Describe the zones of the reef slope, the characteristic coral types, and the physical factors related to zonation.
14. As one moves from the shoreline to the deep-ocean floor, what changes in the physical environment can be expected?
15. What are the major differences between the conditions and biocommunities of the hydrothermal vents and the cold seeps? How are they similar?

R E F E R E N C E S

Chadwick, W. W.; Embley, R. W.; and Fox, C. G. 1991. Evidence for volcanic eruption on the southern Juan de Fuca ridge between 1981 and 1987. *Nature* 350, 416–418.

Childress, J. J.; Fisher, C. R.; Brooks, J. M.; Kennicutt, M. C., II; Bidigare, R.; and Anderson, A. E. 1986. A methanotrophic marine molluscan (Bivalvia, Mytilidae) symbiosis: Mussels fueled by gas. *Science* 233:4770, 1306–1308.

Fisher, C. R. 1990. Chemoautrophic and methanotrophic symbioses in marine invertebrates. *Reviews in Aquatic Sciences* 2:3 and 4, 399–436.

George, D., and George, J. 1979. *Marine life: An illustrated encyclopedia of invertebrates in the sea.* New York: Wiley-Interscience.

Grassle, F. J., and Maciolek, N. J. 1992. Deep-sea species richness: Regional and local diversity estimates from quantitative bottom samples. *American Naturalist* 139:2, 313–341.

Hallock, P., and Schlager, W. 1986. Nutrient excess and the demise of coral reefs and carbonate platforms. *Palaios* 1:389–398.

Hessler, R. R.; Ingram, C. L.; Yayanos, A. A.; and Burnett, B. R. 1978. Scavenging amphipods from the floor of the Philippine Trench. *Deep-Sea Research* 25:1029–1047.

Hessler, R. R., and Lonsdale, P. F. 1991. Biogeography of Mariana Trough hydrothermal vent communities. *Deep-Sea Research* 38:2, 185–199.

Kennicutt, M. C., II; Brooks, J. M.; Bidigare, R. R.; Fay, R. R.; Wade, T. L.; and McDonald, T. J. 1985. Vent-type taxa in a hydrocarbon seep region on the Louisiana slope. *Nature* 317:6035, 351–353.

Klum, L. D.; Suess, E.; Moore, J. C.; Carson, B.; Lewis, B. T.; Ritger, S. D.; Kadko, D. C.; Thornburg, T. M.; Embley, R. W.; Rugh, W. D.; Massoth, G. J.; Langseth, M. G.; Cochrane, G. R.; and Scamman, R. L. 1986. Oregon subduction zone: Venting fauna and carbonates. *Science* 231:4738, 561–566.

MacGinitie, G. E., and MacGinitie, N. 1968. *Natural history of marine animals.* 2nd ed. New York: McGraw-Hill.

Ricketts, E. F.; Calvin, J.; and Hedgpeth, J. 1968. *Between Pacific tides.* Stanford, Calif.: Stanford University Press.

Rona, P. A.; Klinkhammer, G.; Nelson, T. A.; Trefry, J. H.; and Elderfield, H. 1986. Black smokers, massive sulphides and vent biota at the Mid-Atlantic Ridge. *Nature* 321:6065, 33–37.

Thorne-Miller, B., and Catena, J. 1991. *The living ocean: Understanding and protecting marine biodiversity.* Washington, D.C.: Island Press.

Walbran, P. D.; Henderson, R. A.; Jull, A. J. T.; and Head, M. J. 1989. Evidence from sediments of long-term *Acanthaster planci* predation on corals of the Great Barrier Reef. *Science* 245:4920, 847–850.

Williams, E. H., Jr.; Goenaga, C.; and Vicente, V. 1987. Mass bleaching on Atlantic coral reefs. *Science* 238:4830, 877–878.

Yonge, C. M. 1963. *The sea shore.* New York: Atheneum.

Zenevitch, L. A.; Filatove, A.; Belyaev, G. M.; Lukanove, T. S.; and Suetove, I. A. 1971. Quantitative distribution of zoobenthos in the world ocean. *Bulletin der Moskauer Gen der Naturforscher, Abt. Biol.* 76, 27–33.

ZoBell, C. E. 1968. Bacterial life in the deep sea. In Proceedings of the U.S.-Japan Seminar on Marine Microbiology, August 1966, Tokyo. *Bulletin Misaki Marine Biology, Kyoto Inst. Univ.* 12:77–96.

SUGGESTED READING

Sea Frontiers

Alper, J. 1990. The methane eaters. 36:6, 22–29. An account of the exploration of hydrocarbon seep biocommunities in the Gulf of Mexico.

Coleman, N. 1974. Shell-less molluscs. 20:6, 338–342. A description of nudibranchs, gastropods without shells.

George, J. D. 1970. The curious bristle-worms. 16:5, 291–300. The variety of worms belonging to the class *Polychaeta* of phylum Annelida is described. The discussion includes locomotion, feeding, and reproductive habits of the various members of the class.

Gibson, M. E. 1981. The plight of *Allopora* 27:4, 211–218. The reason the author believes the unusual California hydrocoral is headed for the endangered species list is the topic of this article.

Humann, P. 1991. Loving the reef to death. 37:2, 14–21. The many ways divers in particular and humans in general degrade the coral reef environment are discussed.

McClintock, J. 1994. Out of the oyster. 40:3, 18–23. The history of our interest in pearls as jewelry and the development of the industry of culturing pearls are discussed.

Ruggiero, G. 1985. The giant clam: Friend or foe? 31:1, 4–9. The ecology and behavior of the giant clam (*Tridacna gigas*) is discussed in reference to whether it is a danger to divers.

Shinn, E. A. 1981. Time capsules in the sea. 27:6, 364–374. The method by which geologists determine past climatic and environmental conditions by studying coral reefs is discussed.

Viola, F. J. 1989. Looking for exotic marine life? Don't leave the dock. 35:6, 336–341. A description of marine life living on and near pilings for a boat dock are described along with good color photos.

Winston, J. E. 1990. Intertidal space wars. 36:1, 47–51. Florida intertidal life is discussed.

Scientific American

Caldwell, R. L., and Dingle, H. 1976. Stomatopods. 234:1, 80–89. Presents the ecology of these interesting crustaceans that have appendages specialized for spearing and smashing prey.

Feder, H. A. 1972. Escape responses in marine invertebrates. 227:1, 92–100. Discusses the surprisingly rapid movements and other responses made by invertebrates to the presence of predators. Some interesting photographs accompany the text, which describes the escape responses of limpets, snails, clams, scallops, sea urchins, and sea anemones.

Wicksten, M. K. 1980. Decorator crabs. 242:2, 146–157. Describes how species of spider crabs use materials from their environment to camouflage themselves.

Younge, C. M. 1975. Giant clams. 232:4, 96–105. The distribution and general ecology of the tridacnid clams, some of which grow to lengths well over 1 meter (3.3 feet), are investigated.

Appendix I
Scientific Notation

To simplify writing very large and very small numbers, scientists indicate the number of zeros by scientific notation, in which one integer is placed to the left of the decimal and a multiplication times a power of 10 tells which direction and how far the decimal would be moved to write the number out in its long form. For example:

$$2.13 \times 10^5 = 213,000$$
or
$$2.13 \times 10^{-5} = 0.0000213$$

Further examples showing numbers that are powers of 10 are:

$$
\begin{array}{lll}
1,000,000,000 = 1.0 \times 10^9 & \text{or} & 10^9 \\
1,000,000 = 1.0 \times 10^6 & \text{or} & 10^6 \\
1,000 = 1.0 \times 10^3 & \text{or} & 10^3 \\
100 = 1.0 \times 10^2 & \text{or} & 10^2 \\
10 = 1.0 \times 10^1 & \text{or} & 10^1 \\
1 = 1.0 \times 10^0 & \text{or} & 10^0 \\
0.1 = 1.0 \times 10^{-1} & \text{or} & 10^{-1} \\
0.01 = 1.0 \times 10^{-2} & \text{or} & 10^{-2} \\
0.001 = 1.0 \times 10^{-3} & \text{or} & 10^{-3} \\
0.000001 = 1.0 \times 10^{-6} & \text{or} & 10^{-6} \\
0.000000001 = 1.0 \times 10^{-9} & \text{or} & 10^{-9}
\end{array}
$$

To add or subtract numbers written as powers of 10, they must be converted to the same power:

Addition

$$
\begin{array}{ll}
2.1 \times 10^3 & 0.021 \times 10^5 \\
+ 1.0 \times 10^5 & + 1.000 \times 10^5 \\
\hline
 & = 1.021 \times 10^5
\end{array}
$$

Subtraction

$$
\begin{array}{ll}
3.4 \times 10^4 & 3.4 \times 10^4 \\
- 2.0 \times 10^3 & - 0.2 \times 10^4 \\
\hline
 & = 3.2 \times 10^4
\end{array}
$$

To multiply or divide numbers written as powers of 10, the exponents are added or subtracted:

Multiplication

$$
\begin{array}{l}
6.04 \ \times 10^2 \\
\times \ 2.1 \ \ \times 10^4 \\
\hline
= 12.684 \times 10^6
\end{array}
$$

Division

$$
\begin{array}{l}
3.0 \times 10^3 \\
\div 1.5 \times 10^2 \\
\hline
= 2.0 \times 10^1
\end{array}
$$

Appendix II
Latitude and Longitude Determination

A British battle fleet commanded by Sir Cloudesley Shovell ran aground in the Scilly Islands in 1707 with the loss of four ships and 2000 men. This happened because they lost track of their longitude (location east or west) after weeks at sea. To determine longitude at any location in the ocean, it was necessary to know the *time* at a reference meridian (we now use the Greenwich Meridian, running through Greenwich, England). However, clocks in 1707 were driven by pendulums and would not work for long on a ship rocking at sea.

In 1714 the British government offered a £20,000 prize (about $2,000,000 today) for the development of a clock that would work well enough at sea to determine longitude within 0.5° after a voyage to the West Indies. A cabinetmaker in Lincolnshire named John Harrison began working on such a timepiece in 1728. His chronome-

ter was driven by a helical balance spring and remained horizontal regardless of the attitude of the ship. It was complicated, costly, and delicate. In 1736 his first chronometer was successfully tested and he received £500. Eventually, his fourth version was tested in 1761. Upon reaching Jamaica, it was so accurate that the longitude error was only 1.25′ (recall that a minute is only $\frac{1}{60}°$) and it had lost only 5 seconds of time.

This performance greatly exceeded the requirements of the government, and Harrison claimed the prize. The government was slow in paying because they wanted to be convinced that reliable versions of the chronometer could be produced in large numbers. Finally, after the intervention of King George III, Harrison received the balance of his prize in 1773, at age 80.

Figure 1

Determining Latitude. The method of determining latitude used by Pytheas was to measure the angle between the horizon and the North Star (Polaris), which is directly above the North Pole. Latitude north of the equator is the angle between the two sightings. Similar determinations may be made in the Southern Hemisphere by using the Southern Cross, which is directly overhead at the South Pole.

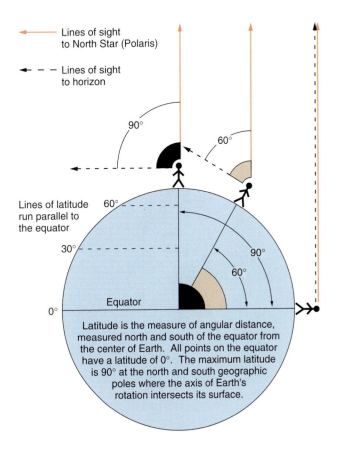

Lines of sight to North Star (Polaris)

Lines of sight to horizon

90°

60°

Lines of latitude run parallel to the equator

60°

30°

90°

60°

Equator

0°

Latitude is the measure of angular distance, measured north and south of the equator from the center of Earth. All points on the equator have a latitude of 0°. The maximum latitude is 90° at the north and south geographic poles where the axis of Earth's rotation intersects its surface.

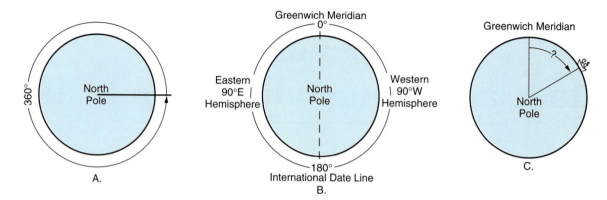

Figure 2

Determining Longitude. View of Earth from outer space, looking down on North Pole.
A: As Earth turns on its rotational axis, it moves through 360° of angle every 24 hours.
B: The meridian that runs from pole to pole through Greenwich, England, was selected as
the reference meridian, dividing Earth into a Western and Eastern hemisphere. After John
Harrison's chronometer was developed, many ships carried it, showing the time on the
Greenwich Meridian—Greenwich time. *C:* Since Earth rotates through 15° of angle, or longitude, per hour (360° ÷ 24 hours = 15°/hour), a ship's captain could easily determine longitude each day at noon. For example, a ship sets sail west across the Atlantic Ocean, checking
its longitude each day at noon (when the sun crosses the meridian running directly overhead). One day when the sun is at the noon position, the captain checks the chronometer.
It reads 16:18 hours (0:00 is midnight and 12:00 is noon). What is the ship's longitude?

Longitude solution:
16:18 hours = 4:18 P.M.
Earth rotates through ¼° (15′) of angle per minute of time.
(One degree of arc is divided into 60 minutes.)
4 hours × 15°/hour = 60° of longitude
18 minutes of time × 15 minutes of angle/minute of time = 270 minutes of arc
270 minutes ÷ 60 minutes/degree (°) = 4.5° of longitude
60° + 4.5° = 64.5°W longitude

Appendix III
Roots, Prefixes, and Suffixes*

a- not, without
ab-, abs- off, away, from
abysso- deep
acanth- spine
acro- the top
aero- air, atmosphere
aigial- beach, shore
albi- white
alga- seaweed
alti- high, tall
alve- cavity, pit
amoebe- change
an- without, not
annel- ring
annu- year
anomal- irregular, uneven
antho- flower
apex- tip
aplysio- sponge, filthiness
aqua- water
arachno- spider
arena- sand
arthor- joint
asthen- weak, feeble
astro- star
auto- self
avi- bird
bacterio- bacteria
balaeno- whale
balano- acorn
barnaco- goose
batho- deep
bentho- the deep sea
bio- living
blast- a germ
botryo- bunch of grapes
brachio- arm
broncho- windpipe
bryo- moss
bysso- a fine thread
calci- limestone (CACO₃)
calori- heat
capill- hair
cara- head

carno- flesh
cartilagi- gristle
caryo- nucleus
cat- down, downward
cen-, ceno- recent
cephalo- head
ceta- whale
chaeto- bristle
chiton- tunic
chlor- green
choano- funnel, collar
chondri- cartilage
cilio- small hair
circa- about
cirri- hair
clino- slope
cnido- nettle
cocco- berry
coelo- hollow
cope- oar
crusta- rind
cteno- comb
cyano- dark blue
cypri- Venus, lovely
-cyst bladder, bag
-cyte cell
deca- ten
delphi- dolphin
di- two, double
diem- day
dino- whirling
diplo- double, two
dolio- barrel
dors- back
-dram- run, a race
echino- spiny
eco- house, abode
ecto- outside, outer
edrio- a seat
en- in, into
endo- inner, within
entero- gut
epi- upon, above
estuar- the sea

eu- good, well
exo- out, without
fauna- animal
fec- dregs
fecund- fruitful
flacci- flake
flagell- whip
flora- flower
fluvi- river
fossili- dug up
fuco- red
geno- birth, race
geo- Earth
giga- very large
globo- ball, globe
gnatho- jaw
guano- dung
gymno- naked, bare
halo- salt
haplo- single
helio- the sun
helminth- worm
hemi- half
herbi- plant
herpeto- creeping
hetero- different
hexo- six
holo- whole
homo-, homeo- alike
hydro- water
hygro- wet
hyper- over, above, excess
hypo- under, beneath
ichthyo- fish
-idae members of the animal family of
infra- below, beneath
insecti- cut into
insula- an island
inter- between
involute- intricate
iso- equal
ite rock
juven- young
juxta- near to

*Source: D. J. Borror, 1960. *Dictionary of Word Roots and Combining Forms*. Palo Alto, Calif.: National Press Books.

kera- horn
kilo- one thousand
lacto- milk
lamino- layer
larvi- ghost
latent- hidden
latero- side
lati- broad, wide
lemni- water plant
limno- marshy lake
lipo- fat
litho- stone
litorial near the seashore
lopho- tuft
lorica- armor
luci- light
luna- moon
lux- light
macro- large
mala- jaw, cheek
mamilla- teat
mandibulo- jaw
mantle- cloak
mari- the sea
masti- chewing
mastigo- whip
masto- breast, nipple
maxillo- jaw
madi- middle
medus- a jellyfish
mega- great, large
meio- less
meridio- noon
meros- part
meso- middle
meta- after
meteor- in the sky
-meter measure
-metry science of measuring
mid- middle
milli- thousandth
mio- less
moll- soft
mono- one, single
-morph form
myo- muscle
myst- mustache
nano- dwarf

necto- swimming
nemato- thread
neo- new
neph- cloud
nerito- sea lymph
noct- night
-nomy the science of
nucleo- nucleus
nutri- nourishing
o-, oo- egg
ob- reversed
ocellus- little eye
octa- eight
oculo- eye
odonto- teeth
oiko- house, dwelling
oligo- few, scant
-ology science of
omni- all
ophi- a serpent
opto- the eye, vision
orni- a bird
-osis condition
oto- hear
ovo- egg
pan- all
para- beside, near
pari- equal
pecti- comb
pedi- foot
penta- five
peri- all around
phaeo- dusky
pholado- lurking in a hole
-phore carrier of
photo- light
phyto- plant
pinnati- feather
pisci- fish
plani- flat, level
plankto- wandering
pleisto- most
pleuro- side
plio- more
pluri- several
pneuma- air, breath
pod- foot

poikilo- variegated
poly- many
polyp- many footed
poro- channel
post- behind, after
-pous foot
pre- before
pro- before, forward
procto- anus
proto- first
pseudo- false
ptero- wing
pulmo- lung
pycno- dense
quadra- four
quasi- almost
radi- radial
rhizo- rot
sali- salt
schizo- split, division
scyphi- cup
semi- half
septi- partition
sessil- sedentary
siphono- tube
-sis process
spiro- spiral, coil
stoma- mouth
strati- layer
sub- below
supra- above
symbio- living together
taxo- arrangement
tecto- covering
terra- earth
terti- third
thalasso- the sea
trocho- wheel
tropho- nourishment
tunic- cloak, covering
turbi- disturbed
un- not
vel- veil
ventro- underside
xantho- yellow
xipho- sword
zoo- animal

Glossary

Abiotic environment   The nonliving components of an ecosystem.

Absolute dating   The use of radioisotope half-lives to determine the age of rock units in years within 2 or 3 percent.

Abyssal clay   Deep-ocean (oceanic) deposits containing less than 30 percent biogenous sediment.

Abyssal hill   Volcanic peaks rising less than 1 kilometer (0.6 miles) above the ocean floor.

Abyssal hill province   Deep-ocean regions, particularly in the Pacific Ocean, where oceanic sedimentation rates are so low that abyssal plains do not form and the ocean floor is covered with abyssal hills.

Abyssal plain   A flat depositional surface extending seaward from the continental rise or oceanic trenches.

Abyssal zone   The benthic environment between 4000 and 6000 meters (13,000 and 20,000 feet).

Abyssopelagic   Open-ocean (oceanic) environment below 4000 meters in depth.

Adiabatic   Pertaining to a change in the temperature of a mass resulting from compression or expansion. It requires no addition of heat to or loss of heat from the substance.

Agulhas Current   A warm current that carries Indian Ocean water around the southern tip of Africa and into the Atlantic Ocean.

Algae   One-celled or many-celled plants that have no root, stem, or leaf systems. Simple plants.

Amino acid   One of more than 20 naturally occurring compounds that contain NH_2 and COOH groups. They combine to form proteins.

Amphidromic point   A nodal or no-tide point in the ocean or sea around which the crest of the tide wave rotates during one tidal period.

Amphipoda   Crustacean order containing laterally compressed members such as the "sand hoppers."

Anadromous   Pertaining to a species of fish that spawns in fresh water and then migrates into the ocean to grow to maturity.

Anaerobic respiration   Respiration carried on in the absence of free oxygen (O_2). Some bacteria and protozoans carry on respiration this way.

Annelida   Phylum of elongated segmented worms.

Anomalistic month   The time required for the moon to go from perigee to perigee, 27.5 days.

Anoxic   Without oxygen.

Antarctic bottom water   A water mass that forms in the Weddell Sea, sinks to the ocean floor, and spreads across the bottom of all oceans.

Antarctic Circle   The latitude of 66.5°S.

Antarctic Circumpolar Current   The eastward-flowing current that encircles Antarctica and extends from the surface to the deep-ocean floor. The largest volume current in the oceans.

Antarctic Convergence   The zone of convergence along the northern boundary of the Antarctic Circumpolar Current where the southward-flowing boundary currents of the subtropical gyres converge on the cold Antarctic waters.

Antarctic Divergence   The zone of divergence separating the westward-flowing East Wind Drift and the easterward-flowing West Wind Drift.

Antarctic Intermediate Water   Antarctic zone surface water that sinks at the Antarctic convergence and flows north at a depth of about 900 meters beneath the warmer upper-water mass of the South Atlantic subtropical gyre.

Antilles Current   This warm current flows north seaward of the Lesser Antilles from the north equatorial current of the Atlantic Ocean to join the Florida Current.

Antinode   Zone of maximum vertical particle movement in standing waves where crest and trough formation alternate.

Aphelion   The point in the orbit of a planet or comet where it is farthest from the sun.

Aphotic zone   Without light. The ocean is generally in this state below 1000 meters (3280 feet).

Apogee   The point in the orbit of the moon or an artificial satellite that is farthest from Earth.

Aragonite   A form of $CaCO_3$ that is less common and less stable than calcite. Pteropod shells are usually composed of aragonite.

Archipelagic apron   A gently sloping sedimentary feature surrounding an oceanic island or seamount.

Arctic Circle   The latitude of 66.5°N.

Arrowworm   A member of the phylum Chaetognatha. This organism averages about 1 centimeter (0.4 inch) in length and is an important member of the plankton.

Aschelminthes   Phylum of wormlike pseudocoelomates.

Aspect ratio   The index of propulsive efficiency obtained by dividing the square of the height by fin area.

Asthenosphere   A plastic layer in the upper mantle 80 to 200 kilometers deep that may allow lateral movement of lithospheric plates and isostatic adjustments.

Atlantic-type margin   The passive trailing edge of a continent that is subsiding due to lithospheric cooling and increasing sediment load.

Atoll   A ring-shaped coral reef growing upward from a submerged volcanic peak. It may have low-lying islands composed of coral debris.

Autolytic decomposition   Decomposition of organic matter that is achieved by enzymes present in the tissue. The enzymes are triggered to begin their work at death of the tissue.

Autotroph   Plants and bacteria that can synthesize organic compounds from inorganic nutrients.

Autumnal equinox   The passage of the sun across the equator as it moves from the Northern Hemisphere into the Southern Hemisphere, approximately September 23.

Backshore   The inner portion of the shore, lying landward of the mean spring tide high water line. Acted upon by the ocean only during exceptionally high tides and storms.

Bacterioplankton   Bacteria that live in plankton.

Bacteriovore   Organisms that feed on bacteria.

Bar-built estuary A shallow estuary (lagoon) separated from the open ocean by a bar deposit such as a barrier island. The water in these estuaries usually exhibits vertical mixing.

Barnacle *See* Cirripedia.

Barrier flat Lying between the salt marsh and dunes of a barrier island, it is usually covered with grasses and even forests if protected from overwash for sufficient time.

Barrier island A long, narrow, wave-built island separated from the mainland by a lagoon.

Barrier reef A coral reef separated from the nearby landmass by open water.

Basalt A dark-colored volcanic rock characteristic of the ocean crust. Contains minerals with relatively high iron and magnesium content.

Bathyal zone The benthic environment between the depths of 200 and 4000 meters (660 and 13,000 feet). It includes mainly the continental slope and the oceanic ridges and rises.

Bathymetry The study of ocean depth.

Bathypelagic zone The pelagic environment between the depths of 1000 and 4000 meters (3300 and 13,000 feet).

Bay barrier A marine deposit attached to the mainland at both ends and extending entirely across the mouth of a bay, separating the bay from the open water.

Beach Sediment seaward of the coastline through the surf zone that is in transport along the shore and within the surf zone.

Benguela Current The cold eastern boundary current of the South Atlantic subtropical gyre.

Benthic Pertaining to the ocean bottom.

Benthos The forms of marine life that live on the ocean bottom.

Bioerosion Erosion of reef or other solid bottom material by the activities of organisms.

Biogenous sediment Sediment containing material produced by plants or animals, such as coral reefs, shell fragments, and housings of diatoms, radiolarians, Foraminifera, and coccolithophores.

Biogeochemical cycles The natural cycling of compounds among the living and nonliving components of an ecosystem.

Biological pump The movement of CO_2 that enters the ocean from the atmosphere through the water column to the sediment on the ocean floor by biological processes—photosynthesis, secretion of shells, feeding, and dying.

Bioluminescence Light produced by chemical reaction. Found in bacteria, phytoplankton, and metazoans.

Biomass The total mass of a defined organism or group of organisms in a particular community or the ocean as a whole.

Biotic community The living organisms that inhabit an ecosystem.

Body wave A longitudinal or transverse wave that transmits energy through a body of matter.

Boiling point The temperature at which a substance changes state from a liquid to a gas at a given pressure.

Bore A steep-fronted tide crest that moves up a river in association with high tide.

Bosphorus A narrow strait between the Black Sea and the Sea of Marmara through which Mediterranean and Black Sea water may mix.

Boundary current The northward- or southward-flowing currents that form the western and eastern boundaries, respectively, of the subtropical circulation gyres.

Brazil Current The warm western boundary current of the South Atlantic subtropical gyre.

Breaker zone Region where waves break at the seaward margin of the surf zone.

Breakwater Any artificial structure constructed to protect a coastal region from the force of ocean waves.

Bryozoa Phylum of colonial animals that often share one coelomic cavity. Encrusting and branching forms secrete a protective housing (zooecium) of calcium carbonate or chitinous material. Possess lophophore feeding structure.

Buoyancy The ability or tendency to float or rise in a liquid.

Calcareous Containing calcium carbonate.

Calcite The most common form of $CaCO_3$.

Calcium carbonate, CaCO3 A chalk-like substance secreted by many organisms in the form of coverings or skeletal structures.

Calorie Unit of heat, defined as the amount of heat required to raise the temperature of 1 gram of water 1°C.

Canary Current The cold eastern boundary current of the North Atlantic subtropical gyre.

Capillarity The action by which a fluid, such as water, is drawn up in small tubes as a result of surface tension.

Capillary wave An ocean wave whose wavelength is less than 1.74 centimeters (0.69 inches). The dominant restoring force for such waves is surface tension.

Carapace Chitinous or calcareous shield that covers the cephalothorax of some crustaceans. Dorsal portion of a turtle shell.

Carbohydrate An organic compound containing the elements carbon, hydrogen, and oxygen with the general formula $(CH_2O)_n$.

Carbonate compensation depth The depth at which carbonate particles falling from above are dissolved.

Caribbean Current The warm current that carries equatorial water across the Caribbean Sea into the Gulf of Mexico.

Carnivore An animal that depends on other animals solely or chiefly for its food supply.

Carotin A red to yellow pigment found in plants.

Catadromous Pertaining to a species of fish that spawns in the sea and then migrates into freshwater to grow to maturity.

Celsius temperature scale 0°C = 273.16 K; 0°C = freezing point of water; 100°C = boiling point of water.

Centripetal force A center-seeking force that tends to make rotating bodies move toward the center of rotation.

Cephalopoda A class of the phylum Mollusca with a well-developed pair of eyes and a ring of tentacles surrounding the mouth. The shell is absent or internal on most members. The class includes the squid, octopus, and nautilus.

Cetacea An order of marine mammals that includes the whales.

Chaetognatha A phylum of elongate transparent wormlike pelagic animals commonly called *arrowworms*.

Chemical energy A form of potential energy stored in the chemical bonds of compounds.

Chemosynthesis A process by which bacteria synthesize organic molecules from inorganic nutrients using chemical energy released from the bonds of some chemical compound by oxidation.

Chesapeake Bay The largest coastal plain estuary in the United States; created by the flooding of the river valleys of the Susquehanna River and its tributaries. It opens into the Atlantic Ocean at Norfolk, Virginia.

Chloride ion, Cl− A chlorine atom that has become negatively charged by gaining one electron.

Chlorinity The amount of chloride ion and ions of other halogens in ocean water expressed in parts per thousand by weight, ‰.

Chlorophyll A group of green pigments that make it possible for plants to carry on photosynthesis.

Chlorophyta Green algae. Characterized by the presence of chlorophyll and other pigments.

Chrysophyta An important phylum of planktonic algae, including the diatoms. The presence of chlorophyll is masked by the pigment carotin that gives the plants a golden color.

Cilium Short, hairlike structures common on lower animals. Beating in unison, they may create water currents that carry food toward the mouth of an animal or may be used for locomotion.

Circadian rhythm   Behavioral and physiological rhythms of organisms related to the 24-hour day. Sleeping and waking patterns are an example.

Circumpolar Current   Eastward-flowing current that extends from the surface to the ocean floor and encircles Antarctica.

Clastic   A rock or sediment composed of broken fragments of preexisting rocks. Two common examples are beach deposits and sandstone.

Clay   A term relating to particle size between silt and colloid. Clay minerals are hydrous aluminum silicates with plastic, expansive, and cation exchange properties.

Cnidoblast   Stinging cell of phylum Coelenterata that contains the stinging mechanism (nematocyst) used in defense and capturing prey.

Coast   A strip of land that extends inland from the coastline as far as marine influence is evidenced in the landforms.

Coastal plain estuary   An estuary formed by rising sea level flooding a coastal river valley.

Coastal upwelling   The movement of deeper nutrient-rich water into the surface water mass as a result of windblown surface water moving offshore.

Coastline   Landward limit of the effect of the highest storm waves on the shore.

Cobalt-rich manganese crust   Hydrogenous deposits found on the flanks of volcanic islands and seamounts.

Coccolith   Tiny calcareous discs averaging about 3 micrometers in diameter that form the cell wall of coccolithophores.

Coccolithophore   A microscopic planktonic form of algae encased by a covering composed of calcareous discs (coccoliths).

Coelenterata   Phylum of radially symmetrical animals that includes two basic body forms, the medusa and the polyp. Includes jellyfish (medusoid) and sea anemones (polypoid). Preferred name is now Cnidaria.

Cold front   A weather front in which a cold air mass moves into and under a warm air mass. It creates a narrow band of intense precipitation.

Colonial animals   Animals that live in groups of attached or separate individuals. Groups of individuals may serve special functions.

Columbia River estuary   An estuary at the border between the states of Washington and Oregon that has been most adversely affected by the construction of hydroelectric dams.

Comb jelly   Common name for members of the phylum Ctenophora. (*See* Ctenophora.)

Commensalism   A symbiotic relationship in which one party benefits and the other is unaffected.

Compensation depth, CaCO3   The depth at which the amount of $CaCO_3$ produced by the organisms in the overlying water column is equal to the amount of $CaCO_3$ the water column can dissolve. There will be no $CaCO_3$ deposition below this depth.

Compensation depth, O2   The depth where the oxygen produced by photosynthesis is equal to the oxygen requirements of plant respiration. A plant population cannot be sustained below this depth, which will be greater in the open ocean than near the shore due to the relatively deeper light penetration in the open ocean.

Condensation   The conversion of water from the vapor to the liquid state. When it occurs, the energy required to vaporize the water is released into the atmosphere. This is about 585 calories/gram of water at 20°C.

Conduction   The transmission of heat by the passage of energy from particle to particle.

Conjunction   The apparent closeness of two heavenly bodies. During the new moon phase, Earth and the moon are in conjunction on the same side of the sun.

Conservative property   A property of ocean water that the water attains at the surface and is changed only by mixing and diffusion after the water sinks below the surface.

Constancy of composition, Rule of   The major constituents of ocean-water salinity are found in the same relative concentrations throughout the ocean-water volume.

Constructive interference   A form of wave interference in which two waves come together in phase, for example, crest to crest, to produce a greater displacement from the still-water line than that produced by either of the waves alone.

Consumers   The animal populations within an ecosystem that consume the organic mass produced by the producers.

Continent   About one-third of Earth's surface that rises above the deep-ocean floor to be exposed above sea level. Continents are composed primarily of granite, an igneous rock of lower density than the basaltic oceanic crust.

Continental borderland   A highly irregular portion of the continental margin that is submerged beneath the ocean and is characterized by depths greater than those characteristic of the continental shelf.

Continental drift   A term applied to early theories supporting the possibility the continents are in motion over Earth's surface.

Continental rise   A gently sloping depositional surface at the base of the continental slope.

Continental shelf   A gently sloping depositional surface extending from the low water line to the depth of a marked increase in slope around the margin of a continent or island.

Continental slope   A relatively steeply sloping surface lying seaward of the continental shelf.

Convection   In a fluid being heated unevenly, the warmer part of the mass will rise and the cooler portions will sink. If the heat source is stationary, cells may develop as the rising warm water cools and sinks in regions on either side of the axis of rising.

Convergence   There are polar, tropical, and subtropical regions of the oceans where water masses with different characteristics come together. Along these lines of convergence, the denser mass will sink beneath the others.

Convergent plate boundary   A lithospheric plate boundary where adjacent plates converge, producing ocean trench-island arc systems, ocean trench-continental volcanic arcs, or folded mountain ranges.

Copepoda   An order of microscopic to nearly microscopic crustaceans that are important members of the zooplankton in temperate and subpolar waters.

Coral   A group of benthic anthozoans that exist as individuals or in colonies and secrete $CaCO_3$ external skeletons. Under the proper conditions corals may produce reefs composed of their external skeletons and the $CaCO_3$ material secreted by varieties of algae associated with the reefs.

Coral reef   A calcareous organic reef composed significantly of solid coral and coral sand. Algae may be responsible for more than half of the $CaCO_3$ reef material. Found in waters where the minimum average monthly temperature is 18°C.

Core   The core of Earth is composed primarily of iron and nickel. It has a liquid outer portion 2270 kilometers (1410 miles) thick and a solid inner core with a radius of 1216 kilometers (756 miles).

Coriolis effect   A small force resulting from Earth's rotation causes particles in motion to be deflected to the right in the Northern Hemisphere and to the left in the Southern Hemisphere.

Cosmogenous sediment   All sediment derived from outer space.

Cotidal lines   Lines connecting points where high tide occurs simultaneously.

Crest (wave)   The portion of an ocean wave that is displaced above the still-water line.

Cruiser   Fish, such as the bluefin tuna, that constantly cruise the pelagic waters in search of food.

Crust   Unit of Earth's structure that is composed of basaltic ocean crust and granitic continental crust. The total thickness of the crustal units may range from 5 kilometers (3 miles) beneath the ocean to 50 kilometers (30 miles) beneath the continents.

Crustacea   A class of phylum Arthropoda that includes barnacles, copepods, lobsters, crabs, and shrimp.

Crystalline rock   Igneous or metamorphic rocks. These rocks are made up of crystalline particles with orderly molecular structures.

Ctenophora   A phylum of gelatinous organisms that are more or less spheroidal with biradial symmetry. These exclusively marine animals have eight rows of ciliated combs for locomotion and most have two tentacles for capturing prey.

Current   A horizontal movement of water.

Cutaneous artery   The artery that runs down both sides of some cruiser-type fish to help maintain a constant elevated temperature in the myomere musculature used for swimming.

Cutaneous vein   The vein that runs down both sides of some cruiser-type fish to help maintain a constant elevated temperature in the myomere musculature used for swimming.

Davidson Current   A northward-flowing current along the Washington-Oregon coast that is driven by geostrophic effects on a large freshwater runoff.

DDT   An insecticide that caused damage to marine bird populations in the 1950s and 1960s. Its use is now banned throughout most of the world.

Declination   The angular distance of the sun or moon above or below the plane of Earth's equator.

Decomposers   Primarily bacteria that break down nonliving organic material, extract some of the products of decomposition for their own needs, and make available the compounds needed for plant production.

Deep boundary current   Relatively strong deep currents flowing across the continental rise along the western margin of ocean basins.

Deep scattering layer   A layer of marine organisms in the open ocean that scatter signals from an echo sounder. It migrates daily from depths of slightly over 100 meters (330 feet) at night to more than 800 meters (2600 feet) during the day.

Deep-sea fan   A large fan-shaped deposit commonly found on the continental rise seaward of such sediment-laden rivers as the Amazon, Indus, or Ganges-Brahmaputra.

Deep-sea system   Includes all benthic environments beneath the littoral (sublittoral, bathyal, abyssal, and hadal).

Deep water   The water beneath the permanent thermocline (pycnocline) that has a uniformly low temperature.

Deep-water wave   Ocean wave traveling in water that has a depth greater than one-half the average wave length. Its speed is independent of water depth.

Delta   A low-lying deposit at the mouth of a river.

Denitrifying bacteria   Bacteria that reduce oxides of nitrogen to produce free nitrogen (N_2).

Density   Mass per unit volume of a substance. Usually expressed as *grams per cubic centimeter*. For ocean water with a salinity of 35‰ at 0°C, the density is 1.028 g/cm³.

Density, in situ (σ or sigma)   Density of ocean water *in place*.

Density, potential (σθ or sigma theta)   Density of ocean water with the adiabatic effect removed. It is always less than in situ density except at the surface where the adiabatic effect is zero.

Desalination   The removal of salt ions from ocean water to produce pure water.

Destructive interference   A form of wave interference in which two waves come together out of phase, for example, crest to trough, and produce a wave with less displacement than the larger of the two waves would have produced alone.

Detritus   Any loose material produced directly from rock disintegration. (Organic: material resulting from the disintegration of dead organic remains.)

Diatom   Member of the class Bacillariophyceae of algae that possesses a wall of overlapping silica valves.

Diatomite   A deposit composed primarily of the frustules of diatoms.

Dipolar   Having two poles. The water molecule possesses a polarity of electrical charge with one pole being more positive and the other more negative in electrical charge.

Diffraction   Any blending of a wave around an obstacle that cannot be interpreted as refraction or reflection.

Diffusion   A process by which fluids move through other fluids from areas of high concentration to areas in which they are in lower concentrations by random molecular movement.

Dinoflagellates   Single-celled microscopic organisms that may possess chlorophyll and belong to the plant phylum Pyrrophyta (autotrophic) or may ingest food and belong to the class Mastigophora of the animal phylum Protozoa (heterotrophic).

Discontinuity   An abrupt change in a property such as temperature or salinity at a line or surface.

Disphotic zone   The dimly lit zone, corresponding approximately to the mesopelagic, in which there is not enough light to carry on photosynthesis; sometimes called the twilight zone.

Dissolved oxygen   Oxygen that is dissolved in ocean water.

Distillation   A method of purifying liquids by heating them to their boiling point and condensing the vapor.

Distributary   A small stream flowing away from a main stream. Such streams are characteristic of deltas.

Diurnal inequality   The difference in the heights of two successive high waters or two successive low waters during a lunar (tidal) day.

Diurnal tide   A tide with one high water and one low water during a tidal day.

Divergence   A horizontal flow of water from a central region, as occurs in upwelling.

Divergent plate boundary   A lithospheric plate boundary where adjacent plates diverge, producing an oceanic ridge or rise (spreading center).

Doldrums   A belt of light variable winds within 10° to 15° of the equator, resulting from the vertical flow of low-density air within this equatorial belt.

Dolphin   1. A brilliantly colored fish of the genus *Coryphaena*. 2. The name applied to the small, beaked members of the cetacean family Delphinidae.

Dorsal   Pertaining to the back or upper surface of most animals.

Dorsal aorta   For most fish, this is the only major artery that runs the length of the fish through openings in the vertebrae and supplies blood. Some pelagic cruisers also have cutaneous arteries.

Downwelling   In the open or coastal ocean where Ekman transport causes surface waters to converge or impinge on the coast, surface water will be carried down beneath the surface.

Drifts   Thick sediment deposits on the continental rise produced where the deep boundary current slows and loses sediment when it changes direction to follow the base of the continental slope.

Drowned beach   An ancient beach now beneath the coastal ocean because of rising sea level.

Dunes   Coastal deposits of sand lying landward of the beach and deriving their sand from onshore winds that transport beach sand inland.

Dynamical tide theory   The theory of tidal behavior that takes into account friction between the ocean water and the ocean floor, the effects of changing depth of the ocean floor, and the interference of the continents on the passage of tidal waves.

Dynamic topography   A surface configuration resulting from the geopotential difference between a given surface and a reference surface of no motion. A contour map of this surface is useful in estimating the nature of geostrophic currents.

Earthquake   A sudden motion or trembling in Earth, caused by the sudden release of slowly accumulated strain by faulting (movement along a fracture in Earth's crust) or volcanic activity.

East Pacific Rise   A fast-spreading divergent plate boundary extending southward from the Gulf of California through the eastern South Pacific Ocean.

East Wind Drift   The coastal current driven in a westerly direction by the polar easterly winds blowing off of Antarctica.

Eastern boundary current   Equatorward-flowing cold drifts of water on the eastern side of all subtropical gyres.

Ebb current   During a decrease in the height of the tide, the ebb current flows seaward.

Echinodermata   Phylum of animals that have bilateral symmetry in larval forms and usually a five-sided radial symmetry as adults. Benthic and possessing rigid or articulating exoskeletons of calcium carbonate with spines, this phylum includes sea stars, brittle stars, sea urchins, sand dollars, sea cucumbers, and sea lilies.

Echosounder   A device that transmits sound from a ship's hull to the ocean floor where it is reflected back to receivers. The speed of sound in the water is known, so the depth can be determined from the travel time of the sound signal.

Ecliptic   The plane of the center of the Earth-moon system as it orbits around the sun.

Ecosystem   All the organisms in a biotic community and the abiotic environmental factors with which they interact.

Eddy   A current of any fluid forming on the side of a main current. It usually moves in a circular path and develops where currents encounter obstacles or flow past one another.

Ekman spiral   A theoretical consideration of the effect of a steady wind blowing over an ocean of unlimited depth and breadth and of uniform viscosity. The result is a surface flow at 45° to the right of the wind in the Northern Hemisphere. Water at increasing depth will drift in directions increasingly to the right until at about 100 meters (330 feet) depth it is moving in a direction opposite to that of the wind. The net water transport is 90° to the wind, and speed decreases with depth.

Ekman transport   The net transport of surface water set in motion by wind. Due to the Ekman spiral phenomenon, it is theoretically in a direction 90° to the right and 90° to the left of the wind direction in the Northern Hemisphere and Southern Hemisphere, respectively.

El Niño   A southerly flowing warm current that generally develops off the coast of Ecuador shortly after Christmas. Occasionally it will move farther south into Peruvian coastal waters and cause the widespread death of plankton and fish.

El Niño-Southern Oscillation   The correlation of El Niño events with an oscillatory pattern of pressure change in a persistent high-pressure cell in the southeastern Pacific Ocean and a persistent low-pressure cell over the East Indies.

Electrolysis   A separation process by which salt ions are removed from saltwater through water-impermeable membranes toward oppositely charged electrodes.

Electromagnetic energy   Energy that travels as waves or particles with the speed of light. Different kinds possess different properties based on wavelength. The longest wavelengths belong to radio waves, up to 100 kilometers in length. At the other end of the spectrum are cosmic rays with greater penetrating power and wavelengths of less than 0.000001 micrometers.

Electromagnetic spectrum   The spectrum of radiant energy emitted from stars and ranging between cosmic rays with wavelengths of less than 10 to 11 centimeters (4 to 4.3

inches) and very long waves with wavelengths in excess of 100 kilometers (60 miles).

Element   One of a number of substances, each of which is composed entirely of like particles—atoms—that cannot be broken into smaller particles by chemical means.

Emergent shoreline   A shoreline resulting from the emergence of the ocean floor relative to the ocean surface. It is usually rather straight and characterized by marine features usually found at some depth.

Endothermic reaction   A chemical reaction that absorbs energy. For example, energy is stored in the organic products of the chemical reaction photosynthesis.

Entropy   A thermodynamic quantity that reflects the degree of randomness in a system. It increases in all natural systems.

Environment   The sum of all physical, chemical, and biological factors to which an organism or community is subjected.

Epicenter   The point on Earth's surface that is directly above the focus of an earthquake.

Epifauna   Animals that live on the ocean bottom, either attached or moving freely over it.

Epipelagic zone   A subdivision of the oceanic province that extends from the surface to a depth of 200 meters (660 feet).

Equatorial countercurrent   Eastward-flowing currents found between the north and south equatorial currents in all oceans, but particularly well developed in the Pacific Ocean.

Exclusive Economic Zone (EEZ)   A coastal zone 200 nautical miles wide over which the coastal nation has jurisdiction over mineral resources, fishing, and pollution. If the continental shelf extends beyond 200 miles, the EEZ may be up to 350 miles (about 560 kilometers) in width.

Exothermic reaction   A chemical reaction that liberates energy. For example, the energy stored in the products of photosynthesis is released by the chemical reaction respiration.

Extrusive rocks   Igneous rocks that flow out onto Earth's surface before cooling and solidifying (lavas).

Fahrenheit temperature scale (°F)   Freezing point of water is 32°; boiling point of water is 212°.

Falkland Current   A northward-flowing cold current found off the southeastern coast of South America.

Fan   A gently sloping, fan-shaped feature normally located near the lower end of a canyon.

Fast ice   Sea ice that is attached to the shore and therefore remains stationary.

Fat   An organic compound formed from alcohol, glycerol and one or more fatty acids; a lipid, it is a solid at atmospheric temperatures.

Fathom   A unit of depth in the ocean, commonly used in countries using the English system of units. It is equal to 1.83 meters, or 6 feet.

Fault   A fracture or fracture zone in Earth's crust along which displacement has occurred.

Fault block   A crustal block bounded on at least two sides by faults. Usually elongate; if it is down-dropped, it produces a graben; if uplifted, it is a horst.

Fauna   The animal life of any particular area or of any particular time.

Fecal pellet   Excrement of planktonic crustaceans that assist in speeding up the descent rate of sedimentary particles by combining them into larger packages.

Ferromagnetism   Minerals rich in iron and magnesium.

Fetch   1. Pertaining to the area of the open ocean over which the wind blows with constant speed and direction, thereby creating a wave system. 2. The distance across the fetch (wave-generating area) measured in a direction parallel to the direction of the wind.

Fjord   A long, narrow, deep, U-shaped inlet that usually represents the seaward end of a glacial valley that has become partially submerged after the melting of the glacier.

Flagellum   A whiplike living process used by some cells for locomotion.

Floe   A piece of floating ice other than fast ice or icebergs. May range in dimension from about 20 centimeters (about 8 inches) across to more than a kilometer.

Flood current   A tidal current associated with increasing height of the tide, generally moving toward the shore.

Flora   The plant life of any particular area or of any particular time.

Florida Current   A warm current flowing north along the coast of Florida. It becomes the Gulf Stream.

Folded mountain range   Mountain ranges formed as a result of the convergence of lithospheric plates. They are characterized by masses of folded sedimentary rocks that formed from sediments deposited in the ocean basin that was destroyed by the convergence.

Food chain   The passage of energy materials from producers through a sequence of a herbivore and a number of carnivores.

Food web   A group of interrelated food chains.

Foraminifera   An order of planktonic and benthic protozoans that possess protective coverings, usually composed of calcium carbonate.

Forced wave   A wave that is generated and maintained by a continuous force such as the gravitational attraction of the moon.

Foreshore   The portion of the shore lying between the normal high and low water marks—the intertidal zone.

Fortnight   Half a synodic month (29.5 days), or about 14.75 days. Normally used in reference to a period of time equal to two weeks. The time that elapses between new moon and full moon.

Fossil   Any remains, print, or trace of an organism that has been preserved in Earth's crust.

Fracture zone   An extensive linear zone of unusually irregular topography of the ocean floor, characterized by large seamounts, steep-sided or asymmetrical ridges, troughs, or long, steep slopes. Usually represents ancient, inactive transform fault zones.

Free wave   A wave created by a sudden rather than a continuous impulse that continues to exist after the generating force is gone.

Freezing   The process by which a liquid is converted to a solid at its freezing point.

Freezing point   The temperature at which a liquid becomes a solid under any given set of conditions. The freezing point of water is 0°C under atmospheric pressure.

Fringing reef   A reef that is directly attached to the shore of an island or continent. It may extend more than 1 kilometer from shore. The outer margin is submerged and often consists of algal limestone, coral rock, and living coral.

Fucoxanthin   The reddish-brown pigment that gives brown algae its characteristic color.

Full moon   The phase of the moon that occurs when the sun and moon are in opposition, on opposite sides of Earth.

Fully developed sea   The maximum average size of waves that can be developed for a given wind speed when it has blown in the same direction for a minimum duration over a minimum fetch.

Galápagos Rift   A divergent plate boundary extending eastward from the Galápagos Islands toward South America. The first deep-sea hydrothermal vent biocommunity was discovered here in 1977.

Galaxy   One of the billions of large systems of stars that make up the universe.

Gaseous state   A state of matter in which molecules move by translation and only interact through chance collisions.

Gastropoda   A class of mollusks, most of which possess an asymmetrical, spiral one-piece shell and a well-developed flattened foot. A well-developed head will usually have two eyes and one or two pairs of tentacles. Includes snails, limpets, abalone, cowries, sea hares, and sea slugs.

Geostrophic current   A current that grows out of Earth's rotation and is the result of a near balance between gravitational force and the Coriolis effect.

Gill   A thin-walled projection from some part of the external body or the digestive tract used for respiration in a water environment.

Glacial epoch   The Pleistocene epoch, the earlier of two divisions of the Quaternary period of geologic time. During this time high-latitude continental areas now free of ice were covered by continental glaciers.

Glacier   A large mass of ice formed on land by the recrystallization of old, compacted snow. It flows from an area of accumulation to an area of wasting where ice is removed from the glacier by melting.

Glauconite   A group of green hydrogenous minerals consisting of hydrous silicates of potassium and iron.

Global plate tectonics   The process by which lithospheric plates are moved across Earth's surface to collide, slide by one another, or diverge to produce the topographic configuration of Earth.

Gondwanaland   A hypothetical protocontinent of the Southern Hemisphere named for the Gondwana region of India. It included the present continental masses Africa, Antarctica, Australia, India, and South America.

Graded bedding   Stratification in which each layer displays a decrease in grain size from bottom to top.

Gradient   The rate of increase or decrease of one quantity or characteristic relative to a unit change in another. For example, the slope of the ocean floor is a change in elevation (a vertical linear measurement) per unit of horizontal distance covered. Commonly measured in meters/kilometer.

Gran method   A system involving the observation of the change in dissolved oxygen within paired transparent and opaque bottles containing phytoplankton and suspended in the ocean surface water to determine the base of the euphotic zone.

Granite   A light-colored igneous rock characteristic of the continental crust. Rich in nonferromagnesian minerals such as feldspar and quartz.

Gravitational force   The force of attraction that exists between any two bodies in the universe that is proportional to the product of their masses and inversely proportional to the distance between the centers of their masses.

Gravity wave   A wave for which the dominant restoring force is gravity. Such waves have a wavelength of more than 1.74 centimeters (0.69 inches), and their speed of propagation is controlled mainly by gravity.

Gray whale   Pacific baleen whales of the species *Rhachianectes glaucus* that feed in the Chukchi Sea and Bering Sea and breed and calve in the warm waters of lagoons in Baja California and Japan.

Greenhouse effect   The heating of Earth's atmosphere that results from the absorption by components of the atmosphere such as water vapor and carbon dioxide of infrared radiation from Earth's surface.

Groin   A low artificial structure projecting into the ocean from the shore to interfere with longshore transportation of sediment. It usually has the purpose of trapping sand to cause the buildup of a beach.

Gross ecological efficiency   The amount of energy passed on from a trophic level to the one above it divided by the amount it received from the one below it.

Gross primary production   The total carbon fixed into organic molecules through photosynthesis or chemosynthesis by a discrete autotrophic community.

Gulf Stream The high-intensity western boundary current of the North Atlantic Ocean subtropical gyre that flows north off the east coast of the United States.

Guyot A tablemount, a conical volcanic feature on the ocean floor that has had the top truncated to a relatively flat surface.

Gyre A circular spiral form. Used mainly in reference to the circular motion of water in each of the major ocean basins centered in subtropical high pressure regions.

Habitat A place where a particular plant or animal lives. Generally refers to a smaller area than environment.

Hadal Pertaining to the deepest ocean environment, specifically that of ocean trenches deeper than 6 kilometers.

Hadal zone Pertaining to the deepest ocean benthic environment, specifically that of ocean trenches deeper than 6 kilometers (3.7 miles).

Half-life The time required for half the atoms of a sample of a radioactive isotope to decay to an atom of another element.

Halocine A layer of water in which a high rate of change in salinity in the vertical dimension is present.

Headland A steep-faced irregularity of the coast that extends out into the ocean.

Heat Energy moving from a high temperature system to a lower temperature system. The heat gained by the one system may be used to raise its temperature or to do work.

Heat budget The equilibrium that exists on the average between the amount of heat absorbed by Earth and its atmosphere in one year and the amount of heat radiated back into space in one year.

Heat capacity Usually defined as the amount of heat required to raise the temperature of 1 gram of a substance 1°C.

Heat energy Energy of molecular motion. The conversion of higher forms of energy such as radiant or mechanical energy to heat energy within a system increases the heat energy within the system and the temperature of the system.

Heat flow (flux) The quantity of heat flow to Earth's surface per unit of time.

Hemoglobin A red pigment found in red blood corpuscles that carries oxygen from the lungs to tissue and carbon dioxide from tissue to lungs.

Herbivore An animal that relies chiefly or solely on plants for its food.

Hermatypic coral Reef-building corals that have symbiotic algae in their ectodermal tissue. They cannot produce a reef structure below the euphotic zone.

Heterotroph Animals and bacteria that depend on the organic compounds produced by other animals and plants as food. Organisms not capable of producing their own food by photosynthesis.

High tide zone That portion of the littoral zone that lies between the lowest high tides and highest high tides that occur in an area. It is, on the average, exposed to desiccation for longer periods each day than it is covered by water.

High water (HW) The highest level reached by the rising tide before it begins to recede.

Higher high water (HHW) The higher of two high waters occurring during a tidal day where tides are mixed.

Higher low water (HLW) The higher of two low waters occurring during a tidal day where tides are mixed.

Highly stratified estuary A relatively deep estuary in which a significant volume of marine water enters as a subsurface flow. A large volume of freshwater stream input produces a widespread low surface-salinity condition which produces a well-developed halocline throughout most of the estuary.

Holoplankton Organisms that spend their entire life as members of the plankton.

Homeotherm An animal that maintains a precisely controlled internal body temperature using its own internal heating and cooling mechanisms.

Horse latitudes The latitude belts between 30° and 35° north and south where winds are light and variable, since the principal movement of air masses at these latitudes is one of vertical descent. The climate is hot and dry.

Hot spot The relatively stationary surface expression of a persistent jet of molten mantle material rising to the surface.

Hurricane A tropical cyclone in which winds reach speeds in excess of 120 kilometers/hour (73 miles/hour). Generally applied to such storms in the North Atlantic Ocean, eastern North Pacific Ocean, Caribbean Sea, and Gulf of Mexico. Such storms in the western Pacific Ocean are called *typhoons*.

Hydrocarbon Any organic compound consisting only of hydrocarbon and carbon. Crude oil is a mixture of hydrocarbons.

Hydrocarbon seep biocommunity Deep bottom-dwelling community associated with a hydrocarbon seep from the ocean floor. The community depends on methane and sulfur-oxidizing bacteria as producers. The bacteria may live free in the water, on the bottom, or symbiotically in the tissues of some of the animals.

Hydrogen bond An intermolecular bond that forms within water because of the dipolar nature of water molecules.

Hydrogenous sediment Sediment that forms from precipitation from ocean water or ion exchange between existing sediment and ocean water. Examples are manganese nodules, phosphorite, glauconite, phillipsite, and montmorillonite.

Hydrologic cycle The cycle of water exchange among the atmosphere, land, and ocean through the processes of evaporation, precipitation, runoff, and subsurface percolation.

Hydrothermal spring Vents of hot water found primarily along the spreading axes of oceanic ridges and rises.

Hydrothermal vent Ocean water that percolates down through fractures in recently formed ocean floor is heated by underlying magma and surfaces again through these vents. They are usually located near the axis of spreading on oceanic ridges and rises.

Hydrothermal vent biocommunity Deep bottom-dwelling community associated with a hydrothermal vent. The hot water vent is usually associated with the axis of a spreading center, and the community is dependent on sulfur-oxidizing bacteria that may live free in the water, on the bottom, or symbiotically in the tissue of some of the animals of the community.

Hypersaline lagoon Shallow lagoons such as Laguna Madre, which may become hypersaline due to little tidal flushing and seasonal variability in freshwater input. High evaporation rates and low freshwater input can result in very high salinities.

Hypersaline seep At the base of the Florida Escarpment there is a biocommunity that depends on methane and sulfur-oxidizing bacteria as producers. The bacteria may live free in the water, on the bottom, or symbiotically in the tissue of some of the animals within the biocommunity.

Hypertonic Pertaining to the property of an aqueous solution having a higher osmotic pressure (salinity) than another aqueous solution from which it is separated by a semipermeable membrane that will allow osmosis to occur. The hypertonic fluid will gain water molecules through the membrane from the other fluid.

Hypotonic Pertaining to the property of an aqueous solution having a lower osmotic pressure (salinity) than another aqueous solution from which it is separated by a semipermeable membrane that will allow osmosis to occur. The hypotonic fluid will lose water molecules through the membrane from the other fluid.

Ice floe *See* floe.

Ice shelf A thick layer of ice with a relatively flat surface that is attached to and nourished by a continental glacier from one side. The shelf, which is for the most part afloat, may extend above water level by more than 50 meters along its seaward cliff formed by the breakoff of larger tabular chunks of ice that become icebergs.

Iceberg   A massive piece of glacier ice that has broken from the front of the glacier (calved) into a body of water. It floats with its tip at least 5 meters (16 feet) above the water's surface and at least four-fifths of its mass submerged.

Igneous rock   One of the three main classes into which all rocks are divided (igneous, metamorphic, and sedimentary). Rock that forms from the solidification of molten or partly molten material (magma).

In situ   In place; in situ density of a sample of water is its density at its original depth.

Inertia   Newton's first law of motion. It states that a body at rest will stay at rest and a body in motion will remain in a uniform motion in a straight line unless acted on by some external force.

Infauna   Animals that live buried in the soft substrate (sand or mud).

Infrared radiation   Electromagnetic radiation lying between the wavelengths of 0.8 micrometers and about 1000 micrometers. It is bounded on the shorter wavelength side by the visible spectrum and on the long side by microwave radiation.

Inner sublittoral   Pertaining to the inner continental shelf, above the intersection with the euphotic zone, where attached plants grow.

Insolation   The rate at which solar radiation is received per unit of surface area at any point at or above Earth's surface.

Interface   A surface separating two substances of different properties, such as density, salinity, or temperature. In oceanography, it usually refers to a separation of two layers of water with different densities caused by significant differences in temperature and/or salinity.

Interface wave   An orbital wave that moves along an interface between fluids of different density. An example is ocean surface waves moving along the interface between the atmosphere and the ocean, which is one thousand times more dense.

Intermolecular bond   A relatively weak bond that forms between molecules of a given substance. The hydrogen bond and the van der Waals bonds are intermolecular bonds.

Internal wave   A wave that develops below the surface of a fluid, the density of which changes with increased depth. This change may be gradual or occur abruptly at an interface.

Intertidal zone   Littoral zone, the foreshore. The ocean floor covered by the highest normal tides and exposed by the lowest normal tides and the water environment of the tide pools within this region.

Intertropical Convergence Zone   Zone where northeast trade winds and southeast trade winds converge. Averages about 5°N in the Pacific and Atlantic oceans and 7°S in the Indian Ocean.

Intrusive rocks   Igneous rocks such as granite that cool slowly beneath Earth's surface.

Invertebrate   Animal without a backbone.

Ion   An atom that becomes electrically charged by gaining or losing one or more electrons. The loss of electrons produces a positively charged cation, and the gain of electrons produces a negatively charged anion.

Ionic bond   A chemical bond formed as a result of the electrical attraction.

Irminger Current   A warm current that branches off from the Gulf Stream and moves up along the west coast of Iceland.

Island arc system   A linear arrangement of islands, many of which are volcanic, usually curved so that the concave side faces a sea separating the islands from a continent. The convex side faces the open ocean and is bounded by a deep-ocean trench.

Island mass effect   As surface current flows past an island, surface water is carried away from the island on the downcurrent side. This water is replaced in part by upwelling of water on the downcurrent side of the island.

Isohaline   Of the same salinity.

Isopoda   An order of dorsoventrally flattened crustaceans that are mostly scavengers or parasites on other crustaceans or fish.

Isostasy   A condition of equilibrium, comparable to buoyancy, by which Earth's brittle crust floats on the plastic mantle.

Isotherm   Line connecting points of equal temperature.

Isothermal   Of the same temperature.

Isotonic   Pertaining to the property of having equal osmotic pressure. If two such fluids were separated by a semipermeable membrane that will allow osmosis to occur, there would be no net transfer of water molecules across the membrane.

Isotope   One of several atoms of an element that has a different number of neutrons, and therefore a different atomic mass, than the other atoms, or isotopes, of the element.

Jellyfish   1. Free-swimming, umbrella-shaped medusoid members of the coelenterate class, Scryphozoa. 2. Also frequently applied to the medusoid forms of other coelenterates.

Jet stream   An easterly moving air mass at an elevation of about 10 kilometers (about 6 miles). Moving at speeds that can exceed 300 kilometers/hour (185 miles/hour), the jet stream follows a wavy path in the midlatitudes and influences how far polar air masses may extend into the lower latitudes.

Jetty   A structure built from the shore into a body of water to protect a harbor or a navigable passage from being shoaled by deposition of longshore (littoral) drift material.

Juan de Fuca Ridge   A divergent plate boundary off the Oregon-Washington coast.

Kelp   Large varieties of Phaeophyta (brown algae).

Kelvin temperature scale (K)   0 K = −273.16°C. One degree on the Kelvin scale equals the same temperature range as one degree on the Celsius scale. 0 K is the lowest temperature possible.

Key   A low, flat island composed of sand or coral debris that accumulates on a reef flat.

Kinetic energy   Energy of motion. It increases as the mass or speed of the object in motion increases.

Knot (kt)   Unit of speed equal to 1 nautical mile per hour, approximately 51 centimeters/second, or 1.67 feet/second.

Krill   A common name frequently applied to members of crustacean order Euphausiacea (euphausids).

La Niña   An event where the surface temperature in the waters of the eastern South Pacific fall below average values. It usually occurs at the end of an El Niño-Southern Oscillation event.

Labrador Current   A cold current flowing south along the coast of Labrador in the northeastern Atlantic Ocean.

Lagoon   A shallow stretch of seawater party or completely separated from the open ocean by an elongate narrow strip of land such as a reef or barrier island.

Laguna Madre   A hypersaline lagoon behind Padre Island along the south Texas coast.

Laminar flow   Flow in which water, or any fluid, flows in parallel layers or sheets. The direction of flow at any point does not change with time. Nonturbulent flow.

Langmuir circulation   A cellular circulation set up by winds that blow consistently in one direction with speeds in excess of 12 kilometers/hour. Helical spirals running parallel to the wind direction are alternately clockwise and counterclockwise.

Larva   An embryo that has a different form before it assumes the characteristics of the adult of the species.

Latent heat   The quantity of heat gained or lost per unit of mass as a substance undergoes a change of state (such as liquid to solid) at a given temperature and pressure.

Latent heat of evaporation   The heat energy that must be added to one gram of a liquid substance to convert it to a vapor at a given temperature below its boiling point. For water, it is 585 calories at 20°C.

Latent heat of melting   The heat energy that must be added to one gram of a substance at its melting point to convert it to a liquid. For water, it is 80 calories.

Latent heat of vaporization   The heat energy that must be added to one gram of a substance at its boiling point to convert it to a vapor. For water, it is 540 calories.

Lateral line system   A sensory system running down both sides of fishes to sense subsonic pressure waves transmitted through ocean water.

Latitude   Location on Earth's surface based on angular distance north or south of the equator. Equator = 0°; North Pole = 90°N; South Pole = 90°S.

Laurasia   A hypothetical protocontinent of the Northern Hemisphere. The name is derived from Laurentia, pertaining to the Canadian Shield of North America, and Eurasia, of which it was composed.

Lava   Fluid magma coming from an opening in Earth's surface, or the same material after it solidifies.

Law of gravitation   *See* gravitational force.

Leeuwin Current   A warm current that flows south out of the East Indies along the western coast of Australia.

Leeward   Direction toward which the wind is blowing or waves are moving.

Levee   1. Natural levees are low ridges on either side of river channels that result from deposition during flooding. 2. Artificial levees are built by human beings.

Light-year   The distance traveled by light during one year at a speed of 300,000 kilometers/second (186,000 miles/second). It equals 9.8 trillion kilometers (6.2 trillion miles).

Limestone   A class of sedimentary rocks composed of at least 80 percent carbonates of calcium or magnesium. Limestones may be either biogenous or hydrogenous.

Limpet   A mollusk of the class Gastropoda that possesses a low conical shell that exhibits no spiraling in the adult form.

Liquid state   A state of matter in which a substance has a fixed volume but no fixed shape.

Lithogenous sediment   Sediment composed of mineral grains derived from the rocks of continents and islands and transported to the ocean by wind and running water.

Lithosphere   The outer layer of Earth's structure, including the crust and the upper mantle to a depth of about 200 kilometers. It is this layer that breaks into the plates that are the major elements of the theory of plate tectonics.

Lithothamnion ridge   A feature common to the windward edge of a reef structure, characterized by the presence of the red algae, *Lithothamnion*.

Littoral zone   The benthic zone between the highest and lowest spring tide shorelines; the intertidal zone.

Lobster   Large marine crustacean used as food. *Homarus americanus* (American lobster) possesses two large chelae (pincers) and is found off the New England coast. *Panulirus* sp. (spiny lobsters or rock lobsters) have no chelae but possess long, spiny antennae effective in warding off predators. *P. argue* is found off the coast of Florida and in the West Indies, whereas *P. interruptus* is common along the coast of southern California.

Longitude   Location on Earth's surface based on angular distance east or west of Greenwich Meridian (0° longitude). 180° longitude is the International Date Line.

Longitudinal wave   A wave phenomenon where particle vibration is parallel to the direction of energy propagation.

Longshore current   A current located in the surf zone and running parallel to the shore as a result of waves breaking at an angle on the shore.

Longshore drift   The load of sediment transported along the beach from the breaker zone to the top of the swash line in association with the longshore current.

Lophophore   Horseshoe-shaped feeding structure bearing ciliated tentacles characteristic of the phyla Bryozoa, Brachiopoda, and Phoronidea.

Low tide zone   That portion of the intertidal zone that lies between the lowest low-tide shoreline and the highest low-tide shoreline.

Low water (LW)   The lowest level reached by the water surface at low tide before the rise toward high tide begins.

Lower high water (LHW)   The lower of two high waters occurring during a tidal day where tides are mixed.

Lower low water (LLW)   The lower of two low waters occurring during a tidal day where tides are mixed.

Lunar day   The time interval between two successive transits of the moon over a meridian, approximately 24 hours and 50 minutes of solar time.

Lunar hour   One-twenty-fourth of a lunar day; about 62.1 minutes.

Lunar tide   The part of the tide caused solely by the tide-producing force of the moon.

Lunger   Fish such as groupers that sit motionless on the ocean floor waiting for prey to appear. A quick burst of speed over a short distance suffices to capture the prey.

Macroplankton   Plankton larger than 2 centimeters (0.8 inches) in their smallest dimension.

Magma fluid   Rock material from which igneous rock is derived through solidification.

Magnetic anomaly   Distortion of the regular pattern of Earth's magnetic field resulting from the various magnetic properties of local concentrations of ferromagnetic minerals in Earth's crust.

Magnetic dip   The dip of magnetite particles in rock units of Earth's crust relative to sea level. It is approximately equivalent to latitude.

Manganese nodules   Concretionary lumps containing oxides of iron, manganese, copper, and nickel found scattered over the ocean floor.

Mangrove swamp   A marshlike environment that is dominated by mangrove trees. They are restricted to latitudes below 30°.

Mantle   The relatively plastic zone rich in ferromagnesian minerals between the core and crust of Earth.

Marginal sea   A semienclosed body of water adjacent to a continent and floored by submerged continental crust.

Mariculture   The application of the principles of agriculture to the production of marine organisms.

Marine terrace   Wave-cut benches that have been exposed above sea level by a drop in sea level.

Marsh   An area of soft, wet land. Flat land periodically flooded by salt water, common in portions of lagoons.

Mean high water (MHW)   The average height of all the high waters occurring over a 19-year-period.

Mean low water (MLW)   The average height of the low waters occurring over a 19-year period.

Mean sea level (MSL)   The mean surface water level determined by averaging all stages of the tide over a 19-year period, usually determined from hourly height observations along an open coast.

Mean tidal range   The difference between mean high water and mean low water.

Meander   A sinuous curve, bend, or turn in the course of a current.

Mechanical energy   Energy manifested as work being done; the movement of a mass some distance.

Mediterranean circulation   Circulation characteristic of bodies of water with restricted circulation with the ocean that results from an excess of evaporation as compared to precipitation and runoff. Surface flow is into the restricted body of water with a subsurface counterflow as exists between the Mediterranean Sea and the Atlantic Ocean.

Medusa   Free-swimming, bell-shaped coelenterate body form with a mouth at the end of a central projection and tentacles around the periphery. Reproduces sexually.

Meiofauna   Small species of animals that live in the spaces among particles in a marine sediment.

Meridian of longitude   Great circles running through the north and south poles.

Meroplankton   Planktonic larval forms of organisms that are members of the benthos or nekton as adults.

Mesopelagic   That portion of the oceanic province 200 to 1000 meters deep. Corresponds approximately with the disphotic (twilight) zone.

Metamorphic rock   Rock that has undergone recrystallization while in the solid state in response to changes of temperature, pressure, and chemical environment.

Metaphyta   Kingdom of many-celled plants.

Metazoa   Kingdom of many-celled animals.

Microplankton   Net plankton. Plankton not easily seen by the unaided eye, but easily recovered from the ocean with the aid of a silk-mesh plankton net.

Mid-Atlantic Ridge   A slow-spreading divergent plate boundary running north-south and bisecting the Atlantic Ocean.

Middle tide zone   That portion of the intertidal zone that lies between the highest low-tide shoreline and the lowest high-tide shoreline.

Migration   Long journeys undertaken by many marine species for the purpose of successful feeding and reproduction. The stimuli that cause the animals to initiate their migrations are for the most part still unknown.

Minamata Bay, Japan   The site of the 1953 occurrence of human poisoning by mercury contained in marine food the victims consumed.

Mineral   An inorganic substance occurring naturally on Earth and having distinctive physical properties and a chemical composition that can be expressed by a chemical formula. The term is also sometimes applied to organic substances such as coal and petroleum.

Mixed interference   A pattern of wave interference in which there is a combination of constructive and destructive interference.

Mixed layer   The surface layer of the ocean water mixed by wave and tide motions to produce relatively isothermal and isohaline conditions.

Mixed tide   A tide having two high waters and two low waters per tidal day with a marked diurnal inequality. Such a tide may also show alternating periods of diurnal and semidiurnal components.

Mixotroph   An organism that depends on a combination of autotrophic and heterotrophic behavior to meet its energy requirements. Many coral reef species exhibit such behavior.

Mohorovicic discontinuity   A sharp compositional discontinuity between the crust and mantle of Earth. It may be as shallow as 5 kilometers below the ocean floor or as deep as 60 kilometers beneath some continental mountain ranges.

Molecular motion   Molecules move in three ways: vibration, rotation, and translation.

Mollusca   Phylum of soft, unsegmented animals usually protected by a calcareous shell and having a muscular foot for locomotion. Includes snails, clams, chitons, and octopi.

Monera   Kingdom of organisms that do not have nuclear material confined within a sheath but spread throughout the cell. Bacteria and blue-green algae.

Mononodal   Pertaining to a standing wave with only one nodal point or nodal line.

Monsoons   A name for seasonal winds derived from the Arabic word for season, *mausim*. The term was originally applied to

winds over the Arabian Sea that blow from the southwest during summer and the northeast during winter.

Moraine   A deposit of unsorted material deposited at the margins of glaciers. Many such deposits have become important economically as fishing banks after being submerged by the rising level of the ocean.

Mud   Sediment consisting of silt and clay-sized particles smaller than 0.06 millimeters. Actually, small amounts of larger particles will also be present.

Mutualism   A symbiotic relationship in which both participants benefit.

Mycota   The kingdom of fungi. In the marine environment they are found living symbiotically with algae as lichen in the intertidal zone and as decomposers of dead organic matter in the open sea.

Myoglobin   A red, oxygen-storing pigment found in muscle tissue.

Myomere   A muscle fiber.

Mysticeti   The baleen whales.

Nadir   The point on the celestial sphere directly opposite the zenith and directly beneath the observer.

Nanoplankton   Plankton less than 50 micrometers in length that cannot be captured in a plankton net and must be removed from the water by centrifuge or special microfilters.

Nansen bottle   A device used by oceanographers to obtain samples of ocean water from beneath the surface.

Nauplius   A microscopic free-swimming larval stage of crustaceans such as copepods, ostracodes, and decapods. Typically has three pairs of appendages.

Neap tide   Tides of minimal range occurring when the moon is in quadrature, first and third quarters.

Nearshore   That zone from the shoreline seaward to the line of breakers.

Nektobenthos   Those members of the benthos that can actively swim and spend much time off the bottom.

Nekton   Pelagic animals such as adult squids, fish, and mammals that are active swimmers to the extent that they can determine their position in the ocean by swimming.

Nematocyst   The stinging mechanism found within the cnidoblast of members of the phylum Cnidaria (Coelenterata).

Neritic province   That portion of the pelagic environment from the shoreline to where the depth reaches 200 meters (660 feet).

Neritic sediment   That sediment composed primarily of lithogenous particles and deposited relatively rapidly on the continental shelf, continental slope, and continental rise.

Net primary production   The primary production of plants after they have removed what is needed for their metabolism.

New moon   The phase of the moon that occurs when the sun and the moon are in conjunction, on the same side of Earth.

New production   Photosynthetic production supported by nutrients supplied from outside the immediate ecosystem by upwelling or other physical transport.

Niche   The ecological role of an organism and its position in the ecosystem.

Niigata, Japan   The site of mercury poisoning of humans in the 1960s by mercury-contaminated seafood.

Nitrogen fixation   Conversion by bacteria of atmospheric nitrogen (N_2) to oxides of nitrogen (NO_2, NO_3) usable by plants in primary production.

Node   The point on a standing wave where vertical motion is lacking or minimal. If this condition extends across the surface of an oscillating body of water, the line of no vertical motion is a nodal line.

Nonconservative property   A property of ocean water attained at the surface and changed by processes other than mixing and diffusion after the water sinks below the surface.

For example, dissolved oxygen content will be altered by biological activity.

Nonferromagnesian Pertaining to a group of common igneous rock-forming minerals that do not contain iron and magnesium.

North Atlantic Deep Water A deep-water mass that forms primarily at the surface of the Norwegian Sea and moves south along the floor of the North Atlantic Ocean.

Northeast Monsoon A northeast wind that blows off the Asian mainland onto the Indian Ocean during the winter season.

Norwegian Current A warm current that branches off from the Gulf Stream and flows into the Norwegian Sea between Iceland and the British Isles.

Nudibranch Sea slug. A member of the mollusk class Gastropoda that has no protective covering as an adult. Respiration is carried on by gills or other projections on the dorsal surface.

Nutrients Any number of organic or inorganic compounds used by plants in primary production. Nitrogen and phosphorus compounds are important examples.

Ocean acoustical tomography A method by which changes in water temperature may be determined by changes in the speed of transmission of sound. It has the potential to help map ocean circulation patterns over large ocean areas.

Ocean beach Beach on the open-ocean side of a barrier island.

Ocean Drilling Program In 1983, this program replaced the Deep Sea Drilling Project. It focuses more on drilling the continental margins using the drill vessel *JOIDES Resolution.*

Oceanic crust A mass of rock with a basaltic composition that is about 5 kilometers thick and may or may not extend beneath the continents.

Oceanic province That division of the pelagic environment where the water depth is greater than 200 meters (660 feet).

Oceanic ridge A linear, seismic mountain range that extends through all the major oceans, rising 1 to 3 kilometers above the deep-ocean basins. Averaging 1500 kilometers in width, rift valleys are common along the central axis. Source of new oceanic crustal material.

Oceanic sediment The inorganic abyssal clays and the organic oozes that accumulate on the deep-ocean floor slowly, particle by particle.

Oceanic spreading center The axes of oceanic ridges and rises that are the locations at which new lithosphere is added to lithospheric plates. The plates move away from these axes in the process of seafloor spreading.

Ocelli Light-sensitive organ around the base of many medusoid bells.

Odonticeti Toothed whales.

Offshore The comparatively flat submerged zone of variable width extending from the breaker line to the edge of the continental shelf.

Oligotrophic Areas such as the midsubtropical gyres where there are low levels of biological production.

Omnivore An animal that feeds on both plants and animals.

Oolite A deposit formed of small spheres from 0.25 to 2 millimeters in diameter. They are usually composed of concentric layers of calcite.

Ooze A pelagic sediment containing at least 30 percent skeletal remains of pelagic organisms, the balance being clay minerals. Oozes are further defined by the chemical composition of the organic remains (siliceous or calcareous) and by their characteristic organisms (diatom ooze, foraminifera ooze, radiolarian ooze, pteropod ooze).

Opal An amorphous form of silica ($SiO_2 \cdot nN_2O$) that usually contains from 3 to 9 percent water. It forms the shells of radiolarians and diatoms.

Opposition The separation of two heavenly bodies by 180° relative to Earth. The sun and moon are in opposition during the full moon phase.

Orbital wave A wave phenomenon in which energy is moved along the interface between fluids of different densities. The wave form is propagated by the movement of fluid particles in orbital paths.

Orthogonal lines Lines drawn perpendicular to wave fronts and spaced uniformly so that equal amounts of energy are contained by the segments of the wave front lying between any two orthogonal lines in a series. The areas where energy is concentrated as the waves break on the shore can be identified by the convergence of the orthogonal lines.

Orthophosphate Phosphoric oxide (P_2O_5) can combine with water to produce orthophosphates ($3H_2O \cdot P_2O_5$ or H_3PO_4) that may be used by plants as nutrients.

Osmosis Passage of water molecules through a semipermeable membrane separating two aqueous solutions of different solute concentration. The water molecules pass from the solution of lower solute concentration into the other.

Osmotic pressure A measure of the tendency for osmosis to occur. It is the pressure that must be applied to the more concentrated solution to prevent the passage of water molecules into it from the less concentrated solution.

Osmotic regulation Physical and biological processes used by organisms to counteract the osmotic effects of differences in osmotic pressures of their body fluids and the water in which they live.

Ostracoda An order of crustaceans that are minute and compressed within a bivalve shell.

Otocyst Gravity-sensitive organs around the bell of a medusa.

Outer sublittoral The continental shelf below the intersection with the euphotic zone where no plants grow attached to the bottom.

Oviparous Referring to embryological development in an egg that is deposited outside the mother's body where nutrition for development is provided by the yolk of the egg.

Ovoviviparous Referring to embryological development in the female reproductive tract where nutrition for development is provided by the yolk of the egg.

Oxygen compensation depth The depth in the ocean at which marine plants receive just enough solar radiation to meet their basic metabolic needs. It marks the base of the euphotic zone.

Oxygen utilization rate (OUR) The rate at which oxygen is being used by the demands of respiration and bacterial decomposition of dead organic material.

Pacific-type margin Leading edge of a continent that undergoes tectonic uplift as a result of lithospheric plate convergence.

Pack ice Any area of sea ice other than fast ice. Less than 3 meters thick, it covers the ocean sufficiently that navigation is possible only by icebreakers.

Paleoceanography The study of the physical and biological changes of the oceans brought about by the changing shapes and positions of the continents.

Paleogeography The study of the historical changes of shapes and positions of the continents and oceans.

Pancake ice Circular pieces of newly formed sea ice from 30 centimeters (1 foot) to 3 meters (10 feet) in diameter that form in early fall in polar regions.

Pangaea A hypothetical supercontinent of the geologic past that contained all the continental crust of Earth.

Panthalassa A hypothetical proto-ocean surrounding Pangaea.

Parasitism A symbiotic relationship between two organisms in which one benefits at the expense of the other.

PCBs A group of industrial chemicals used in a variety of products; responsible for several episodes of ecological damage in coastal waters.

Pelagic environment The open ocean environment, which is divided into the neritic province (water depth 0 to 200 m) and the oceanic province (water depth greater than 200 m).

Pelecypoda A class of mollusks characterized by two more or less symmetrical lateral valves with a dorsal hinge. These filter feeders pump water through the filter system and over gills through posterior siphons. Many possess a hatchet-shaped foot used for locomotion and burrowing. Includes clams, oysters, mussels, and scallops.

Perigee The point on the orbit of an Earth satellite (moon) that is nearest Earth.

Perihelion That point on the orbit of a planet or comet around the sun that is closest to the sun.

Permeability Capacity of a porous rock or sediment for transmitting fluid.

Petroleum A naturally occurring liquid hydrocarbon.

Phaeophyta Brown algae characterized by the carotinoid pigment fucoxanthin. Contains the largest members of the marine plant community.

Phosphorite A sedimentary rock composed primarily of phosphate minerals.

Photic zone The upper ocean in which the presence of solar radiation is detectable. It includes the euphotic and disphotic zones.

Photophore One of several types of light-producing organs found primarily on fishes and squids inhabiting the mesopelagic and upper bathypelagic zones.

Photosynthesis The process by which plants produce carbohydrate food from carbon dioxide and water in the presence of chlorophyll, using light energy and releasing oxygen.

Phycoerythrin A red pigment characteristic of the Rhodophyta (red algae).

Phytoplankton Plant plankton. The most important community of primary producers in the ocean.

Picoplankton Small plankton within the size range of 0.2 to 2.0 micrometers in size. Composed primarily of bacteria.

Pinniped A suborder of marine mammals that includes the sea lions, seals, and walruses.

Plankton Passively drifting or weakly swimming organisms that are not independent of currents. Includes mostly microscopic algae, protozoa, and larval forms of higher animals.

Plankton bloom A very high concentration of phytoplankton, resulting from a rapid rate of reproduction as conditions become optimal during the spring in high latitude areas. Less obvious causes produce blooms that may be destructive in other areas.

Plankton net Plankton-extracting device that is cone-shaped and typically of a silk material. It is towed through the water or lifted vertically to extract plankton down to a size of 50 micrometers.

Plume Rising jets of molten mantle material that create hot spots when they penetrate the crust of Earth.

Plunging breaker Impressive curling breakers that form on moderately sloping beaches.

Pneumatic duct An opening into the swim bladder of some fishes that allows rapid release of air into the esophagus.

Poikilotherm An organism whose body temperature varies with and is largely controlled by its environment.

Polar Pertaining to the polar regions.

Polar easterly winds Cold air masses that move away from the polar regions toward lower latitudes.

Polar emergence The emergence of low- and mid-latitude temperature-sensitive deep-ocean benthos onto the shallow shelves of the polar regions where temperatures similar to that of their deep-ocean habitat exist.

Pollution (marine) The introduction of substances that result in harm to the living resources of the ocean or humans who use these resources.

Polychaeta Class of annelid worms that includes most of the marine segmented worms.

Polynya A nonlinear opening in sea ice.

Polyp A single individual of a colony or a solitary attached coelenterate.

Population A group of individuals of one species living in an area.

Porifera Phylum of sponges. Supporting structure composed of $CaCO_3$ or SiO_2 spicules or fibrous spongin. Water currents created by flagella-waving choanocytes enter tiny pores, pass through canals, and exit through a larger osculum.

Precession Regarding the moon's orbit around Earth, the axis of this orbit slowly changes its direction and describes a complete cone every 18.6 years. This is accompanied by a clockwise rotation of the plane of the moon's orbit that is completed in the same time interval.

Precipitation In a meteorological sense, the discharge of water in the form of rain, snow, hail, or sleet from the atmosphere onto Earth's surface.

Primary productivity The amount of organic matter synthesized by organisms from inorganic substances within a given volume of water or habitat in a unit of time.

Prime meridian The meridian of longitude 0° used as a reference for measuring longitude. The Greenwich Meridian.

Producer The autotrophic component of an ecosystem that produces the food that supports the biocommunity.

Progressive wave A wave in which the waveform progressively moves.

Propagation The transmission of energy through a medium.

Protein A very complex organic compound made up of large numbers of amino acids. Proteins make up a large percentage of the dry weight of all living organisms.

Protista A kingdom of organisms that includes all one-celled forms with nuclear material confined to a nuclear sheath. Includes the animal phylum Protozoa and the phyla of algal plants.

Protoplasm The complicated self-perpetuating living material making up all organisms. The elements carbon, hydrogen, and oxygen constitute more than 95 percent; water and dissolved salts make up 50 to 97 percent of most plants and animals; and carbohydrates, lipids (fats), and proteins constitute the remainder.

Protozoa Phylum of one-celled animals with nuclear material confined within a nuclear sheath.

Pseudopodia An extension of protoplasm in a broad, flat, or long needlelike projection used for locomotion or feeding. Typical of amoeboid forms such as Foraminifera and Radiolaria.

Pteropoda An order of pelagic gastropods in which the foot is modified for swimming and the shell may be present or absent.

Purse seine A curtainlike net that can be used to encircle a school of fish. The bottom is then pulled tight much the way a purse string is used to close a baglike purse.

Pycnocline A layer of water in which a high rate of change in density in the vertical dimension is present.

Pyrrophyta A phylum of microscopic algae that possesses flagella for locomotion—the dinoflagellates.

Quadrature The first and third quarter moon phases occur when the sun and moon are in quadrature, at right angles to one another relative to Earth.

Quarter moon First and third quarter moon phases, which occur when the sun and moon are in quadrature one week after the new moon and full moon phases, respectively.

Radiata A grouping of phyla with primary radial symmetry—phyla Coelenterata and Ctenophora.

Radioactive dating *See* absolute dating.

Radioactivity The spontaneous breakdown of the nucleus of an atom resulting in the emission of radiant energy in the form of particles or waves.

Radiolaria An order of planktonic and benthic protozoans that possess protective coverings usually made of silica.

Ray A cartilaginous fish in which the body is dorsoventrally flattened, eyes and spiracles are on the upper surface, and gill slits are on the bottom. The tail is reduced to a whiplike appendage. Includes electric rays, manta rays, and stingrays.

Red muscle fiber Fine muscle fibers rich in myoglobin that are abundant in cruiser-type fishes.

Red tide A reddish-brown discoloration of surface water, usually in coastal areas, caused by high concentrations of microscopic organisms, usually dinoflagellates. It probably results from increased availability of certain nutrients for various reasons. Toxins produced by the dinoflagellates may kill fish directly; or large populations of animal forms that spring up to feed on the plants, along with decaying plant and animal remains, may rise up the oxygen in the surface water to cause asphyxiation of many animals.

Reef A consolidated rock (a hazard to navigation) with a depth of 20 meters or less.

Reef flat A platform of coral fragments and sand on the lagoonal side of a reef that is relatively exposed at low tide.

Reef front The upper seaward face of a reef from the reef edge (seaward margin of reef flat) to the depth at which living coral and coralline algae become rare, 16 to 30 meters (50 to 100 feet).

Reflection The process in which a wave has part of its energy returned seaward by a reflecting surface.

Refraction The process by which the part of a wave in shallow water is slowed down to cause the wave to bend and tend to align itself with the underwater contours.

Regenerated production The portion of gross primary production that is supported by nutrients recycled within an ecosystem.

Relative dating The determination of whether certain rock units are older or younger than others by the use of fossil assemblages. It was not possible to tell the actual age of rocks by use of fossils until radiometric dating was developed.

Relict beach A beach deposit laid down and submerged by a rise in sea level. It is still identifiable on the continental shelf, indicating that no deposition is presently taking place at that location on the shelf.

Relict sediment A sediment deposited under a set of environmental conditions that still remains unchanged although the environment has changed, and it remains unburied by later sediment. An example is a beach deposited near the edge of the continental shelf when sea level was lower.

Residence time The average length of time a particle of any substance spends in the ocean. It is calculated by dividing the total amount of the substance in the ocean by the rate of its introduction into the ocean or the rate at which it leaves the ocean.

Respiration The process by which organisms utilize organic materials (food) as a source of energy. As the energy is released, oxygen is used and carbon dioxide and water are produced.

Restoring force A force such as surface tension or gravity that tends to restore the ocean surface displaced by a wave to that of a still-water level.

Reverse osmosis A method of desalinating ocean water that involves forcing water molecules through a water-permeable membrane under pressure.

Reversing current The tide current as it occurs at the margins of landmasses. The water flows in and out for approximately equal periods of time separated by slack water where the water is still at high and low tidal extremes.

Rhodophyta Phylum of algae composed primarily of small encrusting, branching, or filamentous plants that receive their characteristic red color from the presence of the pigment phycoerythrin. With a worldwide distribution, they are found at greater depths than other algae.

Right whales Whales of the genus *Eubalaena* that were the favorite target of early whalers.

Rip current A strong narrow surface or near-surface current of short duration (up to 2 hours) and high speed (up to 4 kilometers/hour) flowing seaward through the breaker zone at nearly right angles to the shore. It represents the return to the ocean of water that has been piled up on the shore by incoming waves.

Rise A long, broad elevation that rises gently and rather smoothly from the deep-ocean floor.

Rorqual whales Include the minke, Baird's, Bryde's, sei, fin, blue, and humpback whales. They are baleen whales with many ventral grooves.

Rotary current Tidal current as observed in the open ocean. The tidal crest makes one complete rotation during a tidal period.

Salinity A measure of the quantity of dissolved solids in ocean water. Formally, it is the total amount of dissolved solids in ocean water in parts per thousand by weight (‰) after all carbonate has been converted to oxide, the bromide and iodide to chloride, and all the organic matter oxidized. It is normally computed from conductivity, refractive index, or chlorinity.

Salinometer An instrument that is used to determine the salinity of seawater by measuring its electrical conductivity.

Salpa Genus of pelagic tunicates that are cylindrical, transparent, and found in all oceans.

Salt Any substance that yields ions other than hydrogen or hydroxyl. Salts are produced from acids by replacing the hydrogen with a metal.

Salt marsh A relatively flat area of the shore where fine sediment is deposited and salt-tolerant grasses grow. One of the most biologically productive regions of Earth.

Salt wedge estuary A very deep river mouth with a very large volume of freshwater flow beneath which a wedge of saltwater from the ocean invades. The Mississippi River is an example.

San Andreas Fault A transform fault that cuts across the state of California from the northern end of the Gulf of California to Pt. Arena north of San Francisco.

Sand Particle size of 0.0625 to 2 millimeters. It pertains to particles that lie between silt and granules on the Wentworth scale of grain size.

Sargasso Sea A region of convergence in the North Atlantic lying south and east of Bermuda where the water is a very clear, deep blue color, and contains large quantities of floating *Sargassum*.

Sargassum A brown alga characterized by a bushy form, substantial holdfast when attached, and a yellow-brown, green-yellow, or orange color. Two species, *S. fluitans* and *S. natans,* make up most of the macroscopic vegetation in the Sargasso Sea.

Scarp A linear steep slope on the ocean floor separating gently sloping or flat surfaces.

Scavenger An animal that feeds on dead organisms.

Schooling Well-defined social organizations of fish, squid, and crustaceans that aid them in survival. Often the precise benefit to the schooling species eludes the human observer.

Scyphozoa A class of coelenterates that includes the true jellyfish in which the medusoid body form predominates and the polyp is reduced or absent.

Sea 1. A subdivision of an ocean. Two types of seas are identifiable and defined. They are the *mediterranean seas,* where a number of seas are grouped together collectively as one sea, and *adjacent seas,* which are connected individually to the ocean. 2. A portion of the ocean where waves are being generated by wind.

Sea anemone A member of the class Anthozoa whose bright color, tentacles, and general appearance make it resemble flowers.

Sea arch An opening through a headland caused by wave erosion. Usually develops as sea caves are extended from one or both sides of the headland.

Sea cave A cavity at the base of a sea cliff formed by wave erosion.

Sea cow An aquatic, herbivorous mammal of the order Sirenia that includes the dugong and manatee.

Sea cucumber A common name given to members of the echinoderm class Holotheuroidea.

Seafloor spreading A process producing the lithosphere when convective upwelling of magma along the oceanic ridges moves the ocean floor away from the ridge axes at rates of from 1 to 10 centimeters (0.4 to 4 inches) per year.

Sea ice Any form of ice originating from the freezing of ocean water.

Sea otter A seagoing otter that has recovered from near extinction along the North Pacific coasts. It feeds primarily on abalone and sea urchins.

Sea snake A reptile belonging to the family Hydrophiidae with venom similar to that of cobras. They are found primarily in the coastal waters of the Indian Ocean and the western Pacific Ocean.

Sea turtle Any of the reptilian order Testudinata found widely in warm water.

Sea urchin An echinoderm belonging to the class Echinoidea possessing a fused test (external covering) and well-developed spines.

Seamount An individual peak extending over 1000 meters above the ocean floor.

Seasonal thermocline A thermocline that develops due to surface heating of the oceans in mid- to high latitudes. The base of the seasonal thermocline is usually above 200 meters (656 feet).

Seawall A wall built parallel to the shore to protect coastal property from the waves.

Sediment Particles of organic or inorganic origin that accumulate in loose form.

Sediment maturity A condition in which the roundness and degree of sorting increase and clay content decreases within a sedimentary deposit.

Sedimentary rock A rock resulting from the consolidation of loose sediment, or a rock resulting from chemical precipitation, such as sandstone and limestone.

Seiche A standing wave of an enclosed or semienclosed body of water that may have a period ranging from a few minutes to a few hours, depending on the dimensions of the basin. The wave motion continues after the initiating force has ceased.

Seismic Pertaining to an earthquake or Earth vibration, including those that are artificially induced.

Seismic sea wave *See* tsunami.

Seismic surveying The use of sound-generating techniques to identify features on or beneath the ocean floor.

Semidiurnal tide Tide having two high and two low waters per tidal day with small inequalities between successive highs and successive lows. Tidal period is about 12 hours and 25 minutes solar time. Semidaily tide.

Sessile Permanently attached to the substrate and not free to move about.

Shallow water wave A wave on the surface having a wavelength of at least 20 times water depth. The bottom affects the orbit of water particles and speed is determined by water depth: Speed (meters/second) = 3.1 $\sqrt{\text{Water depth}}$ (meters).

Shelf break The depth at which the gentle slope of the continental shelf steepens appreciably. It marks the boundary between the continental shelf and continental rise.

Shelf ice Thick shelves of glacial ice that push out into Antarctic seas from Antarctica. Large tabular icebergs calve at the edge of these vast shelves.

Shoal Shallow.

Shore Seaward of the coast, extends from highest level of wave action during storms to the low water line.

Shoreline The line marking the intersection of water surface with the shore. Migrates up and down as the tide rises and falls.

Shoreline of emergence Shorelines that indicate a lowering of sea level by the presence of stranded beach deposits and marine terraces above it.

Shoreline of submergence Shorelines that indicate a rise in sea level by the presence of drowned beaches or submerged dune topography.

Side-scan sonar A method of mapping the topography of the ocean floor along a strip up to 60 kilometers (37 miles) wide using computers and sonar signals that are directed away from both sides of the survey ship.

Silica Silicon dioxide (SiO_2).

Siliceous A condition of containing abundant silica (SiO_2).

Sill A submarine ridge partially separating bodies of water such as fjords and seas from one another or from the open ocean.

Silt A particle size of 1/128 to 1/16 millimeter. It is intermediate between sand and clay.

Siphonophora An order of hydrozoan coelenterates that forms pelagic colonies containing both polyps and medusae. Examples are *Physalia* and *Velella*.

Sirenia An order of vegetarian marine mammals that includes the dugong and manatee.

Slack water Occurs when a reversing tidal current changes direction at high or low water. Current speed is zero.

Slick A smooth patch on an otherwise rippled surface caused by a monomolecular film of organic material that reduces surface tension.

Slightly stratified estuary An estuary of moderate depth in which marine water invades beneath the freshwater runoff. The two water masses mix so that the bottom water is slightly saltier than the surface water at most places in the estuary.

SOFAR channel Sound fixing and ranging channel. This is a low-velocity sound travel zone that coincides with the permanent thermocline in low and midlatitudes.

Solar humidification A process by which ocean water can be desalinated by evaporation and the condensation of the vapor on the cover of a container. The condensate then runs into a separate container and is collected as freshwater.

Solar system The sun and the celestial bodies, asteroids, planets, and comets that orbit around it.

Solar tide The partial tide caused by the tide-producing forces of the sun.

Solid state A state of matter in which the substance has a fixed volume and shape. A crystalline state of matter.

Solstice The time during which the sun is directly over one of the tropics. In the Northern Hemisphere the summer solstice occurs on June 21 or 22 as the sun is over the Tropic of Cancer, and the winter solstice occurs on December 21 or 22 when the sun is over the Tropic of Capricorn.

Solute A substance dissolved in a solution. Salts are the solute in saltwater.

Solution A state in which a solute is homogeneously mixed with a liquid solvent. Water is the solvent for the solution that is ocean water.

Solvent A liquid that has one or more solutes dissolved in it.

Somali Current This current flows north along the Somali coast of Africa during the southwest monsoon season.

Sonar An acronym for *sound navigation and ranging*. A method by which objects may be located in the ocean.

Sounding Measuring the depth of water beneath a ship.

Southwest Monsoon A southwest wind that develops during the summer season. It blows off the Indian Ocean onto the Asian mainland.

Southwest Monsoon Current   During the southwest monsoon season, this eastward-flowing current replaces the west-flowing North Equatorial Current in the Indian Ocean.

Species diversity   The number or variety of species found in a subdivision of the marine environment.

Specific gravity   The ratio of density of a given substance to that of pure water at 4°C and at atmospheric pressure.

Specific heat   The quantity of heat required to raise the temperature of 1 gram of a given substance 1°C. For water it is 1 calorie.

Spermatophyta   Seed-bearing plants.

Spicule   A minute, needlelike calcareous or siliceous form found in sponges, radiolarians, chitons, and echinoderms that acts to support the tissue or provide a protective covering.

Spilling breaker   A type of breaking wave that forms on a gently sloping beach, which gradually extracts the energy from the wave to produce a turbulent mass of air and water that runs down the front slope of the wave.

Spit   A small point, low tongue, or narrow embankment of land commonly consisting of sand deposited by longshore currents and having one end attached to the mainland and the other terminating in open water.

Sponge   *See* Porifera.

Spray zone   The shore zone lying between the high-tide shoreline and the coastline. It is covered by water only during storms.

Spreading center   A divergent plate boundary.

Spreading rate   The rate of divergence of plates at a spreading center.

Spring tide   Tide of maximum range occurring every fortnight when the moon is new or full.

Stack   An isolated mass of rock projecting from the ocean off the end of a headland from which it has been detached by wave erosion.

Standard seawater   Ampules of ocean water for which the chlorinity has been determined by the Institute of Oceanographic Services in Wormly, England. The ampules are sent to laboratories all over the world so that equipment and reagents used to determine the salinity of ocean water samples can be calibrated by adjustment until they give the same chlorinity as is shown on the ampule label.

Standing wave   A wave, the form of which oscillates vertically without progressive movement. The region of maximum vertical motion is an *antinode*. On either side are *nodes*, where there is no vertical motion but maximum horizontal motion.

Stenohaline   Pertaining to organisms that can withstand only a small range of salinity change.

Stenothermal   Pertaining to organisms that can withstand only a small range of temperature change.

Storm surge   A rise above normal water level resulting from wind stress and reduced atmospheric pressure during storms. Consequences can be more severe if it occurs in association with high tide.

Strait of Gibraltar   The narrow opening between Europe and Africa through which the waters of the Atlantic Ocean and Mediterranean Sea mix.

Stranded beach   An ancient beach deposit found above present sea level because of a lowering of sea level.

Streamlining   The shaping of an object so it produces the minimum of turbulence while moving through a fluid medium. The teardrop shape displays a high degree of streamlining.

Subduction   The process by which one lithospheric plate descends beneath another as they converge.

Subduction zone seep biocommunity   Animals that live in association with seeps of pore water squeezed out of deeper sediments. They depend on sulfur-oxidizing bacteria that act as producers for the ecosystem.

Sublittoral zone   That portion of the benthic environment extending from low tide to a depth of 200 meters (660 feet); considered by some to be the surface of the continental shelf.

Submarine canyon   A steep, V-shaped canyon cut into the continental shelf or slope.

Submerged dune topography   Ancient coastal dune deposits found submerged beneath the present shoreline because of a rise in sea level.

Submergent shoreline   Shoreline formed by the relative submergence of a landmass in which the shoreline is on landforms developed under subaerial processes. It is characterized by bays and promontories and is more irregular than a shoreline of emergence.

Subneritic province   The benthic environment extending from the shoreline across the continental shelf to the shelf break. It underlies the neritic province of the pelagic environment.

Suboceanic province   Benthic environments seaward of the continental shelf.

Subpolar   Pertaining to the oceanic region that is covered by sea ice in winter. The ice melts away in summer.

Substrate   The base on which an organism lives and grows.

Subsurface current   A current usually flowing below the pycnocline, generally at slower speed and in a different direction from the surface current.

Subtropical   Pertaining to the oceanic region poleward of the tropics (about 30° latitude).

Subtropical Convergence   The zone of convergence that occurs within all subtropical gyres as a result of Ekman transport driving water toward the interior of the gyres.

Subtropical gyre   The trade winds and westerly winds initiated in the subtropical regions of all ocean basins, with the influence of the Coriolis effect, set large regions of ocean water in motion. They rotate clockwise in the Northern Hemisphere and counterclockwise in the Southern Hemisphere, and they are centered in the subtropics.

Sulfur   A yellow mineral composed of the element sulfur. It is commonly found in association with hydrocarbons and salt deposits.

Sulfur-oxidizing bacteria   Bacteria that support many deep-sea hydrothermal vent and cold-water seep biocommunities by using energy released by oxidation to synthesize organic matter chemosynthetically.

Summer solstice   In the Northern Hemisphere, it is the instant when that sun moves north to the Tropic of Cancer before changing direction and moving southward toward the equator, approximately June 21.

Supralittoral zone   The splash or spray zone above the spring high-tide shoreline.

Surf zone   The region between the shoreline and the line of breakers where most wave energy is released.

Surface tension   The tendency for the surface of a liquid to contract owing to intermolecular bond attraction.

Swash   A thin layer of water that washes up over exposed beach as waves break at the shore.

Swell   A free ocean wave by which energy put into ocean waves by wind in the sea is transported with little energy loss across great stretches of ocean to the margins of continents where the energy is released in the surf zone.

Swim bladder   A gas-containing, flexible, cigar-shaped organ that aids many fishes in attaining neutral buoyancy.

Symbiosis   A relationship between two species in which one or both benefit or neither or one is harmed. Examples are commensalism, mutualism, and parasitism.

Tectonic estuary   An estuary, the origin of which is related to tectonic deformation of the coastal region.

Tectonics   Deformation of Earth's surface by forces generated by heat flow from Earth's interior.

Temperate Pertaining to the oceanic region where pronounced seasonal change occurs (about 40° to 60° latitude).

Temperature A direct measure of the average kinetic energy of the molecules of a substance.

Temperature of maximum density The temperature at which a substance reaches its highest density. For water, it is 4°C.

Territorial sea A strip of ocean, 12 nautical miles wide, adjacent to land over which the coastal nation has control over the passage of ships.

Tethered-float breakwater Floating hollow spheres tethered to the bottom and acting as an upside-down pendulum to extract energy from waves near the shore without interfering with the longshore drift of sediment.

Tethys Sea An ancient body of water that separated Laurasia to the north and Gondwanaland to the south. Its location was approximately that of the present Alpine-Himalayan mountain system.

Thermocline A layer of water beneath the mixed layer in which a rapid change in temperature can be measured in the vertical dimension.

Thermohaline circulation The vertical movement of ocean water driven by density differences resulting from the combined effects of variations in temperature and salinity.

Tidal bore A steep-fronted wave that moves up some rivers when the tide rises in the coastal ocean.

Tidal range The difference between high tide and low tide water levels over any designated time interval, usually one lunar day.

Tide Periodic rise and fall of the surface of the ocean and connected bodies of water resulting from the gravitational attraction of the moon and sun acting unequally on different parts of Earth.

Tide-generating force The magnitude of the centripetal force required to keep all particles of Earth having identical mass moving in identical circular paths required by the movements of the Earth-moon system is identical. This required force is provided by the gravitational attraction between the particles and the moon. This gravitational force is identical to the required centripetal force only at the center of Earth. For ocean tides, the horizontal component of the small force that results from the difference between the required and provided forces is the tide-generating force on that individual particle. These forces are such that they tend to push the ocean water into bulges toward the tide-generating body on one side of Earth and away from the tide-generating body on the opposite side of Earth.

Tide wave The long-period gravity wave generated by tide-generating forces described above and manifested in the rise and fall of the tide.

Tissue An aggregate of cells and their products developed by organisms for the performance of a particular function.

Tombolo A sand or gravel bar that connects an island with another island or the mainland.

Topography The configuration of a surface. In oceanography it refers to the ocean bottom or the surface of a mass of water with given characteristics.

Trade winds The air masses moving from subtropical high pressure belts toward the equator. They are northeasterly in the Northern Hemisphere and southeasterly in the Southern Hemisphere.

Transform fault A fault characteristic of oceanic ridges along which they are offset.

Transform plate boundary The shear boundary between two lithospheric plates formed by a transform fault.

Transitional crust Thinned section of continental crust at the trailing edge of a continent created by the breaking apart of an ancient continent over a newly formed spreading center.

Transitional wave A wave moving from deep water to shallow water that has a wavelength more than twice the water depth but less than 20 times the water depth. Particle orbits are beginning to be influenced by the bottom.

Transverse wave A wave in which particle motion is at right angles to energy propagation.

Trench A long, narrow, and deep depression on the ocean floor, with relatively steep sides.

Trophic level A nourishment level in a food chain. Plant producers constitute the lowest level, followed by herbivores and a series of carnivores at the higher levels.

Tropic of Cancer The latitude of 23.5°N.

Tropic of Capricorn The latitude of 23.5°S.

Tropical Pertaining to the regions of the tropics (at or near 23.5° latitude).

Tropical tide A tide occurring twice monthly when the moon is at its maximum declination north and south of the equator. It is in the tropical regions where tides display their greatest diurnal inequalities.

Trough The part of an ocean wave that is displaced below the still-water line.

Tsunami Seismic sea wave. A long-period gravity wave generated by a submarine earthquake or volcanic event. Not noticeable on the open ocean but builds up to great heights in shallow water.

Tunicates Members of the chordate subphylum Urochordata, which includes sacklike animals. Some are sessile (sea squirts), whereas others are pelagic (salps).

Turbidite A sediment or rock formed from sediment deposited by turbidity currents characterized by both horizontally and vertically graded bedding.

Turbidity A state of reduced clarity in a fluid caused by the presence of suspended matter.

Turbidity current A gravity current resulting from a density increase brought about by increased water turbidity. Possibly initiated by some sudden force such as an earthquake, the turbid mass continues under the force of gravity down a submarine slope.

Turbulent flow Flow in which the flow lines are confused heterogeneously due to random velocity fluctuations.

Typhoon A severe tropical storm in the western Pacific.

Ultraplankton Plankton for which the greatest dimension is less than 5 micrometers. Very difficult to separate from the water.

Ultrasonic Sound frequencies above those that can be heard by humans (above 20,000 Hertz).

Ultraviolet radiation Electromagnetic radiation shorter than visible radiation and longer than X rays. The approximate range is 1 to 400 nanometers.

Upper water Includes the mixed layer and the permanent thermocline. It is approximately the top 1000 meters of the ocean.

Upwelling The process by which deep, cold, nutrient-laden water is brought to the surface, usually by diverging equatorial currents or coastal currents that pull water away from the coast.

Valence The combining capacity of an element measured by the number of hydrogen atoms with which it will combine.

van der Waals force Weak attractive force between molecules resulting from the interaction of one molecule and the electrons of another.

Vector A physical quantity that has magnitude and direction. Examples are force, acceleration, and velocity.

Ventral Pertaining to the lower or under surface.

Vernal equinox The passage of the sun across the equator as it moves from the Southern Hemisphere into the Northern Hemisphere, approximately March 21.

Vertebrata Subphylum of chordates that includes those animals with a well-developed brain and a skeleton of bone or cartilage; includes fish, amphibians, reptiles, birds, and animals.

Vertically mixed estuary Very shallow estuaries such as lagoons in which freshwater and marine water are totally mixed from top to bottom so that the salinity at the surface and the bottom is the same at most places within the estuary.

Viviparous Referring to embryonic development inside the mother's reproductive tract with the provision of nutrition to the embryo by the mother.

Viscosity A property of a substance to offer resistance to flow. Internal friction.

Walker Circulation The pattern of atmospheric circulation that involves the rising of warm air over the East Indies low-pressure cell and its descent over the high-pressure cell in the southeastern Pacific Ocean off the coast of Chile. It is the weakening of this circulation that accompanies an El Niño event, which has led to the development of the term El Niño–Southern Oscillation event.

Warm front A weather front in which a warm air mass moves into and over a cold air mass producing a broad band of gentle precipitation.

Water mass A body of water identifiable from its temperature, salinity, or chemical content.

Wave A disturbance that moves over the surface or through a medium with a speed determined by the properties of the medium.

Wave-cut beach A gently sloping surface produced by wave erosion and extending from the base of the wave-cut cliff out under the offshore region.

Wave-cut cliff A cliff produced by landward cutting by wave erosion.

Wave dispersion The separation of waves as they leave the sea area by wave size. Larger waves travel faster than smaller waves and thus leave the sea area first, to be followed by progressively smaller waves.

Wave frequency The number of waves that pass a fixed point in a unit of time, usually one second.

Wave height Vertical distance between a crest and the preceding trough.

Wave period The elapsed time between the passage of two successive wave crests past a fixed point.

Wave steepness Ratio of wave height to wavelength.

Wave train A series of waves from the same direction.

Wavelength Horizontal distance between two corresponding points on successive waves, such as from crest to crest.

Weathering A process by which rocks are broken down by chemical and mechanical means.

West Australian Current This cold current forms the eastern boundary current of the Indian Ocean subtropical gyre. It is separated from the coast by the warm Leeuwin Current except during El Niño-Southern Oscillation events when the Leeuwin Current weakens.

West Wind Drift The surface portion of the Antarctic Circumpolar Current driven in an easterly direction around Antarctica by the strong westerly winds.

Westerly winds The air masses moving away from the subtropical high pressure belts toward higher latitudes. They are southwesterly in the Northern Hemisphere and northwesterly in the Southern Hemisphere.

Western boundary current Poleward-flowing warm currents on the western side of all subtropical gyres.

Western boundary undercurrent (WBUC) A bottom current that flows along the base of the continental slope eroding sediment from it and redepositing the sediment on the continental rise. It is confined to the western boundary of deep-ocean basins.

Westward intensification Pertaining to the intensification of the warm western limb of the subtropical gyre currents that is manifested in higher velocity and deeper flow compared with the cold eastern boundary currents that drift leisurely toward the equator.

Wetlands Biologically productive regions bordering estuaries and other protected coastal areas. They are usually salt marshes at latitudes greater than 30° and mangrove swamps at lower latitudes.

White muscle fiber Thick muscle fibers with relatively low concentrations of myoglobin that make up a large percentage of the muscle fiber in lunger-type fishes.

Wind-driven circulation Any movement of ocean water that is driven by winds. This includes most horizontal movements in the surface waters of the world's oceans.

Windrows Rows of floating debris aligned parallel to the direction of the wind that result from Langmuir circulation.

Windward The direction from which the wind is blowing.

Winter solstice The instant the southward-moving sun reaches the Tropic of Cancer before changing direction and moving north back toward the equator, approximately December 21.

Zenith That point on the celestial sphere directly over the observer.

Zooplankton Animal plankton.

Zooxanthellae A form of algae that lives as a symbiont in the tissue of corals and other coral reef animals and provides varying amounts of their required food supply.

Index

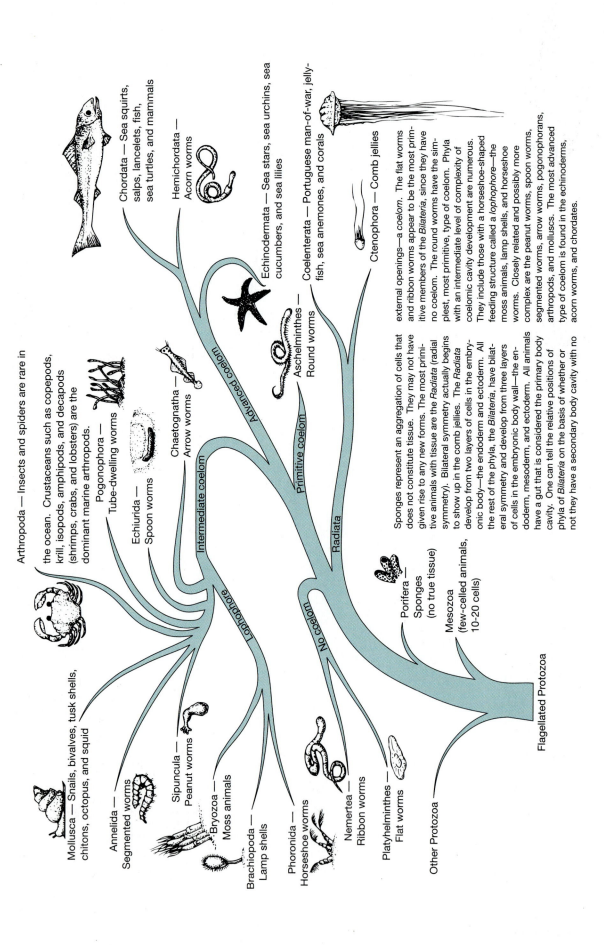

Arthropoda — Insects and spiders are rare in the ocean. Crustaceans such as copepods, krill, isopods, amphipods, and decapods (shrimps, crabs, and lobsters) are the dominant marine arthropods.

Pogonophora — Tube-dwelling worms

Echiurida — Spoon worms

Chaetognatha — Arrow worms

Chordata — Sea squirts, salps, lancelets, fish, sea turtles, and mammals

Hemichordata — Acorn worms

Echinodermata — Sea stars, sea urchins, sea cucumbers, and sea lilies

Coelenterata — Portuguese man-of-war, jelly-fish, sea anemones, and corals

Ctenophora — Comb jellies

Advanced coelom

Intermediate coelom

Primitive coelom

Radiata

Lophophore

No coelom

Mollusca — Snails, bivalves, tusk shells, chitons, octopus, and squid

Annelida — Segmented worms

Sipuncula — Peanut worms

Bryozoa — Moss animals

Brachiopoda — Lamp shells

Phoronida — Horseshoe worms

Nemertea — Ribbon worms

Platyhelminthes — Flat worms

Other Protozoa

Porifera — Sponges (no true tissue)

Mesozoa (few-celled animals, 10–20 cells)

Aschelminthes — Round worms

Flagellated Protozoa

Sponges represent an aggregation of cells that does not constitute tissue. They may not have given rise to any new forms. The most primitive animals with tissue are the *Radiata* (radial symmetry). Bilateral symmetry actually begins to show up in the comb jellies. The *Radiata* develop from two layers of cells in the embryonic body—the endoderm and ectoderm. All the rest of the phyla, the *Bilateria*, have bilateral symmetry and develop from three layers of cells in the embryonic body wall—the endoderm, mesoderm, and ectoderm. All animals have a gut that is considered the primary body cavity. One can tell the relative positions of phyla of *Bilateria* on the basis of whether or not they have a secondary body cavity with no external openings—a *coelom*. The flat worms and ribbon worms appear to be the most primitive members of the *Bilateria*, since they have no coelom. The round worms have the simplest, most primitive, type of coelom. Phyla with an intermediate level of complexity of coelomic cavity development are numerous. They include those with a horseshoe-shaped feeding structure called a *lophophore*—the moss animals, lamp shells, and horseshoe worms. Closely related and possibly more complex are the peanut worms, spoon worms, segmented worms, arrow worms, pogonophorans, arthropods, and molluscs. The most advanced type of coelom is found in the echinoderms, acorn worms, and chordates.

CPSIA information can be obtained
at www.ICGtesting.com
Printed in the USA
LVHW012056270520
656718LV00001B/1

Begin Relaxation Exercise

... Yoga for Beginners

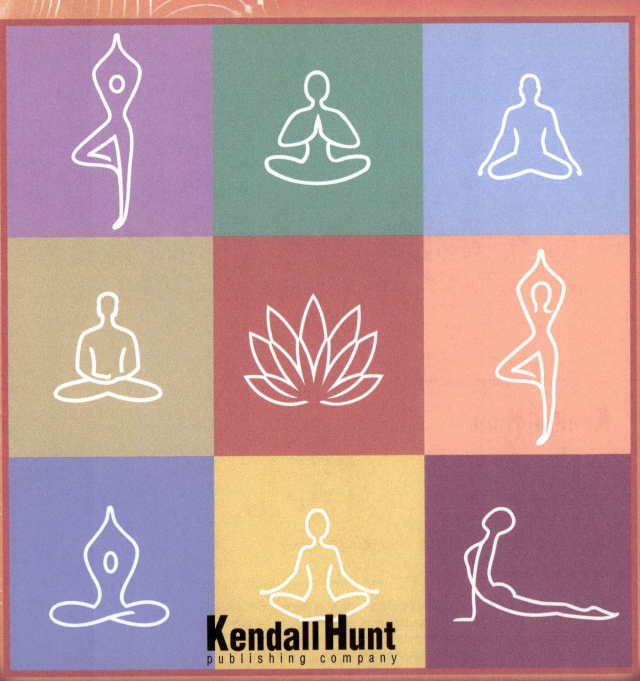

Kendall Hunt
publishing company

Kimberly Hansen-Johnson
Baylor University

Cover image © Shutterstock.com

www.kendallhunt.com
Send all inquiries to:
4050 Westmark Drive
Dubuque, IA 52004-1840

Published in the United States of America

Contents

Preface

As an educator of physical fitness for 28 years, it is my goal with this book to help a yoga beginner learn the basics of the activity for the physical, mental, and relaxation benefits gained from the poses, breathwork, and meditation.

During my first few months of taking yoga classes, I spent most of my time looking around to see what everyone else was doing. I wasn't familiar with yoga terminology or the poses. I want my students to be able to read this book and have no fear or reservation going to a yoga studio.

From this book, the first time yogi, as we call ourselves, will be introduced to yoga terminology. Additionally, you will be well versed in the "ins and outs" of a studio including etiquette while at the same time feeling comfortable and safe doing the poses, breathwork, and meditation. By the time you finish this book, you will be able to attend any yoga class and be able to participate fully or create your own home practice program.

Like most things worth mastering, yoga takes practice and patience. At the same time, it is my belief that yoga is for everyone and available to all generations and fitness levels while providing unlimited benefits.

Source: Kimberly Hansen-Johnson

Acknowledgments

First of all I want to thank God for allowing me to even be here. After experiencing a brain aneurysm rupture in September of 2018 that happened in front of my yoga students, by God's grace, I am alive and well. I want to thank my family, especially my husband Mark. They have supported me through all my training and indulged my crazy ideas about writing a book. I could not have done this without them. Thanks to my brother-in-law, William Lamb, for pushing me to actually write this book and inspiring me by taking me along on his awesome bike rides. Thank you to my sister-in-law, Cheryl Lamb, for helping me through ideas and giving her honest opinions. Thank you to my ZTA sisters, Bush and Reed, who also helped with ideas and inspiration. I want to thank my yoga instructors at The Studio in Waco, Texas for giving me the necessary education and training. Lastly, thank you Baylor University for allowing me to teach your wonderful students. I am truly blessed with amazing friends, family, and community.

Source: Kimberly Hansen-Johnson

Chapter 1: What Is yoga?

1 What Is Yoga?

The word **Yoga** means "to yoke" or "to make one" or "unite."
We are uniting the mind, body, and heart into one with each other.
We accomplish this through 3 main components or limbs of yoga:

1. Breath Control (Pranayama)
2. Postures (Asana)
3. Meditation (Dhyāna)

When practicing the poses and breath control, it will help the student to meditate, which allows the mind and the body to become one.

Yoga is around 5,000 years old, but many believe it is over 10,000 years old. Patanjali, the author of the book *Yoga Sutras,* broke yoga down into **8 limbs:**

1. Abstinence (Yama)
2. Observance (Niyama)
3. *Posture (Asana)*
4. *Breath Control (Pranayama)*
5. Sense Withdrawal (Pratyahara)
6. Concentration (Dharana)
7. *Meditation (Dhyana)*
8. Reflection (Samadhi)

© bc21/Shutterstock.com

In this beginner's book, we will be focusing on these 3 limbs: asana (**poses** or **postures**), pranayama (**breath control**), and dhyana (**meditation**). We will incorporate all 3 limbs into every class period.

Types of Yoga

There are many types of yoga . . . physical, relaxing, restorative, meditative. You will find variations depending on the teacher and studio. It is best to try many different types, styles, and studios, to see what best fits your personality and your daily life. Then switch it around so it does not get boring, repetitive, or a chore.

Hatha yoga

This type of yoga is usually slower and focuses on breath and the physical poses. It is a gentle introduction to yoga and beginner postures.

Iyengar yoga

Focuses on alignment and breath control. The poses are held for a longer time placing emphasis on the alignment and precise movement.

Kundalini yoga

This type of yoga practices uses meditation, breath work, poses, and mantras and is considered more spiritual.

Ashtanga yoga

Always starts with 5 sun salutations A's and 5 sun salutations B's. Then starts the standing and floor sequences while linking the breath to the movement called vinyasa. It is considered more physically demanding.

Yin yoga

Yin Yoga is a slow paced with seated postures that are held for up to 1 to 5 minutes long. Gravity helps push the student deeper into the poses. As one of my yoga instructors once said, "With these poses you hold them with the 'Goldilocks effect . . . not too hard, not too soft, but just right' and still get a good stretch."

Restorative yoga

Props such as blankets, bolsters, blocks, and eye pillows are used to help the student into a deeper relaxed state and the poses are held for a longer time. The difference between restorative and yin is that in restorative you are 100 percent relaxed.

Hot yoga

Bikram hot yoga is a sequence that includes a series of 26 basic poses, with each one performed twice. It is performed in a sauna-like room—usually set to 105 degrees and 40 percent humidity.

Health Benefits

Benefits of yoga include the obvious such as flexibility, strength, balance, posture, relaxation, and stress relief, but it can also help with anxiety, focus, insomnia, lower back pain, migraines, digestive issues, and many more. The physical and mental endurance that yoga provides can help in many different types of situations such as health scares that a person needs the mental and physical toughness to persevere.

Safety

First, this is your practice and your body. Yoga is a great exercise to start listening to your body. There is a fine line of challenging yourself or pushing yourself to a feeling a tightness or injury. If you have an injury, let the instructor know. There is ALWAYS a modification to a pose, so don't push it. Always work within your own range of limits and abilities. If you have any medical concerns, talk with your doctor before practicing yoga. In short, if it hurts do not do it.

Common Terms

Asana = pose or posture.
Pranayama = breath control.
Namaste = the light in me honors the light in you.

Sanskrit = it is the ancient language of Hinduism, Buddhism, and Jainism.
Block = tool to deepen practice usually made of foam, cork, or wood.
Strap = tool to deepen practice usually made of belt material. Can help deepen stretches.
Blanket = tool to deepen practice usually used in a restore class.
Bolster = tool to deepen practice to sit, lay, or prop the body on and used in a restore class.
Flow = linking poses together.
Savasana = corpse pose, final resting pose.
Sun salutation A, B, C = a common sequence of 12 to 17 poses or flow to warm up the body.
Ujjayi breathing = ocean breathing or yogic breathing or victory breath or Darth Vader breath.
Crown of head = top most part of the head.
Sacrum = bone at base of spine.
Sit bones = located under the buttocks, easily feel when sitting on hard surface.
Vinyasa = "link" pose to pose with breath. (Flow).
Example: "Take a vinyasa" = plank, chaturanga, upward facing dog, downward facing dog or ½ plank, to belly, cobra, downward facing dog.
Alignment = ideal position of body, so bone and muscle can work effectively preventing injury.
Example: "Head over heart over pelvis"= bodily alignment.
Flointed = flex of the foot leading from the ball of the foot instead of the toes.
Ground down = all 4 points of the feet and knuckles and palm heels of hands into earth.
Mantra = repeating words, phrases, or sounds for meditation to inspire not to distract.
Set intention = a word, statement, or mantra to bring to your mat during the practice.
Mindful or mindfulness = being present and aware . . . focusing on the present moment.
Mudras = hand positions/postures (hand yoga).
Chakras = it is believed that the 7 chakras are the energy centers in our bodies in which energy flows through. If a chakra is blocked, illnesses or other issues can develop. It is important to learn how to unblock the chakras that are blocked to keep all energy flowing.
Essential oils = essential oils are natural oils from plant extracts that can be used by themselves or a blend of oils together. You might smell them, rub them on your skin, put them in your bath or diffuser.

Common Sayings

Frequently when you go to a yoga class, you will hear the instructor talk about "*setting an intention for the practice*." The act of setting an intention is not a goal. Intentions are the here and now and goal is the future. For instance, If you are a person who likes to take it easy, your intention might be to challenge yourself this class. If you are a person on the go, your intention could be to be present in the here and now and enjoy and just breathe.

Examples of intentions could be:

- Let it go ... whatever happened before let it go or what is happening after let it go.
- Grace ... give yourself grace
- Be present or live in the present
- Forgive and forget
- Smile and laugh
- Love and believe in myself
- Stay positive
- Challenge or push myself
- Open my heart
- Connecting to breath

Oftentimes when coming from corpse pose (Savasana) to a seated position, the instructor might say **"roll to your right side"** and come to a seated position (Sukhasana). Rolling to the side helps to not injure the back, but it also has to do physically and symbolically. Physically the heart (for most people) is on the left and when you roll to your right side the heart remains open and free of pressure. Symbolically, it is advantageous to enter a holy place with your right foot hence the importance of rolling to the right side.

At the end of the practice, you will hear the instructor often say "**Namaste**," with origins from India. Namaste, with hands in prayer at heart center, is considered a traditional greeting. In the western world, it is the handshake. Nowadays in the western world, we use Namaste at the end of a yoga practice, meaning "*The light in me honors the light in you,*" which is considered a respectful ending of class between the student and the teacher.

Etiquette

1. A student should arrive 10 to 15 minutes before the class starts to sign in, get mat situated, and the needed props for that class. Many studios will shut the door at class time and you will not be allowed to enter.
2. Shoes are not worn on the studio floors. Leave shoes at front or shoe area.
3. Dress being mindful of others and having your rear end in the air.
4. For a first timer, it is always a good idea to let the instructor know you are new. Do your research on the studio, classes, and so on.
5. Quiet sanctuary (no talking). For some students who practice, this is their only time to have quiet for meditation. Be mindful of their quiet time.
6. NEVER step on another student's mat. Yuck toe jam ... enough said.
7. Some classes are packed. Hygiene is important so others can have a meditative experience. (On both spectrums, too much perfume is just as bad body odor.)
8. Usually a sign is posted on what tools (equipment) are needed for that particular class that day.
9. ALWAYS disinfect your mat/block if you use the studio's equipment.
10. Put tools/props such as blocks, straps, and blankets away orderly and folded.
11. No phones. Leave your phone in the locker, car, or backpack sound off and turn smart watch, if you need to wear it, on; do not disturb.

Attire and Equipment

Attire: Dress comfortably in exercise attire. Loose clothing can sometimes interfere with certain poses. Yoga is done without shoes and usually without socks due to slippage.

Equipment: No equipment is needed; however, a mat or a towel can be used to perform the yoga postures.

Tools: Block, strap, blanket, bolster are known as props or tools and can *enhance* your practice. These are not crutches but tools to help deepen your practice no matter what level of student. These tools can help you get into a pose and maintain proper alignment but can also deepen the stretch.

What Is Yoga Chapter Test

1. What does the word yoga mean?

   _____

2. How many limbs of yoga are in the Yoga Sutras?

   _____

3. What are some benefits of yoga?

   _____

4. Name the 3 limbs we will go over in this class.

   A. _____

   B. _____

   C. _____

5. What is the English term for asana?

   _____

6. What is the English term for pranayama?

   _____

7. What are 4 tools you can use in yoga?

   A. _____

   B. _____

   C. _____

   D. _____

Chapter 2: Breath Work

2 Breath Control (Pranayama) (4th Limb of Yoga)

In general, breath control can help reduce worry, anxiety, and calm the mind. Breath control can also improve attention, concentration, or focus and help energize the mind and body.

Victory Breath (Ujjayi)

Also known as *ocean breathing or Darth Vader breath or yogic breathing*, start by inhaling through your nose and exhaling saying "ha" through your mouth. Do this 3 times. Then do the same thing and close your lips feeling the same sensation of the "ha" but with lips closed in the inhale and exhale. Quiets the brain, slows and smooths the breath. Be careful not to tighten throat. This is the main breath control we will use throughout the practice.

Three-Part Breath (Dirga)

This is a 3-part breathing practice. It can be done sitting or lying (corpse pose). Inhale deeply through your nose, filling your chest (lungs) so that your belly expands for the count of 2 and pause for a moment. Continue to expand your belly as you fill the next third of your lungs to another count of 2. Continue to expand your belly as you fill the final third of your lungs to another count of 2, pause, and then exhale slowly as you can for a count of 6. Repeat this practice 5 times (1/3 lower belly, 1/3 upper chest, 1/3 throat). This helps you to remain calm by slowing the breath and allowing you to focus more clearly during stressful situation. Stop if you become faint or dizzy.

Sun Piercing Breath/Moon Piercing Breath (Single Nostril Breathing)

Our right nostril associated with our body's heating energy, symbolized by the "Sun" and our left nostril with our body's cooling energy, symbolized by the "Moon."

Do this practice 1 to 3 minutes. For *sun piercing*, block your left nostril and inhale through your right. Then close the right and exhale through the left. Continue in this manner, inhale right, exhale left, for 1 to 3 minutes. For *moon piercing*, simply reverse by inhaling through your left nostril, exhaling through your right. Again continue for 1 to 3 minutes. Sun Piercing = stimulates brain and increases body heat. Moon piercing = quiets brain and cools the body. Do not do if you have high blood pressure or heart disease.

Don't do both breaths on same day.

Humming Bee Breath (Bhramari)

Place your index fingers on the cartilage that is in between the cheek and the ear. On the inhale, press the cartilage in and on the exhale hum a low or high pitch sound. Practice with low pitch and high pitch and see what works best for you. Continue for 3 or 4 times. Helps calm an overactive mind. Also helps with headaches, concentration, and memory.

Cooling Breath (Sitali)

Roll the tongue (also called taco tongue). If you can't roll your tongue, then act like you are drinking through a straw. Inhale through the tongue or straw mouth and exhale through the nose. 5 to 10 times. This breath is calming and cooling. If you start feeling dizzy stop the practice.

Alternate Nostril Breathing (Nadi Shodhana)

With your right hand, start by closing the left nostril lightly with your ring finger and inhale through the right nostril. Close the right nostril with your thumb and open left nostril for an exhale and then an inhale. Close left nostril and open right nostril for exhale and an inhale. Continue 1 to 5 minutes. Calms your mind and relieves stress. This is the beginner's level version.

Breath of Fire (Bhastrika)

Close your eyes and mouth and relax abdominal muscles. Keeping your mouth closed throughout practice inhale and exhale normally. Inhale halfway, and begin to exhale while contracting your abdominal muscles. Continue doing this inhaling passively and exhaling sharply by pumping the abdominal muscles in and out sharp and rapid clearing any waste from air passages. Do 20 rounds. When you are done with the cycles, exhale all your breath. Then, inhale once normally and then exhale. Then inhale and hold as long as you comfortably can and then exhale all your breath. This practice cleanses your sinuses and is a breathing exercise that helps to oxygenate your body while strengthening the muscles of your abdomen. Should never be practiced when an asthmatic attack is in progress. If you feel dizzy or pain stop the practice.

Breath of Joy

A 3-part breath. While standing with legs shoulder width apart, inhale through the nose ⅓ while swinging the arms in front, inhale ⅓ while swinging arms to the side, inhale ⅓ while swinging arms above the head, and on the exhale through the mouth saying "ha" bend over at the waist and swing arms down. Repeat 9 times and feel the joy.

Helps to awaken the sympathetic nervous system and increases oxygen levels in the bloodstream. If you have head/eye injury refrain from doing this practice.

4, 7, 8 Breathing

This breath is also a 3-part breath that includes inhaling, retention, and exhaling. Exhale completely out of your mouth making a whooshing sound. Then inhale through the nose for the count of 4, hold for the count of 7, and exhale through the mouth making a whooshing sound for the count of 8. Repeat 4 full cycles. This breath calms the nerves and can help with anxiety and stress.

Even Breath

For this breath, inhale for a count of 4. Take a moment at the top of the inhale and then exhale for a count of 4. You can experiment with the count, for example, if 5 is a better breath pattern or 6 is better breath pattern for your stress. Continue for 1 to 2 minutes. This breath technique helps to reduce stress and anxiety.

Breath Control Chapter Test

1. What is Sanskrit for breath control?

     _____

2. What are some general benefits of breath control?

     _____

3. What breath control will we be using most in our class?

     _____

4. What is a calming breath?

     _____

5. What is a breath for focus?

     _____

Chapter 3: Alignment and Beginner Poses

© IrinaKrivoruchko/Shutterstock.com

3 Alignment and Poses (3rd Limb of Yoga) (Asana)

Alignment

Alignment is adjusting the body so the muscles can work more effectively and in turn be in the best position for strengthening and stretching so as not to put undue pressure on joints, ligaments, muscles, and bones. When aligned, it keeps the body safe from injury. When working on alignment, start with the base of the posture and work your way up by staking the joints. Always work within your own range of limits and abilities. For example, start with the feet and work your way up stacking ankles, knees, hips, shoulders, and head.

Just remember, the perfect pose and alignment is what is best for you at that given moment. Variations in body types is why nothing is ever set in stone. Listen to your body. How a pose feels is more important than how it will look to you or to others.

The instructor will say verbal cues to remind of alignment and adjustments of the poses. The instructor at some yoga studios will have a hands-on approach to help with alignment and final adjustments.

Seated, Seated Twist, Seated Forward Bend, and Kneeling Poses

Staff Pose (Dandasana)

> Alignment: Sit with legs extended in front of you. Tall spine and palms on ground. Toes spread and toward head.
> Benefits: Strengthens back muscles, stretches the shoulders and chest. Improves posture.
> Caution: Low back or wrist injury.

© fizkes/Shutterstock.com

Easy Pose (Sukhasana)

Alignment: Sit bones on floor or prop, knees same height as hips, spine vertical.
Benefits: Opens hips, strengthens spine, stretches ankles and knees, calming.
Caution: Knee injuries, low back injury.

© LStockStudio/Shutterstock.com

Hero Pose (Virasana)

Alignment: Kneeling with knees hip width apart, sacrum on floor, tops of feet flat on floor and besides the hips, spine tall.
Benefits: Opens thighs and psoas.
Caution: Knee injury, ankle injury.

© fizkes/Shutterstock.com

Diamond Pose (Vajrasana)

Alignment: Sit on legs with bent knees, keep knees close together and form a bowl with the soles of the feet with heels beside buttocks and sitting on them.
Benefits: Strengthens leg, knees, and toes.
Caution: Knee and ankle injuries.

© fizkes/Shutterstock.com

Bound Angle (Baddha Konasana)

Alignment: Start in staff pose. Then bend both knees and bring the soles of the feet together. Hold your ankles and press the small toes side together while opening the inside of the feet (like opening a book).
Benefits: Stretches inner and outer thigh and groin.
Caution: Groin injuries.

© fizkes/Shutterstock.com

Gate (Parighasana)

Alignment: Kneel on the floor with knees stacked under hips, stretch the right leg out, and press the foot to the floor with kneecap toward ceiling. Bring arms up and slide right arm down to rest on shin, calf, or floor while side bending your left arm up and over. Gaze is up toward ceiling.

Benefits: Stretches the sides of the torso and spine, stretches hamstring, and opens shoulder.

Caution: Knee injuries.

© fizkes/Shutterstock.com

Cat (Marjaryasana)

Alignment: Start in tabletop and as you exhale, round the spine toward the ceiling like a mad cat, and release your head to the floor.

Benefits: Massages spine and stomach.

Caution: Neck injury.

© fizkes/Shutterstock.com

Bird Dog Pose (Parsva Balasana)

Alignment: Start in tabletop. Keep spine in neck and back long. Extend the right arm straight and then the left leg straight back.

Benefits: Strengthens and stabilizes the core and strengthens the lower back and increases balance.

Caution: Shoulder or wrist pain.

© fizkes/Shutterstock.com

Easy Pose Forward Fold (Adho Mukha Sukhasana)

Alignment: Sit in easy pose. On an inhale, raise arms above head, and on the exhale, fold forward resting forehead on ground arms extended on ground. If forehead does not touch, bend elbows and stack hands and place forehead on them or use a prop.

Benefits: Stretches back, shoulders, hips, knees, and ankles. Calms the mind and reduces anxiety.

Caution: Knee injury or if tight hips.

© V.S.Anandhakrishna/
Shutterstock.com

Child's Pose (Balasana)

Alignment: Start in tabletop. Bring your big toes together and your knees hip distance apart. Sit your hips back on your heels. Extend the torso forward and let your shoulders round forward allowing the forehead to rest on the floor. Arms lay by legs on each side of hips palms up.

Benefits: Stretches ankle, back, and hips; relieves back pain; reduces stress; and calms the mind.

Caution: Knee injuries.

Extended Child's Pose (Utthita Balasana)

Alignment: Start in tabletop. Bring your big toes together and your knees hip distance apart. Sit your hips back on your heels. Extend the torso forward and let your shoulders round forward allowing the forehead to rest on the floor. Extend the arms forward on the floor.

Benefits: Stretches shoulders, ankles, back, and hips; relieves back pain; reduces stress; and calms the mind.

Caution: Knee injuries.

Extended Puppy Pose (Uttana Shishosana)

Alignment: Start in tabletop. Walk your hands forward keeping hips stacked above knees. Forehead or chin to the floor.

Benefits: Stretches the spine and shoulders.

Caution: Knee Injuries.

Seated Forward Bend (Paschimottanasana)

Alignment: Start in staff pose, and on an exhale bend forward from the hip not the waist. Lengthen the spine and rest hands on shin, ankles, or feet.
Benefits: Improves digestion, reduces stress, and calms the mind.
Caution: Back and knee injuries.

© Ivan Veselinovic/Shutterstock.com

Seated Wide Legged Forward Fold (Upavistha Konasana)

Alignment: Start in staff pose. Slightly lean back opening legs up wide. Rotating thighs outward so kneecaps face the ceiling. Forward fold bending from the hips not the waist.
Benefits: Stretches the insides and back sides of legs, strengthens the spine, stimulates the abdomen, calms the mind.
Caution: Low back injuries.

© fizkes/Shutterstock.com

Seated Twist (Parivrtta Sukhasana)

Alignment: Sit in easy pose; on the inhale bring hands extended above the head and on the exhale twist to the left. Right hand to left thigh and left hand behind you at the base of your spine. Gaze toward the back of the room.
Benefits: Stretches knees, hips, and ankles; stimulates abdomen; reduces stress and anxiety.
Caution: Knee injuries.

© fizkes/Shutterstock.com

Half Lord of the Fishes (Ardha Matsyendrasana)

Alignment: Start in staff pose. Bend your left knee and place left foot on the floor with knee pointing up. Then bend your right leg in where the heel is close to your left hip. Left hand on the floor behind left hip and bring right elbow to outside of left thigh. Gaze toward the back of room.
Benefits: Stretches shoulders, hips, and neck. Stimulates the liver, kidneys, and spine.
Caution: Back or spine injuries.

© fizkes/Shutterstock.com

Head to Knee Forward Bend or Seated Tree (Janu Sirsasana)

Alignment: Start in staff pose. Bend right leg out to the right and rotating hip outward. Slightly turn torso to the left and hinge forward at the hips keeping a flat back and reaching for your left foot with your arms.
Benefits: Stretches spine, hips, groin, and hamstrings. Calming.
Caution: Knee injuries.

Half Monkey (Ardha Hanumanasana)

Alignment: Start in low lunge with left leg forward and right knee on ground. Shift your hips back over your knee and straighten front leg where you feel a stretch but not a strain. Left knee cap pointing up and flex toes toward the face. Fingertips on the ground pressing to keep length in spine.
Benefits: Stretches hips, hamstrings, lower back, and calves. Strengthens hamstrings and stimulates internal organs.
Caution: Back, spine, hamstring, hip, and groin injuries.

Thread the Needle (Parsva Balasana)

Alignment: Begin in tabletop. Slide your left hand under your right arm palm up. Rest your right shoulder, ear, and cheek to the mat. Gaze to the right. Keep hips stacked above the knees.

Benefits: Stretches and opens the shoulders, chest, arms, and neck. Releases tension in shoulders and neck. Calming.
Caution: Knees, shoulders, back, or neck injuries.

Marichi Pose 1 (Sage) (With Bind) (Marichyasana I)

Alignment: Start in staff pose. Bend your left knee. Lean forward and wrap the left arm around the knee and clasp the fingers of the right hand behind the back.
Benefits: Strengthens and stretches spine and stretches shoulders. Stimulates abdominal organs.
Caution: Back or spine injury.

Marichi Pose 2 (Without Bind) (Marichyasana II)

Alignment: Start in staff pose. Bend your left knee. Twist torso to the left and cross your body with your right elbow to the outside left thigh.
Benefits: Strengthens and stretches spine and stretches shoulders. Stimulates abdominal organs.
Caution: Back and spine injuries.

© fizkes/Shutterstock.com

Pigeon (Eka Pada Rajakapotasana) Modified

Alignment: Start in downward facing dog. Lift right knee into your chest and lower your body so that your right knee in on the floor in front of you facing mat, foot facing left and right shin and foot on the floor. Right shin parallel to the front of the mat. Extend the left leg back with the top of foot and knee on the mat and toes pointing straight back.

Benefits: Lengthens hip flexors.

Caution: knee, ankle, and low back injuries.

© Ivan Veselinovic/Shutterstock.com

Sleeping Pigeon (Eka Pada Rajakapotasana) Forward Fold Version

Alignment: Start in downward facing dog. Lift right knee into your chest and lower your body so that your right knee is on the floor in front of you facing mat, foot facing left, and right shin and foot on the floor. Right shin parallel to the front of the mat. Extend the left leg back with the top of foot on the mat and toes pointing straight back. Walk the arms forward hinging from the hips until forehead rests on the mat.

Benefits: Lengthens hip flexors.

Caution: Knee, ankle, and low back injuries.

Standing, Standing Forward Bend Poses, and Standing Twist

Mountain (Tadasana)

Alignment: Feet hip distance apart or big toe mound together. Toes spread with all 4 corners of feet rooted to the mat. Aligned head over heart over pelvis over knees and over heels. Pull belly in and chin parallel to mat. Gaze straight ahead.
Benefits: Strengthens feet, legs, core, and upper back.
Caution: Low back injuries.

Extended Mountain or Upward Salute (Urdhva Hastasana)

Alignment: Start in mountain pose. Feet hip distance apart or big toe mound together. Toes spread with all 4 corners of feet rooted to the mat. Aligned head over heart over pelvis over knees and over heels. Pull belly in and chin parallel to mat. Gaze straight ahead. On the inhale extend arms overhead.

Benefits: Strengthens feet, legs, core, and upper back. Opens chest.
Caution: Low back and neck injuries.

© fizkes/Shutterstock.com

Wide-Legged Forward Fold A (Prasarita Padottanasana A)

Alignment: Start in mountain pose. Step feet apart. Hinge forward from the hips. Hands on floor in line with feet or on blocks. Arms bent with upper arms parallel. Step the feet apart wider than the hips.
Benefits: Stretches hamstrings and spine. Tones abdominals.
Caution: Low back injuries.

© Dmitry Rukhlenko/Shutterstock.com

Wide-Legged Forward Fold B (Prasarita Padottanasana B)

Alignment: Start in mountain pose. Step feet apart wider than the hips. Hinge forward from the hips with hands on the hips.
Benefits: Stretches hamstrings and spine. Tones abdominals.
Caution: Low back injuries.

© Dmitry Rukhlenko/Shutterstock.com

Wide-Legged Forward Fold C (Prasarita Padottanasana C)

Alignment: Start in mountain pose. Step feet apart wider than the hips. Hinge forward from the hips. Interlace hands behind back.
Benefits: Stretches hamstrings and spine. Tones abdominals.
Caution: Low back and shoulder injuries.

© Dmitry Rukhlenko/Shutterstock.com

Wide-Legged Forward Fold D (Prasarita Padottanasana D)

Alignment: Start in mountain pose. Step feet apart wider than the hips. Hinge forward from the hips. Grab big toes.
Benefits: Stretches hamstrings and spine. Tones abdominals.
Caution: Low back injuries.

© Dmitry Rukhlenko/Shutterstock.com

Standing Forward Fold (Uttanasana)

Alignment: Start in extended mountain pose. On an exhale, fold forward hinging at the hips, touching fingers to the floor. Keep a slight bend in knees if hamstrings are tight.
Benefits: Stretches hamstrings, hips, and spine; strengthens thighs and knees; relieves stress.
Caution: Low back injuries.

© fizkes/Shutterstock.com

Standing Half Lift (Ardha Uttanasana)

Alignment: From mountain pose, exhale and hinge from the hips forward until your fingertips touch the floor. On the next inhale, straighten your arms and lift the chest forming a straight back. Fingertips can come to shins or thighs if you cannot touch the floor with a straight back. Press heels into the floor and lift your tailbone up.
Benefits: Stretches hamstrings, hips, and spine; strengthens thighs and knees; relieves stress.
Caution: Low back injuries.

Warrior 1 (Virabhadrasana I)

Alignment: Start in mountain pose. Step back with left leg (like on two skis). Turn your left toes 45 degrees. Keep your left leg straight and bend your right knee. Lift your arms above your head, so that your upper body and arms form a straight line. Palms face each other.

Benefits: Stretches groin, belly, chest, and shoulders. Strengthens shoulders, arms, thighs, ankles, and calves.

Caution: Knee, lower back, and shoulder injuries.

Warrior 2 (Virabhadrasana II)

Alignment: Start in mountain pose. Step back with your right foot turning your foot parallel to the back of your mat. Align your left heel with the arch of your right foot. Keep your left knee bent and torso straight in line head over shoulders over hips. Extend both arms out to the sides and gaze over the left middle fingers.
Benefits: Strengthens thighs and arms. Stretches shoulders, chest, and groin.
Caution: Knee injuries.

© fizkes/Shutterstock.com

Reverse Warrior (Viparita Virabhadrasana)

Alignment: From Warrior II pose (with the left knee bent), bring the right hand down to rest on the right leg. Extend the right arm up toward the ceiling, and reach the fingers away from each other. Look straight ahead or up at the ceiling.
Benefits: Strengthens thighs and arms. Stretches shoulders, chest, and groin.
Caution: Knee or back injuries.

© fizkes/Shutterstock.com

Humble Warrior (Baddha Virabhadrasana)

Alignment: Start in Warrior I with the right leg in front. Gently step your front foot 1 to 2 steps over to the right with your toes slightly turning right to keep the groin open and protect your knee. Interlace your fingers behind your back. Continue to keep your heart open and gently bow forward. Your right shoulder can nudge your right leg but do not collapse into the leg.

Benefits: Strengthens the legs and ankles. Stimulates the abdominal organs. Stretches the shoulders, arms, legs, back, and neck. Opens the hips.

Caution: Knee, hip, and low back injuries.

© Paul Hakimata Photography/Shutterstock.com

Extended Side Angle (Utthita Parsvakonasana)

Alignment: Start in warrior 2 pose with left leg bent and right leg straight. On an exhale hinge and extend your torso to the left resting your left forearm on your thigh. Extend your right hand straight drawing a line behind your ear and lengthen your torso. Right palm is down and gaze toward the ceiling or palm.

Benefits: Strengthens and stretches thighs, knees, ankles, and core. Stretches hips, groin, and side of the body.

Caution: Knee and shoulder injuries.

© Dmitry Rukhlenko/Shutterstock.com

Triangle (Utthita Trikonasana)

Alignment: Start in warrior 2 pose right knee bent and left leg straight. On an exhale straighten the right leg and hinge forward from the hip extending the torso and rotating to the left to keep heart toward ceiling. Rest your hand on your shin, ankle, or the floor next to your right foot. Stretch your left arm toward the ceiling. Gaze toward ceiling or palm.

Benefits: Stretches and strengthens thighs, knees, and ankles. Stretches hips, groin, hamstrings, calves, shoulders, spine, and chest.

Caution: Low blood pressure, headache, neck issues.

© Paul Hakimata Photography/Shutterstock.com

Chair (Utkatasana)

Alignment: Start in mountain pose. On the inhale raise your arms in extended mountain and on the exhale bend your knees. Knees should be touching, weight shifted to your heels, and knees shifted back above your ankles.

Benefits: Strengthens thighs, ankles, spine, and arms. Stretches shoulders and chest.

Caution: Knee injuries.

© soul_studio/Shutterstock.com

Revolved Chair or Twisted Chair (Parivrtta Utkatasana)

Alignment: Start in chair pose. On an inhale, bring arms down to heart center. On the exhale, twist to the left, keeping hips squared and bring your right elbow to the outside of your left thigh.

Benefits: Strengthens thighs, ankles, spine, and arms. Stretches spine and tones abdomen.

Caution: Knee injuries.

© fizkes/Shutterstock.com

Low Lunge (Anjaneyasana)

Alignment: Start in downward facing dog. On an inhale, step your left leg through and forward between your hands. Drop your right knee to the mat. Keep knee over ankle and press pelvis forward for a deeper stretch. Gaze is forward.

Benefits: Stretches hips, thighs, chest, arms, and abdomen. Strengthens thighs. Detoxifying.

Caution: Knee and low back injuries.

© Dmitry Rukhlenko/Shutterstock.com

High Lunge (Prasarita Padottanasana)

Alignment: Start in downward facing dog. On an inhale, step your right leg through and forward between your hands while resting the ball of your left foot on the mat. Gaze is forward keeping spine long.

Benefits: Stretches hips, shoulders, and chest. Strengthens thighs.

Caution: Knee injuries.

© Mehmet Dilsiz/Shutterstock.com

Crescent Warrior or Crescent Lunge (Parivrtta Anjaneyasana)

Alignment: Start in downward facing dog. On an inhale, step your left leg through and forward between your hands while resting the ball of your right foot on the mat. Gain your balance and on the next inhale bring your arms above your head parallel to each other. Gaze is either straight forward or to the ceiling.

Benefits: Strengthens legs, upper back muscles, shoulders, and arms. Stretches hips and upper leg muscles. Increases balance.

Caution: Knee and ankle injuries.

© fizkes/Shutterstock.com

Revolved Crescent Warrior or Revolved Crescent Lunge or Twisted Crescent (Parivrtta Anjaneyasana)

Alignment: Start in crescent warrior with right leg in front and left leg back. On the inhale, bring hands to heart center. On the exhale, twist from the waist and take the left elbow to outside of the right thigh. Gaze to the right or to the ceiling.

Benefits: Strengthens legs and abdomen. Stretches hips and upper leg muscles. Increases balance. Detoxifying.

Caution: Knee, ankle, and back injuries.

© Prostock-studio/Shutterstock.com

Pyramid or Intense Side Stretch (Parsvottanasana)

Alignment: Start in warrior 1 pose right leg bent and left leg straight. On an exhale, straighten your right leg. Hinge from the hips, keeping them squared, and fold forward over your right leg while leading with the heart. Press fingertips to the floor on either side of the right foot.

Benefits: Strengthens legs and spine. Stretches legs.

Caution: Hamstring and spinal injuries.

© fizkes/Shutterstock.com

Garland (Malasana)

Alignment: Step your feet a little more than hip distance apart and squat where your hips are lower than your knee. Leaning torso forward press your elbows against your inner thighs and bring hands to heart center.
Benefits: Stretches ankles, groin, and back.
Caution: Knee and low back injuries.

Backbends

Cow (Bitilasana)

Alignment: Start in tabletop. On the inhale, lift your chest and allow the belly to sink to the floor. Lift your head to look straight forward.
Benefits: Stretches neck, chest, and spine.
Caution: Neck injuries.

Upward Facing Dog (Urdhva Mukha Svanasana)

Alignment: Lie prone on the floor with the tops of your feet touching the ground. Bring your hands under your shoulders on the floor palms down. On an inhale, press through the tops of your feet and while at the same time pressing up with your hands and straightening your arms lifting your thighs and knees off the ground. Keep shoulders over wrist and press into finger pads so as not to collapse into the wrist. Gaze is straight ahead or lift head slightly and gaze up.

Benefits: Strengthens wrist, arms, and spine. Stretches shoulders, chest, and abdomen.

Caution: Shoulder, lower back, and wrist injuries.

© fizkes/Shutterstock.com

Cobra (Bhujangasana)

Alignment: Lie prone on the floor with the tops of your feet touching the ground. Bring your hands under your shoulders on the floor palms down. On an inhale, straighten your arms into lifting your chest and shoulders off the floor.

Benefits: Stretches and strengthens spine. Stretches chest, shoulders, and abdomen.

Caution: Lower back and wrist injuries and pregnancy.

© Dmitry Rukhlenko/Shutterstock.com

Locust (Salabhasana)

Alignment: Lie prone on the floor with the arms and legs extended behind you. On an inhale, lift your head, chest, arms, and legs off the ground.
Benefits: Strengthens the spine, buttocks, and back of the arms and legs. Stretches shoulders, chest, belly, and thighs. Gaze can be slightly forward.
Caution: Back injuries and pregnancy.

© fizkes/Shutterstock.com

Sphinx (Salamba Bhujangasana)

Alignment: Lie prone on the floor with the tops of your feet touching the ground. On an inhale, bring your elbows under your shoulders on the floor and extend forearm palms down with fingers spread. Lift the kneecaps and press the tops of the feet into the mat. Gaze is forward.
Benefits: Stretches chest, shoulders, and abdomen. Strengthens the back and abdomen.
Caution: Back or shoulder injuries. Pregnancy.

© fizkes/Shutterstock.com

Bow (Dhanurasana)

Alignment: Lie prone on the ground with forehead on the floor and arms and legs extended behind you. On an inhale, bring ankles over knees and reach back grabbing ankles on the outside of foot. Keep arms straight while kicking feet into the hands and lifting the chest. Gaze is forward.
Benefits: Strengthens the spine and stretches shoulders, chest, abdomen, and thighs.
Caution: Lower back, knee, and shoulder injuries. Pregnancy.

© fizkes/Shutterstock.com

Camel Modified

Alignment: Start in a kneeling position aligning the hips over the knees. Bend your elbows and place hands on the lower back. Fingers can point up or down. Draw the elbows together, opening up the chest. Keep hips over knees. Gaze to the ceiling or straight ahead.

Benefits: Strengthens spine and core. Opens shoulders and chest.

Caution: Low back and sacroiliac (SI) joint injuries.

© fizkes/Shutterstock.com

Camel (Ustrasana)

Alignment: Start in a kneeling position aligning the hips over the knees. Bend from your upper back and place hands on the heels. Keep hips over knees. Gaze to the ceiling or straight ahead. Modification could be to tuck the toes and raise the heels closer to the hands or use two blocks besides the ankles.

Benefits: Strengthens spine and core. Opens shoulders and chest.

Caution: Low back and SI joint injuries.

© fizkes/Shutterstock.com

Inversions

Downward Facing Dog (Adho Mukha Svanasana)

Alignment: Start in tabletop. On an exhale, tuck the toes and lift the hips up as you straighten the legs. Draw the heels toward the floor. If tight hamstrings, heels can stay up. Pull the chest toward the thighs and keep the ears between the biceps. Keep shoulders away from the ears.

Benefits: Stretches the spine, hamstrings, and calves. Strengthens arms and legs.

Caution: Shoulder or hamstring injuries. Low blood pressure.

© fizkes/Shutterstock.com

Plow (Halasana)

Alignment: Start in supine position with legs bent and arms by the sides palms down. Lift your legs vertical and behind the head. Toes spread and feet flointed. Weight is in back of head and in shoulders, not the neck. Legs are together and straight. Hands can support the mid-back. Lift chin slightly and gaze toward the belly.

Benefits: Strengthens the neck. Stretches the spine.

Caution: Neck or back injuries. Pregnancy.

© Paul Hakimata Photography/ Shutterstock.com

Supported Shoulder Stand (Salamba Sarvangasana)

Alignment: Start in supine position with legs bent. Lift your legs vertical aligning hips over shoulders. Toes spread and feet flointed. Weight is in back of head and in shoulders not the neck. Lift chin slightly. Hands support mid-back. Legs are together and straight. Gaze at the feet.

Benefits: Stretches spine, shoulders, and neck. Strengthens the core.

Caution: Neck injuries. Pregnancy.

© guruXoX/Shutterstock.com

Legs Up the Wall (Viparita Karani)

Alignment: On an inhale, bring knees to chest and extend legs straight up the wall. If you do not have a wall, use a block or bolster at your sacrum for support if needed.
Benefits: Stretches legs, torso, and back of neck. Relieves tired legs and feet. Calms the mind.
Caution: Neck, back, or eye injuries.

© Catalin Petolea/Shutterstock.com

Bridge (Setu Bandha Sarvangasana)

Alignment: Start in supine position. Bend your knees so that your ankles are aligned under your knees and your feet are hip distance apart. Extend your arms along your sides palms down. Press your heels down and lift your pelvis up pushing your chest toward your chin allowing the heart above your head.
Benefits: Strengthens legs. Stretches neck, chest, and spine. Calming.
Caution: Lower back and neck injuries.

© Dmitry Rukhlenko/Shutterstock.com

Bridge Pose Variation (Setu Bandha Sarvangasana)

Alignment: Start in supine position. Bend your knees so that your ankles are aligned under your knees and your feet are hip distance apart. Extend your arms along your sides palms down. Press your heels down and lift your pelvis up pushing your chest toward your chin allowing the heart above your head. Now shimmy your shoulders underneath you and interlace your fingers opening up your chest.
Benefits: Strengthens legs. Stretches neck, chest, and spine. Calming.
Caution: Lower back and neck injuries.

© Dmitry Rukhlenko/Shutterstock.com

Dolphin (Catur Svanasana)

Alignment: Start in tabletop and bring the forearms to the mat with the elbows aligned under the shoulders. Tuck your toes and on the exhale lift your knees off the mat while pressing forearms into the floor and palms together. Keep knees bent if the back starts to round.
Benefits: Strengthens the legs, core, and arms. Stretches shoulders, arms, legs, and calves. Calming.
Caution: Shoulder and neck injuries.

© Dmitry Rukhlenko/Shutterstock.com

Headstand (Sirsasana)

Alignment: Beginners start against a wall preferably a corner. Start in dolphin pose interlacing your fingers. Bring legs together, walk toward chest, allow heels to lift off the floor, and lift up legs bringing one knee at a time. Then begin to straighten the legs toward the ceiling. Big toe mounds touch, toes spread, and feet are flointed. Gaze at the floor. Modification is the tripod headstand (see figures below).
Benefits: Strengthens the arms, legs, abdomen, and spine. Calming.
Caution: Back, neck, and eye injuries, high blood pressure, pregnancy.

© fizkes/Shutterstock.com

© fizkes/Shutterstock.com

Fish Pose (Matsyasana)

Alignment: Start in supine position with knees bent. Lift your hips and tuck your hands under each buttocks while aligning elbows under the shoulders. On an inhale, lift your chest toward the ceiling while the crown of the head reaches for the floor. It does not need to touch. Extend the legs straight.
Benefits: Stretches shoulders, chest, and throat.
Caution: Lower back, neck, and spinal injuries.

© fizkes/Shutterstock.com

Arm Supports

Tabletop

Alignment: Wrist aligned under shoulders. Knees hip distance apart and aligned under the hips. Tops of feet pressing into the mat. Neck is long and gaze is down.
Benefits: Helps to lengthen and realign the spine.
Caution: Knee or wrist injuries.

© Dmitry Rukhlenko/Shutterstock.com

Reverse Tabletop

Alignment: Start in staff pose. Bend your knees and on the exhale, rise up lifting hips up high. Fingers should face the heels while wrists are aligned under the shoulders. Keep spine long without letting head drop back. Gaze is up toward ceiling.
Benefits: Strengthens arms, legs, and core. Opens chest.
Caution: Shoulder, back, or knee injuries. If pain in wrist, turn fingers behind you.

© Dmitry Rukhlenko/ Shutterstock.com

Knees, Chest, and Chin Pose (Ashtanga Namaskar)

Alignment: Start in downward facing dog. On the exhale, lower knees to mat. Keeping elbows hugging in and hip points to the ceiling let chest and then chin touch the floor. Gaze is forward.
Benefits: Strengthens shoulders and core. Chaturanga prep.
Caution: Back, shoulder, and wrist injuries.

© Dmitry Rukhlenko/ Shutterstock.com

Plank

Alignment: Start in downward facing dog, shift your weight forward aligning your wrist under shoulders. Keep arms straight and lengthen tailbone to the heels.
Benefits: Strengthens arms and core.
Caution: Wrist injuries.

© fizkes/Shutterstock.com

Reverse Plank or Upward Plank (Purvottanasana)

Alignment: Start in staff pose with hands a few inches behind shoulders and fingers toward heels. On the exhale, bend knees and lift up into reverse tabletop. On the inhale, extend one leg out at a time and push hips higher. Keep spine long and gaze up toward the ceiling.

Benefits: Strengthens arms, wrist, and legs. Stretches shoulder, chest, and ankles.

Caution: Neck, shoulder, and wrist injuries.

© fizkes/Shutterstock.com

Side Plank (Vasisthasana)

Alignment: Start in plank pose. Shift your weight to the left side of the body stacking your fight foot on top of your left. On the exhale, bring your right arm up so it is pointed to the ceiling. The body is in one straight line from the head to the feet. Gazing toward the right hand.

Benefits: Strengthens arms, legs, core, and wrists. Improves balance.

Caution: Shoulder, wrist, and elbow injuries.

© fizkes/Shutterstock.com

Forearm Plank or Dolphin Plank (Makara Adho Mukha Svanasana)

Alignment: Start in tabletop. Lower elbows to floor aligning them under the shoulders. Interlace your fingers and extend one leg out at a time making your body parallel to the ground. Press heels toward the back of the room and lift the kneecaps. Engage the core so as not to dump the belly low.

Benefits: Strengthens the core and legs. Stretches calves, hamstrings, and shoulders.

Caution: Shoulder and elbow injuries.

© Dmitry Rukhlenko/ Shutterstock.com

Forearm Side Plank or Dolphin Side Plank

Alignment: Start in forearm plank. Step your feet together pressing down into the left forearm and left hand. Roll your body to the left and at the same time allow your left forearm and hand to turn toward the side. Body is in one long line. Do not dump hips toward the floor. Rest your left hand on hip or raise it up to the ceiling. Gaze to the side or to the left hand.

Benefits: Strengthens forearms, arms, shoulders, and core. Improves balance.

Caution: Shoulder, back, arm, or neck injuries.

© Dmitry Rukhlenko/Shutterstock.com

Four Limb Pose (Chaturanga Dandasana)

Alignment: Start in plank pose. On the exhale, shift forward on your toes aligning elbows over the wrist and lower down where shoulders are in line with the elbows.

Benefits: Strengthens spine, abdomen, arms, and wrists.

Caution: Shoulder and wrist injuries. Pregnancy.

© fizkes/Shutterstock.com

Supine Poses

Corpse Pose or Final Resting Pose (Savasana)

Alignment: Lie in supine position. Arms extended to the side palms up. Allow the feet to fall normal to the side and relaxed. Special note: Relax with attention without falling asleep.
Benefits: Calming, lowers blood pressure, decreases muscle tension.
Caution: Back injuries.

© fizkes/Shutterstock.com

Wind Relieving Pose (Pavanamuktasana)

Alignment: Start in supine position. On an exhale, bring both knees to the chest. While holding on to your left knee, extend your right leg straight heel pressing into the ground. To deepen stretch bring nose to knee.
Benefits: Relieves abdominal pain, indigestion, flatulence, bloating, and constipation.
Caution: Spinal or abdominal injuries.

© Dmitry Rukhlenko/
Shutterstock.com

Happy Baby (Ananda Balasana)

Alignment: Start in a supine position. On an exhale, bring knees to chest. Inhale, grab the inside or the outside of your feet allowing the knees to open slightly wider than the torso, bring them toward the mat and armpits. Keep ankles aligned over the knees.
Benefits: Stretches groin and spine. Calming.
Caution: Knee and neck injuries. Pregnancy.

© fizkes/Shutterstock.com

Supine Twist or Reclined Twist (Jathara Parivartanasana)

Alignment: Start in supine position. Bend the right knee to the chest and place arms in a T position. Rotate the right knee over to the left side placing your left hand on right knee. Gaze to the right.
Benefits: Stretches spine. Opens shoulders. Improves digestion. Calming.
Caution: Lower back injuries.

© Dmitry Rukhlenko/
Shutterstock.com

Reclined Figure 4 or Reclined Pigeon (Supta Kapotasana)

Alignment: Start in supine position. Bend knees keeping soles of feet on the mat. Bring your right ankle to the left thigh. For a deeper stretch lift the left leg up and clasp the hands behind left thigh or shin. Head and shoulders release to the ground while bringing knees toward chest.

Benefits: Stretches thighs, groin, and psoas. Opens hips and shoulders.

Caution: Knee injuries and SI issues.

Knees to Chest Pose (Apanasana)

Alignment: Start in supine position. On an exhale, bring both knees to chest clasping hands around the knee. Pull knees closer into chest with each exhale. For a deeper stretch bring head to knee and then relax the head and shoulders back to mat on the exhale.

Benefits: Aids in digestion. Relieves lower back pain.

Caution: Knees injuries. Pregnancy.

Reclined Bound Angle (Supta Baddha Konasana)

Alignment: Start in supine position with knees bent and soles of feet on the mat. On the exhale, open the knees letting them gently fall to the mat using hands on the outside of the thighs to help with this motion. Bring soles of the feet together. Bring arms down by the side or behind the head palms up.

Benefits: Stretches inner thigh and groin.

Caution: Low back and groin injuries.

© wavebreakmedia/Shutterstock.com

Yogi Bicycle (Dwichakrikasan)

Alignment: Start in supine position and hands behind head. On an exhale, bend the left knee to chest and cross right elbow across body toward left knee. Keep the right leg straight and heel off the ground. On the inhale, straighten legs keeping heels off the ground.
Benefits: Strengthens the abdominals.
Caution: Neck or back injuries.

© fizkes/Shutterstock.com

Balancing Poses

Tree (Vriksasana)

Alignment: Start in mountain pose. Bend your left knee and rotate out to the side allowing left knee to point to the side and left toes pointing down. Placement of left

foot can be on thigh, calf, or ankle (kick stand). Arms at heart center, extended up or cactus arms.

Benefits: Strengthens legs, ankles, and feet. Stretches groin and thighs. Improves balance and concentration.

Caution: Low back and groin injuries. Avoid placing foot on knee.

© Fizkes/Shutterstock.com

Eagle (Garudasana)

Alignment: Start in extended mountain pose. On an exhale, lower into chair pose. Lift right leg up and over crossing either wrapped around calf, hovering or kickstand at ankle. Right arm under and cross and wrap around palms together. Keep elbows up even with the shoulders.

Benefits: Stretches hips. Strengthens thighs, knees, and ankles. Releases tension in upper back. Improves balance and concentration.

Caution: Knee, groin, and shoulder injuries.

© Prostock-studio/Shutterstock.com

Warrior 3 (Virabhadrasana III)

Alignment: Start in extended mountain pose. Placing weight on the left foot. Lift right leg up and back and slowly hinging forward from the hips. Keep both hips parallel to the mat. Extend spine into a long line with the foot flexed toes down and kick through the heel. Gaze is forward.

Benefits: Strengthens shoulders, core, spine, thighs, calves, and ankles. Improves balance and concentration. Stretches thighs.

Caution: Low back injuries.

Half Moon (Ardha Chandrasana)

Alignment: Start in triangle pose. Bend your left knee and shift more weight on the left foot. Straighten the left leg while lifting your right leg, staking the hips and left fingertips on the mat. Flex the right foot. Extend right arm to the ceiling staking the shoulders. Gaze toward the right hand.

Benefits: Strengthens thighs, calves, and ankles. Improves balance and concentration.

Caution: Low blood pressure.

Dancer Pose (Natarajasana)

Alignment: Start in mountain pose. Bend your right knee and bringing the heel to the buttocks. Keeping knees close together, grab the big toe side of the foot. Kick into the hand for a deeper backbend but keeping hips square. Extend the left hand up and slightly forward while lifting the right leg higher.

Benefits: Strengthens the legs and ankles. Stretches chest, abdomen, and shoulders. Improves balance and concentration.

Caution: Lower back, foot, and ankle injuries.

© Denis Pepin/Shutterstock.com

Half Boat

Alignment: Start in staff pose. Bend the knees and palms of feet on the floor. Balance behind pelvic bones and not on tailbone. Lift your feet keeping knees bent and shins parallel to the mat. Arms parallel to the mat. Shoulders, back, and heart lift. Keep neck long and lengthen the spine.

Benefits: Strengthens core, back, psoas, and spine.

Caution: Pregnancy. Do not round the back.

© Dmitry Rukhlenko/Shutterstock.com

Boat (Navasana)

Alignment: Start in staff pose. Bend the knees and palms of feet on the floor. Balance behind pelvic bones and not on tailbone. Lift your feet 90 degrees pointing toes and toes spread. Arms parallel to the mat. Shoulders, back, and heart lifted. Keep neck long and lengthen the spine.
Benefits: Strengthens core, back, psoas, and spine.
Caution: Pregnancy. Do not round the back.

© Dmitry Rukhlenko/Shutterstock.com

Alignment and Poses Chapter Test

1. What is alignment in yoga?

    _____

2. Name a seated pose, seated twist, and a seated forward bend.

    A. _____

    B. _____

    C. _____

3. Name a standing pose, standing forward bend, and a standing twist.

    A. _____

    B. _____

    C. _____

4. Name an inversion pose.

    _____

5. Name a backbend pose.

    _____

6. Name an arm support pose.

    _____

7. Name a balance pose.

    _____

Chapter 4: Sun Salutations <Moon Salutations<Dancing Warriors

4 Sun Salutation

Surya Namaskar is Sanskrit and means to honor the sun. A warm up sequence in which we get the blood flowing and the muscles warmed up for the rest of the practice. Sun salutations build heat, improve circulation, improves overall body strength, and improves balance. Sun salutation variations can be used as a warm up or a quick practice if you don't have the time or can be used as an energizer. Sun salutations can be used all through the practice.

½ Salutation

Start: Mountain or standing with hands at heart center
Inhale: Extended mountain
Exhale: Forward fold
Inhale: ½ lift
Exhale: Forward fold
Inhale: Extended mountain
Exhale: Mountain or standing with hands at heart center

Sun Salutation A

© Baleika Tamara/Shutterstock.com

Start = Mountain or standing with hands at heart center
Inhale = Extended mountain
Exhale = Forward fold
Inhale = ½ lift
Exhale and Inhale = Stepping back into plank or ½ plank
Exhale = Chaturanga or to belly
Inhale = Upward facing dog or cobra
Exhale = Downward facing dog
Inhale = Gaze at the top of mat and step or hop forward into ½ lift
Exhale = Forward fold
Inhale = Extended mountain
Exhale = Mountain or standing with hands at heart center

Sun Salutation B

© Baleika Tamara/Shutterstock.com

Start = Mountain or standing with hands at heart center
Inhale = Chair
Exhale = Forward fold
Inhale = ½ lift
Exhale and Inhale = Step back into plank or ½ plank
Exhale = Chaturanga or knees, chest, chin, or to belly
Inhale = Upward facing dog or cobra
Exhale = Downward facing dog
Inhale = 3-legged dog right

Exhale = Warrior prep right
Inhale = Warrior 1 right
Exhale = Chaturanga or knees, chest, chin, or to belly
Inhale = Upward facing dog or cobra
Exhale = Downward facing dog
Inhale = 3-legged dog left
Exhale = Warrior prep left
Inhale = Warrior 1 left
Exhale = Chaturanga or knees, chest, chin, or belly
Inhale = Upward facing dog or cobra
Exhale = Downward facing dog
Inhale = Gaze at the top of the mat and step or hop forward into ½ lift
Exhale = Forward fold
Inhale = Chair
Exhale = Mountain or standing with hands at heart center

Sun Salutation C (Classical Sun Salutation)

© Baleika Tamara/Shutterstock.com

Start = Mountain or standing with hands at heart center
Inhale = Extended mountain
Exhale = Forward fold
Inhale and Exhale = Low lunge right
Inhale = Plank

Exhale = Knee, chest, chin pose
Inhale = Cobra
Exhale = Downward facing dog
Inhale = Low lunge left
Inhale = Gaze at the top of the mat and hop or step to hands
Exhale = Forward fold
Inhale = Extended mountain
Exhale = Mountain or standing with hands at heart center

Dancing Warriors

Dancing Warriors = A beautiful flow with the breath moving from warrior to warrior.

Inhale = Warrior 1
Exhale = Warrior 2
Inhale = Reverse warrior
Exhale = Extended side angle
Inhale = Warrior 2
Exhale = Triangle pose
Inhale = Crescent lunge
Exhale = Chaturanga or belly
Inhale = Upward facing dog or cobra
Exhale = Downward facing dog

Repeat on other side

Moon Salutation

Moon Salutation = Chandra Namaskar is Sanskrit to honor the moon. It is a cooling, quieting, and calming.

Start = Mountain
Inhale = Extended mountain
Exhale = Standing side bend (crescent moon) right steeple fingers
Inhale = Extended mountain
Exhale = Standing side bend (crescent moon) left steeple fingers
Inhale = Extended mountain
Exhale = ¼ turn to the left
Inhale = 5-pointed star
Exhale = Goddess pose
Inhale = Triangle left
Exhale = Pyramid left
Inhale = Half-moon left*
Exhale = low lunge left
Inhale = Side lunge left
Exhale = Up transition
Inhale= Side lunge right

Exhale = Low lunge right
Inhale = Half-moon right*
Exhale = Pyramid right
Inhale = Triangle right
Exhale = Goddess
Inhale = 5-pointed star
Exhale = ½ turn to right front
Inhale = Extended mountain
Exhale = Crescent moon left
Inhale = Extended mountain
Exhale = Crescent moon right
Inhale = Extended mountain

Repeat on the other side to do the full moon

*I added half-moon pose to this sequence.

Sun Salutations Chapter Test

1. What is the Sanskrit term for sun salutation?

    _____

2. Name the 3 sun salutations we will be doing in our class?

    _____

3. What are the benefits of sun salutations?

    _____

4. What is the Sanskrit term for moon salutation?

    _____

5. What are the benefits of the moon salutation?

    _____

Chapter 5: Meditation

5 Meditation (7th Limb of Yoga) Dhyana

Meditation means free of thoughts, but you cannot stop thinking; however, you can acknowledge the thought and then proceed back to your meditation. Slowing down the mind. Trying to block everything else out in the world and to just focus on one thing. It means awareness. Listening to the rain can be meditation. Prayer can be meditation. Walking can be meditation. Whatever has your attention and are free from other thoughts and distractions of the mind can be meditation.

The act of regular meditation can have many benefits on a person's mental and physical health. Meditation can help reduce stress, control anxiety, increase focus and improve sleep.

© AVA Bitter/Shutterstock.com

Sometimes we can get bored with a meditation and it becomes a challenge. If this happens, change it up by using a different type of meditation or venues. Use different resources to change up your meditation routine. The web is a wonderful treasure box!

If you feel scattered and cannot be still to meditate, do a few yoga poses or sun salutations to calm the mind. Along with warming up the body, utilizing breathing techniques such as even breath or 4, 7, 8 breathing before you start your meditation can deepen your practice and start to calm your mind.

Types of Meditations

1. **Guided Meditations**
   There are hundreds of resources online that have a huge supply of guided meditations and music to help soothe your soul. (YouTube, iTunes, Google)
   Guided Visualization—you are guided through a relaxation meditation.
   Progressive Muscle Relaxation—starting from your head or your toes and progressively relaxing these muscles by flexing and releasing the body part.
2. **Candle Staring**
   If you have trouble focusing, you can light a candle and stare at it. Your attention will be held. If your mind races, just observe what the mind is doing and bring your thoughts back to the flame.
3. **Mantra**
   Repeating words, statements, or sounds.
4. **Visualization**
   Focus on the picture and allow yourself to be in it and enjoy.

5. **Present Moment Meditation**
   Meditation that starts with focusing on the breath, then the body and its sensations, and then focusing on your senses. Then reverse the process slowly ending with your breath.

6. **Prayer Meditation**
   "Sitting with God"—a silent meditation, by first reading scripture or story, in which we focus all our mind, heart, and soul on the presence of God.

7. **Alternative Ways of Meditation**
   - Music—Listening to music that speaks to you helps empty the mind, while allowing the music to consume you.
   - Nature—Unlike our mind, our body and senses are always in the present. Being present in nature makes it much easier for us to go within our body and delve deep into our senses.
   - Walking—A walking meditation is a simple and quick way for developing calm and reduce anxiety and bring awareness.
   - Coloring—Adult coloring books are fun. Coloring or painting can calm your mind. (See cover page for each chapter.)

Meditation Chapter Test

1. What is meditation?

   _____

2. Name 2 benefits of meditation.

   A. _____

   B. _____

3. Name 4 different types of meditation

   A. _____

   B. _____

   C. _____

   D. _____

4. Name 2 alternative types of meditation.

   A. _____

   B. _____

Chapter 6: Warriors
<Balance Poses<Twists

6 Warriors Poses

History

The warriors are considered some of the most iconic poses in yoga. In fact, a whole chapter could go into the myth and twisted story of the warriors. In short, Shiva, a Hindu god, was avenging the death of his wife, Sati. Sati's father, Daksha, did not approve of Shiva. Daksha had tried many attempts of shaming Sati and trying to convince her to leave Shiva. Finally, Daksha held a party purposely not inviting Shiva. While Sati was at the party, she burst into flames due to the shame, and humility that her father caused her. When Shiva heard the news he was devastated. He tore out one of his dreadlocks and threw it on the ground. The dreadlock snaked its way through the earth, springing up from the energy and **Virabhadra** was created. It was a huge and ferocious being. Shiva ordered Virabhadra to kill and Daksha's head was cut off. When Shiva saw the bloody aftermath, he felt remorse and placed Daksha's head on a stake. Daksha and the other gods honored Shiva for this, calling him kind. Shiva left with his lifeless wife and became a recluse.

Warrior 1 = represents Virabhadra emerging from the ground, arms reaching up and gaze forward. (Dreadlock snaking it was through the Earth and emerging from the ground.)

© Dmitry Rukhlenko/Shutterstock.com

Warrior 2 = represents Virabhadra drawing his sword and slicing off Daksha's head.

© Dmitry Rukhlenko/Shutterstock.com

Warrior 3 = represents Virabhadra as he picked up the head to place it on a stake (arms reaching forward).

Benefits

The Warriors increase flexibility in the hips and strengthens and tones the legs, ankles, and feet, while at the same time builds balance and core strength.

Importance of Balance Poses

Balance poses help promote stability, coordination, and strength. These poses can also help with focus and concentration and to calm the mind. By working on balance poses, you are strengthening your equilibrium to help you with everyday activities. Building balance helps prevent injury from falling as we age.

Tips for Balance pose:

1. Find your Dishti (focal point). Place your focus on an object ahead of you that is not moving.
2. Squeezing energy in and up. For example, in tree pose press foot against thigh and thigh against foot with equal pressure. In Eagle, press inner thighs together.

Benefits of Twists

There are 3 types of twist: Supine, Standing, and Seated.

The purpose of twist is to rotate the spine. Twist can be used for poor digestion by helping move food and waste along the digestive system. Releasing the twist allows the fresh oxygenated blood and nutrients to rush back into the digestive organs detoxifying the body. Twist can improve the range of motion and flexibility, especially for those of us who enjoy golf, tennis, baseball, and softball.

© Dmitry Rukhlenko/Shutterstock.com

© Prostock-studio/Shutterstock.com

© fizkes/Shutterstock.com

Warriors Poses Chapter Test

1. Name 3 benefits of the balance poses.

   A. _____

   B. _____

   C. _____

2. What is Dishti?

   _____

3. Name 3 benefits of the warrior poses.

   A. _____

   B. _____

   C. _____

4. Name a benefit of twists.

   _____

5. Name 3 types of twists.

   A. _____

   B. _____

   C. _____

Chapter 7: Mudras, Mantras, Mindfulness

7 Mudras

Mudras are energetic seals, hand positions or hand yoga that direct or retain energy flows in the body. If your practice is more meditative, mudras can bring a physicality to your practice. Mudras can be used for healing, mood changes, grounding, spiritual awareness, and balancing many of the 5 elements provided in the following list. Our fingers take on these characteristics of the elements. When a finger is in contact with a thumb, the energy is redirected or retained to help balance. Each finger has an element associated with it.

Air = Index Finger
Fire = Thumb
Water = Little Finger
Earth = Ring Finger
Space = Middle Finger

There are hundreds of mudras. Some common mudras you might see in yoga classes are as follows:

Prana Mudra = Awakens and energizes the body.

Gyan or Chin Mudra = A meditative mudra meaning consciousness.

Surya Mudra = Increases the fire element and improves metabolism.

© SvetaZi/Shutterstock.com

Apan Mudra = Detoxifying and energizing.

© SvetaZi/Shutterstock.com

Anjali Mudra (Prayer hands at heart center) = To honor or divine offering.

© SvetaZi/Shutterstock.com

Mantras

Mantras are word, phrases, or sounds repeated silently over and over again during a meditation or yoga practice that resonate with you. Examples are as follows:

- Let it go
- I am enough
- Forgive
- Be a warrior, not a worrier
- Breathe in love, breathe out kindness.
- I give myself permission to slow down

Mindfulness

Sometimes we spend so much time worrying about the future that the here and the now, the present, comes and goes without us even realizing it is gone. When we are being mindful, we might think about the present moment rather rehashing the past or thinking about the future. Mindfulness is being fully present while being aware of or conscious of something. It can be sitting, walking, eating, listening to others when they are speaking to us, listening to music, or being in nature. Being mindful is nonjudgmental to ourselves and others. Mindfulness is acceptance. We pay attention to our thoughts and feelings without judging them.

Mudras Chapter Test

1. What are mudras?

   _____

   _____

2. What are mantras?

   _____

   _____

3. Name a mantra that has meaning for you.

   _____

   _____

4. What is mindfulness?

   _____

   _____

Chapter 8: Chakras

© Katika/Shutterstock.com

8   Chakras

Chakras are the 7 energy centers of the body. The belief is to have all energy centers open and not blocked. For instance in simplest of terms, if you have a sore throat, your throat chakra is blocked.

1. Root Chakra = Safety, grounding = base of spine= color red
2. Sacral Chakra = Emotions, creativity, sexuality = lower abdomen = color orange
3. Solar Plexus Chakra= Will, social self = upper abdomen = color yellow
4. Heart Chakra = Compassion, love = center of chest = color green
5. Throat Chakra = Personal truth, expression, speech = throat = color blue
6. Third Eye Chakra = Perception, intuition = forehead between eyes = color indigo
7. Crown Chakra = Wisdom, universality = very top of head = color violet

Each chakra has a color, poses, mantras, mudras, coordinating foods, essential oils, and healing crystals to help open up and balance these energy fields.

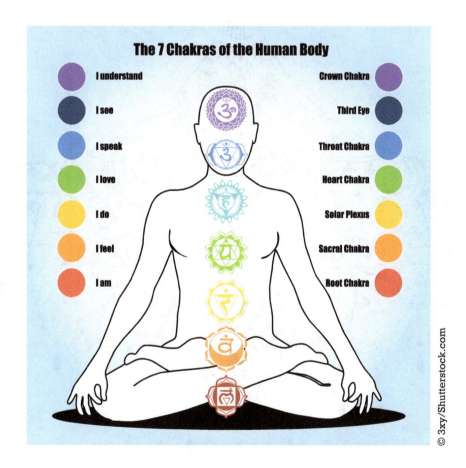

The 7 Chakras of the Human Body

I understand	Crown Chakra
I see	Third Eye
I speak	Throat Chakra
I love	Heart Chakra
I do	Solar Plexus
I feel	Sacral Chakra
I am	Root Chakra

© 3xy/Shutterstock.com

Chakras Chapter Test

1. What are the chakras?

   _____

2. How many chakras are there?

   _____

3. What color is the root chakra?

   _____

4. What chakra is green?

   _____

5. What chakra can affect your speech?

   _____

6. Name three things that can help balance your chakras?

   A. _____

   B. _____

   C. _____

Chapter 9: Brain Break

9   Brain Break

Sometimes as college students and in the work world, we need a quick pick me up to give us energy or ways to help us focus during a long lecture, meeting, or a long night of studying. We might not have time to go to the gym or on a long run. This chapter gives natural suggestions to help with energy and concentration.

Poses for Energizing

If you need to be energized during a slump of studying, perform Sun Salutations A, B, or C.

Standing and Balancing Poses for Focus

Chair, Tree, Warrior 2 and 3, Eagle.

Inversions for Focus

Use gravity to pull oxygen-rich blood to the brain reversing the blood flow in the body to improve concentration and memory.

Half expressions: Downward facing dog, plow, supported shoulder stand, bridge, legs up the wall.

Full expressions: Headstand and handstand and forearm balance.

Mudras for Focus

The great thing about mudras is that you can do them anytime anywhere and still be inconspicuous. You can do them in your lap or under a desk while in a lecture or test.

Ajna Chakra—clarifies the mind and improves concentration and intuition.

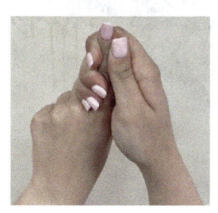

Anjali—calming and centering in order to concentrate.

Chin—sharpens the intellect, reduces daydreaming.

Panchamukha—balances left and right sides of brain, improves concentration.

Sahasrara—improves brain function, reduces headaches, increases concentration.

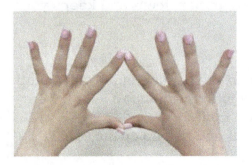

Sumukha—balances left and right sides of brain, increases concentration.

Vajra—improves concentration.

Pranayama for Focus or Energy

In Chapter 2 we learned many types of breath control. These breaths in this chapter can help for focus: even breath, ocean breath, alternate nostril breath, interrupted inhales, and interrupted exhales, humming bee breath. For energy, both breath of joy and breath of fire generate heat and release natural energy (see Chapter 2 pranayama).

Brain Yoga

Brain yoga activates acupuncture points on the earlobe that helps balances the left and right side of the brain. This practice can improve concentration, focus and keeps your mind sharp.

- Stand tall with legs apart hands by your side.
- Take you left hand to your right earlobe with thumb and pointer finger with thumb in front and then cross your right arm OVER your left arm and grab your left earlobe with thumb and pointer finger with thumb in front.
- On an inhale squat down. Hold squat for 2 seconds and rise up on the exhale.
- Repeat 15 times during the day.

Eye Yoga

Eye yoga helps to improve eyesight, peripheral vision, and concentration. It also helps reduce dry eyes, relieves eye strain.

Warm up the palms by rubbing hands together for 30 seconds. After warming up the palms place palms over eyes. Not pressing hard but blocking out the light. This is called palming. Then start eye yoga keeping the face still and only moving the eyes.

Look up and down quickly for 10 times, palming (see picture)
Look left and right quickly for 10 times, palming (see picture)
Look diagonal top left low right 10 times, palming (see picture)
Look diagonal top right low left 10 times, palming (see picture)
Rolling the eyes 10 full circles, palming

© Inspiring/Shutterstock.com

Test Taking Strategies

Inversion

- Yes you read right . . . inversions during class or a test. No I am not talking about going to the floor, but I am talking about getting blood to your head to help you focus.
- Accidentally drop your pencil/pen to the floor. Take a few seconds there fumbling around trying to find your pencil while at the same time allowing the blood to run to your head.
- Untie your shoe before the lecture or test starts. When you need to focus, just bend down and let the blood flow to your head and slowly tie your shoe.

Meditation for Brain Power

Any type of meditation from guided, visualization, breathing, and so on may improve concentration, focus, and memory retention. It doesn't hurt to try it and at the same time help with relaxation, stress relief, and sleep.

Essential Oils or Blends to Help with Focus

Basil, frankincense, patchouli, rosemary, sage, ylang ylang, blends.

ADHD = Lavender, lemon, blends.

Brain Break Chapter Test

1. How do inversions help with focus?

   _____

   _____

2. Name 2 balance poses for focus.

   A. _____

   B. _____

3. Name 2 mudras for focus or calming.

   A. _____

   B. _____

4. Name 2 benefits for eye yoga.

   A. _____

   B. _____

5. Name 2 benefits of brain yoga.

   A. _____

   B. _____

Chapter 10: Self-Care

© Katika/Shutterstock.com

10   Self-Care

Nutrition

Of course, a balanced diet is must for a physically active person. The proper diet includes the nutrients needed for your body to function properly. Without proper nutrition, your body is more prone to diseases, infections, and fatigue. Majority of the calories should be consumed in vegetables, fruit, legumes, nuts, whole grains, and lean proteins.

Not eating too close to yoga class time is important since some of the poses can put pressure on the stomach organs. However, going to class starving is not a good idea due to low blood sugar.

© ImageFlow/Shutterstock.com

Hydration

Water is another key element in exercise performance. 7 to 8 glasses a day is ideal for optimal organ function and health. However it is not all about drinking water, you can eat your water too! Fruits, vegetables, soups, and grains have a high water content such as celery, lettuce, bell peppers, watermelon, pineapples, and oranges.

Hydration for preyoga is important so as not to cramp, be stiff, or sore. Make sure you rehydrate after class too. Thirst is one of the first signals of dehydration. Adding fruit to your water can change it up a little.

TRANSPARENT
vector water
eps10

© Julia-art/Shutterstock.com

Rest

Sleep is important to your physical health. Your body repairs itself while you are a sleep. Exercise such as yoga can aid in sleep. Stick to a bedtime ritual and schedule. Limit your caffeine and alcohol intake. As for rest, your body needs rest between physical exercise in order to repair itself and get stronger. Listen to your body (see below in injury).

sleep tight!

© Orange Vectors/Shutterstock.com

Injury

Listen to your body, an emotional and physical check in. *If it doesn't feel right don't do it.* When we are emotionally not present, that is when we tend to get hurt. Be mindful of any pose and if you feel a twinge, stop. Do what works best for you. The alignments for the poses are a general rule of thumb.

Mental Health

Mental health is just as important as physical health. It is easy to neglect your mental health because some feel that they are an indulgence. If you are mentally healthy, you tend to feel more motivated to do school work, you are more productive, and you have good relationships. Mental health can be in a form of meditation, face masks, movie, music, exercise, or nature. Whatever lights you up and you enjoy can be a mental health break.

Self-Care Chapter Test

1. Without proper nutrition your body is more prone to

   A. _____

   B. _____

   C. _____

2. Hydration for preyoga is important so as not to

   A. _____

   B. _____

3. What is one of the first signs of dehydration?

   _____

4. _____ can aid in sleep.

5. Listen to your body. "If it doesn't feel right

   _____   _____   _____."

6. What is your favorite mental health self-care?

   _____

Chapter 11: Purpose of Home Journaling

© An Vino/Shutterstock.com

11   Purpose of Home Journaling

Journaling can help know yourself better, reduce stress, and can clarify your thoughts and feelings. The purpose of these 5 journal prompts are to promote deeper thinking into these parts of your yoga practice. You might have fond thoughts or ill thoughts. It is your journal and your writing. It is all accepting and nonjudgmental.

Journal 1 Prompt Intention Setting

The purpose of this journal is to reflect on setting intentions (see Chapter 1). However, this journal is not about a yoga practice for a certain day but in general.

First, journal on what intention you are setting for this course? And how are some ways your intention might support you in your yoga practice?

Lastly, journal on ways you might be able to bring your intention into your everyday life?

Journal 2 Prompt Sun Salutations

The purpose of this journal entry is to have you reflect on the purpose and benefits of sun salutations (see Chapter 4).

First, journal ways to improve your alignment. And maybe ways to incorporate props.

Second, journal on the sun salutation best suited for your body . . . a, b, or c.

Lastly, journal about how you can utilize the sun salutations in your daily life.

Journal 3 Prompt Mindfulness

The purpose of this journal entry is to reflect on the word mindfulness and being mindful (see Chapter 7).

First, reflect and journal about what it means to you to be present? And what does it mean to you to do something mindfully?

Second, when do you have a hard time just letting go and being here in the present moment? What parts of your life are most distracting? Do you have any zoning out behaviors? (Instagram, Snapchat, Email, etc.)

Journal 4 Prompt Love/Hate Poses

This journal entry is to reflect on our poses. You are not alone if you have a pose or poses you just hate and poses that you love.

The poses we hate are the most valuable for our practice. Go inward and ask yourself why you have negative feelings toward or are struggling with a pose. However, Yoga poses that we are fond of are just as important to us. These are the poses we go to when we are feeling low or sick that can comfort us.

Then journal about your "nemesis pose." Also journal about a modified version of the pose or the use of props, the wall, or ways to improve your alignment. Do you use the pose as a challenge or a goal to take the negative feelings away or do you just stay away from it?

Now reflect on your favorite pose or your "Go to" pose. Journal about why it is your favorite pose? Do you feel strong in this pose? Is it relaxing? Is it a goal achieved and you are proud? Does it have special meaning to you? Do you use this pose to boost your mood?

Journal 5 Self-Care

The purpose of this journal is to reflect on self-care and its importance. Think of your yoga and meditation practice.

Journal on if you loved yourself unconditionally, how would you treat yourself? How can you act on that feeling now?

Now journal about how you can approach your daily routine. What can you do to add more ease and relaxation to your night time routine? When I'm really busy, how can I find 10 minutes of time for myself? What can I do in that time?

Finally, does your self-care practice feel like a chore or an item on your to-do list?

Chapter 12: Skills Test

© Luba_kistochka/Shutterstock.com

12   Skills Test

1. Sun salutation A, B, or C
2. Poses (Choose 3 from each category)
    - Seated, Seated twist, Seated Forward Bend, and Kneeling Poses
    - Standing, Standing Forward Bend, and Standing Twist
    - Backbends
    - Supine
    - Arm Supports
    - Inversions

*There are modified versions for a pose. If you are not sure, research it or ask the yoga teacher.

© Radharani/Shutterstock.com

Skills Test Worksheet

Sun Salutation Test

I will be performing sun salutation _____.

Poses Skills Test

Seated, Seated twist, Seated Forward Bend, and Kneeling Poses

> **1.**
> **2.**
> **3.**

Standing, Standing Forward Bend, and Standing Twist

> **1.**
> **2.**
> **3.**

Backbends

> **1.**
> **2.**
> **3.**

Inversions

> **1.**
> **2.**
> **3.**

Arm Supports

> **1.**
> **2.**
> **3.**

Supine

> **1.**
> **2.**
> **3.**

Chapter 13: Home Practice

13   Home Practice

Starting a home practice is very easy and inexpensive. Taking advantage of free YouTube yoga channels and meditation channels is a must to freshen up your practice. However, you have the tools now to do your own home sequence. Below is a sequence builder. It can give you the flexibility to do whatever, whenever, and wherever you need to have your practice (in your home, in the park, at the gym, at the beach, etc.).

Yoga Sequence Builder

Depending on the length you want your practice, start with 1 to 3 poses from each category and then when you feel comfortable and you have more time, add more poses to each category. Be creative and change out the poses so your practice does not become stagnant.

- Adding different breathing techniques (Chapter 2), mantras and mudras (Chapter 7) to your sequence can add more depth to your practice.
- Hold each pose for 30 seconds to 1 minute (5–10 breaths)

Warm-up

During this time, you are going inward and preparing your mind and body for your practice. These poses are good to start with in your practice, but are not limited to these.

- Seated meditation and breath work
- Seated head and shoulders stretches, side and forward fold body stretches
- Seated twists
- Child's pose
- Mountain pose

Movement

Some like to start with sun salutations after the warm up and that is fine. Do 3 or 4 rounds of the salutations and then move to the standing poses. Be creative. Do what makes you feel good. Some examples are as follows:

- Sun salutations A, B, C
- Kneeling poses
- Warriors 1, 2, 3, reverse warrior, humble warrior, crescent
- Extended side angle, pyramid, triangle, revolved triangle
- Chair, revolved chair

Have fun and challenge yourself

It is always good to build on your practice with challenge poses. They can be any pose that you feel challenges you. It is important to build up to that pose and make sure you are doing poses beforehand that warmup that part of the body. For example, if you really want to try full bow pose, make sure you are doing poses to have the back warmed up such as core work like boat and yogi bicycle and of course, some back stretches such as bridge, cobra, upward-facing dog, sphinx, and locust. Of course, any of the poses in the movement part of your practice can be your challenge poses. Some examples are as follows:

- Balance poses = tree, eagle, dancer, half moon, warrior 3
- Floor poses = bow, camel, wheel, monkey pose
- Arm support poses = crow, crane, headstands, handstands

Cooldown

Cooldown allows your body to restore and reset. Using bolsters (pillows) and blocks (books) can help with the cool down. Here are some examples of restorative poses:

- Pigeon
- Seated forward bend
- Reclined bound angle
- Supine twist
- Seated twist
- Side bends
- Legs up the wall
- Happy baby
- Child's pose

Meditate

Take at least 1 to 5 minutes to relish the stillness and let the nutrients of your practice soak in.

- Savasana (use props to help the relaxation.)

Yoga Sequence Worksheet

- What is my yoga today? Morning, bedtime, challenge pose, fitness, restorative, and so on.
- Do I have a mantra today?
- What pranayama (breath control) do I want to do today besides Ujjayi breathing?
- Do I have a mudra today?

Warm-up

1.	6.
2.	7.
3.	8.
4.	9.
5.	10.

Movement

1.	6.
2.	7.
3.	8.
4.	9.
5.	10.

Challenge

1.
2.
3.

Restore

1.	6.
2.	7.
3.	8.
4.	9.
5.	10.

Meditate

Helpful Sequences

Back/Spine Health

The health of your spine is key to the health of your whole body. Use these yoga poses to improve flexibility and _decrease back pain_. Hold each pose for 8 breaths. Make sure do both right and left sides. Warm up with 3+ sun salutations (see Chapter 3 to put names to poses).

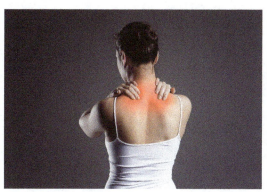

© Chingyunsong/Shutterstock.com

- Easy pose with victory breath technique
- Seated spinal twist (both sides)
- Cow/cat
- Thread the needle(both sides)
- Cobra OR baby cobra
- Sphinx
- Bow
- Extended child's pose OR child's pose
- Puppy pose
- Camel
- Rabbit
- Seated forward bend
- Bridge pose and supported bridge
- Fish pose
- Corpse pose (Savasana)

Upper Back Pain

We keep a lot of stress in our upper back, shoulders, and neck. The way we sit (our posture) at the computer or in class can also cause tightness in these areas. Here are some poses you can do at home to help alleviate some of the pain. Hold each pose 8 breaths. Make sure to do both right and left sides. Warm up with 3+ sun salutations (see Chapter 3 to put names to poses).

© Staras/Shutterstock.com

- Easy pose with victory breath technique
- Easy pose with ear to shoulder (both sides) and neck circles both directions
- Easy pose with shoulder rolls forward and backward
- Easy pose with eagle arms (both sides)
- Easy pose with interlace fingers behind back
- Easy pose with arm across chest stretch

- Easy pose with hand to elbow tricep stretch
- Cow/cat (8 rounds)
- Thread the needle (both sides)
- Extended child's pose with side stretch (both sides)
- Extended puppy pose
- Rabbit pose
- Supine twist
- Supported fish pose with bolster or pillow

Bedtime Yoga

Ease into bedtime with a quiet yoga practice focused on deep breathing to calm your mind and release physical tension. Hold each pose 5 to 10 breaths. Make sure to do both right and left sides. Warm up with 3+ sun salutations (see Chapter 3 to put names to poses).

- Easy pose with 4, 7, 8 breath 4 cycles
- Easy pose with head and shoulder rolls
- Easy pose with seated cat/cow
- Easy pose with side bend
- Seated forward bend with blocks/books for forehead
- Supported child's pose; allow your heart, mind, and soul to relax using bolster/pillow under chest. *Offer yourself grace and focus on your breathing*.
- Tabletop
- Cat/cow
- Downdog
- Lizard (using block on forearms/book) right and left and/or Pigeon
- Plank
- Locust
- Rabbit
- Bound angle with foot massage (ankle, heels, toes)
  - Interlace fingers under toes and inhale look up exhale chin to chest
- Supported bridge with block, pillow, or blanket
- Reclined bound angle hands on belly connecting to your breath.
- Supported fish pose using bolster, pillow, or blanket
- Supported spinal twist right and left bolster, pillow, and blanket
- Knees to chest circle them both directions
- Wind relieving pose right and left
- Supported legs up the wall
- Happy baby
- Supported savasana; *Choose to let go of the day*. Bolster, pillow, and blanket
  - 4 cycles of the 4, 7, 8 breath technique

Quick Yin Wall Yoga

If you are short on time, here is a quick yin _wall_ sequence. Hold each pose 1 to 3 minutes depending on your time. Make sure to do both right and left sides.

- Easy pose with seated meditation with back against wall. Checking your posture. Even breath technique or 3-part breath technique.
- Easy pose right shoulder next to wall. Right hand on wall fingers up with hand head height and little past shoulder.
  - Left ear to left shoulder
  - Turn around and do other side
- Reclined bound angle
  - Facing wall with hips close to the wall. Start in legs up the wall. Place palms of feet touching and slid feet down.
  - Palms against legs to help stretch.
  - Invite heels to drop down into groin.
- Knees to chest
- Head to knees
- Reclined Figure 4 (reclined pigeon pose)
  - Allowing the right palm of foot to slide down the wall while the left ankle is on top of right thigh.
  - Arms relax to the side
  - Let gravity do its job
  - Do both sides.
- Straddle V
  - Hips close to the wall
  - Let gravity do its job
- Knee to chest
- Head to knees
- Legs up the wall

© Coka/Shutterstock.com

© Coka/Shutterstock.com

Chapter 14: Yoga Games

14 Yoga Games

It is always good to switch up your routine to make it fresh and fun. Adding games such as the following can challenge you and relieve stress because it's entertaining, especially with friends.

  **Head or Tails**

Directions: Flip a quarter and choose a pose from the side it lands on. Make sure to hold pose 3 to 5 breaths and do both right and left sides of the body if applicable.

Heads	Tails
Upward Facing Dog	Downward Facing Dog
Bound Angle	Reclined Bound Angle
Seated Wide Leg Forward Fold	Standing Wide Leg Forward Fold
Half Lord of the Fishes	Seated Twist
Cobra	Sphinx
Plank	Side Plank
Forearm Plank	Forearm Side Plank
Warrior 1	Crescent Lunge (Warrior)
Warrior 2	Warrior 3
Half Moon	Dancer
Tree	Eagle
Triangle	Pyramid
Extended Side Angle	Revolved Triangle
Chair	Revolved Chair
Revolved Crescent Lunge	Bow
Bridge	Locust
Camel	Plow
Supported Shoulder Stand	Frog
Boat	Pigeon
Reverse Warrior	Humble Warrior
Rabbit	Fish

 A to Z Yoga

Directions: Spell your first and last name and do the corresponding pose. Spell your hometown, and so on.

A = Abdomen...Any core strengthening pose

B = Boat or bow or bound angle or bridge

C = Cobra or cat or camel or chair

D = Downward facing dog or dancer or dolphin or diamond

E = Eagle or extended side angle pose or extended child's pose

F = Frog or fish

G = Garland or goddess or gate

H = Humble warrior or half lord of the fishes or hero or happy baby

I = Intense side stretch pose (also known as pyramid)

J = Janusirsasana is Sanskrit for seated tree pose

K = Kneeling poses...examples (hero, diamond)

L = Locust or lizard

M = Marichi's or monkey pose (splits) or mountain pose

N = Noose pose

O = Oh so favorite pose

P = Pigeon or plow or pyramid or plank

Q = Quadriceps work...chair pose or warrior 1 or warrior 2

R = Rabbit or revolved chair or revolved triangle or reverse warrior or reclined bound angle

S = Sphinx or seated twist or standing wide leg forward fold = side plank

T = Tree or three legged dog or triangle

U = Upward facing dog

V = Virabhadrasana is Sanskrit for warriors (reverse or crescent or humble)

W = Warriors 1, 2, or 3 or wild thing

X = X-factor means a special talent or quality. Do the pose that you do best!

Y = Yoga favorite pose

Z = Pick any pose from A to Z that you have not done.

Search modifications for a pose that you cannot perform or lead up poses.

Hold for 3 breaths and perform on both sides if applicable.

What's in the Cards?

Directions: If a number card is drawn perform that suit for that many breaths (Example: 3 of diamonds do standing pose for 3 breaths). Face cards do not do suit, but do what the face card key says.

Arm Supports

© Martial Red/Shutterstock.com

Backbends/Inversions

© ahmad agung wijayanto/Shutterstock.com

Standing or Standing Forward Bend Pose

© Benton Frizer/Shutterstock.com

Seated or Seated Twist Poses

© Flipser/Shutterstock.com

= Sun Salutation

© Francesco Abrignani/Shutterstock.com

= Pranayama (Breath)

© Francesco Abrignani/Shutterstock.com

= Mudra

© PandaWild/Shutterstock.com

= Favorite Pose

© elbud/Shutterstock.com

Partner Pose Challenge

Directions: Find a partner and attempt these fun yoga partner challenges.

Benefits: Stress reliever, study break, mood booster.

© Denys Kurbatov/Shutterstock.com

© MediaGroup_BestForYou/Shutterstock.com

© fizkes/Shutterstock.com

© Joyja_Lee/Shutterstock.com

© Joyja_Lee/Shutterstock.com

© Prostock-studio/Shutterstock.com

© ZephyrMedia/Shutterstock.com

© Joyja_Lee/Shutterstock.com

Useful Online Resources

Abide Christian Meditation App
Calm.com
Dartmouth College Meditations http://dartmouth.edu/-healthed/relax/downloads.html

Headspace.com
UCLA Mindful Meditations http://marc.ucla.edu/mindful-meditations

Yogajournal.com
YouTube Yoga With Adriene

Glossary

Alignment—adjusting the bones and muscles in positions so the pose can be at the best position for strengthening so as not to cause injury.

Asana—pose or posture, 3rd limb of yoga.

Benefits—how and where a pose helps the body.

Blanket—towel like tool or prop in yoga.

Block—a rectangle or egg shaped tool or prop in yoga.

Bolster—a pillow or cushion used as a tool or prop in yoga.

Chakras—7 energy centers inside the body.

Cautions—to be careful if the body has these injuries.

Crown of Head—top of head.

Flointed—extension of the foot, leading with the ball instead of the toe.

Flow—linking poses together.

Gaze—to stare or where the eyes look.

Ground Down—all four points of the feet, knuckles, and palm heels of hands into the earth.

Hamstring—muscle on back side of leg above the knee.

Intention Setting—a word or mantra to bring to your mat during the practice.

Mantra—repeating words, phrases, or sounds for meditation.

Meditation—free of thoughts and slowing down the mind.

Mindfulness—being present and aware of the present moment.

Mudra—hand positions or gestures to aid in meditation.

Namaste—the light in me honors the light in you.

Pranayama—breath control, 4th limb of yoga.

Prone—lying flat, facing downward.

Psoas—hip flexor.

Quadriceps—front leg muscles above the knee.

Sacrum—bone at the base of the spine.

Sacroiliac (SI) Joint—the joint between the sacrum and the ilium bones of the pelvis.

Sanskrit—ancient language of Hinduism and Buddhism.

Savasana—final resting pose or corpse pose.

Sun Salutation—set of poses put into a flow that warm up the body.

Sequence—yoga poses or postures in a particular order.

Sit Bones—located under the buttocks.

Strap—belt like tool or prop.

Supine—lying flat on one's back.

Torso—front upper body.

Ujjayi—a type of breath control.

Vinyasa—link pose with breath.

Yoga—to yoke or unite.

Pose Index